Signal Processing and Communications

Advanced  Magnetic
Image in Resonance
Processing Imaging

edited by

Luigi Landini
Vincenzo Positano
Maria Filomena Santarelli

Taylor & Francis
Taylor & Francis Group
Boca Raton   London   New York

A CRC title, part of the Taylor & Francis imprint, a member of the
Taylor & Francis Group, the academic division of T&F Informa plc.

Published in 2005 by
CRC Press
Taylor & Francis Group
6000 Broken Sound Parkway NW, Suite 300
Boca Raton, FL 33487-2742

© 2005 by Taylor & Francis Group, LLC
CRC Press is an imprint of Taylor & Francis Group

Library of Congress Cataloging-in-Publication Data

Advance image processing in magnetic resonance imaging / edited by Luigi Landini, Vicenzo Positno, Maria Santarelli.
    p. cm. -- (Signal processing and communications ; 26)
  Includes bibliographical references and index.
  ISBN 0-8247-2542-5
  1. Magnetic resonance imaging. 2. Image processing. I. Landini, Luigi. II. Positano, Vicenzo. III. Santarelli, Maria. IV. Series.

RC78.7.N83A377 2005
616.07'548--dc22   2005043995

Taylor & Francis Group
is the Academic Division of T&F Informa plc.

Visit the Taylor & Francis Web site at
http://www.taylorandfrancis.com

and the CRC Press Web site at
http://www.crcpress.com

Preface

Magnetic Resonance (MR) imaging produces images of the human tissues in a noninvasive manner, revealing the structure, metabolism, and function of tissues and organs. The impact of this image technique in diagnostic radiology is impressive, due to its versatility and flexibility in joining high-quality anatomical images with functional information.

Signal and image processing play a decisive role in the exploitation of MR imaging features, allowing for the extraction of diagnostic and metabolic information from images. This book attempts to cover all updated aspects of MR image processing, ranging from new acquisition techniques to state-of-art imaging techniques. Because the textbook provides the tools necessary to understand the physical and chemical principles, and the basic signal and image processing concepts and applications, it is a valuable reference book for scientists, and an essential source for upper-level undergraduate and graduate students in these disciplines.

The book's 18 chapters are divided into five sections. The first section focuses on MR signal and image generation and reconstruction, the basics of MR imaging, advanced reconstruction algorithms, and the parallel MRI field. In the second section, the state-of-art techniques for MR images filtering are described. In particular, the second section covers the signal and noise estimation, the inhomogeneities correction and the more advanced image filtering techniques, taking into account the peculiar features of the noise in MR images. Quantitative analysis is a key issue in MR diagnostic imaging. The third section's topics range from image registration to integration of EEG and MEG techniques with MR imaging. Two chapters cover the cardiac image quantitative analysis issue. In the fourth section, MR spectroscopy is described, from both the signal generation and the data analysis point of view. Diffusion tensor MR imaging and MR elastography are also examined. Finally, in the last section, the functional MR image processing is described in detail. Fundamentals and advanced data analysis (as exploratory approach), bayesian inference, and nonlinear analysis are also depicted.

Experts from all fields of MRI have cooperated in preparing this book. We would like to thank all authors for their excellent and up-to-date chapters. Some overlap was unavoidable, due to the multiple connections in MR image processing fields.

The outline of this book was born at the MR Laboratory at the CNR Institute of Clinical Physiology in Pisa, Italy. The MR Laboratory is a multidisciplinary arena for biomedical engineers, physicists, computer scientists, radiologists, cardiologists, and neuroscientists, and it shares experts from CNR and university departments. For this textbook, we are sincerely grateful to professor Luigi Donato, the chief of the CNR Institute of Clinical Physiology, and the pioneer of the multidisciplinary approach to biomedicine.

Contributors

Antonio Benassi
Institute of Clinical Physiology, CNR
Pisa, Italy

Alessandra Bertoldo
Department of Information
  Engineering
University of Padova
Padova, Italy

Sergio Cerutti
Department of Biomedical
  Engineering
Polytechnic University
Milano, Italy

Claudio Cobelli
Department of Information
  Engineering
University of Padova
Padova, Italy

Arjan J. den Dekker
Delft Center for Systems and
  Control
Delft University of Technology
Delft, The Netherlands

Francesco Di Salle
Department of Neurological
  Sciences
University of Naples, Italy
Naples, Italy

Elia Formisano
Department of Cognitive
  Neurosciences
Faculty of Psychology
University of Maastricht
Maastricht, The Netherlands

Alejandro Frangi
C.N.R. Institute of Clinical
  Physiology
Department of Technology
Pompev Fabra University
Barcelona, Spain

Karl J. Friston
The Wellcome Department of Imaging
  Neuroscience
University College London
London, U.K.

Rainer Goebel
Department of Cognitive
  Neurosciences
Faculty of Psychology
University of Maastricht
Maastricht, The Netherlands

Steven Haker
Harvard University School of
  Medicine
Brigham and Women's Hospital
Boston, Massachusetts

Yaroslav O. Halchenko
Psychology Department
Rutgers University
Newark, New Jersey

Stephen Jose Hanson
Rutgers University
Psychology Department
Newark, New Jersey

W. Scott Hoge
Harvard University School of
  Medicine
Brigham and Women's Hospital
Boston, Massachusetts

Filip Jiru
Section of Experimental MR of the
  CNS
Department of Neuroradiology
Tübingen, Germany

Uwe Klose
Section of Experimental MR of the
  CNS
Department of Neuroradiology
Tübingen, Germany

Walid Elias Kyriakos
Harvard University School of
  Medicine
Brigham and Women's Hospital
Boston, Massachusetts

Luigi Landini
Department of Information
  Engineering
University of Pisa
Pisa, Italy

Koen Van Leemput
HUS Helsinki Medical Imaging
  Center
Helsinki, Finland

Boudewijn P.F. Lelieveldt
Laboratory for Clinical and
  Experimental Image Processing
Department of Radiology, Leiden
  University Leiden, The Netherlands

Zhi-Pei Liang
Department of Electrical Engineering
  & Computer Engineering
University of Illinois at Urbana
Urbana, Illinois

Massimo Lombardi
MRI Laboratory
Institute of Clinical Physiology, CNR
Pisa, Italy

Luca T. Mainardi
Department of Biomedical
  Engineering
Polytechnic University
Milano, Italy

Armando Manduca
Mayo Clinic and Foundation
Rochester, Minnesota

Wiro J. Niessen
Image Sciences Institute
Utrecht University
Utrecht, The Netherlands

Lauren O'Donnell
Massachusetts Institute of Technology
Harvard Division of Health, Science
  and Technology
Boston, Massachusetts

Barakk A. Pearlmutter
Hamilton Insitute and Department of
  Computer Science
Kildare, Ireland

William D. Penny
The Wellcome Department of Imaging
  Neuroscience
University College London
London, U.K.

Vincenzo Positano
Institute of Clinical Physiology, CNR
Pisa, Italy

Maria Filomena Santarelli
Institute of Clinical Physiology, CNR
Pisa, Italy

Jan Sijbers
Vision Laboratory, Campus
  Middelheim
University of Antwerp
Antwerp, Belgium

Martin Styner
Department of Computer Science
University of North Carolina
Chapel Hill, North Carolina

Nicola Vanello
Department of Electrical Systems
  and Automation
University of Pisa
Pisa, Italy

M.A. Viergever
Image Sciences Institute
Utrecht University
Utrecht, The Netherlands

Yongmei Michelle Wang
Departments of Statistics, Psychology
  Bioengineering
University of Illinois at
  Urbana–Champaign
Champaign, Illinois

Carl-Fredrik Westin
Laboratory of Mathematics in Imaging
Harvard Medical School
Brigham and Women's Hospital
Boston, Massachusetts

Leslie Ying
Department of Electrical Engineering
  and Computer Science
University of Wisonsin
Milwaukee, Wisconsin

Alistair A. Young
Department of Anatomy with
  Radiology
University of Auckland
Auckland, New Zealand

Francessca Zanderigo
Department of Information Engineering
University of Padova
Padova, Italy

Contents

Part I

Signal and Image Generation and Reconstruction

1 Basic Physics of MR Signal and Image Generation

Maria Filomena Santarelli

CONTENTS

1.1 INTRODUCTION

The nuclear magnetic resonance (NMR or MR) phenomenon in bulk matter was first demonstrated by Bloch and associates [1] and Purcell and associates [2] in 1946. Since then, MR has developed into a sophisticated technique with applications in a wide variety of disciplines that now include physics, chemistry, biology, and medicine. Over the years, MR has proved to be an invaluable tool for molecular structure determination and investigation of molecular dynamics in solids and liquids. In its latest development, application of MR to studies of living systems has attracted considerable attention from biochemists and clinicians alike. These studies have progressed along two parallel and perhaps complementary paths. First, MR is used as a spectroscopic method to provide chemical information from selected regions within an object (magnetic resonance spectroscopy [MRS]). Such information from a localized area in living tissue provides valuable metabolic data that are directly related to the state of health of the tissue and, in principle, can be used to monitor tissue response to therapy. In the second area of application, MR is used as an imaging tool to provide anatomic and pathologic information.

The rapid progress of MR to diverse fields of study can be attributed to the development of pulse Fourier transform techniques in the late 1960s [3]. Additional impetus was provided by the development of fast Fourier transform algorithms, advances in computer technology, and the advent of high-field superconducting magnets. Then, the introduction of new experimental concepts such as two-dimensional MR has further broadened its applications [4] to the magnetic resonance imaging (MRI) technique.

MRI is a tomographic imaging technique that produces images of internal physical and chemical characteristics of an object from externally measured MR signals. Tomography is an important area in the ever-growing field of imaging science. The Greek term *tomos* (τομοσ) means "cut," but tomography is concerned with creating images of the internal (anatomical or functional) organization of an object without physically cutting it open.

Image formation using MR signals is made possible by the spatial information encoding principles, originally named *zeugmatography* [4,5]. As it will be briefly described in this chapter, these principles enable one to uniquely encode spatial information into the activated MR signals detected outside an object.

As with any other tomographic imaging device, an MRI scanner outputs a multidimensional data array (or image) representing the spatial distribution of some measured physical quantity. But unlike many of them, MRI can generate two-dimensional sectional images at any orientation, three-dimensional volumetric images, or even four-dimensional images representing spatial–spectral or spatial–temporal (i.e., in cardiac imaging) distributions. In addition, no mechanical adjustments to the imaging machinery are involved in generating these images.

Another peculiarity of the MRI technique is that MR signals used for image formation come directly from the object itself. In this sense, MRI is a form of emission tomography similar to position emission tomography (PET) and single-photon computed tomography (SPECT). But unlike PET or SPECT, no injection of radioactive isotopes into the object is needed for signal generation in MRI. There are other forms of tomography in use, including transmission tomography and diffraction tomography. X-ray computer-assisted tomography (CT) belongs to the first category, while most acoustic tomography is of the diffraction type. In both cases, an external signal source is used to "probe" the object being imaged.

MRI operates in the radio frequency (RF) range, as shown in Figure 1.1. Therefore, the imaging process does not involve the use of ionizing radiation and does not have the associated potential harmful effects. However, because of the unique imaging scheme used, the resulting spatial resolution of MRI is not limited by the probing (or working) frequency range as in other remote-sensing technologies.

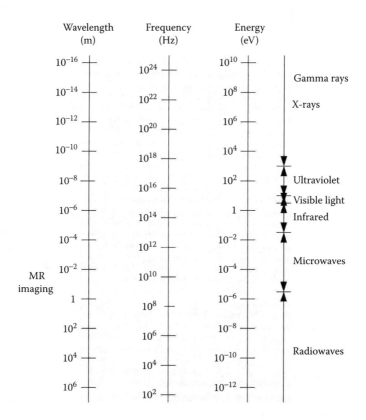

FIGURE 1.1 Electromagnetic wave range for MR and MRI.

Images are extremely rich in information content. The image pixel value is in general dependent on a lot of intrinsic parameters, including the nuclear spin density (ρ), the spin-lattice relaxation time (T1), the spin-spin relaxation time (T2), molecular motions (such as diffusion and perfusion), susceptibility effects, and chemical shift differences. The imaging effects of these parameters can be suppressed or enhanced in a specific experiment by another set of operator-selectable parameters, such as repetition time (TR), echo time (TE), and flip angle (α). Therefore, an MR image obtained from the same anatomical site can look drastically different with different data acquisition protocols.

In the present chapter, the basic physical principles of MRI are presented. The objective is to allow readers to understand and interpret the MR signal and image generation in order to introduce them to processing analysis: why and how signal processing theories and methods, described in the following chapters of the book, can be applied on MR signals and images.

For a deeper study of MRI physical principles, extensive literature exists [6–13], which the reader is encouraged to consult; part of the material of this chapter has been extracted from these texts.

1.2 NUCLEAR SPIN

The basis of NMR lies in a property possessed by certain nuclei, called the *spin angular momentum* (**p**). The spin angular momentum of the nucleus can be considered as an outcome of the rotational or spinning motion of the nucleus about its own axis. For this reason, nuclei having spin angular momentum are often referred to as nuclear spins. The spin angular momentum of a nucleus is defined by the nuclear *spin quantum number* I, and is given by the relationship

$$|\mathbf{p}| = \hbar \cdot \sqrt{[I(I+1)]} \qquad (1.1)$$

where $\hbar = h/2\pi$ and h is the Planck's constant. The value of spin quantum number depends on the structure of the nucleus — the number of protons and neutrons — and can be an integer, half-integer, or zero. In Table 1.1, the spin quantum number of some selected nuclei is given. Hydrogen (^{1}H, with I = 1/2), the most abundant element in nature and in the body, is most receptive to NMR experiments. On the other hand, the most common isotopes of carbon (^{12}C) and oxygen (^{16}O) have nuclei with I = 0 and hence cannot be observed by magnetic resonance experiments.

Because the nucleus is a charged particle, spin angular momentum is accompanied by a *magnetic moment* (**μ**) given by

$$\mathbf{\mu} = \gamma \mathbf{p} \qquad (1.2)$$

where γ is called the *gyromagnetic ratio*.

Note that both **μ** and **p** are vector quantities having magnitude and direction. The gyromagnetic ratio is characteristic of a particular nucleus (see Table 1.1), and it is proportional to the charge-to-mass ratio of the nucleus.

TABLE 1.1
NMR Properties of Some Selected Nuclei

Nucleus	Nuclear Spin	Gyromagnetic Ratio (MHz/T)	Natural Abundance (%)	Relative Sensitivity*
1H	1/2	42.58	99.98	1
13C	1/2	10.71	1.11	0.016
19F	1/2	40.05	100	0.870
31P	1/2	17.23	100	0.066
23Na	3/2	11.26	100	0.093

* calculated at constant field for an equal number of nuclei

1.3 NUCLEI IN A MAGNETIC FIELD

In MR experiments we are concerned with the behavior of nuclei placed in an external magnetic field. According to the *classical model*, the presence of a magnetic moment means that the nucleus behaves like a tiny bar magnet, with a north and a south pole, and it will therefore interact with a magnetic field.

In the case of a bar magnet, application of an external magnetic field would cause the magnet simply to align with or against the direction of the field. However, a nucleus has angular momentum and, consequently, precesses about the direction of the applied field, just as a spinning top precesses in the Earth's gravitational field. The interaction between the magnetic moment $\boldsymbol{\mu}$ and the field $\mathbf{B_0}$ tries to align the two, according to the formula:

$$\mathbf{L} = \boldsymbol{\mu} \times \mathbf{B_0} \tag{1.3}$$

where $\mathbf{L}$ is the torque or turning force.

This precessional motion is schematized in Figure 1.2 as a rotation of the magnetic moment vector $\boldsymbol{\mu}$ about the direction of the external magnetic field $\mathbf{B_0}$ in addition to the nuclear spin about its own axis.

The torque force causes the nucleus to precess about $\mathbf{B_0}$, altering the angular momentum $\mathbf{p}$; in fact,

$$d\mathbf{p}/dt = \mathbf{L} = \boldsymbol{\mu} \times \gamma \mathbf{B_0} \tag{1.4}$$

But $\mathbf{p} = \boldsymbol{\mu}/\gamma$, so that

$$d\boldsymbol{\mu}/dt = \boldsymbol{\mu} \times \gamma \mathbf{B_0} = \boldsymbol{\omega_0} \times \boldsymbol{\mu} \tag{1.5}$$

where $\boldsymbol{\omega_0} = -\gamma\,\mathbf{B_0}$
is the *frequency of precession*, i.e., the rate at which $\boldsymbol{\mu}$ rotates about $\mathbf{B_0}$, and is called the *Larmor frequency*; the minus sign describes the rotation direction.

As said before, this result can be deduced from the laws of classical physics. However, in order to obtain a complete description of the behavior of nuclei in a magnetic field, quantum mechanical theory must be considered.

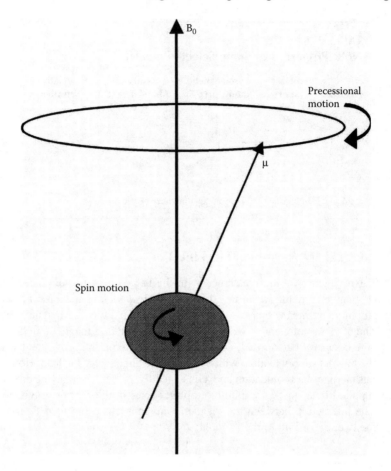

FIGURE 1.2 Precession of nuclear magnetic moment about the magnetic field $\mathbf{B_0}$.

According to the *quantum mechanical model*, the orientation of the spin angular momentum vector, and hence the magnetic moment, of the nucleus with respect to the direction of $\mathbf{B_0}$ cannot be arbitrary and it is subjected to certain restrictions. In fact, the nuclear magnetic moment can only have $2I + 1$ orientations in a magnetic field, corresponding to $2I + 1$ allowed energy levels, so that the proton (^1H) nucleus can either align with or against the applied field. These two orientations are similar to the two directions in which a bar magnet may orient in a magnetic field and are consequently called the parallel (spin-up) and the anti-parallel (spin-down) orientations. These orientations correspond to low and high energy states respectively, as schematically shown in Figure 1.3.

The difference in energy, $\Delta\mathbf{E}$, between the two states is proportional to the strength of the magnetic field $\mathbf{B_0}$, and it is given by

$$\Delta\mathbf{E} = \gamma\mathbf{B_0} \tag{1.6}$$

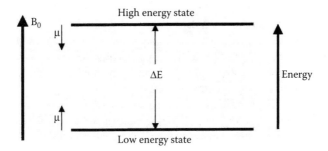

FIGURE 1.3 Energy level diagram for I = 1/2.

So far, the behavior of a single isolated nucleus has been considered. However, in practice, the mean result due to a large number of similar nuclei is observed. When an ensemble of nuclei is subjected to an external magnetic field, nuclei distribute themselves in the allowed orientations. At equilibrium, the population of nuclei in the parallel orientation (lower energy state) exceeds that in the antiparallel orientation (higher energy state) by a small amount, according to Boltzman statistics. The excess population in the lower energy state, ΔN, is dependent on the energy difference between the spin states and the absolute temperature, T:

$$\Delta N \cong N_0 \frac{\Delta E}{2kT} \qquad (1.7)$$

where N_0 is the total number of nuclei in the sample, and k is the Boltzman constant. The fractional excess of population ($\Delta N/N_0$) in the lower energy state is extremely small; for example, for hydrogen nuclei at body temperature (310 K) in a magnetic field $B_0 = 1$ T, $\Delta N/N_0 = 3.295 \times 10^{-6}$.

1.3.1 NOTES ON LARMOR FREQUENCY: CHANGES DUE TO DISHOMOGENEITIES

We have seen that the precession frequency of μ experiencing a $\mathbf{B_0}$ field is given by $\omega_0 = -\gamma \, \mathbf{B_0}$ and, considering the modulus:

$$\omega_0 = \gamma B_0 \qquad (1.8)$$

This relation, commonly called the *Larmor equation*, is important because the Larmor frequency is the natural resonance frequency of a spin system. Equation 1.8 shows that the resonance frequency of a spin system is linearly dependent on both the strength of the external magnetic field $\mathbf{B_0}$ and the value of the gyromagnetic ratio γ. This relationship describes the physical basis for achieving nucleus specificity. In fact, as shown in Table 1.1, the nuclei of ^{1}H and

^{31}P in an object resonate at 42.58 MHz and 11.26 MHz, respectively, when the object is placed in a magnetic field $\mathbf{B_0} = 1$ T; this difference in resonance frequency enables us to selectively image one of them without interacting with the other.

Actually, a specific spin system (i.e., hydrogen protons) may have a range of resonance frequencies. In this case, each group of nuclear spins that share the same resonance frequency is called an *isochromat*.

There are two main reasons for a magnetized spin system to have multiple isochromats: (1) the presence of inhomogeneities in the $\mathbf{B_0}$ field and (2) the chemical shift effect, which is exploited for chemical components studies.

When $\mathbf{B_0}$ is not homogeneous, spin with the same γ value will have different Larmor frequencies at different spatial locations. Such a condition can be usefully exploited, as we will see in MR image generation, when the inhomogeneity of $\mathbf{B_0}$ is known. However, if $\mathbf{B_0}$ inhomogeneities are not known, they are considered as bringing a negative effect, that is artifacts.

The *chemical shift effect* is due to the fact that nuclei in a spin system are part of different molecules in a chemically heterogeneous environment. Because each nucleus of a molecule is surrounded by orbiting electrons, these orbiting electrons produce their own weak magnetic fields, which "shield" the nucleus to varying degrees depending on the position of the nucleus in the molecule. As a result, the effective magnetic field that a nucleus "sees" is

$$\hat{B}_0 = B_0(1 - \delta) \tag{1.9}$$

where δ is a shielding constant taking on either positive or negative values. Based on the Larmor relationship, the resonance frequency for the nucleus is

$$\hat{\omega}_0 = \omega_0 - \Delta\omega = \omega_0(1 - \delta) \tag{1.10}$$

Equation 1.10 shows that spins in different chemical environments will have relative shifts in their resonance frequency even when B_0 is homogeneous. The frequency shift $\Delta\omega$ depends on both the strength of the magnetic field B_0 and the shielding constant δ. Usually, the value of δ is very small, on the order of a few parts per million (ppm), and it depends on the local chemical environment in which the nucleus is embedded. Knowledge of these chemical shift frequencies and the corresponding spin densities is of great importance for determining the chemical structures of an object, which is the subject of MR spectroscopy.

1.3.2 BULK MAGNETIZATION

To describe the collective behavior of a spin system, we introduce a macroscopic magnetization vector $\mathbf{M}$, which is the vector sum of all the microscopic magnetic moments in the object:

$$M = \sum_{i=1}^{N_s} \mu_i \tag{1.11}$$

where $\boldsymbol{\mu}_i$ is the magnetic moment of the i-th nuclear spin, and N_s is the total number of spins.

Although there is a microscopic transverse component for each magnetic moment vector, the transverse component of **M** is zero at equilibrium because the precessing magnetic moments have random phases. The macroscopic effect of an external magnetic field $\mathbf{B_0}$ on an ensemble of nuclei with nonzero spins is the generation of an observable bulk magnetization vector **M** pointing along the direction of $\mathbf{B_0}$.

The magnitude of the equilibrium magnetization **M** is equal to that of the spin excess predicted by the quantum model. **M** itself behaves like a large magnetic dipole moment, and if perturbed from its equilibrium state, it will precess at ω_0 about $\mathbf{B_0}$. By analogy with Equation 1.5, we can write:

$$d\mathbf{M}/dt = \boldsymbol{\omega_0} \times \mathbf{M} = \gamma \mathbf{M} \times \mathbf{B_0} \tag{1.12}$$

1.4 RF EXCITATION FOR THE RESONANCE PHENOMENON GENERATION

It is the precessing bulk magnetization that we detect in an MR experiment. In order to detect it, we must somehow perturb the system from its equilibrium state, and get **M** to process about $\mathbf{B_0}$. This is done by applying a second magnetic field $\mathbf{B_1}$, perpendicular to $\mathbf{B_0}$, rotating about $\mathbf{B_0}$ at ω_0 in synchronism with the precessing nuclear magnetic moments.

The $\mathbf{B_1}$ field causes **M** to tilt away from $\mathbf{B_0}$ and to execute a spiral path, as schematically described in Figure 1.4.

The term *RF pulse* is a synonym of the $\mathbf{B_1}$ field generation, so called because $\omega_0/2\pi$ is normally between 1MHz and 500 MHz, corresponding to radio waves. The field is usually turned on for a few microseconds or milliseconds. Also, in contrast to the static magnetic field $\mathbf{B_0}$, the $\mathbf{B_1}$ field is much weaker (i.e., $\mathbf{B_1}$ = 50 mT while $\mathbf{B_0}$ = 1.5 T).

A typical $\mathbf{B_1}$ field takes the following form:

$$\mathrm{B}_1(t) = \mathrm{B}_{1,x}(t) + i\mathrm{B}_{1,y}(t) = \mathrm{B}_1(t)e^{-i(\omega_0 t + \varphi)} \tag{1.13}$$

where $\mathbf{B_1}(t)$ is the envelope function, ω_0 is the excitation carrier frequency, and φ is the initial phase angle.

Equation 1.13 describes a circularly polarized RF pulse, perpendicular to the z axis, and hence to the $\mathbf{B_0}$ field. The initial phase angle φ, if it is a constant, has no significant effect on the excitation result so that we assume it is equal to zero. The excitation frequency ω_0 can be considered as a constant for almost all RF pulses, and it is determined by the resonance conditions. The envelope function $\mathbf{B_1}(t)$ is the heart of an RF pulse. It uniquely specifies the shape and duration of an RF pulse and thus its excitation property. In fact, many RF pulses are named

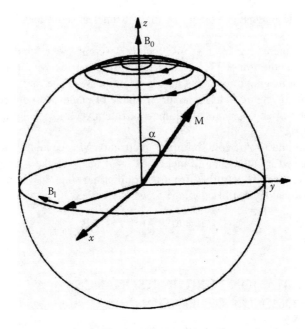

FIGURE 1.4 Trajectory of **M** after an RF pulse.

based on the characteristics of this function. For example, the envelope function of the widely used *rectangular RF pulse* (see Figure 1.5a) is given by:

$$B_1(t) = B_1 \prod \left(\frac{t - \tau/2}{\tau} \right) = \begin{cases} B_1 & 0 \le t \le \tau \\ 0 & \text{otherwise} \end{cases} \tag{1.14}$$

where τ is the pulse time width. Another frequently used RF pulse is the following *sinc pulse* (see Figure 1.5b):

$$B_1(t) = \begin{cases} B_1 \sin c[\pi f_\omega (t - \tau/2)] & 0 \le t \le \tau \\ 0 & \text{otherwise} \end{cases} \tag{1.15}$$

When $\mathbf{B_1}$ is switched off, **M** continues to precess, describing a cone at some angle α to $\mathbf{B_0}$ as shown in Figure 1.4. This is called the *flip angle* (FA); it depends on the strength of the $\mathbf{B_1}$ field and how long it is applied.

We can adjust the value of the FA by changing the duration τ or the amplitude of the envelope function $\mathbf{B_1}(t)$. If we turn off the transmitter RF after **M** has precessed down into the transverse plane, this is called a $\pi/2$ (or 90°) pulse. A $\pi/2$ pulse is commonly used as an excitation pulse, because it elicits the maximum signal from a sample that is in equilibrium. An RF pulse twice as long, a

Rectangular RF pulse

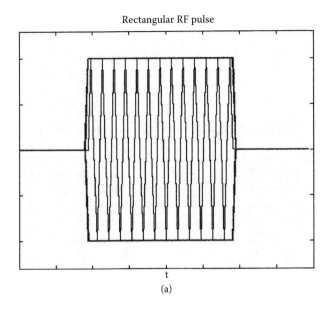

t

(a)

Sinc RF pulse

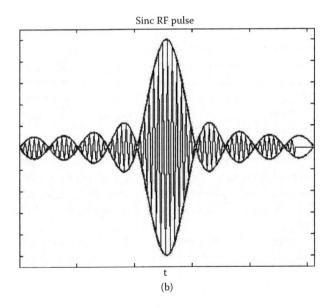

t

(b)

FIGURE 1.5 Typical RF pulse shapes: (a) rectangular pulse, (b) sinc pulse.

π (or 180°) pulse, will place **M** along the negative z axis. This is often called an *inversion pulse*. In general, for a constant **B**$_1$ amplitude, an α pulse causes the magnetization vector **M** to precess through the flip angle:

$$\alpha = \gamma \mathbf{B}_1 \tau \tag{1.16}$$

More generally, if we allow the amplitude of the applied RF to vary with time, the FA α is given by:

$$\alpha = \gamma \int_0^\tau B_1(t)dt \tag{1.17}$$

1.5   MR SIGNAL GENERATION AND ACQUISITION

Once displaced from the z axis by the RF pulse, the net magnetization **M** is no longer at equilibrium. We denote this nonequilibrium magnetization vector by **M**, and the magnitudes of its components along the x, y, and z axes will be denoted by M_x, M_y, and M_z, respectively. The magnitude of the component of **M** in the transverse xy plane (i.e., the resultant of M_x and M_y) will be denoted as M_{xy}. The equilibrium magnetization M_0 represents the situation in which **M** is aligned along the z axis corresponding to the case of $M_z = M_0$ and $M_{xy} = 0$.

During the period following the pulse, **M** experiences a torque due to the B_0 field and thus precesses about the B_0 field at the Larmor frequency. The effect of the precessing magnetization is similar to that of a rotating bar magnet and hence is equivalent to producing a periodically changing magnetic field in the transverse plane. If the sample investigated is surrounded by a suitable oriented coil of wire, an alternating voltage will be induced in the coil according to Faraday's law of induction. The coil used to receive the signal can be the same as or different from that used to produce the B_1 field so that precession of **M** corresponds to reorientation of M_{xy} in the xy plane at a rate equal to the Larmor frequency (i.e., M_x and M_y are oscillatory with time at a frequency equal to the Larmor frequency) while M_z is unchanged.

It can be shown that the amplitude of the alternating voltage induced in the receiver coil is proportional to the transverse magnetization component M_{xy}. Hence, a maximum amplitude voltage signal is obtained following a $\pi/2$ pulse because such a pulse creates a maximum M_{xy} component equal to M_0. In general, for RF pulse of flip angle α, the amplitude of the alternating voltage is proportional to $M_0 \sin(\alpha)$.

As a result of relaxation (such phenomena will be described later), M_{xy} (i.e., the amplitudes of oscillation of M_x and M_y) decays to zero exponentially; thus, the voltage signal observed in practice corresponds to an oscillating signal at the Larmor frequency, with exponentially decaying signal amplitude. This type of decaying signal, obtained in the absence of B_1, is called a *free induction signal* or *free induction decay* (FID).

1.5.1   FREE INDUCTION DECAY AND THE FOURIER TRANSFORM

The FID induced in the receiver coil is extremely weak and has a frequency in the RF range: very-high-frequency signal information cannot easily be stored in a computer, and direct amplification without introducing distortion is a difficult task. Therefore, low-level amplitude of the MR signal at the Larmor frequency is converted to a low-frequency signal by subtracting a frequency component equal to the frequency of a chosen reference signal (RF detection). The latter usually is chosen to be equal to that of $\mathbf{B_1}$ (i.e., ω_0); the resulting low-frequency signal has a frequency of $\Delta\omega$ in the Hz to KHz range. This signal is amplified again to the required level. The low-frequency signal (we will call it *base-band FID*) contains all the information, previously present in the RF signal, which is required to generate the MR spectrum.

In order to store the base-band FID in a computer, the analog signal is sampled at specific times (analog-to-digital conversion) and an array of numbers representing the sampled voltages is stored in the computer memory. The process of analog signal sampling must be performed in accordance with the Nyquist sampling theorem to ensure that the analog signal is correctly represented in digital form; the time duration in which the FID is sampled is referred to as the *acquisition* or *readout time*.

When a single group of equivalent nuclei (i.e., ^{1}H in a water sample) is considered, the FID represents a simple decay oscillation at a particular frequency, as described before. This frequency can be determined simply by measuring the period of the oscillation, t_{osc}, and calculating the value of $1/t_{osc}$. However, if the sample being examined contains different chemical shift frequencies, the observed FID represents the composite of several individual FID signals, with slightly different frequencies. The individual frequency components of any FID are most conveniently identified by subjecting the FID to Fourier transformation. The Fourier relationship between the FID and the MR spectrum is shown in Figure 1.6. All the characteristics of the FID are represented in the spectrum, but in a different format: the frequency of oscillation of the FID is indicated by the horizontal scale of the spectrum, the rate of decay of the FID is inversely related to the width of the spectral line at its half maximum, and the height (maximum amplitude) of the spectral line is directly proportional to

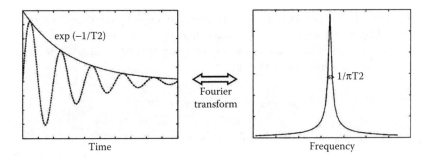

FIGURE 1.6 Fourier transform relationship between FID in the time domain and frequency domain.

the initial amplitude of the FID or, more exactly, it is equal to the area under the FID envelope.

1.6 MR SIGNAL CHARACTERISTICS

1.6.1 RELAXATION

Following the perturbation by the RF pulse, the spin system returns to the equilibrium population distribution between the energy levels by releasing excess energy into the surroundings. In the classical vector model, this corresponds to the return of the magnetization **M** to the equilibrium position along the z axis. Thus, during the relaxation period, any transverse magnetization component M_{xy} created by the RF pulse decays to zero, and, at the same time, the longitudinal magnetization component M_z returns to the equilibrium value of $\mathbf{M_0}$. The decay of M_{xy} and the recovery of M_z are two distinct processes, and they are referred to as *spin-spin* and *spin-lattice relaxation*, respectively.

1.6.1.1 The Decay of Transverse Magnetization: T2

The decay of transverse magnetization following an α-degree pulse, due to spin-spin relaxation can be described with the equation:

$$M_{xy}(t) = M_0 \sin \alpha e^{i(\omega_0 t + \varphi)} e^{-t/T2} \qquad (1.18)$$

where the parameter $T2$ is called the *transverse relaxation time* or *spin-spin relaxation time*; it represents the time interval required for the transverse magnetization to decay to 36.7% of its initial value M_0.

1.6.1.2 The Recovery of Longitudinal Magnetization: T1

After the application of an α-degree RF pulse, the longitudinal magnetization M_z recovers back to equilibrium at a rate that is linearly proportional to the difference between its current value and the equilibrium value; this rate is also characteristic of the sample. It can be derived that:

$$M_z(t) = M_0(1 - (1 - \cos \alpha)e^{-t/T1}) \qquad (1.19)$$

where the parameter $T1$ is called the *longitudinal relaxation time* or *spin-lattice relaxation time*, and it represents the time interval needed for the longitudinal magnetization to recover to a value of 63.2% of the equilibrium value $\mathbf{M_0}$.

This process of recovery of longitudinal magnetization is very important when calculating the contrast between tissues with different T 1 values and determining the imaging method that obtains the greatest signal-to-noise ratio (SNR).

The relaxation times T 1 and T 2 are determined by the molecular environment and thus are dependent on the sample. If the sample contains chemically shifted resonances, the nuclei in different chemical environments exhibit T 1 and T 2 values characteristic of the particular environment. Knowledge of T 1 and T 2 values is obtained experimentally by perturbing the equilibrium magnetization in appropriated multiple RF pulses. For pure liquids T 1 = T 2 and for biological samples T 2 < T 1. Molecules in a mobile liquid environment have T 1 and T 2 values in the range of tens of hundreds of milliseconds. For tissues in the body, the relaxation times are in the ranges 250 msec < T 1 < 2500 msec and 25 msec < T 2 < 250 msec and, usually, $5T 2 < = T 1 < = 10T 2$.

1.6.1.3 Pseudo-Relaxation: T2*

In practice, variations in the value of the magnetic field throughout the sample also cause dephasing of the magnetic moments, mimicking the decay of transverse magnetization caused by relaxation. It is convenient to define another relaxation time, T2*, describing the "observed" rate of decay of the FID:

$$\frac{1}{T2^*} = \frac{1}{T2} + \frac{1}{T2'} \qquad (1.20)$$

where T2′ describes the decay in signal due to the magnetic field inhomogeneity:

$$\frac{1}{T2'} = \gamma \Delta B_0 \qquad (1.21)$$

where ΔB_0 is the extent of variation of the applied magnetic field strength over the region occupied by the sample. This causes the nuclei in different regions of the sample to experience slightly different magnetic fields and hence to precess at different Larmor frequencies. The FID obtained in such a field is seen to decay faster than that determined by T 2 (T 2* < T 2).

1.6.2 PROTON DENSITY

Most of the hydrogen atoms in the tissues are within water molecules; it is these protons that we detect in an MR experiment. The term *proton density* (PD) simply refers to the number of protons per unit volume and is effectively proportional to the density of water in the tissue. Thus, for example, bone has very low proton density, and liver has high proton density, while blood has a very high proton density.

It is easy to see that we can identify the proton density with M_0, the equilibrium magnetization.

1.6.3 THE BLOCH EQUATIONS

From the preceding description of the observed behavior of $\mathbf{M}(t)$ the following equation can be postulated for the complete description of this motion:

$$\frac{d\mathbf{M}(t)}{dt} = \gamma \mathbf{M}(t) \times \mathbf{B}(t) - R\{\mathbf{M}(t) - \mathbf{M}_0\} \qquad (1.22)$$

where $\mathbf{B}(t)$ is composed of the static field and the RF field i.e., $\mathbf{B}(t) = \mathbf{B}_0 + \mathbf{B}_1(t)$ and R is the relaxation matrix:

$$R = \begin{bmatrix} 1/T2 & 0 & 0 \\ 0 & 1/T2 & 0 \\ 0 & 0 & 1/T1 \end{bmatrix} \qquad (1.23)$$

and the vector $\mathbf{M}_0 = [0, 0, M_0]$.

This is the set of equations used when constructing models of MRI. For most applications, these equations are transformed into the rotating frame of reference.

1.6.3.1 Rotating Frame of Reference

Usually, MR experiments involve the application of a sequence of RF pulses. Viewed from fixed coordinates (stationary frame), a description of the motion of the magnetization results can be complicated and difficult to visualize, especially when two or more RF pulses are applied. Hence, we consider the motion of the magnetization from the point of view of an observer rotating about an axis parallel with $\mathbf{B}_0$, in synchronism with the precessing nuclear magnetic moments: this is the so-called *rotating frame of reference*.

A rotating frame is a coordinate system whose transverse plane is rotating clockwise at an angular frequency ω. To distinguish it from the conventional stationary frame, we use x', y', and z' to denote the three orthogonal axes of this frame, and correspondingly, $\mathbf{i}'$, $\mathbf{j}'$, and $\mathbf{k}'$ as their unit directional vectors. Mathematically, this frame is related to the stationary frame by the following transformations:

$$\begin{cases} \mathbf{i}' = \cos(\omega t)\mathbf{i} - \sin(\omega t)\mathbf{j} \\ \mathbf{j}' = \sin(\omega t)\mathbf{i} + \cos(\omega t)\mathbf{j} \\ \mathbf{k}' = \mathbf{k} \end{cases} \qquad (1.24)$$

1.7 MULTIPLE RF PULSES

Up to now, we have described single-RF-pulse effects and subsequent detection of FID (one-pulse experiments). Now we consider the effects of multiple pulses on the magnetization.

Interrogation of the nuclear spin system with multiple pulses provides information that is not accessible via one-pulse experiment. In fact, multipulse experiments, such as multiple gradient-echo, inversion-recovery, and spin-echo sequences, enable experimental determination of proton density, T1, and T2, respectively.

1.7.1 GRADIENT ECHO

Gradient-echo pulse (GE) includes an α-degree RF pulse (usually a low flip angle is defined for fast imaging) followed by a time-varying gradient magnetic field in order to generate (and then acquire) an echo signal instead of a decaying FID. For physical concepts of magnetic field gradients, see Section 1.8. The key concept underlying GE formation is that a gradient field can dephase and rephase a signal in a controlled fashion so that one or multiple echo signals can be created. After the application of an α-degree RF pulse, a negative gradient (for example, along x axis) is switched on; as a result, spins in different x positions will acquire different phases, which can be expressed as:

$$\varphi(x,t) = \gamma \int_0^t -xG_x d\tau = -\gamma xG_x t \qquad (1.25)$$

so that the loss of spin phase coherence becomes progressively worse as time elapses after the excitation pulse. When the signal decays to zero, a positive gradient of the same strength is applied; the transverse magnetization components will gradually rephase, resulting in a regrowth of the signal. The spin phase angle is now given by:

$$\varphi(x,t) = -\gamma xG_x t' + \gamma \int_{t'}^t -xG_x d\tau \qquad (1.26)$$

The phase dispersal introduced by the negative gradient is gradually reduced over time after positive gradient is switched on at $t = t'$. After a time t', the spin phase φ is zero for any x value, which means that all the spins have rephased, and therefore an echo signal is formed. Time t' is called *echo time*, TE.

1.7.2 INVERSION RECOVERY

In the inversion-recovery (IR) pulse, the equilibrium magnetization $\mathbf{M_0}$ is initially perturbed by a 180° pulse. Following a short time period TI (time of inversion), a second perturbation is introduced in the form of a 90° pulse. This can be written as a 180°-TI-90°-FID sequence. In order to examine the effects of this pulse sequence, assume that the RF pulses are applied along the x′ axis of the rotating frame and are on-resonance. Prior to the 180° pulse the magnetization is at equilibrium (Figure 1.7a). The 180° pulse causes the magnetization to rotate by 180° about the x′ axis, and at the end of the pulse it will be oriented along the negative z axis (Figure 1.7b). Because, in this case, the 180° pulse

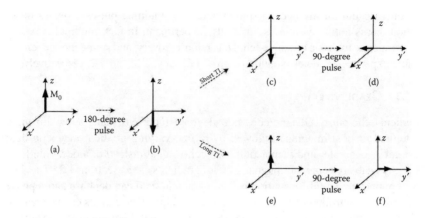

FIGURE 1.7 Rotating frame view of a magnetization subjected to an IR sequence.

inverts the magnetization from positive z axis to negative z axis, it is sometimes referred to as an inversion pulse. During the period TI, magnetization relaxes exponentially to the equilibrium position at a rate determined by T1; the magnitude of magnetization along the z axis (M_z) at a time TI after the 180° pulse is given by

$$M_z = M_0[1 - 2e^{(-TI/T1)}] \qquad (1.27)$$

Hence, the magnetization present at the end of the period TI is dependent on the ratio TI/T1. The magnetization following relatively short and long TI periods is shown in Figure 1.7c and Figure 1.7e, respectively. During the period TI, the magnetization, though not at equilibrium, is completely oriented along the z axis (i.e., $M_x = M_y = 0$); hence, an MR signal cannot be observed during this delay.

Application of a 90° pulse causes the magnetization present at that time to rotate by 90° about the x axis. Thus, at the end of the 90° pulse, the magnetization will be directed along the negative y′ axis or the positive y′ axis, depending on the value of TI chosen (Figure 1.7d and Figure 1.7f, respectively). In the subsequent time period, the transverse magnetization generated by the 90° pulse allows a FID to be observed. The initial amplitude of the FID will be proportional to M_z, remaining at the end of the delay period TI. In Figure 1.8 it is shown how the z component of magnetization changes as a function of the parameter TI for different T1 values.

1.7.3 Spin Echo

The spin-echo (SE) pulse sequence consists of an initial 90° pulse followed by a 180° pulse after a period TE/2 (time of echo), that is a 90°-TE/2-180°-echo sequence. The rotating frame view of an on-resonance magnetization subjected

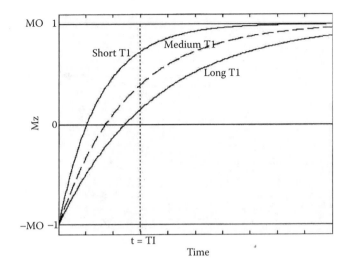

FIGURE 1.8 Plot of Mz vs. time in the IR pulse sequence.

to this sequence is shown in Figure 1.9. The initial 90° pulse rotates the equilibrium magnetization $\mathbf{M_0}$ by 90° about the x′ axis, creating transverse magnetization oriented along the y′ axis (Figure 1.9a and Figure 1.9b). During the following period, τ = TE/2, this transverse magnetization decays because of spin-spin relaxation and the inhomogeneity of the $\mathbf{B_0}$ magnetic field.

As previously said, inhomogeneity of the field causes the nuclei in different regions of the sample to have different Larmor frequencies; this means that the isochromats are off-resonance from the $\mathbf{B_1}$ field to different extents and hence will precess in the xy plane at slightly different frequencies. Therefore, individual isochromats will be seen to dephase in the xy plane following the 90° pulse. The position of a few isochromats at a time TE/2 after the pulse is shown in Figure 1.9c; the isochromats are displayed spread out on both sides of the y′ axis, because the Larmor frequency of some isochromats will be greater than the nominal value ($\Delta\omega > 0$) whereas that of others will be lower ($\Delta\omega < 0$). Consequently, isochromats with $\Delta\omega$ greater and less than zero will be seen to precess in opposite directions in the rotating frame. At sufficiently long TE/2 values, isochromats will be completely dephased in the xy plane and M_{xy} will be reduced to zero. In addition to spin-spin relaxation, inhomogeneity of the field contributes to the gradual reduction of the net magnetization in the xy plane. However, the additional decay of the magnetization due to field inhomogeneity can be reversed, applying a 180° pulse. Application of a 180° pulse at time TE/2 following the initial 90° pulse causes all the isochromats to rotate by 180° about the x′ axis; this brings the isochromats to their mirror-image position (Figure 1.9d). Following the 180° pulse, the frequency and direction of precession of isochromats in the xy plane remain same as prior to the 180° pulse as the Larmor frequency of each isochromat is unchanged. Hence, immediately

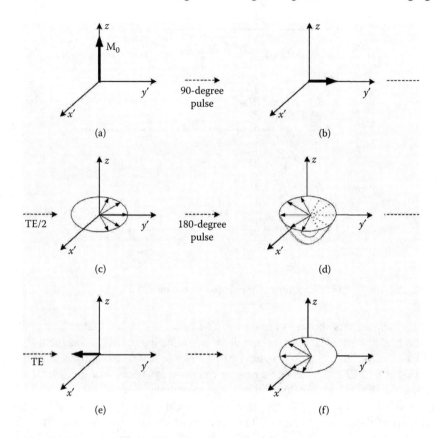

FIGURE 1.9 Rotating frame view of a magnetization subjected to an SE sequence.

after the 180° pulse, the isochromats precessing faster are seen to lag behind the slower ones. Precession of isochromats for a period TE/2 after the 180° pulse allows the faster isochromats to rephase with the slower ones, and at this instant all isochromats will be refocused along the negative y′ axis (Figure 1.9e). Further precession of isochromats following refocusing causes them to dephase in the xy plane (see Figure 1.9f). Therefore: following the 180° pulse, the net magnetization along the y′ axis ($M_{y'}$) increases, until a maximum is reached at time TE; after reaching the maximum, the magnetization decreases in a similar manner to the decay following the initial 90° pulse.

The net magnetization $M_{y'}$ at this time is determined by the decay due to spin-spin relaxation only and is given by:

$$M_{y'} = M_0 e^{(-TE/T2)} \qquad (1.28)$$

The maximum amplitude of the echo is proportional to M_y, given by Equation 1.28.

Equation 1.28 is applicable when the effect of molecular diffusion is negligible. For complete refocusing of isochromats, each nucleus must experience the same field during period TE. Movement of nuclei in an inhomogeneous field because of diffusion causes the echo amplitude to be reduced.

1.8 MAGNETIC FIELD GRADIENTS

Magnetic field gradients allow spatial information to be obtained from analysis of the MR signal. A field gradient is an additional magnetic field in the same direction as B_0, whose amplitude varies linearly with position along a chosen axis. The application of a field gradient G_x in the x direction, for example, causes the magnetic field strength to vary according to:

$$B_z(x) = B_0 + xG_x \qquad (1.29)$$

Magnetic field gradients are produced by combining the magnetic fields from two sources: the main homogeneous B_0 field, plus a smaller magnetic field directed primarily along the z axis; such a secondary magnetic field is produced by current-carrying coils (gradient coils). The design of a gradient coil is such that the strength of the magnetic field produced by it varies linearly along a certain direction. When such a field is superimposed on the homogeneous field B_0, it either reinforces or opposes B_0 to a different degree, depending on the spatial coordinate. This results in a field that is centered on B_0.

In Figure 1.10, the gradient field is added to the uniform B_0 field, obtaining the total magnetic field that varies linearly along the x axis.

It is important to note that the total magnetic field is always along the B_0 axis (for convention, z axis), while the gradient can be along any axis, x, y, or z:

$$\begin{aligned} G_x &= dB_z/dx \\ G_y &= dB_z/dy \\ G_z &= dB_z/dz \end{aligned} \qquad (1.30)$$

The units of magnetic field gradient are Tesla per meter (Tm^{-1}).

To understand how magnetic field gradients are utilized in MRI, consider a large sample of water placed in such a field. If the sample is subjected to a G_x gradient, ^1H nuclei at different x coordinates possess different Larmor frequencies:

$$\omega(x) = \gamma(B_0 + xG_x) \qquad (1.31)$$

Hence, in the presence of G_x, planes of constant field strength also become planes of constant resonance frequency. Equation 1.31 describes that a relationship between the position and the resonance frequency can be established by application of a magnetic field gradient. It follows that the ^1H spectrum obtained

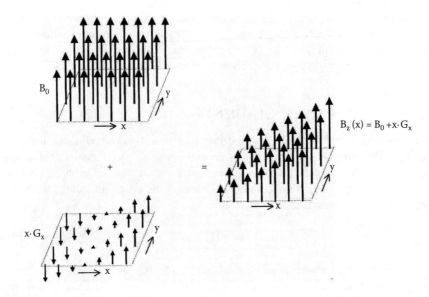

FIGURE 1.10 Magnetic field gradient along x axis. Up: uniform magnetic field **B₀**; down: the gradient field; right: the total magnetic field.

from the water sample subjected to a gradient magnetic field will show a distribution of resonance frequencies.

The spectral amplitude at a particular frequency is proportional to the number of nuclei in the given constant-frequency plane. It can be seen that the MR spectrum of an object placed in linear magnetic field gradient corresponds to the projection of the object onto the gradient direction (Figure 1.11).

1.9 SPATIAL LOCALIZATION OF MR SIGNALS

There are three main methods of spatial discrimination, all of which use field gradients, and that are combined in the *imaging pulse sequence*. The techniques are called *slice selection, frequency encoding*, and *phase encoding*.

1.9.1 SLICE SELECTION

Slice selection is the method by which the RF excitation, and therefore the signal, is limited to a chosen slice within the sample. It is achieved by applying the excitation pulse simultaneously with a gradient perpendicular to the desired slice. So, if we want to image a slice in the xy plane, the selection gradient will be applied orthogonal to this plane, i.e., in the z direction. The effect of the gradient

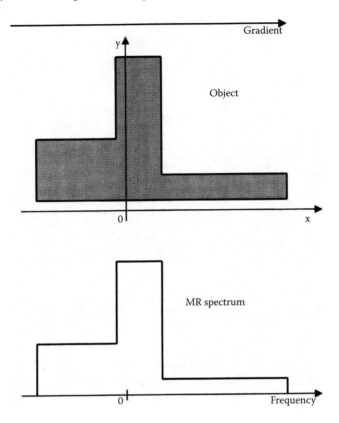

FIGURE 1.11 Relationship between the spatial extent of the sample of water and the MR spectrum, in the presence of a linear gradient: the MR spectrum corresponds to the projection of the object in the gradient direction.

is to make the resonance frequency a function of position along the z direction:

$$\omega(z) = \gamma B_z(z) = \gamma (B_0 + zG_z) \qquad (1.32)$$

As seen before, the RF pulse is not just an on-off pulse of radio waves, but it is shaped (i.e., amplitude modulated) so that it contains a narrow spread of frequencies close to the fundamental resonance frequency of the magnet ω_0.

Then, the width of the slice, Δz, is given by:

$$\Delta z = \Delta \omega / \gamma G_z \qquad (1.33)$$

as shown in Figure 1.12, where $\Delta \omega$ is the bandwidth of frequencies contained within the selective RF pulse.

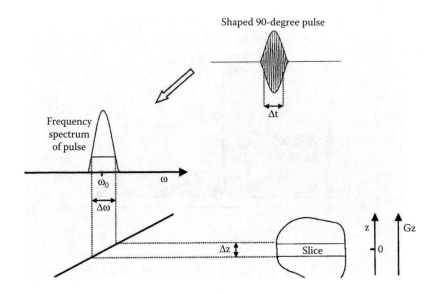

FIGURE 1.12 Selective excitation of a thin slice of the sample.

$\Delta\omega$ is related to the shape and the duration of the pulse; if we consider a Gaussian-shaped pulse, $\Delta\omega = 2\pi/\Delta t$, where $\Delta\omega$ is the full width at half maximum (FWHM) of the pulse's frequency spectrum, and Δt is the relevant FWHM of the pulse envelope in seconds. From Equation 1.33, the slice can be made thinner by decreasing the spectral bandwidth of the pulse (i.e., by making the pulse longer in time) or by increasing the strength of the slice selection gradient G_z.

The slice profile is determined by the spectral contents of the selective pulse, and it is approximately given by the Fourier transform of the RF pulse envelope. Thus, a Gaussian-shaped 90° pulse gives a roughly Gaussian slice profile.

The slice selection pulse sequence can be represented by a *pulse timing diagram*, as shown in Figure 1.13, showing the RF pulse and selection gradient as a function of time. The selection gradient is followed by a negative gradient pulse in order to bring the spins back into phase across the slice.

After a signal has been activated by a selective or nonselective pulse, spatial information can be encoded into the signal during the free precession period. We have essentially two ways to encode spatial information: frequency encoding and phase encoding.

1.9.2 FREQUENCY ENCODING

Frequency encoding makes the oscillation frequency of an MR signal linearly dependent on its spatial origin. Let us consider an idealized one-dimensional object with spin distribution $\rho(x)$. If the magnetic field that the object experiences

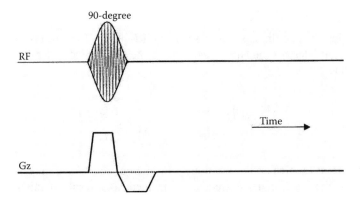

FIGURE 1.13 Example of a pulse timing diagram relevant to the slice selection operation.

after an excitation is the homogeneous $\mathbf{B_0}$ field plus another linear gradient field (xG_x), the Larmor frequency at position x is

$$\omega(x) = \omega_0 + \gamma x G_x \qquad (1.34)$$

Correspondingly, the FID signal ds(x,t), generated locally from spins in an infinitesimal interval dx at point x without considering the transverse relaxation effect, is

$$ds(x,t) = c\rho(x)dxe^{-i\gamma(B_0+xG_x)t} \qquad (1.35)$$

where the constant of proportionality c is dependent on the flip angle, the main field strength, and so on. The signal in Equation 1.35 is said to be *frequency encoded* because its oscillation frequency is linearly related to the spatial location; consequently, G_x is called a *frequency-encoding gradient*. The signal received from the entire sample in the presence of this gradient is:

$$s(t) = \int_{-\infty}^{\infty} ds(x,t) = \int_{-\infty}^{\infty} c\rho(x)e^{-i\gamma(B_0+xG_x)}dx = \left[\int_{-\infty}^{\infty} c\rho(x)e^{-i\gamma xG_x t}dx\right]e^{-i\omega_0 t}$$

$$(1.36)$$

In the general case, the received frequency-encoded FID signal after demodulation (i.e., after removal of the carrier signal $\exp(-i\omega_0 t)$), is given by

$$s(t) = c\int_{-\infty}^{\infty} \rho(\mathbf{r})e^{-i\gamma r G_{freq} t}d\mathbf{r} \qquad (1.37)$$

where G_{freq} is the frequency-encoding gradient defined by $G_{freq} = (G_x, G_y, G_z)$.

1.9.3 PHASE ENCODING

Considering the one-dimensional case after a RF pulse, if we turn on a gradient G_y for a short interval t_y and then we turn it off, the local signal under the influence of this gradient is:

$$ds(t,y) = \begin{cases} \rho(y)e^{-i\gamma B_0 + yG_y)t} & 0 \le t \le t_y \\ \rho(y)e^{-i\gamma yG_y t_y}e^{-i\gamma B_0 t} & t \ge t_y \end{cases} \tag{1.38}$$

where $\rho(y)$ is the spin distribution along y. From Equation 1.38, during the interval $0 \le t \le t_y$ the local signal is frequency encoded; as a result of this frequency encoding, signals from different y positions accumulate different phase angles after a time interval t_y. Therefore, the signal collected after t_y will bear an initial phase angle

$$\varphi(y) = -\gamma y G_y t_y \tag{1.39}$$

Because $\varphi(y)$ is linearly related to the signal location y, the signal is said to be *phase encoded*, the gradient G_y is called *phase-encoding gradient*, and t_y is the *phase-encoding interval*.

Phase encoding along an arbitrary direction can be also done for a multi-dimensional object by turning on G_x, G_y, and G_z simultaneously during the phase-encoding period $G_{phas} = (G_x, G_y, G_z)$ for $0 \le t \le t_y$; the initial angle is $\varphi(\mathbf{r}) = -\gamma \mathbf{r} G_{phas} t_y$. Similar to frequency encoding, the received signal is the sum of all the local phase-encoded signals and is given by:

$$s(t) = \int_{-\infty}^{\infty} ds(r,t) = c \left[\int_{-\infty}^{\infty} \rho(r)e^{-i\gamma r G_{phas} t_G} dr \right] e^{-i\omega_0 t} \tag{1.40}$$

where the carrier signal $\exp(-i\omega_0 t)$ is removed after signal demodulation.

1.9.4 PHASE HISTORY OF MAGNETIZATION VECTORS DURING PHASE ENCODING

Let us consider the evolution of the phase angle of magnetization vectors in the transverse plane as a function of a different phase-encoding gradient ampli-tude $G'_y = mG_y$ by varying m; we call this a *phase-encoding step*. Referring to the scheme of Figure 1.14, each phase-encoding step corresponds to a different value of m, that is, a different amplitude of the phase-encoding gradient. We can write the sequence of phase shifts added to a magnetization vector at the location y_0 as $\varphi^m(y_0) = -\gamma y_0 m G_y t_y$. Therefore, the expression for a set of different

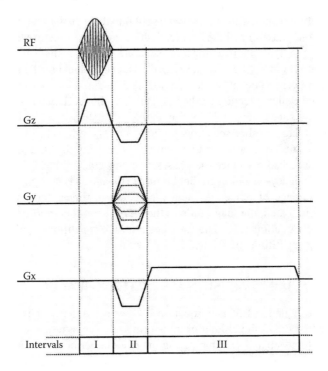

FIGURE 1.14 Schematic representation of a sequence timing diagram.

phase-encoding gradient amplitudes can be written:

$$s^m(t) = c\left[\int_{-\infty}^{\infty} \rho(y)e^{-i\gamma mG_y t_y}dy\right]e^{-i\omega_0 t} \tag{1.41}$$

A Fourier transform applied to the sequence $s^m(t)$ for different m values, can be used to compute the position of the object in the y direction. If the signal is from a collection of point objects in a column with different y offsets, the Fourier transform of the resulting signal will yield a spectrum that is proportional to a profile of the column.

1.9.5 TIMING DIAGRAM OF AN IMAGING SEQUENCE

Image sequence timing diagrams, also called *sequence diagrams*, are commonly used to describe the implementation of a particular MR sequence, and show the magnitude and duration of the three orthogonal magnetic field gradients and the RF pulses. An example of pulse sequence diagram is shown in Figure 1.14. In particular it shows the two-dimensional Fourier transform image formation method, as described in Reference 14, that is a development of the earlier technique of Fourier zeugmatography [5]. It could be considered the basic imaging sequence from which all the hundreds of image sequences existing nowadays are derived.

Generally, a sequence diagram can be split into three distinct sections, namely, slice selection, phase encoding, and readout, according to the previous descriptions. In Figure 1.14 such sections are separated in time (three time intervals) but, especially in more recently developed sequences, some of them overlap.

The first event to occur in the imaging sequence represented in Figure 1.14 is to turn on the slice selection gradient, together with the RF pulse. As previously described, the slice selective RF pulse should be a shaped pulse. Once the RF pulse is complete, the slice selection gradient is turned off, and a phase-encoding gradient is turned on. In order to obtain an MR echo signal, a negative read gradient is switched on. Once the phase-encoding gradient has been turned off, a positive frequency-encoding gradient is turned on and an echo signal is recorded.

This sequence of pulses is usually repeated m times, and each time the sequence is repeated, the magnitude of the phase-encoding gradient is changed according to Equation 1.41. The time between the repetitions of the sequence is called the repetition time, TR.

1.10 ACQUIRING MR SIGNALS IN THE K-SPACE

According to Equation 1.36 and Equation 1.41 we can describe the signals resulting from a two-dimensional Fourier transform sequence as a function of both the phase-encoding step and the time during the readout period. When M frequency-encoded FIDs are obtained, each one experiences a different value of the phase-encoding gradient amplitude; usually both positive and negative amplitudes are applied:

$$s^m(t) = c\left[\int_{-\infty}^{\infty}\int_{-\infty}^{\infty} \rho(x, y)e^{-i\gamma(xG_xt+ymG_yt_G)}dxdy\right]e^{-i\omega_0t} \qquad (1.42)$$

In order to obtain digital MR signals, data acquired during frequency gradient activation are sampled. So that, k-space data are sampled data, memorized in a matrix of $N \times M$ points, if N is the number of samples along reading gradient and M the number of times the phase gradient is activated.

Then, if each FID is sampled and Δt is the sampling time interval, and we consider the demodulated signal, from Equation 1.42 we obtain:

$$s(n,m) = c\int_{-\infty}^{\infty}\int_{-\infty}^{\infty} \rho(x,y)e^{-i\gamma(xG_xn\Delta t+ymG_yt_G)}dxdy$$

$$0 \le n \le N$$
$$-M/2+1 \le m \le M/2 \qquad (1.43)$$

The formula in Equation 1.43 is usually referred to as the *imaging equation*. So that, if M = 256 and the FID sampling points N are 256, then a 256 × 256 data matrix of complex numbers is the result.

Considering the following substitution:

$$k_x = \gamma n \Delta t G_x$$
$$k_y = \gamma m t_G G_y \tag{1.44}$$

the formula in Equation 1.43 can be rewritten as:

$$s(k_x, k_y) = c \int_{-\infty}^{\infty} \int_{-\infty}^{\infty} \rho(x,y) e^{-i(xk_x + yk_y)} dx dy \tag{1.45}$$

This shows that the data matrix $s(k_x, k_y)$ is a sampling of the Fourier coefficients of the function $\rho(x, y)$. Therefore, by applying a two-dimensional inverse Fourier transform to the data $s(n, m)$, the result will be an estimate of the function $\rho(x, y)$.

Several parameters of interest in the k-space can be defined in terms of parameters described in the pulse sequence. The sample spacing and width of the k-space are:

$$\Delta k_x = \frac{\gamma}{2\pi} G_x \Delta t$$

$$\Delta k_y = \frac{\gamma}{2\pi} G_y \Delta y$$

$$W_{kx} = N \Delta k_x = \frac{\gamma}{2\pi} G_x T_x \tag{1.46}$$

$$W_{ky} = M \Delta k_y = \frac{\gamma}{2\pi} 2G_{y,\max} t_G$$

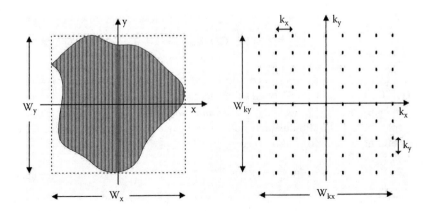

FIGURE 1.15 Sampling parameters of k-space.

where (see also Figure 1.15):

 Δk_x and Δk_y are the frequency interval step sizes in the x and y directions, respectively,

 W_{kx} and W_{ky} are the maximum frequencies that contain the object information

 Δt is the readout sampling time interval and T_x is the readout gradient time duration

 Δy is the phase-encoding gradient step size

 $G_{y,max}$ is the phase gradient maximum amplitude

Creating an MR image requires sampling the two-dimensional k-space with sufficient density (Δk_x, Δk_y) over a specified extent (W_{kx}, W_{ky}). The Nyquist sampling theorem dictates the sampling spacing necessary to prevent spatial aliasing of the reconstructed object (parts of the object can alias to different locations). The unaliased region is known as the *field of view* (FOV). The extent of the acquisition in Fourier space dictates the high-spatial frequency content and, hence, the spatial resolution.

For data acquired on a two-dimensional rectilinear grid in k-space and reconstructed with a two-dimensional Fast Fourier transform (FFT), the spatial resolution and FOV relationships are:

$$FOV_x = 1/\Delta k_x \qquad FOV_y = 1/\Delta k_y$$
$$\Delta x = 1/W_{kx} \qquad \Delta y = 1/W_{ky} \tag{1.47}$$

Up to now, we have described the MR signal equations for MRI for a particular pulse sequence (two-dimensional Fourier) and a homogeneous sample. In general, the value of the MR parameters, ($T1$, $T2$, and $T2^*$) vary with position; this generates the contrast between tissues. So, a more general form of Equation 1.45 is:

$$s(\mathbf{k}(t)) = \int_{sample} \zeta(\mathbf{r},\mathbf{p})e^{-2\pi i \mathbf{k}(t)\cdot\mathbf{r}}d\mathbf{r}$$

$$\mathbf{k}(t) = \gamma \int_0^t \mathbf{G}(\tau)d\tau \tag{1.48}$$

where $\zeta(\mathbf{r}, \mathbf{p})$ is a function of position $\mathbf{r}$, and $\mathbf{p}$ is a parameters vector $\mathbf{p} = (\rho(\mathbf{r}),$ $T1(\mathbf{r}), T2(\mathbf{r}), \ldots,$ TR, TE, $\alpha, \ldots)$ that describes the dependence on the *tissue parameters* ($\rho, T1, T2$, etc.) and the *scanner parameters* (i.e., TR, TE, etc.). For example, in the two-dimensional Fourier sequence, if there are no effects from magnetic field inhomogeneities (i.e., $T2 = T2^*$):

$$\zeta(\mathbf{r},\mathbf{p}) = \rho(\mathbf{r})e^{-TE/T2(\mathbf{r})}(1 - e^{-TR/T1(\mathbf{r})}) \tag{1.49}$$

$\mathbf{k}(t)$, described in Equation 1.48, is the spatial frequency vector as a function of time; the Fourier coefficients of the image $s(\mathbf{k}(t))$ can be sampled along the pathway defined by $\mathbf{k}(t)$.

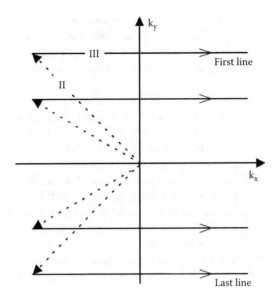

FIGURE 1.16 k-space path for 2-D Fourier transform.

1.10.1 K-Space Trajectories

By applying Equation 1.42 to a field gradient sequence, a path in k-space can be traced out. Consider again the sequence diagram of Figure 1.14; let us trace the paths in the k-space generated by the three time intervals as shown in Figure 1.14.

Figure 1.16 shows the path(s) in k-space corresponding to intervals II and III. Each different phase encode takes us to a new starting point on the left during interval II; during interval III, only the readout gradient is applied, ensuring that **k** travels horizontally from left to right at constant speed. Such trajectory is due to the x-gradient (readout gradient), shape, which has a negative compensation lobe before the sampling period that allows the k-space to be sampled symmetrically with the readout gradient; i.e., during the readout period, **k** goes from $-k_{max}$ to k_{max}.

The important parts of the paths are where the MR signal is being sampled for processing into an image. In a straightforward FT imaging procedure, the whole track in k-space should be a rectilinear scan, preferably with a square aspect ratio, because this implies equal spatial resolution along both axes.

1.11 IMAGING METHODS

There are numerous variations on the basic MRI sequences described earlier. Other than the RF pulse shapes and repetitions (gradient echo, spin echo, inversion recovery, etc.), there are many aspects that distinguish them from one another; for example, data acquisition and image reconstruction velocity. There exist a number of *rapid imaging techniques* that could be grouped under echo-planar imaging (EPI), spiral imaging, and, more recently, parallel imaging. Another important aspect that characterizes groups of imaging sequences is the *k-space sampling method*: it can be homogeneous or not, it can include the partial k-space filling instead of full k-space filling, etc.

In the present section, only a quick description is given: the most recent and efficient methods and their applications will be described in more detail in the rest of the book.

One feature that distinguishes different image acquisition methods is how quickly one can acquire data for an image and how easily the methods can be extended to generate higher spatial resolution. The two-dimensional Fourier transform method described in the preceding text is excellent for generating high-quality, high-spatial resolution images, but it is rather slow because only one line in k-space is acquired for each TR. The EPI method is well known as a very fast method that generally has limited spatial resolution. The EPI pulse sequence is uniquely characterized by a zigzag-like pathway that goes back and forth rapidly in the frequency direction (k_x) and moves in small steps in the phase direction (k_y), as schematically shown in Figure 1.17.

As in the two-dimensional Fourier transform method, we apply the Fourier transform in both frequency and phase directions obtaining the two-dimensional spatial map of MRI data, which is an MR image. As it can be seen, a single RF

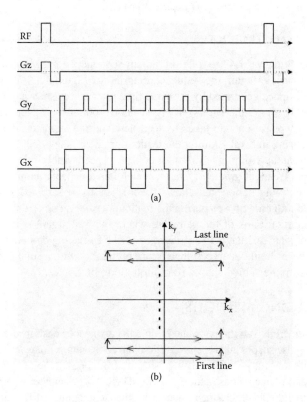

FIGURE 1.17 (a) Timing diagram of EPI sequence and (b) relevant k-space path.

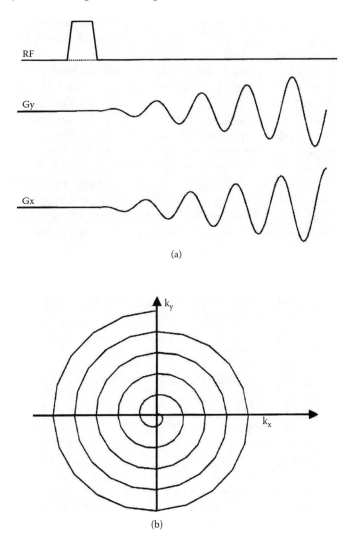

RF

Gy

Gx

(a)

k_y

k_x

(b)

FIGURE 1.18 Timing diagram of (a) spiral sequence and (b) relevant k-space path.

pulse is all that is needed to produce sufficient data for an image. This is called a *single-shot imaging method*, which is very popular in functional MRI (fMRI). There are limits on how fast the gradient fields can be changed and, thus, there are limits on the spatial resolution that can be achieved with this method.

Another common rapid imaging method is *spiral imaging*. It exploits the mathematics of how k-space trajectories are related to gradient waveforms (Equation 1.48) and that allows nearly any continuous trajectory to be acquired. In particular, in spiral imaging, gradients are designed to produce a spiral trajectory in k-space and samples are placed along that pathway (as schematically shown in Figure 1.18).

Then, taking the two-dimensional Fourier transform of data in this plane will produce an image. Similar to EPI, spiral imaging can produce is suitable for use in rapid imaging such as fMRI and cardiac imaging.

Parallel MRI techniques are characterized by multiple RF receiver coils and associated RF receiver electronics. The proposed methods, such as SMASH [15], SENSE [16], and SPACE RIP [17] offer improved temporal and/or spatial resolution, so that they are good candidates for clinical imaging applications that require both high speed and resolution, such as cardiac imaging. Each parallel MRI method uses a unique reconstruction scheme that exploits the independence of the spatial sensitivity profiles of the RF coils. Most recently, they have been successfully combined with other fast acquisition methods, offering further improvements; for example, the UNFOLD method for increasing temporal resolution was recently applied to parallel coil acquisition [18], and non-Cartesian SENSE [19] renders the use of the SENSE reconstruction technique compatible with complicated k-space trajectories, such as spiral imaging.

ACRONYMS

EPI	Echo-planar imaging
FA	Flip angle
FFT	Fast Fourier transform
FID	Free induction decay
fMRI	functional magnetic resonance imaging
FOV	Field of view
GE	Gradient echo
IR	Inversion recovery
MR	Magnetic resonance
MRI	Magnetic resonance imaging
MRS	Magnetic resonance spectroscopy
NMR	Nuclear magnetic resonance
PD	Proton density
RF	Radio frequency
SE	Spin echo
SNR	Signal-to-noise ratio
TE	Echo time
TI	Inversion time
TR	Repetition time

REFERENCES

1. Bloch, F. (1946). Nuclear induction. *Phys. Rev.* 70: 460.
2. Purcell, E.M., Torrey, H.C., and Pound, R.V. (1946). Resonance absorption by nuclear magnetic moments in a solid. *Phys. Rev.* 69: 37.
3. Ernst, R.R. and Anderson, W.A. (1966). Application of Fourier transform spectroscopy to magnetic resonance. *Rev. Sci. Instrum.* 37: 93.

4. Lauterbur, P.C. (1973). Image formation by induced local interactions: examples employing nuclear magnetic resonance. *Nature* 242: 190–191.

5. Kumar, A., Welti, D., and Ernst, R.R. (1975). NMR Fourier Zeugmatography. *J. Magn. Reson.* 18: 69–83.

6. Talagala, S.L. and Wolf, G.L. (1991). Principles of nuclear magnetic resonance. in Melvin, L.M., Schelbert, H.R., Skorton, D.J., and Wolf, G.L. (Eds.). *Cardiac Imaging.* W.B. Saunders Company. pp. 732–743.

7. Talagala, S.L. and Wolf, G.L. (1991). Principles of magnetic resonance imaging. in Melvin, L.M., Schelbert, H.R., Skorton, D.J., and Wolf, G.L. (Eds.). *Cardiac Imaging.* W.B. Saunders Company. pp. 744–758.

8. Morris, P.G. (1987). *Nuclear Magnetic Resonance in Medicine and Biology.* Clarendon Press, Oxford.

9. Chen, C.N. and Hoult, D.I. (1989). *Biomedical Magnetic Resonance Technology.* IOP Publishing Ltd., New York.

10. Liang, Z.-P. and Lauterbur, P.C. (1999). *Principles of Magnetic Resonance Imaging.* SPIE Press–IEEE Press.

11. Stark, D.D. and Bradley, W.G. (1992). *Magnetic Resonance Imaging.* Vol. 1 and 2, Mosby-Year Book, St. Louis.

12. Haacke, E.M., Brown, R.W., Thomson, M.R., and Venkatesan, R. (1999). *Magnetic Resonance Imaging — Physical Principles and Sequence Design.* Wiley-Liss. John Wiley & Sons, New York.

13. Brown, M.A. and Semelka, R.C. (1999). *MRI: Basic Principles and Applications.* Wiley-Liss., New York.

14. Edelstein, W.A., Hutchison, J.M.S., Johnson, G., and Redpath, T. (1980). Spin warp NMR imaging and applications to human whole-body imaging. *Phys. Med. Biol.* 25: 751-756.

15. Sodickson, D.K. and Manning, W.J. (1997). Simultaneous acquisition of spatial harmonics (SMASH): fast imaging with radiofrequency coil arrays. *Magn. Reson. Med.* 38(4):591–603.

16. Pruessmann, K.P., Weiger, M., Scheidegger, M.B., and Boesiger, P. (1999). SENSE: Sensitivity encoding for fast MRI. *Magn. Reson. Med.* 42(5): 952–962.

17. Kyriakos, W.E., Panych, L.P., Kacher, D.F., Westin, C.F., Bao, S.M., Mulkern, R.V., and Jolesz, F.A. (2000). Sensitivity profiles from an array of coils for encoding and reconstruction in parallel (SpaceRIP). *Magn. Reson. Med.* 44(2): 301–308.

18. Madore, B. (2002). Using UNFOLD to remove artifacts in parallel imaging and in partial-fourier imaging. *Magn. Reson. Med.* 48(3): 493–501.

19. Pruessmann, K.P., Weiger, M., Bornert, P., and Boesiger, P. (2001). Advances in sensitivity encoding with arbitrary k-space trajectories. *Magn. Reson. Med.* 46(4): 638–651.

2 Advanced Image Reconstruction Methods in MRI

Leslie Ying[1] and Zhi-Pei Liang

CONTENTS

This chapter provides a tutorial overview of advanced image reconstruction methods used in MRI. The term "advanced" is used loosely to refer to the class of non-Fourier reconstruction methods developed for handling the inverse problem with limited Fourier samples. We will consider two specific cases: (a) the superresolution reconstruction problem (associated with limited Fourier samples collected at the Nyquist rate) and (b) the parallel imaging problem (arising when Fourier samples are collected at sub-Nyquist rates, using multiple nonuniform receiver channels).

For notational convenience, we will consider only the one-dimensional case. The following is a summary of notations used in this chapter.

$\rho(x)$ Desired image
$\hat{\rho}(x)$ Reconstructed image

$\vec{\rho}$ Vector of sample values from $\rho(x)$

$\hat{\vec{\rho}}$ Vector of sample values from $\hat{\rho}(x)$

$\vec{\rho}_r$ Vector of sample values of a regularization image

$D(k_n)$ k-space data sampled at k_n

$d_\ell(x)$ Aliased reconstructed image of the ℓth coil

$\vec{d}$ Vector of $d_\ell(x)$ from the coil array

L Total number of receiver channels (coils)

R Acceleration factor or reduction factor in parallel MRI

$s_\ell(x)$ Sensitivity of lth coil

Δk Nyquist sampling interval (i.e., $\Delta k = 1/W$)

$\Delta\hat{k}$ Increased sampling interval used in parallel MRI

W $\rho(x)$ is assumed to be support-limited to $|x| < W/2$

$\hat{W}$ Reduced FOV due to sub-sampling ($\hat{W} = W/R$)

λ Regularization parameter.

2.1 INTRODUCTION

This chapter is focused on Fourier transform MRI, in which the imaging equation can be written, in general, as

$$D(k_n) = \int_{-w/2}^{w/2} \rho(x)s(x)e^{-i2\pi k_n x}dx, \qquad (2.1)$$

where we explicitly include the sensitivity weighting function $s(x)$ of the receiver coil. In conventional Fourier imaging, $s(x)$ is often ignored because it can be assumed to be a constant over the field of view (FOV), and $D(k_n)$ is usually measured at $k_n = n\Delta k$ for $n = -N/2, -N/2 + 1,...,N/2 - 1$, with N being the total number of encodings acquired. In multichannel Fourier imaging (often known as parallel imaging), an array of receiver channels (or coils) with sensitivity functions $s_\ell(x)$ is used to acquire $D_\ell(k_n)$ simultaneously for $\ell = 1, 2,...,L$. To increase imaging speed, $D_\ell(k_n)$ is measured at $k_n = n\Delta\hat{k}$ for $n = -M/2, -N/2 + 1,...,M/2 - 1$, with $M = N/R$ and $\Delta\hat{k} = R\Delta k$. In other words, the k-space signal is measured at a sub-Nyquist rate in each receiver channel. Before we discuss advanced techniques to handle the image reconstruction problem associated with these two data acquisition schemes, a brief review of the popular Fourier reconstruction method is in order.

2.2 FOURIER RECONSTRUCTION

Given $D(n\Delta k)$ and $s(x) = 1$, it is well known that $\rho(x)$ can be reconstructed using the Fourier series, that is,

$$\rho(x) = \Delta k \sum_{n=-\infty}^{\infty} D(n\Delta k)e^{i2\pi\Delta k x}, \quad |x| < \frac{1}{2\Delta k}. \qquad (2.2)$$

For finite sampling, there are not sufficient data to define this series. The conventional Fourier reconstruction method treats the unknown coefficients as zero and, as a result, we have

$$\hat{\rho}(x) = \Delta k \sum_{n=-N/2}^{N/2-1} D(n\Delta k)e^{i2\pi n\Delta kx}, \quad |x| < \frac{1}{2\Delta k}, \tag{2.3}$$

which can be evaluated efficiently using the fast Fourier transform (FFT) algorithm. Some basic properties of the Fourier reconstruction method are summarized in the following remarks:

Remark 1: Given $D(n\Delta k)$ for $n = -N/2, -N/2 + 1, \ldots, N/2 - 1$, any $\hat{\rho}(x)$ given below with satisfies, Equation 2.1,

$$\hat{\rho}(x) = \Delta k \sum_{n=-N/2}^{N/2-1} D(n\Delta k)e^{i2\pi n\Delta kx} + \sum_{n<-N/2; n\geq N/2} c_n e^{i2\pi n\Delta kx}, \tag{2.4}$$

which is often called a feasible reconstruction of $\rho(x)$.

Remark 2: The Fourier reconstruction, $\hat{\rho}(x)$, given in Equation 2.3 is the minimum-norm feasible solution because

$$\bar{c}_n = \arg\min \int_{-\frac{1}{2\Delta k}}^{\frac{1}{2\Delta k}} |\hat{\rho}(x)|^2 \, dx = 0. \tag{2.5}$$

Remark 3: The Fourier reconstruction, $\hat{\rho}(x)$, is related to the true image $\rho(x)$ by

$$\hat{\rho}(x) = \int_{-\frac{1}{2\Delta k}}^{\frac{1}{2\Delta k}} \rho(\hat{x})h(x - \hat{x})d\hat{x}, \tag{2.6}$$

where $h(x)$, known as the point spread function (PSF), is given by

$$h(x) = \Delta k \frac{\sin(\pi N\Delta kx)}{\sin(\pi\Delta kx)} e^{-i\pi\Delta kx}. \tag{2.7}$$

Note that $h(x)$ is a periodic function, and within each period it is similar to a sinc function. The width of its main lobe, as measured by the interval between the first two zero crossings, is $2/(N\Delta k)$. The effective width W_h of $h(x)$ is often taken to be the width of an approximating rectangular pulse with height $h(0)$ and the same area. It is easy to show that

$$W_h = \frac{1}{h(0)} \int_{-\frac{1}{2\Delta k}}^{\frac{1}{2\Delta k}} h(x)dx = \frac{1}{N\Delta k},$$
(2.8)

which is exactly half the width of the main lobe of $h(x)$.

The right-hand side of Equation 2.8 is known as the Fourier pixel size, in contrast to the usual image pixel size Δx. Note that Δx can be made arbitrarily small using any signal interpolation schemes, but image resolution is fundamentally limited to $1/(N\Delta k)$. Another implication of Equation 2.8 is that W_h and N cannot be reduced simultaneously; in other words, improving image resolution and reducing the number of measured data points cannot be achieved simultaneously.

In addition to a loss of resolution in $\hat{\rho}(x)$, the convolution operation in Equation 2.6 also results in the well-known Gibbs ringing artifact in $\hat{\rho}(x)$. This artifact manifests itself as spurious ringing around sharp edges, as illustrated in Figure 2.1. The maximum undershoot or overshoot of the spurious ringing is about 9% of the intensity discontinuity and is independent of the number of data points used in the reconstruction. The frequency of oscillation, however, increases as more data points are used. For this reason, when a large number of data points is used in practice, the spurious ringing does not cover an appreciable distance in the reconstructed image and thus becomes invisible.

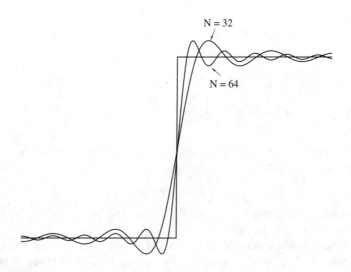

FIGURE 2.1 Gibbs ringing artifacts.

2.3 CONSTRAINED IMAGE RECONSTRUCTION

For years, the belief existed that information beyond the measurement cutoff frequency was not recoverable, thus the Rayleigh resolution limit [1]. Although the information is not apparent in the measured data, we now have learned how to look elsewhere for the additional information required to restore those frequency contents [2,3]. Constrained methods are the mathematical tools developed to accomplish this objective by using *a priori* information to compensate for the lack of high-frequency experimental data in the reconstruction process. Although constrained data processing methods have been used extensively for decades in other fields, application of the constrained reconstruction concept to MRI is very recent. The first successful effort was perhaps due to Smith [4] and, since then, research interest in this area has continued to grow for at least two reasons: first, the rapid development of computing technology has made it possible to use computation-intensive algorithms for practical applications and, second, the advantages of modern constrained reconstruction methods have made them worthwhile. In particular, the ability to reduce data truncation artifacts and improve image resolution is very desirable and can produce effects unmatched by the traditional unconstrained Fourier methods. Nonparametric constraints permit the use of the conventional Fourier series model for image function, and reconstruction methods of this type usually involve explicit data extrapolation to recover some of the unmeasured (presumably lost) high-spatial-frequency data so as to reduce truncation artifact. Parametric modeling methods, on the other hand, represent the image function in terms of a set of parameterized basis functions, rather than the nonparameterized harmonic sinusoidal functions used in the Fourier series. These methods can, in principle, generate images of infinite resolution from the model without explicitly extrapolating the data to the infinite frequency range. In this sense, parametric model constraints are often more powerful than nonparametric constraints, although sometimes they may not be as robust. Explicit data extrapolation is also possible and often used with parametric methods by using the model to generate the unmeasured data.

2.3.1 NONPARAMETRIC METHODS

A popular mathematical algorithm used in many nonparametric reconstruction methods is alternate projection, or projection onto convex sets (POCS). The principle of POCS has been discussed in great detail in the signal processing literature. We review here only the central ideas and give a couple of examples of its use in MRI.

> *Definition:* A subset Ω in the Hilbert space H is said to be convex if together with any x_1 and x_2, it also contains $\mu x_1 + (1 - \mu)x_2$ for all μ, $0 \le \mu \le 1$.
>
> *Definition:* For any $x \in H$, the projection $P_\Omega x$ of x onto Ω is the element in Ω closes to x. If Ω is closed and convex, P_Ω exists and is uniquely determined by x and Ω from the following minimality criterion
>
> $$\| x - P_\Omega x \| = \min_{y \in \Omega} \| x - y \|.$$

Briefly, the method of POCS is simply an iterative algorithm that finds a solution subject to a number of convex-type constraints by alternate projection. The following theorem is central to this technique, which assures the convergence of such an iterative process.

Theorem: Given m closed convex sets Ω_i, $i = 1, 2, \ldots, m$, in H and their corresponding projection operators P_i, if $\Omega_0 = \overset{m}{\underset{i=1}{\cap}} \Omega_i$ is nonempty, the sequence generated by

$$f_{\ell+1} = P_m P_{m-1}, \ldots, P_1 f_\ell, \quad l = 0, 1, \ldots$$

converges (weakly) to an element $f \in \Omega_0$ for any initial value $f_0 \in H$.

For problems with m pieces of *a priori* constraints of which each restricts the solution to a convex set, POCS is an ideal method for finding a solution. Various type of convex-type constraints exist for the reconstruction problem addressed. For example, the following constraints are of convex type:

1. Data-consistency constraint:

$$\Omega_1 = \{\rho(x) : F\{\rho(x)\} = D(k), \quad \text{for all the measured } k \text{ values}\}.$$

2. Limited-support constraint:

$$\Omega_2 = \{\rho(x) : \rho(x) = 0, \quad \text{for } |x| > W/2\}.$$

3. Bounded-magnitude constraint:

$$\Omega_3 = \{\rho(x) : |\rho(x)| \leq B\}.$$

4. Phase constraint:

$$\Omega_4 = \{\rho(x) : \arg\{\rho(x)\}\theta(x)\}.$$

5. Bounded-noise variance constraint:

$$\Omega_5 = \{\rho(x) : \| s - F\{\rho(x)\} \|^2 \leq \sigma_\varepsilon^2 \}.$$

Although POCS has played an important role in nonparametric image reconstruction, it is not without limitations. Three of the most serious limitations are: First, it can handle only convex-type constraints, which prevents some effective but nonconvex-type constraints from being used. Secondly, POCS can be computationally expensive, because convex set projection operators are nonlinear in nature and sometimes involve constrained nonlinear optimization: for example, $P_{\Omega 5}$ for the convex set Ω_5 defined above requires a quadratic programming step. Thirdly, POCS may converge very slowly and not necessarily to a unique solution when the projection operators are not contractive mapping; therefore, POCS can also converge to a "bad" solution.

2.3.2 PARAMETRIC METHODS

Parametric image models often take the following form:

$$\rho(x) = \sum_n c_n \varphi_n(x), \tag{2.9}$$

where $\varphi_n(x)$ are the basis functions used to absorb any *a priori* information and c_n are the series coefficients chosen to match the measured data.

Selecting a set of "good" basis functions is essential for the model in Equation 2.9. A particular set of basis functions is given in the form of weighted complex sinusoids [5,6]:

$$\varphi_n = C(x)e^{i2\pi n\Delta kx}, \tag{2.10}$$

where $C(x)$ is a nonnegative function incorporating *a priori* information. With this set of basis functions, the model, known as the *generalized series* (GS) *model* [5,6], becomes

$$\rho(x) = C(x)\sum_n c_n e^{i2\pi n\Delta kx}. \tag{2.11}$$

This model has several useful properties. Specifically, when no nontrivial *a priori* information is available, namely, $C(x) = 1$, Equation 2.11 automatically reduces to the conventional Fourier series model. This is desirable because the Fourier series model is indeed optimal in this case. On the other hand, if $C(x) = \rho(x)$, the multiplicative Fourier series factor will be forced to unity by the data-consistency constraint, and a perfect reconstruction will result. In general, if $C(x)$ is properly chosen, the new basis functions given in Equation 2.10 enable the GS model to converge faster than the Fourier series model. Therefore, within a certain error bound, fewer terms can be used to represent an image function than are required by the Fourier series method, leading to a reduction of the truncation artifact. The optimality of the GS model in Equation 2.11 can also be justified from the minimum cross entropy principle [7].

Selection of the weighting function $C(x)$ is application dependent. For the limited data reconstruction problem, it was suggested [8] that $C(x)$ be chosen to be a summation of boxcar functions as

$$C(x) = \left| \sum_{m=1}^{M} a_m \prod \left[\frac{x - \frac{1}{2}(\beta_m + \beta_{m+1})}{\beta_{m+1} - \beta_m} \right] \right|, \tag{2.12}$$

where M represents the number of boxcar functions in the model, and β_m and a_m are the edge locations and amplitude of the mth boxcar, respectively. This function is particularly suitable for image functions containing sharp edges because they are explicitly built into the basis functions.

The weighting function $C(x)$ can also be determined experimentally. A typical example is time-sequential imaging, which involves the acquisition of a time series of images, $\rho_1(x), \rho_2(x),..., \rho_L(x)$, from the same anatomical site. For many of this type of imaging experiments, the underlying high-resolution morphology in the desired image sequence does not change from one image to another. As a result, it is not necessary to acquire each of these images independently. Specifically, with the GS model, we first acquire one high-resolution (reference) data set with N encodings, followed by a sequence of reduced data set with M encodings. In the image reconstruction step, the high-resolution reference image $\rho_{ref}(x)$ is used as the weighting function for the GS basis functions. That is, we set

$$C(x) = |\rho_{ref}(x)| \qquad (2.13)$$

for the GS model when it is used for image reconstruction from the reduced data sets. After $C(x)$ is known, the series coefficients c_n are determined by solving a set of linear equations from the data-consistency constraints. That is,

$$D(n\Delta k) = \sum_{m=-N/2}^{N/2-1} c_m D^c[(n-m)\Delta k], \qquad (2.14)$$

where $D^c(n\Delta k) = F\{C(x)\}(n\Delta k)$.

2.3.3 APPLICATION EXAMPLES

Constrained image reconstruction has been successfully used in several practical applications. This section discusses two specific examples: partial Fourier imaging and dynamic imaging.

Example 2.1: Partial Fourier Reconstruction

In partial Fourier imaging, k-space is sampled asymmetrically, say, $D(n\Delta k)$ is measured for $n \in N_{data} = \{-n_0, -n_0 + 1,...,N - 1\}$. Such a sampling scheme arises in MRI when a short echo time is used to avoid spin dephasing due to short T_2^* caused by local susceptibility changes or uncompensated motion effects. It is sometimes also used in the phase-encoding direction when an asymmetric set of phase-encoding measurements is acquired to reduce data acquisition time. Usually, n_0 is much smaller than N, typically, $n_0 = 16$ or 32 with n being on the order of 128. The central k-space data are used first to obtain an phase estimate $\hat{\varphi}(x)$, which is then used as a constraint to get the final reconstruction. The phase-constrained reconstruction problem lends itself nicely to the POCS algorithm. Specifically, let

$$\Omega_1 = \{\rho(x) \mid \angle\rho(x) = \hat{\varphi}(x)\} \qquad (2.15)$$

and

$$\Omega_2 = \{\rho(x) \mid F\{\rho(x)\} = D(n\Delta k), -n_0 \leq n \leq N-1\}. \qquad (2.16)$$

Clearly, Ω_1 contains all the images satisfying the predetermined phase constraint, whereas Ω_2 contains all the images consistent with the measured data. The desired image $\rho(x)$ lies in the intersection of Ω_1 and Ω_2. That is,

$$\rho(x) \in \Omega = \Omega_1 \cap \Omega_2, \qquad (2.17)$$

which can be found by alternating projections of an initial estimate onto these two sets. More specifically,

$$\rho_{m+1}(x) = \wp_1 \wp_2 \{\rho_m(x)\}, \qquad (2.18)$$

where

$$\wp_1\{\rho(x)\} = |\rho(x)| e^{i\hat{\varphi}(x)} \qquad (2.19)$$

and

$$\wp_2\{\rho(x)\} = F^{-1} R F\{\rho(x)\}, \qquad (2.20)$$

in which R is a data replacement operator defined as

$$R\{\hat{D}(n\Delta k)\} = \begin{cases} D(n\Delta k), & -n_0 \le n \le N-1 \\ \hat{D}(n\Delta k), & \text{otherwise.} \end{cases} \qquad (2.21)$$

It is apparent that $\wp_1$ projects any image function $\rho(x)$ onto Ω_1, whereas $\wp_2$ projects it onto Ω_2. The initial condition $\rho_0(x)$ for Equation 2.18 is usually chosen to be the zero-filled Fourier reconstruction.

Example 2.2: Data-Sharing Dynamic Imaging

Constrained image reconstruction finds wide application in dynamic imaging. The keyhole and reduced encoding by generalized series reconstruction (RIGR) techniques [6,9,10] are two typical examples. A common feature of these two methods is that a high-resolution reference image and a sequence of reduced dynamic data sets (usually in central k-space) are collected. Assuming that N encodings are collected for the reference data set and M encodings for each of the dynamic data sets, a factor of improvement N/M in temporal resolution (or imaging efficiency) is gained with this data acquisition scheme as compared to the conventional full-scan imaging method. In image reconstruction, the reference data is used to compensate for the loss of high-frequency data in the dynamic data sets. In keyhole, this is done in a straightforward fashion; that is, the unmeasured encodings of each dynamic data set are replaced directly by the corresponding reference data to create a "full-size" data set. A weakness of this data-sharing method is that any data inconsistency between the dynamic and reference data sets will result in data truncation artifact and, as a result, dynamic image features are produced only at low resolution. With RIGR, image reconstruction is done using the GS model described in Section 2.2, in which the basis functions are determined by the reference data and the coefficients are determined by the dynamic data. This reconstruction algorithm can overcome the limited resolution problem with the keyhole method. It has been shown that with multiple references, RIGR can reconstruct dynamic features in a resolution close to that of the reference image [11].

2.4 REGULARIZED IMAGE RECONSTRUCTION IN PARALLEL MRI

2.4.1 BASIC RECONSTRUCTION METHODS

The Fourier image of the ℓth channel (ignoring the data truncation effects) is given by

$$d_\ell(x) = \sum_{m=0}^{R-1} \rho(x - m\hat{W})s_\ell(x - m\hat{W}),\tag{2.22}$$

for $\ell = 1, 2, ..., L$, and $W/2 - \hat{W} < x < W/2$. Assuming that $R \le L$, we can solve for $\rho(x)$ pixel by pixel from the earlier equations. More specifically, rewriting Equation 2.22 in matrix form

$$\mathbf{S}\vec{\rho} = \vec{d}\tag{2.23}$$

where

$$\mathbf{S} = \begin{bmatrix} s_1(x) & s_1(x - \hat{W}) & \cdots & s_1(x - (R-1)\hat{W}) \\ s_2(x) & s_2(x - \hat{W}) & \cdots & s_2(x - (R-1)\hat{W}) \\ \vdots & \vdots & & \vdots \\ s_L(x) & s_L(x - \hat{W}) & \cdots & s_1(x - (R-1)\hat{W}) \end{bmatrix}$$

$$\vec{\rho} = \begin{bmatrix} \rho(x) \\ \rho(x - \hat{W}) \\ \vdots \\ \rho(x - (R-1)\hat{W}) \end{bmatrix}, \quad \text{and} \quad \vec{d} = \begin{bmatrix} d_1(x) \\ d_2(x) \\ \vdots \\ d_L(x) \end{bmatrix}.$$

Equation 2.23 is known as the sensitivity encoding (SENSE) reconstruction formula [12], which can be derived from Papoulis' generalized sampling theorem [13]. Clearly, perfect reconstruction of $\rho(x)$ requires: (a) precise knowledge of $s_\ell(x)$ to form, (b) to be nonsingular for $W/2 - \hat{W} < x < W/2$, and (c) $d_\ell(x)$ to be noiseless and not corrupted by the data truncation artifact.

In practice, Equation 2.23 is often solved in the least-squares (LS) sense or minimum-variance (MV) sense. The LS solution is given by

$$\hat{\rho}_{LS} = (\mathbf{S}^H \mathbf{S})^{-1} \mathbf{S}^H \vec{d}, \tag{2.24}$$

and the MV solution is given by [12]

$$\vec{\rho}_{MV} = (\mathbf{S}^H \boldsymbol{\psi}^{-1} \mathbf{S})^{-1} \mathbf{S}^H \boldsymbol{\psi}^{-1} \vec{d}, \tag{2.25}$$

where $\boldsymbol{\Psi}$ is the data noise covariance matrix. Some basic properties of the LS and MV solutions are summarized in the following remarks.

Remark 4: When S and $\boldsymbol{\Psi}$ are accurate, the variance of the reconstruction error due to data noise is given by

$$\sigma_{LS}(x) = \sqrt{[(\mathbf{S}^H \mathbf{S})^{-1} \mathbf{S}^H \boldsymbol{\Psi} \mathbf{S} (\mathbf{S}^H \mathbf{S})^{-1}]_x}, \tag{2.26}$$

for the LS solution, and

$$\sigma_{MV}(x) = \sqrt{[\mathbf{S}^H \boldsymbol{\Psi}^{-1} \mathbf{S})^{-1}]_x}, \tag{2.27}$$

for the MV solution, where the subscript x denotes the index of the matrix corresponding to location x.

Remark 5: The LS and the MV solutions are the same if the acceleration factor equals the number of coils or noise is uncorrelated between coils, in which case there is no need to measure the noise covariance matrix.

Remark 6: The SNR of the MV solution is always greater or equal to that of the LS solution. The MV solution minimizes the variance of the reconstruction error vector $E(\Delta\vec{\rho}^H \Delta\vec{\rho}) = \mathrm{trace}E(\Delta\vec{\rho}\Delta\vec{\rho}^H)$ over all possible estimators when the noise is Gaussian and over all linear unbiased estimators for non-Gaussian noise. Therefore, the mean-squared error of the MV solution is less than that of the LS solution.

The earlier results are based on the assumption that both S and $\boldsymbol{\Psi}$ are accurate. In practice, S and $\boldsymbol{\Psi}$ are estimated from experimental data, and any error in S (denoted as ΔS) and/or in $\boldsymbol{\Psi}$ errors (denoted as $\Delta\boldsymbol{\Psi}$) will contribute to $\Delta\vec{\rho}$. Suppose that

$$\| \Delta\mathbf{S} \|_2 < \sigma_{\min}(\mathbf{S}),$$

$$\| \Delta\boldsymbol{\Psi} \|_2 < \sigma_{\min}(\boldsymbol{\Psi}), \tag{2.28}$$

where $\sigma_{\min}(\cdot)$ denotes the minimum singular value of the matrix. It can be shown for the MV solution that [14]

$$\frac{\left\|\Delta\vec{\rho}_{\mathrm{MV}}\right\|}{\left\|\vec{\rho}_{\mathrm{MV}}\right\|} \leq \kappa(\mathbf{S}_\Psi)\left(\frac{\left\|\Delta\mathbf{S}_\Psi\right\|}{\left\|\mathbf{S}_\Psi\right\|} + \frac{\left\|\Delta\vec{d}_\Psi\right\|}{\left\|\vec{d}_\Psi\right\|}\right), \tag{2.29}$$

where the subscript Ψ denotes a matrix or vector premultiplied by $\Psi^{-1/2}$, with $\Psi^{-1/2}$ being defined as $\Psi^{-\frac{1}{2}} = \mathbf{V}\Lambda^{-\frac{1}{2}}\mathbf{V}^H$ assuming that $\Psi = \mathbf{V}\Lambda\mathbf{I}^H$. The earlier result can be easily extended to the LS solution by setting $\Psi = \mathbf{I}$ in Equation 2.29.

2.4.2 REGULARIZED RECONSTRUCTION METHODS

The SENSE reconstruction (either $\vec{\rho}_{\mathrm{LS}}$ or $\vec{\rho}_{\mathrm{MV}}$) is sensitive to $\Delta\vec{d}$, $\Delta\mathbf{S}$, and $\Delta\Psi$, especially when $\mathbf{S}$ is ill-conditioned. To desensitize the solution to data noise and model errors, regularization methods are often used. Tikhonov regularization is perhaps the most common regularization scheme, in which we form a weighted sum of the data misfit term $\|\mathbf{S}\vec{\rho} - \vec{d}\|^2$ and a regularization term $\|\mathbf{A}(\vec{\rho} - \vec{\rho}_r)\|^2$ using a weighting factor λ^2, and find the solution $\vec{\rho}_{\mathrm{reg}}$ that minimizes this sum, i.e.,

$$\vec{\rho}_{\mathrm{reg}} = \arg\min\{\|\mathbf{S}\vec{\rho} - \vec{d}\|^2 + \lambda^2\|\mathbf{A}(\vec{\rho} - \vec{\rho}_r)\|^2\}, \tag{2.30}$$

where λ is often referred to as the regularization parameter and $\vec{\rho}_r$ is a regularization image. A closed-form solution for $\vec{\rho}_{\mathrm{reg}}$ exists for L_2-norm and is given by

$$\vec{\rho}_{\mathrm{reg}} = \vec{\rho}_r + (\mathbf{S}^H\mathbf{S} + \lambda^2\mathbf{A}^H\mathbf{A})^{-1}\mathbf{S}^H(\vec{d} - \mathbf{S}\vec{\rho}_r). \tag{2.31}$$

Selecting "good" values for λ and $\vec{\rho}_r$ is essential for this regularized reconstruction scheme. Although this is still a research problem, several algorithms have been proposed, which find useful practical applications. We will briefly review some of them to illustrate the concept.

2.4.2.1 Construction of $\vec{\rho}_r$

There are basically three schemes to construct $\vec{\rho}_r$: (a) setting $\vec{\rho}_r = 0$, (b) recycling an initial SENSE reconstruction to create $\vec{\rho}_r$, and (c) collecting additional data to generate $\vec{\rho}_r$.

Scheme (a) corresponds to, perhaps, the simplest version of the Tikhonov regularization scheme. It was used in Reference 15 with some success. In scheme (b), the conventional SENSE algorithm is used to obtain an initial reconstruction, which is then filtered by a median filter to suppress any residual aliasing artifacts [16]. However, if the matrix $\mathbf{S}$ is highly ill-conditioned within a large region, the filtering step may not be effective in suppressing the aliasing artifacts. Scheme (c) acquires additional k-space center lines at the Nyquist

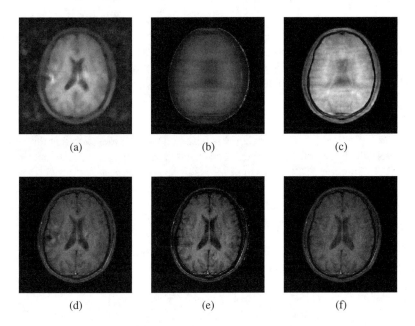

FIGURE 2.2 SENSE reconstructions from a real data set acquired with 4 coils and $R = 4$. (a) Median-filtered SENSE and (d) corresponding regularized reconstructions; (b) low-resolution reconstruction from autocalibration and (e) corresponding regularized reconstruction; and (c) GS reconstruction and (f) corresponding regularized reconstruction.

rate, known as the autocalibration scan [17,18,19] and uses these data to reconstruct a low-resolution regularization image [20]. A high-resolution regularization image can also be created from these data using the GS model. Details of the algorithm can be found in Reference 21.

Figure 2.2 shows a set of regularized reconstructions with different regularization images from real experimental data acquired with four receiver coils and $R = 4$. As can be seen, different regularization images can affect the final reconstruction.

2.4.2.2 Selection of λ

A straightforward way to select the regularization parameter is to set λ heuristically as a constant over the entire image. This method is not effective because the condition of **S** varies at different locations. A more elaborate way is to select λ adaptively using traditional regularization methods such as the L-curve or the generalized cross-validation (GCV) methods [22]. The L-curve method was used in parallel imaging with some success [20]. The GCV method works well in general but sometimes gives biased results if the noise $\Delta \vec{d}$ is highly correlated [22].

A significant weakness of both the conventional L-curve and the GCV methods lies in the fact that they choose $\lambda(x)$ independently for different spatial locations. This problem was addressed in Reference 21 with an algorithm to select $\lambda(x)$ jointly. Specifically, the algorithm first sets $\lambda(x)$ to be within $[\lambda_{min}, \lambda_{max}]$, and then forms $\lambda(x)$ as a linear function of the local condition number of S, i.e.,

$$\lambda(x) = \alpha \kappa(S) + \beta, \tag{2.32}$$

for $W/2 - \hat{W} < x < W/2$. This scheme is based on the consideration that the larger the $\kappa(S)$, the heavier the regularization is needed for Equation 2.31. To determine α and β, Equation 2.32 is rewritten as

$$\lambda(x) = \frac{\kappa(S) - \kappa_{min}}{\kappa_{max} - \kappa_{min}} (\lambda_{max} - \lambda_{min}) + \lambda_{min}, \tag{2.33}$$

where κ_{max} and κ_{min} are the maximum and the minimum condition numbers of all S, and λ_{min} and λ_{max} are determined by

$$\lambda_{min} = \arg\min_{\lambda} \left\{ \frac{\max_i \sigma_i}{\min_i (\sigma_i + \lambda^2 / \sigma_i)} < K \right\}, \tag{2.34}$$

and

$$\lambda_{max} = \arg\max_{\lambda} \left\{ \sum_x \| S\vec{\rho}_{reg}(\lambda) - \vec{d} \| \leq \varepsilon \right\}, \tag{2.35}$$

where σ_i is the ith singular value of S, and K and ε are user-specified constants. Details of the algorithm can be found in [21].

Figure 2.3 shows a set of exemplary regularized reconstructions with different regularization parameters from real experimental data acquired with four receiver coils and an acceleration factor of four. The importance of regularization parameters can be appreciated by comparing the results in (a) to (d).

2.4.2.3 Sensitivity Analysis

An upper bound for the sensitivity of the regularized solution to data noise and model error is given by

$$\frac{\| \Delta\vec{\rho}_{reg} \|}{\| \vec{\rho}_{reg} \|} \leq \frac{\sigma_{max}}{\sigma_q + \frac{\lambda^2}{\sigma_q}} \left(\frac{\| \Delta S \|}{\| S \|} + \frac{\| \Delta\vec{d} \|}{\| \vec{d} \|} \right) + \frac{\lambda^2}{\sigma_n^2 + \lambda^2} \frac{\| \vec{\rho}_r - \vec{\rho} \|}{\| \vec{\rho} \|}, \tag{2.36}$$

where σ_{max} denotes the largest singular value of matrix S and

$$q = \arg\min_j \left(\sigma_j + \frac{\lambda^2}{\sigma_j} \right). \tag{2.37}$$

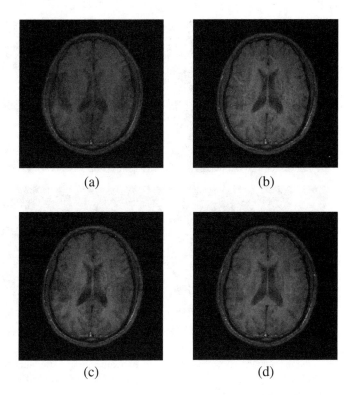

(a)   (b)

(c)   (d)

FIGURE 2.3 Regularized SENSE reconstructions with 4 coils, $R = 4$, $\vec{\rho}_r$ = GS reconstruction, and λ (a) being a constant, (b) using GCV, (c) using L-curve, and (d) using the method proposed in Reference 21.

As expected, the result in Equation 2.29 is a special case of Equation 2.36 when $\lambda = 0$. Comparing Equation 2.36 and Equation 2.29 yields

$$k(S_\lambda) = \frac{\sigma_{max}}{\sigma_q + \frac{\lambda^2}{\sigma_q}},$$ (2.38)

which can be regarded as the effective condition number of the regularized reconstruction. Clearly, $k(S_\lambda)$ is reduced by increasing λ.

2.4.3 APPLICATION EXAMPLE

An example is shown in Figure 2.4, where the data were collected using three coils in a dynamic contrast-enhanced MRI experiment. In addition to the usual SENSE data, eight encodings were collected at the Nyquist rate in central k-space

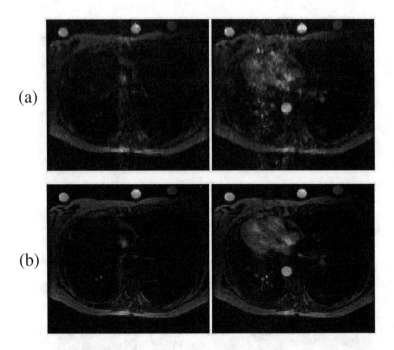

FIGURE 2.4 Dynamic images of a chest tumor at two time points after injection of a contrast agent: (a) SENSE reconstruction ($R = 3$, $L = 3$), and (b) improved SENSE reconstruction by the proposed method.

for each data frame, from which $\vec{\rho}_r$ was derived using the GS model. As can be seen, the regularized SENSE reconstruction (Figure 2.4b) is significantly better than that from the standard SENSE algorithm (Figure 2.4a).

2.5 CONCLUSION

Image reconstruction from limited Fourier data is a classical problem in tomographic imaging. Although a general solution to this problem is not available, a number of practical techniques have emerged, which can provide optimal (or close-to-optimal) solutions to a particular application problem, leading to significant improvements in image quality. This chapter provided a tutorial discussion of some representative techniques, including parametric and nonparametric methods for superresolution image reconstruction from limited Fourier data and regularization methods for image reconstruction from multichannel undersampled Fourier data. The chapter is also intended to provide some basic background knowledge of the area for the reader to apply these techniques to particular problems or to further improve them.

REFERENCES

1. di Francia, G.T. (1955). Resolving power and information. *J. Opt. Soc. Am. A*. 45: 497–501.
2. Kosarev, E.L. (1990). Shannon's superresolution limit for signal recovery. *Inverse Problems*. 6: 4975–76.
3. Pike, E.R., McWhirter, J.G., Bertero, M., and de Mol, C. (1984). Generalized information theory for inverse problems in signal processing. *Proc. IEEE*. 131: 660–667.
4. Smith, M.R., Nichols, S.T., Henkelman, R.M., and Wood, M.L. (1986). Application of auto-regressive moving average parametric modeling in magnetic resonance image reconstruction. *IEEE Trans. Med. Imaging*. MI-5: 132–139.
5. Liang, Z.-P. and Lauterbur, P.C. (1991). A generalized series approach to MR spectroscopic imaging. *IEEE Trans. Med. Imaging*. 10: 132–137.
6. Liang, Z.-P. and Lauterbur, P.C. (1994). An efficient method for dynamic magnetic resonance imaging. *IEEE Trans. Med. Imaging*. 13: 677–686.
7. Shore, J.E. and Johnson, R.W. (1980). Axiomatic derivation of the principle of maximum entropy and the principle of minimum cross-entropy. *IEEE Trans. Inform. Theory*. 26: 26–37.
8. Liang, Z.-P., Boada, F., Constable, T., Haacke, E.M., Lauterbur, P.C., and Smith, M.R. (1992). Constrained reconstruction methods in MR imaging. *Rev. Magn. Reson. Med*. 4: 67–185.
9. Jones, R.A., Haraldseth, O., Muller, T.B., Rinck, P.A., and Oksenda, A.N. (1993). k-space substitution: A novel dynamic imaging technique. *Magn. Reson. Med*. 29: 830–834.
10. van Vaals, J.J., Brummer, M.E., Dixon, W.T., Tuithof, H.H., Engels, H., Nelson, R.C., Gerety, B.M., Chezmar, J.L., and den Boer, J.A. (1993). Keyhole method for accelerating imaging of contrast agent uptake. *J. Magn. Reson. Imaging*. 3: 671–675.
11. Hanson, J.M., Liang, Z.-P., Wiener, E., and Lauterbur, P.C. (1996) Fast dynamic imaging using two reference images. *Magn. Reson. Med*. 36: 172–175.
12. Pruessmann, K.P., Weiger, M., Scheidegger, M.B., and Boesiger, P. (1999). SENSE: Sensitivity encoding for fast MRI. *Magn. Reson. Med*. 42: 952–962.
13. Papoulis, A. (1977). Generalized sampling expansion. *IEEE Trans. Circuits Syst*. 24: 652–654.
14. Xu, D., Ying, L., and Liang, Z.-P. (2004). Parallel MR imaging in the presence of large uncertainty in coil sensitivity functions: Effects, remedies and applications. in *Proceedings of the 26th Annual International Conference of the IEEE Engineering in Medicine and Biology Society*. San Francisco, USA, September 1–5.
15. King, K.F. and Angelos, L. (2001). SENSE image quality improvement using matrix regularization. in *Proceedings of the 9th Annual Meeting of International Society of Magnetic Resonance in Medicine*. April 21–27. Glasgow, Scotland.
16. Liang, Z.-P., Bammer, R., Ji, J., Pelc, N.J., and Glover, G.H. (2002). Improved image reconstruction from sensitivity-encoded data by wavelet denoising and Tikhonov regularization. in *Proceedings of the 1st IEEE International Symposium on Biomedical Imaging*. Washington D.C., USA, July.
17. Jakob, P.M., Griswold, M.A., Edelman, R.R., and Sodickson, D.K. (1998). AUTO-SMASH: a self-calibrating technique for SMASH imaging. *MAGMA* 7: 42–54.
18. Heidemann, R.M., Griswold, M.A., Haase, A., and Jakob, P.M. (2001). VD-AUTO-SMASH imaging. *Magn. Reson. Med*. 45(6): 1066–1074.

19. McKenzie, C.A., Yeh, E.N., Ohliger, M.A., Price, M.D., and Sodickson, D.K. (2002). Self-calibrating parallel imaging with automatic coil sensitivity extraction. *Magn. Reson. Med.* 47(3): 529–538.

20. Lin, F.-H., Kwong, K.K., Belliveau, J.W., and Wald, L.L. (2004). Parallel imaging reconstruction using automatic regularization. *Magn. Reson. Med.* 51: 559–567.

21. Ying, L., Xu, D., and Liang, Z.-P. (2004). On Tikhonov regularization for image reconstruction in parallel MRI. in *Proceedings of the 26th Annual International Conference of the IEEE Engineering in Medicine and Biology Society.* San Francisco, USA, September 1–5.

22. Hansen, P.C. (1998). *Rank-Deficient and Discrete Ill-Posed Problems.* Philadelphia: SIAM Press.

3 Parallel MRI: Concepts and Methods

Walid Elias Kyriakos and W. Scott Hoge

CONTENTS

3.1 INTRODUCTION

Throughout history, human nature's fascination with speed has constituted the main driver for technological advancement. From the wheel to the space shuttle, the pursuit of speed has been a defining attribute of our civilization, as the realization of mortality pushes us to find faster ways to experience, understand, and interact with our reality. Short of being an essay on the philosophy of speed, this chapter pertains to increasing the speed of image acquisition in the medical imaging modality of magnetic resonance imaging (MRI).

In the field of MRI, the goal is to observe a medically significant projection of the state of the human body in a reasonable amount of time. With dynamic structures and processes, such as the beating heart and blood flow, the observed state is in constant change. Being able to resolve yet finer projections of that state onto the time axis has been a constant drive for diagnostic medicine. The ability to follow changes in time would provide a better understanding of the underlying biology and a better ability to detect anomalies and disease. When observing static structures, such as joints, fast imaging can help minimize the amount of time a patient has to lie in the scanner. This is a main consideration for patient comfort and has the economical advantage of allowing more patients access to the scanner. Other considerations such as the inability to maintain long breath holds, as well as claustrophobia, have contributed to making reduced image acquisition time a priority in MRI research and development.

The simultaneous use of multiple detectors in order to increase the speed of imaging has previously been employed in computer-aided tomography (CAT). This chapter describes the techniques that use multiple sensors in parallel in order to increase the speed of imaging in MRI. It is divided into three parts. The first part describes the history of parallel MRI in chronological order. In the second part, each of the techniques is described and analyzed in detail. The third part of this chapter describes results of reconstructions using all the techniques, and shows their comparison.

In MRI, spins are placed in a static magnetic field and manipulated by both radio frequency (RF) electromagnetic fields and a superposition of gradient magnetic fields, in order to detect their spatial distribution and local spatial interactions. Manipulation of the spin relaxation parameters (often referred to as $T1$ and $T2$) is used in order to provide fine levels of image contrast by the precise design of imaging pulse sequences. The MRI signals are received in RF coils placed around the object of interest and contain information relating the spin content to the $T1$ and $T2$ contrast parameters. The data acquisition domain, containing the spatial frequency information in the image, is often referred to as k-space and is related to the image domain by Fourier transformation. The signals received in the RF coils are used to populate k-space whereby their location is related to their spatial frequencies. These frequencies are lowest at the center of k-space and increase as we move away from it.

3.2 HISTORY OF PARALLEL MRI

Within a decade of its advent (1), MRI was gaining recognition as a leading diagnostic imaging modality. Exquisite soft tissue contrast coupled with minimal invasiveness has placed it at the forefront of clinical use. Although many advances have occurred in MRI, there is still a need for further increases in the speed of image acquisition. Speed of image acquisition has constituted a major limitation and goal of technical development. Over a decade was spent on the development of sequential acquisition methods, involving pulse sequence design and fast gradient-switching technology.

These advances have led to ultrafast multiecho pulse sequences that go by names such as RARE, EPI, and BURST but have not been able to address the speed requirements in certain applications. Dynamic imaging applications such as cardiac and interventional imaging would still be greatly served if an order of magnitude reduction in scan time were achieved without sacrificing spatial resolution and signal-to-noise ratio (SNR). During the 1990s, the answer to the imaging speed needs was starting to take shape in the parallel acquisition paradigm. The idea that multiple RF receivers could be used in parallel to speed up the image acquisition, as is the case in computer-aided tomography (CAT), was gaining momentum. Moreover, this new field of parallel imaging can be combined with the previous ultrafast multiecho methods to further increase imaging speed.

The theoretical feasibility of fast data acquisitions using multiple detectors in MRI was first described by Hutchinson and Raff in 1988 (2), and subsequently by Kwiat et al. in 1991 (3). Both groups investigated methods to solve the inverse source problem on MR signals received in multiple RF receiver coils, requiring the use of a number of closely packed RF coils, equal to the number of pixels in the image, as well as greatly increased receiver coil sensitivities, in order to eliminate the requirement of phase encoding by gradient switching. These requirements are quite impractical in conventional MR imaging and have prevented these techniques from being applied in practice.

In 1993, Carlson and Minemura (4) described the use of a two-coil array, using one coil with homogeneous sensitivity over the field of view (FOV) and the other having a linear gradient in sensitivity. Partial data sets were acquired in each coil, and the missing lines in k-space were generated using a series expansion in terms of other phase-encoded lines. This approach yielded twofold image acceleration and was the first technique that produced accelerated images using coil-sensitivity information; however, it remained impractical due to design conditions required of the sensitivity of the coils.

Later that same year, Ra and Rim (6) introduced the first feasible parallel imaging method, which used coil sensitivity as a way to remove the aliasing in regularly undersampled images acquired with multiple coils. Although this technique later constituted the basis for the sensitivity encoding (SENSE) method, which is currently enjoying wide commercial use, Ra and Rim's original paper presented only phantom reconstructions. However, during the time when the technique came out, more research emphasis was being given to sequential fast imaging methods, and little effort was made to develop it.

In 1997, Sodickson et al. introduced the simultaneous acquisition of spatial harmonics (SMASH) method (7), which used the sensitivity profile of receiver coils as a complementary encoding function. In its initial embodiment, SMASH tried to fit these profiles to sinusoidal harmonics in order to emulate the effect of phase encoding. It was the first parallel acquisition method that produced *in vivo* imaging and launched the field of parallel imaging using sensitivity encoding. SMASH has since undergone a large number of changes and adaptations, as a number of practical limitations prevented it from reaching commercial reliability.

Today SMASH remains an elegant theoretical development, although without real widespread applicability.

The SENSE method proposed by Pruessmann et al. (8) was introduced in 1999 and is another parallel imaging technique that relies on the use of 2-D or 3-D sensitivity profile information in order to reduce image acquisition times in MRI. Similar to SMASH, the Cartesian version of SENSE requires the acquisition of equally spaced k-space lines in order to reconstruct sensitivity-weighted, aliased versions of the image. The aliasing is removed with the use of the sensitivity profile information at each pixel. This is done by solving in the space domain the linear system of equations defined by the subsampling pattern. This technique is very similar to the approach described by Ra and Rim in 1993 (6), and only differs in the sensitivity profile estimation, as well as in its use of numerical system regularization strategies.

The general version of SENSE allows for data to be sampled along arbitrary k-space trajectories, and has a number of advantages over the Cartesian version, including the ability to minimize artifact, and maximize the SNR, effectively achieving higher acceleration factors with acceptable quality. A very high computational cost, however, accompanies the arbitrary k-space sampling in generalized SENSE.

In 2000, the SPACE RIP (9) technique was introduced whereby flexibility is allowed in the choice of the k-space lines acquired, given that the frequency-encoded direction is kept unchanged. This allows one to maintain the advantages of SNR and minimized artifacts of generalized SENSE, while having a considerably smaller computational load. Similar to SENSE, SPACE RIP uses 2-D or 3-D sensitivity profile information in order to solve a linear system of equations. In addition, it allows for total flexibility in the positioning of the coil array around the object of interest.

Another technique, partially parallel imaging with localized sensitivities (termed PILS), was described in 2000 (10), which requires the use of coils having localized sensitivities and circumvents the need to estimate the sensitivity profiles. PILS is simple in principle, but the condition that it imposes on the coil sensitivities is impractical for a large number of applications.

Since 2001, the effort shifted toward optimizing the existing techniques and a number of works have surfaced that describe better ways to estimate the coil-sensitivity profiles, as well as to condition the reconstruction in the various techniques. Generalized autocalibrating partially parallel acquisitions (GRAPPA) (11), published in 2002, introduced a generalized autocalibrating approach that can result in more robust SMASH-like reconstructions with computational complexity far below more general SMASH approaches (12). Currently, the trend has shifted toward finding a generalized formalism that encompasses all the preceding techniques, as they all strive to solve the same system of equations. In addition, more effort is spent on the design of coil arrays that are optimized for use in parallel imaging. In the following section, we describe the general problem of parallel MRI, and present the basics of each of the techniques.

3.3 FORMULATION OF THE PROBLEM

The starting point of all parallel imaging techniques in MRI is the equation describing the received signal. When using an array of N coils, the general equation of the signal received in a coil "k" can be written as:

$$s_k(G_y^i, G_z^i, t) = \iiint I(x,y,z)W_k(x,y,z)P^L(x,y,z)e^{j\gamma(G_x^i xt + G_y^i y\tau_1 + G_z^i z\tau_2)}dxdydz \quad (3.1)$$

where $I(x, y, z)$ represents the image that we seek to identify; $W_k(x, y, z)$ represents the 3-D sensitivity profile of coil "k"; $P^L(x, y, z)$ represents the RF selective excitation profile during the L-th excitation; G_x^i, G_y^i, G_z^i represent the values of the gradients applied, respectively, in the x, y, and z directions during the i-th acquisition, assuming x as the frequency-encoded direction; and τ_1 and τ_2 represent the duration of the phase-encoding gradients in the y and z directions, respectively. This equation lumps the T1, T2, and spin-density dependencies into the element designating the image I(x, y, z). In addition, it takes into account the possibility of acquiring multiple echoes (index "i") for each excitation profile (index "L"), and allows for the total flexibility in the design of the excitation profile $P^L(x, y, z)$. In Fourier imaging, $P^L(x, y, z)$ is kept constant for all values of "m," $\tau_1 = \tau_2 = \tau$, unless otherwise specified, and G_x^i is kept constant for all values of "i."

Equation 3.1 is a linear equation relating the time signal of an acquired echo to the image, through the parameters of the imaging pulse sequence. The time signal is sampled, and each time sample provides one linear equation of the image I(x, y, z). If I(x, y, z) is an $M \times N \times P$ matrix, then we need $M \times N \times P$ independent equations to resolve it. For the sake of simplicity, we will describe 2-D imaging in the following text; the same analysis can be extended to the 3-D case by adding the z dimension.

In 2-D imaging, RF excitation is used to select a slice in the volume. This is equivalent to setting the profile $P^L(x, y, z)$ to a value of 1 for the desired slice, and 0, otherwise. For the sake of the discussion, let us assume that the slice selection is performed along the z direction, for a position z_0. This means that $P^L(x, y, z) = 1$ when $z = z_0$, and $P^L(x, y, z) = 0$ for all other values of z. Equation 3.1 would become:

$$s_k(G_y^i, t) = \iint I(x,y)W_k(x,y)e^{j\gamma(G_x xt + G_y^i yt)}dxdy \quad (3.2)$$

for the imaged slice at $z = z_0$. With a change of variables, $k_x = jG_x$ and $k_y^i = jG_y^i\tau$, this equation can be discretized to yield:

$$s_k(k_y^i, T) = \sum_{m=1}^{M}\sum_{n=1}^{N} I(m,n)W_k(m,n)e^{(k_x mT + k_y^i n)} \quad (3.3)$$

Where m, n, and T take discrete values, and the number of time points sampled is equal to the resolution along the frequency-encoded direction M. If N phase-encoding steps are used, Equation 3.3 can be written in matrix form as:

$$
\begin{bmatrix}
s_k(G_y^1,1) \\
s_k(G_y^1,2) \\
\cdot \\
\cdot \\
s_k(G_y^1,M) \\
\\
= \\
\cdot \\
s_k(G_y^N,1) \\
\cdot \\
\cdot \\
s_k(G_y^N,M)
\end{bmatrix}
\begin{bmatrix}
W_k(1,1)e^{(k_x1+k_y1)} & \cdot & \cdot & W_k(M,N)e^{(k_xM+k_yN)} \\
W_k(1,1)e^{(k_x1.2+k_y1)} & \cdot & \cdot & W_k(M,N)e^{(k_xM\cdot2+k_yN)} \\
 & \cdot & \cdot & \cdot \\
 & \cdot & \cdot & \\
W_k(1,1)e^{(k_x1\cdot M+k_y1)} & \cdot & \cdot & W_k(M,N)e^{(k_xM\cdot M+k_yN)} \\
 & \cdot & \cdot & \cdot \\
 & \cdot & \cdot & \cdot \\
W_k(1,1)e^{(k_x1+k_yN)} & \cdot & \cdot & W_k(M,N)e^{(k_xM\cdot1+k_yN)} \\
 & \cdot & \cdot & \cdot \\
 & \cdot & \cdot & \cdot \\
W_k(1,1)e^{(k_x1+k_yN)} & \cdot & \cdot & W_k(M,N)e^{(k_xM\cdot M+k_yN)}
\end{bmatrix}
\bullet
\begin{bmatrix}
I(1,1) \\
I(1,2) \\
\cdot \\
\\
I(1,M) \\
I(2,1) \\
\\
\cdot \\
I(M,1) \\
\cdot \\
\cdot \\
I(M,N)
\end{bmatrix}
$$

$$(3.4)$$

where the left-hand-side vector [s_k] contains the time signals received in the k-th coil for all the phase-encoding steps performed (1 to N), and the right-hand-side vector [I(m, n)] contains the elements of the image to be reconstructed. The size of the system of the reconstruction matrix is ($M \cdot N \times M \cdot N$). A slow way of computing the image I(m, n) is by inverting the system of equations; this is never used in practice due to computational inefficiency. In the case where the sensitivity profile information is unknown, Equation 3.4 can also be written such that the sensitivity profile information, $W_k(m, n)$, is included in the right-hand-side vector. If K coils are used simultaneously, the system can be augmented to ($K \cdot M \cdot N \times M \cdot N$), where all the s_k vectors are included.

3.3.1 FOURIER ENCODING

In the widely used Fourier imaging, the 2DFT relationship between I(m, n) and $s_k(G_y^1, T)$ is exploited in order to perform computationally efficient image reconstructions. Once N different phase-encoding steps are acquired, at the Nyquist rate of the spatial resolution in the y direction, the linear system of equations can be solved very efficiently by applying a 2-D Fourier transform to the acquired matrix yielding the image. When using multiple coils, the resulting images in all the coils {$W_1 I, W_2 I,...,W_n I$} are combined in the least-squares sense to yield a composite image: $(W_1^2 + W_2^2 + \cdots + W_n^2)^{1/2} I$. Figure 3.1 shows images reconstructed from a four-coil array and combined to yield a

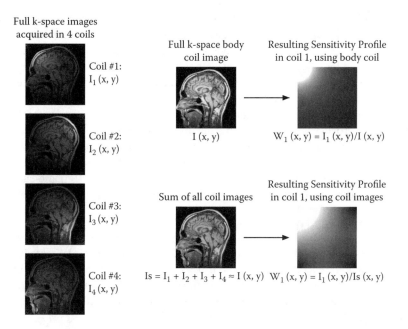

FIGURE 3.1 Coil-sensitivity profile computation from full k-space data.

composite image in the least-squares sense. This was found to be the optimal combination strategy by Roemer et al., in Reference 13.

3.3.2 SAMPLING AT THE NYQUIST RATE AND EQUATION INDEPENDENCE

The time signal $s_k(t)$ described in Equation 3.1 is related by a Fourier transform to the projection of the image onto the frequency-encoded direction. This can be seen by taking the Fourier transform of Equation 3.2 along the x direction:

$$F \cdot T \cdot \left[s_k \left(G_y^i, t \right) \right] = S_k \left(G_y^i, x \right) = \int I(x,y) \cdot W_k(x,y) e^{j\gamma(G_y^i y \tau_1)} dy \qquad (3.5)$$

In order to fully represent the time signal acquired, the sampling rate needs to be at least equal to the Nyquist rate for one period of the signal. From a linear algebraic standpoint, the samples acquired at the Nyquist rate provide an independent and complete set of equations for the specified resolution. Sampling below the Nyquist rate yields an underdetermined linear system, whereas sampling at a higher rate will result in overdetermination for the chosen maximum resolution. Once the FOV and the number of image pixels M along the frequency-encoded direction are determined, image resolution in that direction is set. In practice, the sampling rate of the signal is constant. The effect of varying the sampling rate is obtained by varying the magnitude of the magnetic field gradient in the frequency-encoded direction G_x. This results in M samples acquired at the Nyquist rate of the specified resolution along the x direction. Acquiring more samples than M, however, at the same sampling rate, amounts to adding more dependent equations to the linear system, resulting in increasing the resolution of the image in the x direction, if that resolution is set, and comes at a higher computational cost without real benefit. In practice, the user specifies the size of the FOV as well as the number of pixels in all directions. The gradient magnitude G_x is then automatically computed by the scanner to provide M samples on the time axis. Each acquired signal can therefore be considered as providing M independent equations in the linear system that we seek to solve. In order to get M × N × P independent equations, the values of G_y and G_z need to be varied N and P times respectively. Hence, N × P phase-encoding steps should be performed in order to solve for the image I(m, n, p).

3.3.3 PARALLEL IMAGING METHODS

In parallel imaging, the sensitivity profiles of the receiver coils $\{W_k\}$ are used as complementary encoding functions to phase encoding, a role that is not played in Fourier imaging. Each sensitivity profile, W_k, provides an additional independent view of the image, and hence knowledge of these profiles is necessary in order to resolve the system of equations.

3.3.3.1 Coil-Sensitivity Estimation

A preliminary requirement for almost all parallel imaging techniques is the knowledge of the coil-sensitivity profiles, W_k. A number of approaches have been adopted in order to estimate these profiles, based on solving Equation 3.2 for $W_k(x, y)$.

3.3.3.1.1 Static Estimate

The static estimate assumes that the coil-sensitivity profile is constant during the imaging procedure and is most accurate when the imaging coils are fixed. A calibration scan is done prior to the initiation of parallel imaging whereby a full k-space data set is acquired with all the coils, and the images are reconstructed. The resulting images are weighted by the coil-sensitivity profiles $W_k(x, y) \cdot I(x, y)$. An extra scan is performed using the body coil of the scanner where it is assumed that the sensitivity profile is $W_k(x, y) = 1$. The resulting image is I(x, y). Taking the ratio of the two images yields $W_k(x, y)$ for all the coils. Figure 3.1 shows a four-element array example. Full k-space images are shown.

For the case when coils are arranged around the FOV, it is common to assume that the combined sensitivity profile is homogeneous. This means that adding the images in all the coils results in I(x, y). If this assumption is valid, there is no need to acquire an additional image using the body coil.

3.3.3.1.2 Dynamic Self-Calibrated Estimate

The dynamic self-calibrated sensitivity profile estimates consider that the coil sensitivity varies during the dynamic scans, and seeks to compute it dynamically from a small number of k-space lines acquired during parallel imaging.

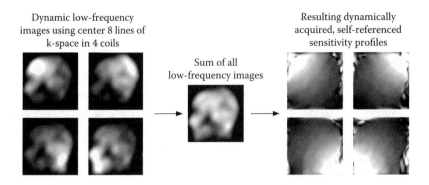

Dynamic low-frequency
images using center 8 lines of
k-space in 4 coils

Sum of all
low-frequency images

Resulting dynamically
acquired, self-referenced
sensitivity profiles

FIGURE 3.2 Sensitivity profile estimation using the center-lines of k-space.

Coil-sensitivity profiles are smoothly varying functions, and can therefore be assumed to contain only low spatial frequencies. A set of contiguous, low-frequency k-space lines is acquired with every acquisition in a dynamic sequence and is used to reconstruct a low-frequency image from each coil (W_1I_{LF}, W_2I_{LF}, W_3I_{LF}, W_4I_{LF}). The images are then added together to form: $(W_1 + W_2 + W_3 + W_4) \cdot I_{LF} \approx I_{LF}$. The resulting sensitivity profiles are computed by taking the ratio of the single-coil low-frequency images, to the sum of the low-frequency coil images. This approach is shown in Figure 3.2. The low-frequency images shown on the left are reconstructed using the eight center lines of k-space.

3.3.3.2 Parallel MR Image Reconstruction Techniques

In this section, we describe how each of the parallel imaging techniques solves the linear system shown in Equation 3.2, assuming that the sensitivity profile information is known.

Two main approaches to solving the linear system of equations shown in Equation 3.2 have been adopted. The k-space approaches, such as SMASH and GRAPPA, seek to use partial data acquired in parallel in all the coils, in order to synthesize a representation of the full k-space of the image. Once this is done, image reconstruction is performed by usual Fourier reconstruction. Image domain approaches such as regular SENSE and SPACE RIP solve the linear system of equations in the image domain.

3.3.3.3 K-Space Approaches

3.3.3.3.1 SMASH

SMASH (7) is dubbed a k-space technique, in the sense that it seeks to estimate a composite k-space of the image from partial data acquired in different coils. Image reconstruction is performed by Fourier transformation of the composite k-space. SMASH operates by using linear combinations of simultaneously acquired signals from multiple surface coils with different spatial sensitivities.

First, the sensitivity profiles $W_k(x, y)$ are linearly combined to approximate composite spatial harmonics in the phase-encoded direction as follows:

$$W_m^{Comp} = \sum_{k=1}^{K} a_k^m \cdot W_k(x, y) \approx A \cdot e^{j(m \cdot \Delta k_y \cdot y)} \tag{3.6}$$

where $k_y = 2/FOV$, A is a complex constant, and K is the total number of coils used. This approximation tries to fit the 2-D coil-sensitivity profiles to a 1-D function in the y direction, rendering the fit challenging and error prone. Once the weights a_k^m are computed, the signals can be combined to form shifted lines of the image k-space as follows:

$$s_m^{Comp} = \sum_{k=1}^{K} a_k^m \cdot s_k(G_y^i, t) = \sum_{k=1}^{K} a_k^m \cdot \iint I(x, y) W_k(x, y) e^{j(k_x xt + k_y y)} dx dy$$

$$= \iint I(x, y) \cdot \sum_{k=1}^{K} a_k^m \cdot W_k(x, y) e^{j(k_x xt + k_y y)} dx dy \tag{3.7}$$

Combining Equation 3.6 and Equation 3.7 results in:

$$s_m^{Comp} = \iint I(x, y) \cdot A \cdot e^{j(m \cdot \Delta k_y \cdot y)} \cdot e^{j(k_x xt + k_y y)} dx dy$$

$$= A \cdot \iint I(x, y) \cdot e^{j[k_x xt + (k_y + m \cdot \Delta k_y) \cdot y]} dx dy \tag{3.8}$$

This shows that shifted k-space lines can be synthesized using a linear combination of the same phase-encoded signals acquired in all the coils in the array. Figure 3.3 shows schematically how the coil sensitivities from eight coils can be

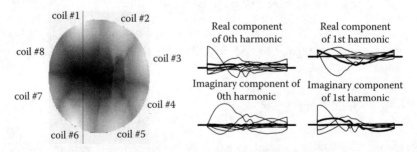

FIGURE 3.3 Combining coil sensitivities in SMASH imaging. On the left, a composite view of all 8 coil estimates is shown. On the right, the real and imaginary components from each coil, for the column marked with the dotted line in the left image. The bold line illustrates the estimated harmonic function formed through a linear combination of the coil components.

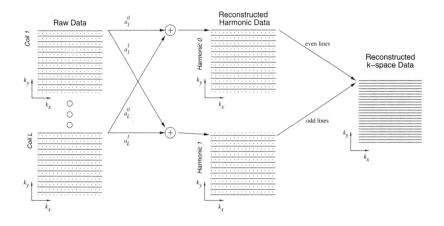

FIGURE 3.4 Data flow and reconstruction in SMASH imaging.

combined to yield spatial harmonics in the y direction. Note that the harmonic fit is approximately, but not exactly, sinusoidal—leading to the problems in reconstruction illustrated in the Examples section.

Figure 3.4 shows the signal flow for the SMASH algorithm. Here each coil is used to collect regularly subsampled k-space sets with a skip factor of two. Reconstructing these sets yields aliased images, weighted by the coil sensitivity profiles. The acquired signals are combined linearly to form the harmonics as described in Equation 3.8. Reconstruction of each harmonic set separately also yields aliased images, where the coil weighting has been removed. Finally, the harmonic sets are combined into a k-space representation, filling odd lines from one harmonic and even lines from the other. Note that the chosen acceleration factor in SMASH imaging determines the number of harmonic fits needed to generate the composite k-space data.

3.3.3.3.1.1 Better Harmonic Fitting in SMASH

The quality of SMASH imaging is a function of the accuracy of the fits computed in Equation 3.6; variations on the design of this fit have been proposed in order to maximize its accuracy. Better fits covering the full FOV of the image can be found when Equation 3.6 is expanded to include the x direction.

$$W_m^{\text{Comp}} = \sum_{k=1}^{K} a_k^m \cdot W_k(x, y) \approx A \cdot B(x) \cdot e^{j(m \cdot \Delta k_y \cdot y)} \qquad (3.9)$$

where each location along the frequency-encoded axis x would have a set of $a_k^m(x)$ parameters associated with it. If N is the resolution along the frequency-encoded direction, then N different SMASH reconstructions can be performed with the different $a_k^m(x)$ sets computed, whereby in each reconstruction, one column location along the x axis is optimal. The N optimal columns are combined in order to get the final image. This procedure increases the computational load by a factor of N, and can be reduced by assuming that multiple locations along the frequency-encoded direction can have the same fitting parameters.

3.3.3.3.2 AUTO-SMASH

One of the significant difficulties with implementing SMASH clinically is that the reconstruction quality is greatly dependent on the coil configuration used. Compounding this difficulty is the fact that coil sensitivities are only approximately known. Thus, more recent k-space methods have focused on bypassing the need to estimate coil-sensitivity maps.

The first such approach was AUTO-SMASH (14), in which one additional line of k-space is acquired in the low-frequency region. The harmonic fit coefficients are then determined by fitting the acquired lines of k-space to this autocalibration line. Specifically, the k-space extrapolation coefficients, $a_k^{(m)}$ are determined via solution of a linear system of equations:

$$S^{comp}(k_y + m\Delta k_y) = \sum_{l=1}^{L} b_l^{(m)} S_l(k_y) = \sum_{l=1}^{L} a_k^{(m)} S_l^{ACS}(k_y + m\Delta k_y) \quad (3.10)$$

This expression can be written in matrix form as

$$S^{comp}(:, k_y + m\Delta k_y) = \begin{bmatrix} b_1^{(m)} & b_2^{(m)} & \cdots & b_L^{(m)} \end{bmatrix} \bullet \begin{bmatrix} S_1(.., k_y + m\Delta k_y) \\ S_2(.., k_y + m\Delta k_y) \\ \vdots \\ S_L(.., k_y + m\Delta k_y) \end{bmatrix} \quad (3.11)$$

where $S^{comp}(:, k_y + m\Delta k_y)$ and $S_l(:, k_y + m\Delta k_y)$ represent vectors from the acquired k-space data set. The reconstructed image is formed using these coefficients in the same fashion as the original SMASH reconstruction equations. A schematic describing the data flow in AUTO-SMASH is shown in Figure 3.5.

This approach can be extended to variable density strategies as well through the use of additional autocalibration lines, VD-AUTO-SMASH (15).

3.3.3.3.3 GRAPPA

In MR, there is significant concern in maximizing SNR in the received signal. This is due to the fact that the ratio of spins contributing to the recorded signal is on the order of 10^{-3}. Accelerating the acquisition by reducing the amount of

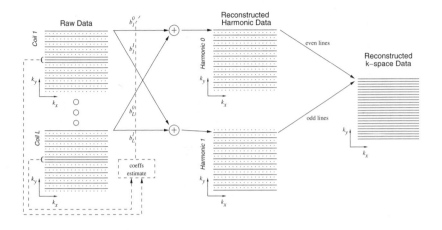

FIGURE 3.5 Data flow diagram for the AUTO-SMASH algorithm.

data acquired necessarily reduces SNR. In considering these issues, the creators of GRAPPA (11) proposed a combination of VD-AUTO-SMASH and coil-by-coil SMASH reconstructions (16) in which a full FOV data set for each coil is produced, and then the final image is constructed by combining the separate coil data using optimal SNR approaches.

The strategy in GRAPPA is to use multiple k-space lines from multiple coils to reconstruct a single k-space estimate in each coil. The reconstruction parameters are again determined by solving a linear system of equations

$$S_j^{\text{acs}}(:, k_y + m\Delta k_y) = \sum_{l=1}^{L} \sum_{b=0}^{N_b-1} n_{l,j}^{(m,b)} S_l(:, k_y - b A \Delta k_y) \qquad (3.12)$$

which can be solved by rewriting in matrix form, as in the AUTO-SMASH case. Here, A represents the acceleration factor, and the coefficients $n_{i,j}^{(m,b)}$ are indexed over both the number of coils and the number of k-space lines used to construct the linear system. A diagram of the data processing flow is given in Figure 3.6.

3.3.3.4.3.4 *k-Space Methods Summary*
Estimation of missing k-space lines has proved to be a successful strategy in parallel MR image reconstruction. However, we must emphasize that these methods are approximations to solving the signal equation. They do not provide an optimal, in the least-squares sense, solution to the signal equation; rather, they attempt to simulate the acquisition of the missing k-space lines through extrapolation.

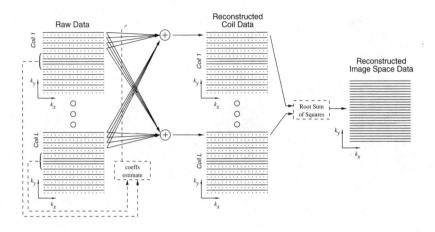

FIGURE 3.6 Data flow diagram for the GRAPPA algorithm.

3.3.3.4   Image Domain Approaches

3.3.3.4.1   SENSE

As originally presented, the Cartesian form of SENSE (8) is founded upon a uniform downsampling pattern. Based on the subsequent aliasing pattern that results from this subsampling choice, one can construct a small system for each spatial-domain pixel in the acquired data reference frame. Solving this small system gives unaliased spatial-domain pixels. This process is then repeated for each pixel in the FOV.

For example, consider a multiple coil acquisition employing a "uniform downsampling by 2" phase-encode k-space acquisition pattern. In the spatial domain, this sampling pattern will alias a pixel $I(j, x)$ with a pixel from the alternate half of the FOV, $I(j + N/2, x)$, as shown in Figure 3.7.

To reconstruct the image, SENSE seeks to identify these two pixel values, $[I(j, x), I(j + N/2, x)]^T$, for each pixel location in the acquired subsampled—and,

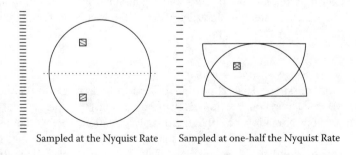

Sampled at the Nyquist Rate   Sampled at one-half the Nyquist Rate

FIGURE 3.7 Diagram of uniform down sampling by 2 and associated aliasing pattern.

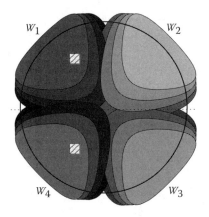

FIGURE 3.8 Local FOV coil-sensitivity encoding is used to untangle separate pixels in aliased acquisitions.

thus, spatially aliased—image. The reconstruction equation is formed using estimates of the coil sensitivity at the location of these aliased pixels.

$$v_1(j,x) = W_1(j,x)I(j,x) + W_1(j+N/2,x)I(j+N/2,x) \qquad (3.13)$$

where $v_l(\mathbf{r}')$ is the spatial-domain representation of acquired data for coil l. Note that the coordinate frame $\mathbf{r}'$ is the aliased version of $\mathbf{r}$. Ideally, each coil "sees" a different portion of the FOV, as in Figure 3.8.

Collecting together the expressions for each coil yields the SENSE linear system of equations

$$\begin{bmatrix} v_1(j,x) \\ v_2(j,x) \\ \vdots \\ v_L(j,x) \end{bmatrix} = \begin{bmatrix} W_1(j,x) & W_1(j+N/2,x) \\ W_2(j,x) & W_2(j+N/2,x) \\ \vdots & \vdots \\ W_L(j,x) & W_L(j+N/2,x) \end{bmatrix} \begin{bmatrix} I(j,x) \\ I(j+N/2,x) \end{bmatrix} \qquad (3.14)$$

which is repeated for each pixel location in the acquired spatial-domain coordinate system. Using an identical framework, this approach has also been used for variable density subsampling patterns (17).

3.3.3.4.2 SPACE RIP

From the signal equation, SPACE RIP (9) takes the Fourier transform of Equation 3.2 along the x direction, when a phase-encoding gradient G_y^i is applied yields:

$$S_k(G_y^i, x) = \int I(x,y)W_k(x,y)e^{j\gamma(G_y^i y \tau)}dy, \qquad (3.15)$$

which is the phase-modulated projection of the sensitivity-weighted image onto the x axis. Equation 3.15 can be discretized along y and written as follows:

$$S_k(G_y^i, x) = \sum_{n=1}^{N} I(x,n) W_k(x,n) e^{j\gamma(G_y n\tau)}.$$
(3.16)

This expression can be converted to matrix form for each position x along the horizontal direction of the image, as follows:

$$
\begin{pmatrix} S_1(G_y^1, x) \\ \\ S_1(G_y^F, x) \\ S_2(G_y^1, x) \\ \\ S_2(G_y^F, x) \\ \\ \\ S_K(G_y^1, x) \\ \\ S_K(G_y^F, x) \end{pmatrix}
=
\begin{pmatrix}
W_1(x,1)e^{j\gamma(G_y^1 1\tau)} \dots\dots W_1(x,N)e^{j\gamma(G_y^1 N\tau)} \\ \\
W_1(x,1)e^{j\gamma(G_y^F 1\tau)} \dots\dots W_1(x,N)e^{j\gamma(G_y^F N\tau)} \\
W_2(x,1)e^{j\gamma(G_y^1 1\tau)} \dots\dots W_2(x,N)e^{j\gamma(G_y^F N\tau)} \\ \\
W_2(x,1)e^{j\gamma(G_y^F 1\tau)} \dots\dots W_2(x,N)e^{j\gamma(G_y^F 1\tau)} \\ \\ \\
W_K(x,1)e^{j\gamma(G_y^1 1\tau)} \dots\dots W_K(x,N)e^{j\gamma(G_y^F 1\tau)} \\ \\
W_1(x,1)e^{j\gamma(G_y^1 1\tau)} \dots\dots W_K(x,N)e^{j\gamma(G_y^F 1\tau)}
\end{pmatrix}
\bullet
\begin{pmatrix} I(x,1) \\ I(x,2) \\ I(x,3) \\ \\ \\ \\ I(x,N) \end{pmatrix}
$$
(3.17)

where F is the number of phase encodes used in the experiment, $[W_1, \dots, W_k]$ represent the sensitivity profiles of the coils, $I(x,y)$ represents the image, and $[S_1, \dots, S_k]$ represent the discrete Fourier transform of the signals received in all the coils. Equation 3.17 is a matrix equation where the term on the left side is a K × F element vector containing the F phase-encoded values for all K coils. The term on the far right is an n-element vector representing the "image" for one column. The middle term in Equation 3.17 is a matrix with K × F rows and N columns; it is constructed based on the sensitivity profiles and phase encodes used. Solving Equation 3.17 for each position along the x axis yields a column-by-column reconstruction of the image. Increasing F results in an increase of the rank of the matrices, yielding system matrices that are better conditioned. SPACE RIP allows for the arbitrary choice of phase encodes, making it well suited to getting real-time sensitivity profile estimations by fully sampling the center of k-space. These estimates are more

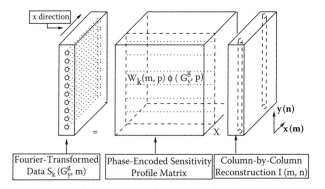

FIGURE 3.9 A full image reconstruction representation. Equation 3.17 is solved independently for each column in the image.

accurate than estimates computed prior to the beginning of the dynamic acquisition, leading to a minimization of artifacts. The inclusion of these center lines in the reconstruction further contributes to a higher SNR in the image, due to the high level of energy present at the center of k-space. Reconstructing the full image amounts to a column-by-column reconstruction whereby each location along the frequency-encoded direction can be computed independently; a schematic description of the full image reconstruction is shown in Figure 3.9.

Note that one can convert the given SPACE RIP linear system of equations to the Generalized SMASH linear system of equations with an inverse- and forward-unitary Fourier transform matrix between the system matrix in Equation 3.17 and the reconstruction vector (12). This effectively converts the objective from a spatial domain reconstruction to a k-space domain reconstruction, while maintaining the basic structure if the linear system.

3.3.3.4.4.2.1 Conditioning of the Reconstruction Matrix

The reconstruction scheme outlined in the SPACE RIP technique is based on matrix inversion. In order to ensure stable and robust reconstruction, the condition number of the inverted matrices (defined as the ratio of the largest eigenvalue to the lowest eigenvalue) should be minimized. Equation 3.17 shows that the condition number is a function of the choice of the phase encodes F acquired per coil; it is also a function of the sensitivity profile estimations of the receiver coil array. Because the sensitivity profiles are coil dependent and generally fixed during a dynamic acquisition, conditioning the reconstruction of the SPACE RIP technique is practically performed by careful selection of the acquired k-space lines as well as minimizing the noise in the sensitivity estimates. The SNR in the resulting images is also a function of the condition number of the reconstruction matrix shown in Equation 3.17. To avoid errors due to numerical propagation, the pseudoinverse of each reconstruction matrix is restricted to those singular values that are greater than a given threshold. This effectively removes any noise amplification due to poor conditioning.

3.3.3.4.4.2.2 Choice of the Phase Modulations

To appropriately cover the *k*-space distribution of the image I(x, y), the choice of the phase modulations used in the inversion matrix should be influenced by the frequency content of the sensitivity profile. In the spatial domain, the image received in a coil having a sensitivity profile $W_c(x, y)$ can be written as $I_c(x, y) =$ I(x, y) $W_c(x, y)$. Therefore, the k-space profile of $I_c(x, y)$ is the convolution of the k-space profile $I(k_x, k_y)$ of the image I(x, y), with the *k*-space profile $W_c(k_x, k_y)$ of the sensitivity profile $W_c(x, y)$. This convolution amounts to a blurring of the k-space data $I(k_x, k_y)$ in an image. Because a different convolution is performed for each coil, a different blurring of $I(k_x, k_y)$ occurs for each coil. Subsampling the convolved k-space data received in different coils, therefore, results in different coverage of the k-space of the image I(x, y). Hence, in order to get the best k-space coverage of I(x, y), it is necessary to optimally sample the k-space data from all the coils. In contrast to SMASH and SENSE, which require the use of equally spaced k-space lines, SPACE RIP is completely flexible in this regard. In the following section, we show how a carefully chosen irregular sampling pattern (whereby the center of k-space is sampled more densely than the periphery) coupled with appropriate matrix conditioning, would better capture the spatial energy distribution, yielding reconstructions with higher SNR and fewer artifacts than in other parallel imaging techniques.

3.4 EXAMPLES

The following examples demonstrate each of the preceding algorithms and illustrate their effectiveness in removing aliasing artifacts from subsampled parallel MR acquisitions. The ACR quality phantom data shown here were acquired using an eight-channel head coil on a GE Signa Lx 1.5-T MR scanner. This phantom slice provides both a large uniform area to easily identify residual aliasing artifacts and a set of boxes with varying resolution through which one can measure blur. For these examples, the full 256 by 256 k-space set was obtained once, then each of the various subsampling patterns were simulated by excluding those k-space lines from the reconstruction. To more accurately reflect the current common practice in parallel MR imaging, the sensitivity maps were estimated using self-referenced data. In this case, between 8 and 20 lines closest to the lowest frequency in k-space were used to estimate the coil-sensitivity maps. A Gaussian envelope was applied to this data to filter the high-spatial-frequency components along the readout direction and to limit ringing along the phase-encode direction, and the estimates were then normalized. An example of these coil-sensitivity estimates are shown in Figure 3.10.

3.4.1 EXAMPLE 1: UNIFORM SUBSAMPLING

Figure 3.11(a) shows a 2x-acceleration uniform subsampling pattern. Figure 3.11(b) shows the resulting uncorrected image reconstruction using the subsampled data. The twofold aliasing pattern that parallel MR reconstruction algorithms aim to suppress is clearly evident.

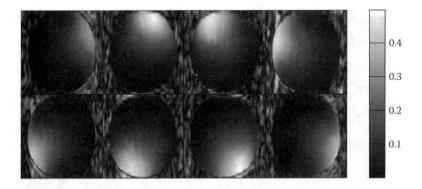

FIGURE 3.10 Self-calibrated 8-channel head coil-sensitivity maps.

Figure 3.12 shows the reconstructions from four separate algorithms on the 2x-accelerated data. Figure 3.12(a) shows a reconstruction following the SMASH approach described in the preceding text. As mentioned earlier, the SMASH reconstruction approach is very sensitive to the coil-sensitivity map estimates. As this example shows, SMASH reconstruction is not well suited for self-calibrated sensitivity estimation. A better approach is to use the AUTO-SMASH reconstruction, shown in Figure 3.12(b). This k-space domain method does not require estimation of the coil-sensitivity maps and produces a much cleaner image reconstruction with significant suppression of the aliasing artifacts. As noted earlier, however, the SMASH and AUTO-SMASH approaches present only an approximate solution to the parallel MR signal equation. The generalization of these approaches in GRAPPA provides enough flexibility in reconstruction parameter selection to effectively suppress the artifacts visible in the other two k-space methods, as the reconstruction with 8 ACS lines in 3.12(c) illustrates. This reconstruction is comparable to the SENSE approach, which does minimize the least-squared error between the acquired data and the signal acquisition model

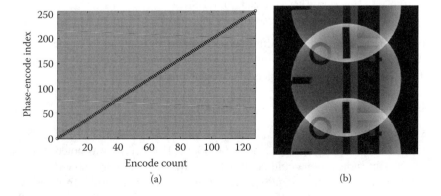

FIGURE 3.11 Uniform 2x sampling pattern and spatial-domain aliasing result.

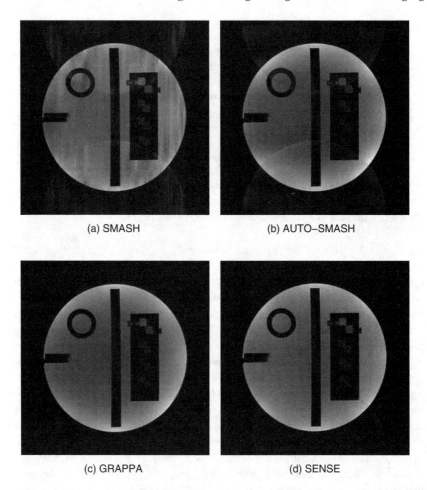

FIGURE 3.12 Reconstructions of uniform 2x-acceleration data using (a) SMASH, (b) AUTO-SMASH, and (c) GRAPPA, and (d) SENSH.

for the particular case of uniform phase-encode subsampling. As seen in the SENSE reconstruction of Figure 3.12(d), this approach successfully suppresses almost all of the aliasing artifacts at 2x acceleration.

3.4.2 EXAMPLE 2: VARIABLE SUBSAMPLING

As the previous example shows, successful reconstruction techniques for uniform downsampling approaches have been developed. However, it was noted early on that artifact suppression could be improved through nonuniform downsampling. Figure 3.13 shows an example of the aliasing pattern that results from using a 3x-"hat" acceleration sampling pattern, and it is significantly different from the uniform-2x aliasing pattern. Here, the hat sampling pattern follows a top-hat

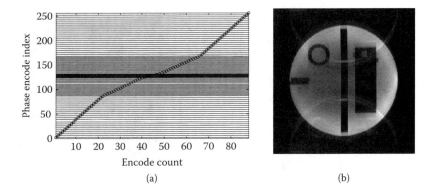

(a)   (b)

FIGURE 3.13 Non-uniform 3x-"hat" sampling pattern and spatial-domain aliasing result.

profile — densely sampled (1x-acceleration) close to the center of k-space (zero frequency), less-densely sampled (2x-acceleration) farther away from the center of k-space, and sparsely sampled (4x-acceleration) at the edges of k-space. The ratio of the total number of selected lines to the number of lines in the FOV determines the final acquisition acceleration.

Note that changing from a uniform to an irregular sampling pattern complicates the spatial-domain aliasing pattern. Consequently, the SENSE equations no longer solve the parallel MR signal equation in the least-squares sense. This led to the development of reconstruction algorithms, e.g., SPACE RIP, Generalized SMASH, and GRAPPA that do consider irregular subsampling patterns.

Figure 3.14 presents both a GRAPPA and SPACE RIP reconstruction of 3x-hat accelerated data. As seen in the figure, the aliasing artifacts visible in Figure 3.14(b) are visibly suppressed. For comparison, a 3x-uniform accelerated SENSE reconstruction is shown in Figure 3.14(c) with significant artifacts in the reconstruction.

While both GRAPPA and SPACE RIP provide accurate reconstructions, there are significant differences between them. On the sub-sampling side, SPACE RIP

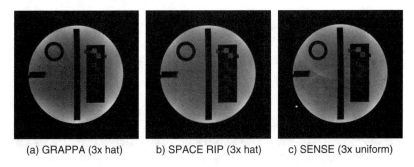

(a) GRAPPA (3x hat)   b) SPACE RIP (3x hat)   c) SENSE (3x uniform)

FIGURE 3.14 Reconstructions of 3x-"hat"-accelerated acquisition data using (a) GRAPPA and (b) SPACERIP. For comparison, a SENSE reconstruction of uniform-3x-acceleration data is shown in (c).

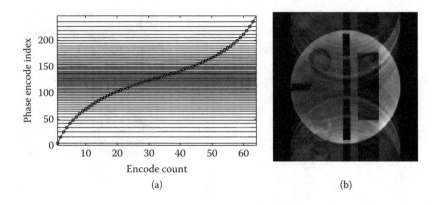

(a)   (b)

FIGURE 3.15 Non-uniform 4x-"exponential" sampling pattern and spatial-domain aliasing result.

(and closely related techniques like Generalized SMASH) provides greater flexibility in sub-sampling selection than GRAPPA. One can choose a variety of sub-sampling patterns, including the exponentially-weighted sampling distribution — densely sampled in the center of k-space (1x), sparsely sampled at the edges of k-space (12x), with the change in step size following an exponential decay. — shown in Figure 3.15. This sub-sampling flexibility allows one to potentially tailor the sub-sampling choice to particular applications.

On the reconstruction side, SPACE RIP and Generalized SMASH provide a minimum least-squares error estimate of the image in the same manner as SENSE, where as GRAPPA provides only at approximate solution in the same manner as SMASH. The effect of this difference becomes more apparent at high acceleration factors. For example, Figure 3.16 shows a 4x-acceleration reconstruction for GRAPPA and SPACE RIP. The GRAPPA reconstruction, in Figure 3.16(a), shows higher spatial noise throughout the field of view and a significant loss of resolution in the boxes to the right. In contrast, although the SPACE RIP reconstruction from identical phase encodes lines in Fig. 3.16(b) shows stronger residual artifacts

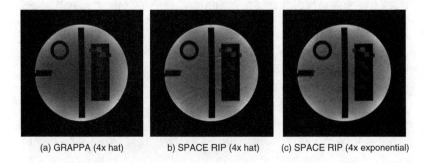

(a) GRAPPA (4x hat)   b) SPACE RIP (4x hat)   (c) SPACE RIP (4x exponential)

FIGURE 3.16 Reconstructions of 4x-exponential-acceleration data using (a) GRAPPA and (b) SPACE-RIP.

than the GRAPPA reconstruction, the resolution of the boxes on the right is maintained. Furthermore, the inherent ability within SPACE RIP to modify the phase encode selection allows one to choose different phase encode lines that distribute the residual artifact more evenly over the field of view while maintaining the required resolution. This is illustrated in Figure 3.16(c), using the 4x-exponentially-weighted sub-sampling distribution given in Figure 3.15.

The cost of this flexibility in sub-sampling is computational complexity. SPACE RIP reconstructions typically take six to ten times longer to compute than GRAPPA. This is because for each image, SPACE RIP must solve a system of equations for each column, with each system of size LP-by-N, where L is the number of coils, P is the number of phase encodes acquired, and N is the size of the FOV in the phase-encode direction. In contrast, GRAPPA needs to solve a system of equations for each missing line of k-space: L coils × M-P lines (size of FOV–number of phase encodes). For each of these systems the matrix equation that solves the GRAPPA parameter fit equation is of size p-by-N, where p is the number of parameters to determine and N is the size of the FOV along the readout direction. With a smaller system matrix size, GRAPPA requires significantly less computational resources to compute.

The ability to provide reconstruction of irregularly sampled data with minimal computational resources has led to rapid adoption of GRAPPA in clinical settings. However, with rapidly improving computational resources, it is anticipated that SPACE RIP will become more prevalent, given its greater flexibility in tailoring sub-sampling strategies for both artifact suppression and resolution improvement.

3.4.3 EXAMPLE 3: IN VIVO APPLICATIONS

Parallel imaging has found wide applicability in clinical imaging, because it can be used to improve both spatial and temporal resolution. Our final examples illustrate two applications in which parallel MR imaging has found strong popularity and a promising future. In cardiac imaging, shown in Figure 3.17, the standard clinical examination currently requires a patient breath hold of

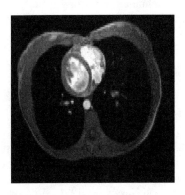

FIGURE 3.17 Axial cardiac image.

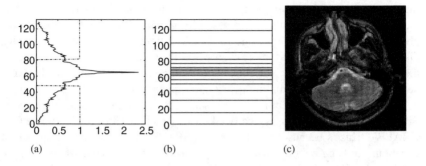

(a)   (b)   (c)

FIGURE 3.18 Axial neurological image acquired using irregular sub-sampling strategy.

approximately 20 sec. This breath hold is certainly possible with a healthy volunteer, but very taxing on patients with cardiac disease. Using parallel imaging to increase the acquisition rate decreases the breath-hold time by the acceleration factor. Thus, with 2x- and 4x-acceleration, the breath-hold time reduces to approximately 11 sec and 6 sec, respectively.

If used to improve spatial resolution, parallel imaging provides an improvement proportional to the acceleration factor. Figure 3.18 shows a demonstration of this with an axial image of the brain. The image was acquired using four coils at 4x-acceleration. The acquired data were subsampled using a set of k-space lines whose density mirrors the Fourier signal energy distribution. This distribution is shown in Figure 3.18(a), with the corresponding sampling pattern is shown Figure 3.18(b). The resulting image reconstructed using SPACE RIP is shown in Figure 3.18(c).

3.5 SUMMARY

This chapter presented a basic overview of the current state of the art in parallel MR image reconstruction techniques. Starting from the signal acquisition model, we reviewed the reconstruction approach given by SMASH, its subsequent improvements leading to GRAPPA, SENSE, and SPACE RIP. A number of reconstruction examples were given to illustrate the effectiveness of each method and a few of the differences between them.

REFERENCES

1. Lauterbur, P.C. (1973). Image formation by induced local interactions: examples employing nuclear magnetic resonance. *Nature* 242: 190–191.
2. Hutchinson, M. and Raff, U. (1998). Fast MRI data acquisition using multiple detectors. *Magn. Reson. Med.* 6(1): 87–91.
3. Kwiat, D., Einav, S., and Navon, G. (1991). A decoupled coil detector array for fast image acquisition in magnetic resonance imaging. *Med. Phys.* 18(2): 251–265.

4. Carlson, J.W. and Minemura, T. (1993). Imaging time reduction through multiple receiver coil data acquisition and image reconstruction. *Magn. Reson. Med.* 29(5): 681–687.

5. Kelton, J.R., Magin, R.L., and Wright, S.M. (1989). In: An algorithm for rapid image acquisition using multiple receiver coils. Proceedings of the SMRM 8th annual meeting, Amsterdam, The Netherlands. p. 1172.

6. Ra, J.B. and Rim, C.Y. (1991). In: Fast imaging method using multiple receiver coils with subencoding data set. Proceedings of the SMRM 10th annual meeting, San Francisco, CA. p. 1240.

7. Sodickson, D.K. and Manning, W.J. (1997). Simultaneous acquisition of spatial harmonics (SMASH): ultra-fast imaging with radio frequency coil arrays. *Magn. Reson. Med.* 38(4): 591–603.

8. Pruessmann, K.P., Weiger, M., Scheidegger, M.B., and Boesiger, P. (1999). SENSE: Sensitivity encoding for fast MRI. *Magn. Reson. Med.* 42(5): 952–962.

9. Kyriakos, W.E., Panych, L.P., Kacher, D.F., Westin, C.F., Bao, S.M., Mulkern, R.V., and Jolesz, F.A. (2000). Sensitivity profiles from an array of coils for encoding and reconstruction in parallel (SPACE RIP). *Magn. Reson. Med.* 44(2): 301–308.

10. Griswold, M.A., Jakob, P.M., Hedemann, R.M., and Haase, A. (2000). Parallel imaging with localized sensitivities (PILS). *Magn. Reson. Med.* 44(4): 602–609.

11. Griswold, M.A., Jakob, P.M., Heidemann, R.M., Nittka, M., Jellus, V., Wang, J., Kiefer, B., and Haase, A. (2002). Generalized autocalibrating partially parallel acquisitions (GRAPPA). *Magn. Reson. Med.* 47(6): 1202–1210.

12. Bydder, M., Larkman, D.J., and Hajnal, J.V. (2002). Generalized SMASH imaging. *Magn. Reson. Med.* 47(1): 160–170.

13. Roemer, P.B., Edelstein, W.A., Hayes, C.E., Souza, S.P., and Mueller, O.M. (1990). The NMR phased array. *Magn. Reson. Med.* 16(2): 192–225.

14. Jacob, P.M., Griswold, M.A., Edelman, R.R., and Sodickson, D.K. (1998). AUTO-SMASH: a self-calibrating technique for SMASH imaging. Simultaneous acquisition of spatial harmonics. *MAGMA* 7: 42–54.

15. Heidemann, R.M., Griswold, M.A., Haase, A., and Jakob, P.M. (2001). VD-AUTO SMASH imaging. *Magn. Reson. Med.* 45(6): 1066–1074.

16. McKenzie, C.A., Ohliger, M.A., Yeh, E.N., Price, M.D., and Sodickson, D.K. (2001). Coil-by-coil phased array image reconstruction with SMASH. *Magn. Reson. Med.* 46(3): 619–623.

17. Modore, B. (2004) UNFOLD-SENSE: A parallel MRI method with self calibration and artifact Suppression. *Magn. Reason. Med.* 52(2): 310–320.

Part II

SNR Improvement and Inhomogeneities Correction

4 Estimation of Signal and Noise Parameters from MR Data

A.J. den Dekker and J. Sijbers

CONTENTS

4.1 INTRODUCTION

Magnetic resonance imaging (MRI) is a magnificent imaging technique in bio-medicine that is able to produce high-quality images containing an abundance of physiological, anatomical, and functional information. Such information is often not extracted by visual interpretation of the images alone; digital data from the scanners are generally processed and analyzed in a quantitative way using advanced digital-data-processing tools.

Most of the current image processing applications applied to MR image data can be formulated as a parameter estimation problem. For example, in the case of noise filtering the parameter to be estimated is given by the true signal component underlying the noise-corrupted data, whereas in the construction of T_1 and T_2 maps the parameters to be estimated are given by the relaxation time constants [1–9].

Nowadays, there exist several estimation procedures, each of which seems to be slightly different from the other. So which one should we use? Which one is optimal with respect to a specified error criterion? Throughout this chapter, we hope to give the reader answers to these questions by analyzing commonly used signal and noise estimation methods as well as the maximum likelihood (ML) method. Each method is described in detail and evaluated in terms of precision and accuracy.

This chapter is organized as follows: Because optimal quantitative analysis requires exploitation of the knowledge of the underlying data statistics, Section 4.2 describes various probability density functions (PDFs) that appear when dealing with MR data. Section 4.3 reviews some results from statistical parameter estimation theory, which are used in the remainder of the chapter. Different performance measures for estimators as well as the so-called Cramér–Rao lower bound (CRLB) and the ML estimator are discussed. In Section 4.4 and Section 4.5 we explain how the various PDFs can be exploited to estimate parameters from MR data. In particular, in Section 4.4 we will focus on the estimation of (noiseless) signal components, whereas in Section 4.5 we will consider the estimation of the image noise variance.

4.2 PDFs IN MRI

Whenever quantitative information needs to be extracted from MRI data, knowledge of the PDF of the data is of vital importance. Indeed, if an incorrect PDF is assumed *a priori*, systematic errors (bias) may be introduced when estimating parameters from these data. Therefore, this section starts with an overview of the various PDFs that would appear when dealing with (processed) MR data [10].

4.2.1 GAUSSIAN PDF

The raw MR data acquired in K-space during an MR acquisition scheme are known to be complex valued. The complex data are composed of noiseless signal components and noise contributions that are assumed to be additive and independent and are characterized by a zero-mean Gaussian PDF [11–13]. An MR reconstruction is then obtained by means of an inverse Fourier transform (FT). Because of the linearity and orthogonality of the FT, the complex data resulting from the transformation are still independent and Gaussian distributed* [14–16]. Hence, the PDF of a raw, complex data point $\underline{c} = (\underline{w}_r, \underline{w}_i)$ is given by

$$p_{\underline{c}}(\omega_r, \omega_i \mid A, \varphi, \sigma) = \frac{1}{2\pi\sigma^2} e^{-\frac{(\omega_r - A\cos\varphi)^2}{2\sigma^2}} e^{-\frac{(\omega_i - A\sin\varphi)^2}{2\sigma^2}}, \qquad (4.1)$$

---

* It is assumed that the MR signals are sampled on a uniform grid in K-space. Furthermore, the variance of the noise is assumed to be equal for each raw data point.

where σ^2 denotes the noise variance, and $(\underline{\omega}_r, \underline{\omega}_i)$ are the real and imaginary variables, respectively, corresponding to the complex observation $(\underline{w}_r, \underline{w}_i)$, with underlying true amplitude and phase value, A and φ, respectively. In Equation 4.1 and in what follows, stochastic (i.e., random) variables are underlined [10]. In general, a Gaussian PDF is described by

$$p_{\underline{x}}(x \mid \mu, \sigma) = \frac{1}{\sqrt{2\pi\sigma^2}} e^{-\frac{(x-\mu)^2}{2\sigma^2}}, \tag{4.2}$$

with μ and σ denoting the mean and standard deviation of the PDF, respectively.

4.2.1.1 Moments of the Gaussian PDF

Analytical expressions for the moments of a Gaussian PDF are given by

$$é[\underline{x}^v] = \sigma^{2(v-1)} \left[\frac{d^{v-1}}{d\mu^{v-1}} \left(\mu e^{\frac{\mu^2}{2\sigma^2}} \right) \right] e^{-\frac{\mu^2}{2\sigma^2}}, \tag{4.3}$$

where $\mathbb{E}[.]$ is the expectation operator and $v \in \mathbb{N}_0$ [17]. For the first four moments we have, explicitly

$$\mathbb{E}[\underline{x}] = \mu, \tag{4.4}$$

$$\mathbb{E}[\underline{x}^2] = \mu^2 + \sigma^2, \tag{4.5}$$

$$\mathbb{E}[\underline{x}^3] = \mu^3 + 3\mu\sigma^2, \tag{4.6}$$

$$\mathbb{E}[\underline{x}^4] = \mu^4 + 6\mu^2\sigma^2 + 3\sigma^4. \tag{4.7}$$

4.2.1.2 Central Moments

For the central moments, we have

$$\mathbb{E}[(\underline{x} - \mu)^v] = \begin{cases} 0 & \text{if } v \text{ is odd} \\ \dfrac{v!\sigma^v}{(v/2)!2^{v/2}} & \text{if } v \text{ is even} \end{cases}. \tag{4.8}$$

4.2.2 RICIAN PDF

During MR data processing, it is a common practice to work with magnitude data instead of real and imaginary data, because magnitude data have the advantage of being immune to the effects of incidental phase variations due to radio-frequency (RF) angle inhomogeneity, system delay, noncentered sampling windows, etc. In this section, the PDF of the magnitude data is discussed.

To construct a magnitude image from the complex data, the magnitude is computed on a pixel-by-pixel basis

$$m = \sqrt{\omega_r^2 + \omega_i^2}, \qquad (4.9)$$

where m is the magnitude variable corresponding to the magnitude observation m. As root extraction is a nonlinear transformation, the PDF of the magnitude data is no longer expected to be Gaussian [18,19].

The PDF of the magnitude data is found by transforming the joint PDF of the real and imaginary data, given in Equation 4.1, into polar coordinates:

$$p_{(m,\phi|A,\varphi,\sigma)} = \frac{1}{2\pi\sigma^2} e^{-\frac{(A\cos\varphi - \cos\phi)^2}{2\sigma^2}} e^{-\frac{(A\sin\varphi - \sin\phi)^2}{2\sigma^2}} m, \qquad (4.10)$$

where ϕ denotes the phase variable corresponding to the phase observation ϕ. The last factor, m, is the Jacobian of the transformation (see Appendix). Integration of Equation 4.10 over a full cycle of $\cos\phi$ leads us to the PDF that characterizes magnitude data:

$$p_{\underline{m}}(|A,\sigma) = \frac{m}{\sigma^2} e^{-\frac{m^2 + A^2}{2\sigma^2}} I_0\left(\frac{A}{\sigma^2}\right) \varepsilon(m), \qquad (4.11)$$

with I_0 denoting the zeroth-order modified Bessel function of the first kind. The unit step Heaviside function $\varepsilon(.)$ is used to indicate that the expression for the PDF of $\underline{m}$ is valid for nonnegative values of m only. The preceding distribution is called the *Rician distribution*, after S. O. Rice, who derived it in the context of communication theory in 1944 [18]. Note that the shape of the Rician PDF depends on the signal-to-noise ratio (SNR), which is here defined as the ratio A/σ. Figure 4.1 shows the Rician PDF as a function of the magnitude variable for various values of the SNR.

4.2.2.1 Asymptotic Approximation of the Rician Distribution

This subsection describes the behavior of the Rician distribution for very low and very high SNR:

- For low SNR, the modified Bessel function is given by

$$I_\nu(z) \sim \left(\frac{z}{2}\right)^\nu \Gamma(\nu+1) \quad \text{for} \quad z \to 0. \qquad (4.12)$$

Hence, the Rician PDF then leads to a Rayleigh distribution:

$$p_{\underline{m}}(m|\sigma) = \frac{m}{\sigma^2} e^{-\frac{m^2}{2\sigma^2}} \varepsilon(m) \quad \text{for} \quad \text{SNR} \to 0. \qquad (4.13)$$

The Rayleigh PDF characterizes the random intensity distribution of nonsignal background areas such as air.

At high SNR, the asymptotic approximation of the modified Bessel function is given by

$$I_v(z) \sim \frac{e^z}{\sqrt{2\pi z}} \quad \text{for} \quad z \to \infty. \qquad (4.14)$$

Then, given that $\sqrt{m/A} \to 1$ for SNR $\to \infty$, we find for the Rician PDF:

$$p_{\underline{m}}(m \mid A\sigma) = \frac{1}{\sqrt{2\pi\sigma^2}} e^{-\frac{(m-A)^2}{2\sigma^2}} \varepsilon(m) \quad \text{for} \quad \text{SNR} \to \infty. \qquad (4.15)$$

Hence, for high SNR, the Rician PDF approaches a Gaussian PDF, with a mean A and a variance σ^2. The transition between the two limits of the Rician PDF can visually be appreciated in Figure 4.1.

4.2.2.2 Moments of the Rician PDF

The moments of the Rician PDF can be analytically expressed as a function of the confluent hypergeometric function of the first kind $_1F_1$

$$\mathbb{E}[\underline{m}^v] = (2\sigma^2)^{v/2}\Gamma\left(1+\frac{v}{2}\right){}_1F_1\left[-\frac{v}{2};1;-\frac{A^2}{2\sigma^2}\right], \qquad (4.16)$$

with Γ representing the Gamma function [20]. The even moments of the Rician distribution (i.e., when v is even) are simple polynomials. The expressions for the

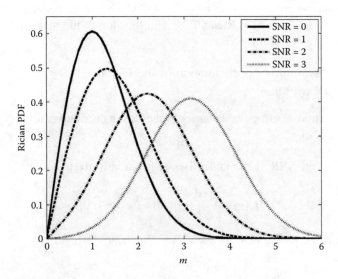

FIGURE 4.1 The Rician PDF as a function of the magnitude variable M, drawn for various values of the SNR where M=1. At SNR = 0, the distribution equals a Rayleigh PDF, whereas at high SNR (SNR > 3), the Rician PDF approaches a Gaussian PDF.

odd moments are much more complex. However, the confluent hypergeometric function can be expressed in terms of modified Bessel functions, from which an analytic expression of the odd moments can be derived. Hence, we have explicitly:

$$\mathbb{E}\big[\underline{m}\big] = \sigma\sqrt{\frac{\pi}{2}}e^{-\frac{A^2}{4\sigma^2}}\left[\left(1+\frac{A^2}{2\sigma^2}\right)I_0\left(\frac{A^2}{4\sigma^2}\right)+\frac{A^2}{2\sigma^2}I_1\left(\frac{A^2}{4\sigma^2}\right)\right], \qquad (4.17)$$

$$\mathbb{E}[\underline{m}^2] = A^2 + 2\sigma^2 , \qquad (4.18)$$

$$\mathbb{E}[\underline{m}^3] = \sigma^3\sqrt{\frac{\pi}{2}}e^{-\frac{A^2}{4\sigma^2}}\left[\left(3+3\frac{A^2}{\sigma^2}+\frac{A^4}{2\sigma^4}\right)I_0\left(\frac{A^2}{4\sigma^2}\right)\right.$$
$$\left.+\left(2\frac{A^2}{\sigma^2}+\frac{A^4}{2\sigma^4}\right)I_1\left(\frac{A^2}{4\sigma^2}\right)\right] \qquad (4.19)$$

$$\mathbb{E}[\underline{m}^4] = A^4 + 8\sigma^2 A^2 + 8\sigma^4, \qquad (4.20)$$

with I_1 denoting the first-order modified Bessel function of the first kind.

4.2.2.3 Moments of the Rayleigh PDF

For completeness, we also mention the general expression for the moments of the Rayleigh PDF, to which the Rician PDF tends at low SNR. For the Rayleigh PDF, we have

$$\mathbb{E}[\underline{m}^\nu] = (2\sigma^2)^{\nu/2}\Gamma\left(1+\frac{\nu}{2}\right), \qquad (4.21)$$

Explicitly, the first four moments are

$$\mathbb{E}[\underline{m}] = \sqrt{\frac{\pi}{2}}\sigma , \qquad (4.22)$$

$$\mathbb{E}[\underline{m}^2] = 2\sigma^2 , \qquad (4.23)$$

$$\mathbb{E}[\underline{m}^3] = 3\sqrt{\frac{\pi}{2}}\sigma^2, \qquad (4.24)$$

$$\mathbb{E}[\underline{m}^4] = 8\sigma^4. \qquad (4.25)$$

4.2.2.4 Generalized Rician PDF

If magnitude data are computed from more than two Gaussian-distributed variables, the underlying PDF is the *generalized Rician PDF*. Such data are found in phased array magnitude MR images, where use is made of multiple receiver coils [21], and in phase-contrast MR (PCMR) images [22,23]. In general, a (random) PCMR pixel variable denoted by $\underline{m}$ can be written as

$$\underline{m} = \sqrt{\sum_{k=1}^{K} \underline{s}_k^2},$$
(4.26)

with K denoting twice the number of orthogonal Cartesian directions in which flow is encoded. The set $\{\underline{s}_k\}$ contains independent Gaussian-distributed variables with mean $\{a_k\}$ and variance σ^2. The deterministic signal component of the PCMR pixel variable is given by

$$A = \sqrt{\sum_{k=1}^{K} a_k^2},$$
(4.27)

The PDF of such a PCMR variable is given by

$$p_{\underline{m}}(\mathsf{m}) = \frac{\mathsf{m}}{\sigma^2} \left(\frac{\mathsf{m}}{A}\right)^{\frac{K}{2}-1} \exp\left(-\frac{\mathsf{m}^2 + A^2}{2\sigma^2}\right) I_{\frac{K}{2}-1}\left(\frac{\mathsf{m}A}{\sigma^2}\right) \varepsilon(\mathsf{m}).$$
(4.28)

When $A \to 0$ the PDF of the magnitude PCMR variable turns into a *generalized Rayleigh PDF*:

$$p_{\underline{m}}(\mathsf{m}) = \frac{2\mathsf{m}^{K-1}}{(\sigma\sqrt{2})^K \Gamma(K/2)} \exp\left(-\frac{\mathsf{m}^2}{2\sigma^2}\right) \varepsilon(\mathsf{m}).$$
(4.29)

Figure 4.2 shows the generalized Rician PDF for SNR = 0 and SNR = 3 and for $K = 2$, 4, and 6 and $\sigma = 1$.

4.2.2.5 Moments of the Generalized Rician PDF

The general expression for the moments is written as

$$\mathbb{E}[\underline{m}^v] = (2\sigma^2)^{v/2} \frac{\Gamma[(K+v)/2]}{\Gamma(K/2)} {}_1F_1\left(-\frac{v}{2}, \frac{K}{2}; -\frac{A^2}{2\sigma^2}\right).$$
(4.30)

Again, the even moments turn out to be simple polynomials:

$$\mathbb{E}[\underline{m}^2] = K\sigma^2 + A^2.$$
(4.31)

$$\mathbb{E}[\underline{m}^4] = K^2\sigma^4 + 2K\sigma^4 + 2A^2K\sigma^2 + 4A^2\sigma^2 + A^4.$$
(4.32)

For $A = 0$, we obtain the moments of the generalized Rayleigh PDF:

$$\mathbb{E}[\underline{m}^v] = (2\sigma^2)^{v/2} \frac{\Gamma[(K+v)/2]}{\Gamma(K/2)}.$$
(4.33)

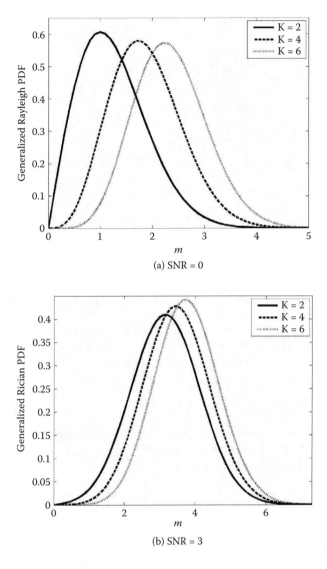

(a) SNR = 0

(b) SNR = 3

FIGURE 4.2 Plots of the generalized Rician PDF as a function of the magnitude variable m for K = 2, 4, and 6 and with $\sigma = 1$.

4.2.2.6 PDF of Squared Magnitude Data

Consider a set of N real and imaginary observations $\{(w_{r,n}, w_{i,n})\}$ with $n = 1, \ldots, N$, where all observations are assumed to be statistically independent and Gaussian distributed with standard deviation σ and arbitrary means $\mu_{r,n}$ and $\mu_{i,n}$, respectively. Next, consider the sum of N squared magnitude variables constructed from these N complex data points:

$$y = \sum_{n=1}^{N} m_n^2 = \sum_{n=1}^{N} \left(\omega_{r,n}^2 + \omega_{i,n}^2 \right). \tag{4.34}$$

Then, it can be shown that the PDF of y is given by

$$p_y(y) = \frac{1}{2\sigma^2} \left(\frac{y}{\mu^2} \right)^{\frac{N-1}{2}} e^{-\frac{y+\mu^2}{2\sigma^2}} I_{N-1} \left(\frac{\sqrt{y}\mu}{\sigma^2} \right), \tag{4.35}$$

where $\mu^2 = \sum_n (\mu_{r,n}^2 + \mu_{i,n}^2)$ [22]. Note that Equation 4.35 is the PDF of the sum of $2N$ Gaussian-distributed variables. Also, note the following:

- If the variance of the Gaussian-distributed components equals one, the PDF given in Equation 4.35 turns into a noncentral chi-squared distribution. The mean is given by $N + \mu^2$. The variance is given by $2(2\mu^2 + N)$.
- If, in addition, the mean of the components equals zero, it turns into the chi-squared distribution with $2N$ degrees freedom and with mean and variance equal to N and $2N$, respectively.

4.2.3 PDF OF PHASE DATA

Phase data, which are commonly obtained during flow imaging, are constructed from the real and imaginary observations $\{(w_{r,n}, w_{i,n})\}$ by calculating the arctangent of their ratio for each complex data point:

$$\phi_n = \arctan \left(\frac{\omega_{i,n}}{\omega_{r,n}} \right). \tag{4.36}$$

The PDF of the phase deviation $\Delta\phi$ from the true phase value is given by [24]

$$p_{\Delta\phi}(\Delta\phi) = \frac{1}{2\pi} e^{-A^2/2\sigma^2} \left[1 + \frac{A}{\sigma} \cos\Delta\phi\, e^{A^2 \cos^2 \Delta\phi/2\sigma^2} \int_{-\infty}^{\frac{A\cos\Delta\phi}{\sigma}} e^{-x^2/2} dx \right]. \tag{4.37}$$

Note that the distribution can be expressed solely in terms of the SNR, defined as A/σ. A graphical representation of the phase difference PDF as a function of the SNR is shown in Figure 4.3. Although the general expression for the distribution of $\Delta\phi$ is complicated, the two limits of the SNR turn out to yield simple distributions:

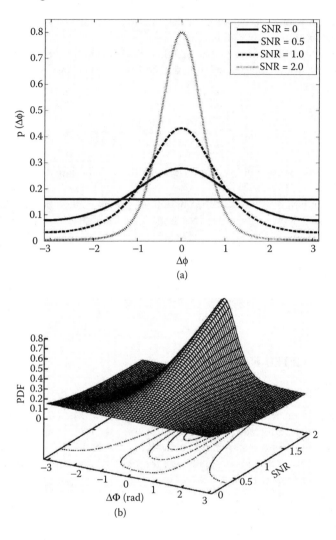

FIGURE 4.3 Plot of the phase error PDF as a function of the SNR. (From Sijbers, J., Van der Linden, A. (2003). *Encyclopedia of Optical Engineering*, chap. Magnetic resonance imaging, pp. 1237–1258. Marcel Dekker, ISBN: 0-8247-4258-3.)

- In regions where there is only noise, (i.e., where the SNR is zero), Equation 4.37 reduces to a uniform PDF:

$$P_{\underline{\Delta\phi}}(\Delta\phi) = \begin{cases} \dfrac{1}{2\pi} & \text{if } -\pi < \Delta\phi < \pi \\[2mm] 0 & \text{otherwise.} \end{cases} \tag{4.38}$$

Stated in another way, complex data that only consist of noise "point in all directions" with the same probability.

- For high SNR, it is easy to see that the probability of observing large values for the phase deviation will be small. In that case, Equation 4.37 reduces to

$$p_{\underline{\Delta\phi}}(\Delta\phi) = \frac{1}{\sqrt{2\pi}} \frac{A}{\sigma} \exp\left(-\frac{\Delta\phi^2 A^2}{2\sigma^2} \right). \tag{4.39}$$

Thus, the phase noise $\underline{\Delta\phi}$ is governed by a Gaussian distribution when SNR $\rightarrow \infty$.

The standard deviations of the phase noise can in general be calculated from Equation 4.37. However, for the SNR limits given in Equation 4.38 and Equation 4.39, it is given by

$$\sigma_{\underline{\Delta\phi}} = \begin{cases} \pi/\sqrt{3} & \text{if SNR} = 0 \\ \sigma/A & \text{if SNR} \gg 1. \end{cases} \tag{4.40}$$

4.3 PARAMETER ESTIMATION

4.3.1 PERFORMANCE MEASURES OF ESTIMATORS

In the remainder of this chapter, several estimators will be considered. In order to compare the performance of these estimators, we require appropriate performance measures. In this subsection, we will introduce two of the most commonly used and widely accepted ones, namely *precision* and *accuracy*. Furthermore, the *mean squared error* (MSE), a measure incorporating both precision and accuracy, will be introduced.

In what follows, $\hat{\boldsymbol{\theta}} = (\hat{\boldsymbol{\theta}}_1,\ldots,\hat{\boldsymbol{\theta}}_K)^T$ represents an estimator of the $K \times 1$ parameter vector $\boldsymbol{\theta} = (\boldsymbol{\theta}_1,\ldots,\boldsymbol{\theta}_K)^T$.

4.3.2 PRECISION

The precision of an estimator concerns the spread of the estimates when the experiment is repeated under the same conditions. It is represented by the variance of the estimator or, equivalently, by the standard deviation of the estimator, which is the square root of the variance. The variance is thus a measure of the nonsystematic error. If the estimator is vector valued, it has a covariance matrix associated with it, which is defined as

$$C_{\hat{\underline{\theta}}} = \mathbb{E}[(\hat{\underline{\theta}} - \mathbb{E}[\hat{\underline{\theta}}])(\hat{\underline{\theta}} - \mathbb{E}[\hat{\underline{\theta}}])^T]. \tag{4.41}$$

The diagonal elements of $C_{\hat{\theta}}$ represent the variances of the elements of $\hat{\underline{\theta}}$, whereas the nondiagonal elements represent the covariances between the elements of the estimator.

4.3.3 ACCURACY

The accuracy of an estimator can be described in terms of its bias. The *bias* of an estimator is defined as the deviation of the expectation value of the parameter (vector) from the true value:

$$b(\hat{\underline{\theta}}) = \mathbb{E}[\hat{\underline{\theta}}] - \underline{\theta}. \tag{4.42}$$

Hence, the bias represents the systematic error. If the expectation of the estimator equals the true value of the parameter, the estimator is said to be unbiased. Otherwise, it is biased.

4.3.4 MSE

A potential measure of the quality of an estimator, taking account of both accuracy and precision, is given by the MSE. The MSE of the kth element of the estimator $\hat{\underline{\theta}}$ is defined as

$$\text{MSE}(\hat{\underline{\theta}}_k) = \mathbb{E}\left[\left(\hat{\underline{\theta}}_k - \underline{\theta}_k\right)^2\right]. \tag{4.43}$$

Note that the MSE can also be written as the sum of the variance of the estimator and its bias squared:

$$\text{MSE}(\hat{\underline{\theta}}_k)\arcsin\theta = b^2(\hat{\underline{\theta}}_k) + \text{Var}(\hat{\underline{\theta}}_k). \tag{4.44}$$

The MSE of the vector estimator $\hat{\underline{\theta}}$ is given by the scalar value:

$$\text{MSE}(\hat{\underline{\theta}}) = \mathbb{E}[(\hat{\underline{\theta}} - \underline{\theta})^T (\hat{\underline{\theta}} - \underline{\theta})] \tag{4.45}$$

$$= \sum_{k=1}^{K} \text{MSE}(\hat{\underline{\theta}}_k). \tag{4.46}$$

4.3.5 CRLB

The same parameter can be estimated using different estimators. Generally, different estimators have different precisions. Then, one might ask what precision might be achieved, or, in other words, is there a lower bound on the attainable variance? The answer is that such a lower bound exists. It can be computed from the joint PDF of the observations (i.e., data points) as follows [25].

Suppose that the joint PDF $p_{\underline{x}}(x; \boldsymbol{\theta})$ of a set of observations $\underline{x} = (\underline{x}_1, \ldots, \underline{x}_N)$ is determined by the parameter vector $\boldsymbol{\theta} = (\boldsymbol{\theta}_1, \ldots, \boldsymbol{\theta}_K)^T$. In addition, define the matrix I by

$$I = -\mathbb{E}\left[\frac{\partial^2 \ln p_{\underline{x}}(\underline{x}; \boldsymbol{\theta})}{\partial \boldsymbol{\theta} \, \partial \boldsymbol{\theta}^T}\right]. \tag{4.47}$$

In this expression, $\partial^2 \ln p_{\underline{x}}(\underline{x}; \boldsymbol{\theta})/\partial\boldsymbol{\theta}\partial\boldsymbol{\theta}^T$ is the $K \times K$ Hessian matrix of $\ln p_{\underline{x}}(\underline{x}; \boldsymbol{\theta})$ defined by its (q, r)th element $\partial^2 \ln p_{\underline{x}}(\underline{x}; \boldsymbol{\theta})/\partial\boldsymbol{\theta}_q\partial\boldsymbol{\theta}_r$. The matrix I is called the *Fisher information matrix*. Next, let $\hat{\boldsymbol{\tau}} = (\hat{\boldsymbol{\tau}}_1, \ldots, \hat{\boldsymbol{\tau}}_L)^T$ be any unbiased estimator of the vector function $\boldsymbol{\tau}(\boldsymbol{\theta}) = (\tau_1(\boldsymbol{\theta}), \ldots, \tau_L(\boldsymbol{\theta}))^T$, that is, $\mathbb{E}[\hat{\boldsymbol{\tau}}] = \boldsymbol{\tau}(\boldsymbol{\theta})$. Then, under a number of conditions that are not too restrictive, the Cramér–Rao inequality states that [26]:

$$\mathrm{Cov}(\hat{\boldsymbol{\tau}}, \hat{\boldsymbol{\tau}}) \geq \left(\frac{\partial \boldsymbol{\tau}}{\partial \boldsymbol{\theta}}\right) I^{-1} \left(\frac{\partial \boldsymbol{\tau}}{\partial \boldsymbol{\theta}}\right)^T. \tag{4.48}$$

In this expression, $\partial\tau/\partial\theta$ is the $L \times K$ Jacobian matrix of the vector τ with respect to θ, and $\mathrm{Cov}(\hat{\boldsymbol{\tau}}, \hat{\boldsymbol{\tau}})$ is the $L \times L$ covariance matrix of the estimator $\hat{\boldsymbol{\tau}}$. Therefore, the diagonal elements of $\mathrm{Cov}(\hat{\boldsymbol{\tau}}, \hat{\boldsymbol{\tau}})$ are the variances of $\hat{\boldsymbol{\tau}}_1, \ldots, \hat{\boldsymbol{\tau}}_L$, respectively.

The inequality in Equation 4.48 expresses that the difference between the positive semidefinite left-hand and right-hand members is positive semidefinite. The right-hand member defines the minimum variance bound or the CRLB on the covariance of any unbiased estimator of $\tau(\theta)$. A property of positive semidefinite matrices is that their diagonal elements cannot be negative. Therefore, the diagonal elements of $\mathrm{Cov}(\hat{\boldsymbol{\tau}}, \hat{\boldsymbol{\tau}})$, that is, the variances of the elements of the estimator $\hat{\boldsymbol{\tau}}$, cannot be smaller than the corresponding elements of the CRLB. Consequently, the diagonal elements of the CRLB are a lower bound on the variances of the elements of the estimator $\hat{\boldsymbol{\tau}}$. The CRLB thus defines the highest attainable precision. Notice that if $\tau(\theta) = (\theta_1, \ldots, \theta_K)^T$, the CRLB simplifies to I^{-1}.

Finally, it is known that if there exists an unbiased estimator having the CRLB as a covariance matrix, it is the ML estimator [25]. In the next subsection, the ML estimator will be discussed.

4.3.6 ML ESTIMATION

Consider a set of N observations $\underline{x} = (\underline{x}_1, \ldots, \underline{x}_N)$ with joint PDF given by

$$p_{\underline{x}}(x|\boldsymbol{\theta}), \tag{4.49}$$

where $\boldsymbol{\theta} = (\boldsymbol{\theta}_1,...,\boldsymbol{\theta}_K)^T$. Notice that for statistically independent observations, the joint PDF is given by the product of the marginal PDFs of the observations:

$$p_{\underline{x}}(\boldsymbol{x} \mid \boldsymbol{\theta}) = \prod_{n=1}^{N} p_{\underline{x}_n}(x_n \mid \boldsymbol{\theta}), \qquad (4.50)$$

with $p_{\underline{x}_n}(x_n \mid \boldsymbol{\theta})$ is the PDF of $\underline{x}_n$.

To construct the ML estimator, we first substitute the available observations $x_1,...,x_N$ for the corresponding independent variables in Equation 4.38. Because these observations are numbers, the resulting expression depends only on the elements of the parameter vector $\boldsymbol{\theta}$. In a second step, we regard the fixed true parameters $\boldsymbol{\theta}$ as variables. The resulting function $L(\boldsymbol{\theta} \mid x_1,...,x_N)$ is called the *likelihood function* of the sample. The ML estimate $\boldsymbol{\theta}$ of the parameters $\boldsymbol{\theta}$ is defined as the value of $\boldsymbol{\theta}$ that, within the admissible range of θ, maximizes the likelihood function [27,28]:

$$\hat{\boldsymbol{\theta}}_{\text{ML}} = \arg\left\{\max_{\theta}(\ln L)\right\}. \qquad (4.51)$$

The most important properties of the ML estimators are the following [25]:

Consistency: Under very general conditions, ML estimators are consistent, i.e.,

$$\forall \varepsilon \in \mathbb{R} \mid \{\Pr(|\hat{\boldsymbol{\theta}}_{\text{ML}} - \boldsymbol{\theta}| < \varepsilon) = 1 \quad \text{if} \quad N \to \infty\}, \qquad (4.52)$$

with $\mathbb{R}^+$ denoting the set of positive real numbers.

Asymptotic efficiency: Under not too restrictive conditions, the covariance matrix of an ML estimator $\underline{\boldsymbol{\theta}}$ equals the CRLB asymptotically.

Asymptotic normality: If the number of data points increases, the probability density function of an ML estimator tends to a normal distribution.

Invariance property: If $\hat{\boldsymbol{\theta}}_{\text{ML}}$ is the ML estimator of the $K \times 1$ parameter vector $\boldsymbol{\theta}$, then $\hat{\boldsymbol{\tau}}(\hat{\boldsymbol{\theta}}_{\text{ML}})$ is the ML estimator of the $L \times 1$ vector $\boldsymbol{\tau}(\boldsymbol{\theta}) = (\tau_1(\boldsymbol{\theta}),...,\tau_L(\boldsymbol{\theta}))^T$ of functions of $\boldsymbol{\theta}$.

4.4 SIGNAL AMPLITUDE ESTIMATION

4.4.1 INTRODUCTION

In this section, the problem of signal amplitude estimation from MR data is addressed. In particular, we focus on the estimation of the magnitude magnetization values from data acquired during an MR imaging procedure.

Raw MR data, directly obtained from an MR scanner, represent the FT of a magnetization distribution of a volume at a certain point in time. Such data are generally complex valued and corrupted by zero-mean Gaussian-distributed noise [29]. An inverse FT, which yields the complex magnetization distribution of the object under study, does not change the type of data PDF because of the linearity and orthogonality of the FT. Hence, these data are corrupted by zero-mean Gaussian-distributed noise as well. The complex data, however, are generally not retained but transformed into magnitude and phase data. This is because the amplitudes and phase values of the magnetizations are of greater interest than the real and imaginary components of the complex data values. Transformation to a magnitude or phase MR image, however, is nonlinear. As a consequence of this transformation, the PDF of the data changes. As discussed in the preceding text, magnitude data are then governed by a Rician PDF.

Nevertheless, conventional estimation techniques usually assume the data to be Gaussian distributed. Whenever other PDFs come into play (such as a Rician distribution), one still tends to use parameter estimation techniques that are based on Gaussian-distributed data [30–32]. The justification for this is that when the SNR is high, the actual data PDF is very similar to a Gaussian PDF (cf. Subsection 4.2.1). In addition, Gaussian PDFs have attractive computational properties.

However, the Rician PDF deviates significantly from a Gaussian PDF when the SNR is low, leading to significantly biased results. To reduce this bias, parameter estimation methods were proposed that exploit the properties of the Rician PDF [24,33–35]. Although these estimators reduce the bias, they are not optimal. In this section, it is shown where the bias appears in the conventional estimation. In addition, the ML estimator for Rician-distributed data is constructed and its performance is compared that of conventional estimators [36].

Furthermore, consider a data-processing application (e.g., noise filtering) that requires estimation of the underlying signal amplitude from a number of noise-corrupted, complex-valued data points. Thereby, for each complex data point belonging to the data set, the underlying signal amplitude is assumed to be the same. This amplitude can be estimated either by first transforming the complex data points into a set of magnitude data points and then estimating the signal amplitude from the so-obtained data set, or by directly estimating the signal amplitude from the original complex-valued data points. Indeed, both data sets contain the signal amplitude to be estimated.

In general, if N complex points are available, the data set consists of $2N$ observations (N real and N imaginary data points) and $N + 2$ unknowns (N true phase values, 1 true signal amplitude, and the noise variance). On the other hand, if the N complex data points are transformed into a set of N magnitude data points, such a data set has only two unknown parameters (the true signal amplitude and the noise variance). Hence, questions such as "Should we use the complex data set or the magnitude data set when estimating the unknown signal amplitude?" and "Does it matter whether or not the true phase values of the complex data points, from which the signal amplitude is estimated, are the same?" may arise. In this section, these questions are addressed. In order to simplify the discussion, we will elaborate on the estimation of the underlying signal amplitude from a set of data points of which this signal amplitude is assumed to be the same (i.e., a constant model). It is, however, clear that similar reasoning is valid for any other underlying (parametric) model of the data points.

For both data sets (complex and magnitude), the ML estimators of the signal amplitude will be derived. The use of the ML estimator is justified by the fact that the ML estimator has a number of favorable statistical properties, as discussed in Section 4.3.6 [25]. First, it is asymptotically precise, i.e., it achieves the so-called CRLB for an infinite number of observations. The CRLB defines a lower bound on the variance of any unbiased estimator of a parameter. Second, the ML estimator is consistent, which means that it converges to the true parameter in a statistically well-defined way if the number of observations (i.e., data points) increases. Third, the ML estimator is asymptotically normally (i.e., Gaussian) distributed, with a mean equal to the true value of the parameter(s) and a (co)variance (matrix) equal to the CRLB. Whether these asymptotic properties also apply when the number of observations is finite depends on the particular estimation problem under concern. In the present case, this can be found out analytically (for complex data points) or by means of simulations (for magnitude data points). It is known that if there exists an estimator that attains the CRLB, it is given by the ML estimator [25]. For both data sets, the performance of the corresponding ML estimators of the signal amplitude will be evaluated in terms of the MSE, a measure of both accuracy (bias) and precision (variance). Moreover, for both complex and magnitude data, the variance of the ML estimator will be compared with the CRLB, which can be computed analytically. In addition, for both types of data sets, the ML estimators of the variance of the noise will be derived, after which their performance will be evaluated in terms of both accuracy and precision [37].

4.4.2 SIGNAL AMPLITUDE ESTIMATION FROM COMPLEX DATA

We start by considering complex, Gaussian-distributed data. The CRLB for unbiased estimation of the underlying amplitude signal as well as the ML estimator of this signal will be derived. This will be done for data with identical underlying phase values, as well as for data with different underlying phase values.

4.4.2.1 Region of Constant Amplitude and Phase

Consider a set of N independent, Gaussian-distributed, complex data points $\underline{c} = \{(\underline{w}_{r,n}, \underline{w}_{i,n})\}$ with underlying true amplitude and phase values A and φ, respectively. This means that $A\cos\varphi$ and $A\sin\varphi$ represent the true real and imaginary values, respectively. As the real and imaginary data are independent, the joint PDF of the complex data, $p_{\underline{c}}$, is simply the product of the marginal PDFs of the Gaussian-distributed real and imaginary data points:

$$p_{\underline{c}}(\{(\underline{w}_{r,n}, \underline{w}_{i,n})\} \mid A, \varphi) = \left(\frac{1}{2\pi\sigma^2}\right)^N \prod_{n=1}^{N} e^{-\frac{(\omega_{r,n} - A\cos\varphi)^2}{2\sigma^2}} e^{-\frac{(\omega_{i,n} - A\sin\varphi)^2}{2\sigma^2}}, \quad (4.53)$$

where σ denotes the standard deviation of the noise, and $\{(\omega_{r,n}, \omega_{i,n})\}$ are the real and imaginary variables corresponding with the complex data $\{(\underline{w}_{r,n}, \underline{w}_{i,n})\}$.

4.4.2.1.1 CRLB

It follows from Subsection 4.3.5 that the CRLB for unbiased estimation of (A, φ) can be computed from the Fisher information matrix I [25]:

$$I = -\mathbb{E}\left[\begin{pmatrix} \dfrac{\partial^2 \ln p_c}{\partial A^2} & \dfrac{\partial^2 \ln p_c}{\partial A \partial \varphi} \\[2ex] \dfrac{\partial^2 \ln p_c}{\partial \varphi \partial A} & \dfrac{\partial^2 \ln p_c}{\partial \varphi^2} \end{pmatrix}\right] = \begin{pmatrix} \dfrac{N}{\sigma^2} & 0 \\[2ex] 0 & \dfrac{NA^2}{\sigma^2} \end{pmatrix} \tag{4.54}$$

with the joint PDF p_c given by Equation 4.42. Applying the inverse operator yields for the CRLB:

$$\mathrm{CRLB} = I^{-1} = \begin{pmatrix} \dfrac{\sigma^2}{N} & 0 \\[2ex] 0 & \dfrac{\sigma^2}{NA^2} \end{pmatrix} \tag{4.55}$$

4.4.2.1.2 ML Estimation

Following the procedure described in Subsection 4.3.6, the likelihood function L is obtained by substituting the available observations $\{(w_{r,n}, w_{i,n})\}$ for $\{(\omega_{r,n}, \omega_{i,n})\}$ in the joint PDF (4.53):

$$L(A, \varphi \mid \{(w_{r,n}, w_{i,n})\}) = \left(\frac{1}{2\pi\sigma^2}\right)^N \prod_{n=1}^{N} e^{-\frac{(w_{r,n} - A\cos\varphi)^2}{2\sigma^2}} e^{-\frac{(w_{i,n} - A\sin\varphi)^2}{2\sigma^2}}. \tag{4.56}$$

Then, the ML estimates of (A, φ) are found by maximizing this function with respect to A and φ. Taking the logarithm yields:

$$\ln L = -N\ln(2\pi\sigma^2) + \frac{1}{2\sigma^2}\sum_{n=1}^{N}[(w_{r,n} - A\cos\varphi)^2 + (w_{i,n} - A\sin\varphi)^2]. \tag{4.57}$$

At the maximum, the first derivative of $\ln L$ with respect to A and φ should be zero:

$$\frac{\partial \ln L}{\partial A} = \frac{NA}{\sigma^2} - \frac{1}{\sigma^2}\sum_{n=1}^{N}(w_{r,n}\cos\varphi + w_{i,n}\sin\varphi), \tag{4.58}$$

$$\frac{\partial \ln L}{\partial \varphi} = \frac{A}{\sigma^2}\sum_{n=1}^{N}(w_{r,n}\sin\varphi - w_{i,n}\cos\varphi). \tag{4.59}$$

Setting Equation 4.48 and Equation 4.59 to zero yields the ML estimators of A and φ:

$$\widehat{\underline{A}}_{ML} = \frac{1}{N} \sqrt{\left(\sum_{n=1}^{N} \underline{w}_{r,n} \right)^2 + \left(\sum_{n=1}^{N} \underline{w}_{i,n} \right)^2}, \tag{4.60}$$

$$\widehat{\underline{\varphi}}_{ML} = \arctan \left(\frac{\sum_{n=1}^{N} \underline{w}_{i,n}}{\sum_{n=1}^{N} \underline{w}_{r,n}} \right). \tag{4.61}$$

Notice that the estimator $\widehat{\underline{A}}_{ML}$ is obtained by taking the square root of the quadratic sum of two Gaussian-distributed variables. Hence, $\widehat{\underline{A}}_{ML}$ is Rician distributed [18].

4.4.2.1.3   MSE

As $\widehat{\underline{A}}_{ML}$ is Rician distributed, we find for its MSE, which is the sum of the bias (b) squared and the estimator's variance (cf. Equation 4.44):

$$MSE = \left[b\left(\widehat{\underline{A}}_{ML} \right) \right]^2 + Var\left(\widehat{\underline{A}}_{ML} \right) \tag{4.62}$$

$$= 2A^2 - 2A\mathbb{E}\left[\widehat{\underline{A}}_{ML} \right] + 2\sigma^2/N \tag{4.63}$$

$$= 2A\left(A - \mathbb{E}\left[\widehat{\underline{A}}_{ML} \right] \right) + 2\sigma^2/N, \tag{4.64}$$

where the first moment of its PDF can be deduced from Equation 4.17:

$$\mathbb{E}\left[\widehat{\underline{A}}_{ML} \right] = \sigma \sqrt{\frac{\pi}{2N}} e^{-\frac{NA^2}{4\sigma^2}} \left[\left(1 + \frac{NA^2}{2\sigma^2} \right) I_0\left(\frac{NA^2}{4\sigma^2} \right) + \frac{NA^2}{2\sigma^2} I_1\left(\frac{NA^2}{4\sigma^2} \right) \right]. \tag{4.65}$$

4.4.2.2   Region of Constant Amplitude and Different Phases

Now assume that the complex data $\underline{c} = \{(\underline{w}_{r,n}, \underline{w}_{i,n})\}$ have an underlying signal amplitude A and arbitrary phase values $\varphi_1, \ldots, \varphi_N$. Then, the joint PDF of the complex data, $p_{\underline{c}}$, is given by

$$p_{\underline{c}}(\{(\underline{w}_{r,n}, \underline{w}_{i,n})\} \mid A, \{\varphi_n\}) = \left(\frac{1}{2\pi\sigma^2} \right)^N \prod_{n=1}^{N} e^{-\frac{(\omega_{r,n} - A\cos\varphi_n)^2}{2\sigma^2}} e^{-\frac{(\omega_{i,n} - A\sin\varphi_n)^2}{2\sigma^2}}. \tag{4.66}$$

4.4.2.2.1 CRLB

The Fisher information matrix of the data with respect to the parameter vector $(A, \varphi_1, \ldots, \varphi_N)$ is given by

$$
\mathbf{I} = -\mathbb{E}
\begin{bmatrix}
\begin{pmatrix}
\dfrac{\partial^2 \ln p_c}{\partial A^2} & \dfrac{\partial^2 \ln p_c}{\partial A \partial \varphi_1} & \cdots & \dfrac{\partial^2 \ln p_c}{\partial A \partial \varphi_N} \\[3mm]
\dfrac{\partial^2 \ln p_c}{\partial \varphi_1 \partial A} & \dfrac{\partial^2 \ln p_c}{\partial \varphi_1^2} & \cdots & \dfrac{\partial^2 \ln p_c}{\partial \varphi_1 \partial \varphi_N} \\[3mm]
\cdots & \cdots & \cdots & \cdots \\[3mm]
\dfrac{\partial^2 \ln p_c}{\partial \varphi_N \partial A} & \dfrac{\partial^2 \ln p_c}{\partial \varphi_N \partial \varphi_1} & \cdots & \dfrac{\partial^2 \ln p_c}{\partial \varphi_N^2}
\end{pmatrix}
\end{bmatrix}
=
\begin{pmatrix}
\dfrac{N}{\sigma^2} & 0 & \cdots & 0 \\[3mm]
0 & \dfrac{A^2}{\sigma^2} & \cdots & 0 \\[3mm]
\cdots & \cdots & \cdots & \cdots \\[3mm]
0 & 0 & \cdots & \dfrac{A^2}{\sigma^2}
\end{pmatrix}
\qquad (4.67)
$$

and the CRLB for unbiased estimation of $(A, \varphi_1, \ldots, \varphi_N)$ is given by

$$
\mathrm{CRLB} =
\begin{pmatrix}
\dfrac{\sigma^2}{N} & 0 & \cdots & 0 \\[3mm]
0 & \dfrac{\sigma^2}{A^2} & \cdots & 0 \\[3mm]
\cdots & \cdots & \cdots & \cdots \\[3mm]
0 & 0 & \cdots & \dfrac{\sigma^2}{A^2}
\end{pmatrix}
\qquad (4.68)
$$

4.4.2.2.2 ML Estimation

The likelihood function for N statistically independent, Gaussian-distributed complex observations $\underline{c} = \{(w_{r,n}, w_{i,n})\}$ with underlying noiseless signal amplitude A and arbitrary phase values $\varphi_1, \ldots, \varphi_N$ is given by

$$
L(A, \varphi_1, \ldots, \varphi_N \mid \{(w_{r,n}, w_{i,n})\}) = \left(\frac{1}{2\pi\sigma^2}\right)^N \prod_{n=1}^{N} e^{-\frac{(w_{r,n} - A\cos\varphi_n)^2}{2\sigma^2}} e^{-\frac{(w_{i,n} - A\sin\varphi_n)^2}{2\sigma^2}}. \qquad (4.69)
$$

Taking the logarithm yields

$$
\ln L = -N \ln(2\pi\sigma^2) - \frac{1}{2\sigma^2} \sum_{n=1}^{N} [(w_{r,n} - A\cos\varphi_n)^2 - (w_{i,n} - A\sin\varphi_n)^2]. \qquad (4.70)
$$

The first derivative of $\ln L$ with respect to A and φ_n are given by

$$
\frac{\partial \ln L}{\partial A} = \frac{1}{\sigma^2} \sum_{n=1}^{N} [w_{r,n} \cos\varphi_n + w_{i,n} \sin\varphi_n] - \frac{NA}{\sigma^2}, \qquad (4.70)
$$

$$
\frac{\partial \ln L}{\partial \varphi_n} = -\frac{A}{\sigma^2}(w_{r,n} \sin\varphi_n - w_{i,n} \cos\varphi_n). \qquad (4.71)
$$

Setting these equations to zero yields the ML estimators of A and φ_n:

$$\hat{\underline{A}}_{ML} = \frac{1}{N} \sum_{n=1}^{N} \sqrt{\underline{w}_{r,n}^2 + \underline{w}_{i,n}^2}, \tag{4.72}$$

$$\hat{\underline{\varphi}}_{n,ML} = \arctan\left(\frac{\underline{w}_{i,n}}{\underline{w}_{r,n}}\right). \tag{4.73}$$

4.4.2.2.3 MSE

The ML estimator of the signal amplitude, given by Equation 4.73, is distributed as the average of N independent, Rician-distributed variables. Therefore, its mean value is simply given by the average of the mean values of the individual Rician-distributed variables, whereas its variance is given by the sum of their variances divided by N^2. Hence, we have for the MSE:

$$\mathrm{MSE}\left(\hat{\underline{A}}_{ML}\right) = \left[b\left(\hat{\underline{A}}_{ML}\right)\right]^2 + \mathrm{Var}\left(\hat{\underline{A}}_{ML}\right), \tag{4.75}$$

$$= \left(A - \mathbb{E}\left[\hat{\underline{A}}_{ML}\right]\right)^2 + \left(A^2 + 2\sigma^2 - \mathbb{E}\left[\hat{\underline{A}}_{ML}\right]^2\right)/N \tag{4.76}$$

where $\mathbb{E}[\hat{\underline{A}}_{ML}]$ is now given by Equation 4.14 or equivalently:

$$\mathbb{E}\left[\hat{\underline{A}}_{ML}\right] = \sigma\sqrt{\frac{\pi}{2}} e^{-\frac{A^2}{4\sigma^2}} \left[\left(1 + \frac{A^2}{2\sigma^2}\right) I_0\left(\frac{A^2}{4\sigma^2}\right) + \frac{A^2}{2\sigma^2} I_1\left(\frac{A^2}{4\sigma^2}\right)\right]. \tag{4.77}$$

Note that Equation 4.77 does not depend on N. Furthermore, Equation 4.77 is identical to Equation 4.65 if $N = 1$.

4.4.3 SIGNAL AMPLITUDE ESTIMATION FROM MAGNITUDE DATA

Though raw MR data are complex valued and Gaussian distributed, it is common practice to transform them into magnitude MR data, because physiological and anatomical information are more closely related to the magnitude of the magnetization vectors. However, as we have seen earlier, computing the magnitude results in a change of the underlying data PDF, which has to be accounted for when extracting quantitative information [38].

As earlier, we try to estimate the underlying, noiseless amplitude signal A from a region of interest (ROI), where A is assumed to be constant. The ROI now consists of N independent, Rician-distributed data points $\underline{m} = (\underline{m}_1, ..., \underline{m}_N)$.

Unlike that from complex data, estimation of the signal amplitude from magnitude data requires either prior knowledge of the noise variance or simultaneous estimation of signal amplitude and noise variance:

- The noise variance may be estimated separately if a background region is available, i.e., a region in which the underlying signal is zero (see

Subsection 4.5.3). Moreover, if large background areas are available, which is often the case, many more data points are available for the estimation of the noise variance than for the estimation of the signal amplitude. Then, the noise variance can be estimated with much higher precision. Hence, it might be a valid assumption to regard the noise variance as known (i.e., to regard the estimated noise variance as the true noise variance).

• If the noise variance cannot be estimated separately (with sufficient precision), it acts as a nuisance parameter that needs to be estimated simultaneously with the signal amplitude.

Both cases are discussed in the following subsections.

4.4.3.1 Region of Constant Amplitude and Known Noise Variance

4.4.3.1.1 CRLB

The Fisher information matrix of the data with respect to the parameter A is given by

$$I = -\mathbb{E}\left[\frac{\partial^2 \ln p_m}{\partial A^2}\right] = \frac{N}{\sigma^2}\left(Z - \frac{A^2}{\sigma^2}\right), \tag{4.78}$$

with

$$Z = \mathbb{E}\left[\frac{m^2}{\sigma^2}\frac{I_1^2\left(\frac{Am}{\sigma^2}\right)}{I_0^2\left(\frac{Am}{\sigma^2}\right)}\right], \tag{4.79}$$

and m a Rician-distributed random variable with true parameters (A, σ) [5]. The expectation value in Equation 4.79 can be evaluated numerically. Note that I is in fact a scalar, from which the CRLB can easily be obtained by applying the inverse operator:

$$\text{CRLB} = \frac{\sigma^2}{N}\left(Z - \frac{A^2}{\sigma^2}\right)^{-1}. \tag{4.80}$$

4.4.3.1.2 Conventional Estimation

Usually, Equation 4.18 is exploited for the estimation of the underlying signal A. Thereby, $\mathbb{E}[m^2]$ is estimated from a simple spatial average of the squared pixel values of the ROI [35–39]:

$$\widehat{\mathbb{E}[m^2]} = \langle m^2 \rangle = \frac{1}{N}\sum_{n=1}^{N} m_n^2. \tag{4.81}$$

Note that this estimator is unbiased because $\mathbb{E}[\langle \underline{m}^2 \rangle] = E[\underline{m}^2] = A^2 + 2\sigma^2$. If the noise variance σ^2 is assumed to be known, an unbiased estimator of A^2 is given by

$$\underline{\hat{A}}_c^2 = \langle \underline{m}^2 \rangle - 2\sigma^2. \tag{4.82}$$

Taking the square root of Equation 4.64 gives the conventional estimator of A [39,40,34,35]:

$$\underline{\hat{A}}_c = \sqrt{\langle \underline{m}^2 \rangle - 2\sigma^2}. \tag{4.83}$$

4.4.3.1.3 Discussion

The parameter to be estimated is the signal A. Obviously, A is known *a priori* to be real valued and nonnegative. However, this *a priori* knowledge has not been incorporated into the conventional estimation procedure. Consequently, the conventional estimator $\underline{\hat{A}}_c$, given in Equation 4.83, may reveal estimates that violate the *a priori* knowledge and are therefore physically meaningless. This is the case when $\hat{A}_c^2$ becomes negative. Therefore, $\underline{\hat{A}}_c$ cannot be considered a useful estimator of A if the probability that $\hat{A}_c^2$ is negative differs from zero significantly. To determine this probability, the PDF of $\underline{\hat{A}}_c^2$ is required. It can be derived from

$$p_{\underline{y}}(y) = \frac{1}{2\sigma^2} \left(\frac{y}{(NA)^2} \right)^{\frac{N-1}{2}} \exp\left(-\frac{y + (NA)^2}{2\sigma^2} \right) I_{N-1}\left(\frac{\sqrt{y}(NA)}{\sigma^2} \right) \varepsilon(y), \tag{4.84}$$

where y is given by the sum of N real and N imaginary, independent, squared, Gaussian-distributed variables (as discussed in Subsection 4.2.4) i.e.,

$$\underline{y} = \sum_{n=1}^{N} \left(\underline{w}_{r,n}^2 + \underline{w}_{i,n}^2 \right) = \sum_{n=1}^{N} \underline{m}_n^2. \tag{4.85}$$

The deterministic signal component of $\underline{y}$ is given by NA^2. From Equation 4.81 and Equation 4.82, we have that $\hat{A}_c^2 = y/N - 2\sigma^2$, and with (cf. Appendix)

$$p_{\underline{\hat{A}}_c^2}\left(\hat{A}_c^2 \right) = N p_{\underline{y}}\left(N\left(\hat{A}_c^2 + 2\sigma^2 \right) \right), \tag{4.86}$$

the PDF of $\hat{A}_c^2$ becomes explicitly

$$p_{\underline{\hat{A}}_c^2}\left(\hat{A}_c^2 \right) = \frac{N}{2\sigma^2} \left(\frac{\hat{A}_c^2 + 2\sigma^2}{A^2} \right)^{\frac{N-1}{2}} \exp\left(-N \frac{\left(\hat{A}_c^2 + 2\sigma^2 \right) + A^2}{2\sigma^2} \right)$$

$$\times I_{N-1}\left(\frac{NA\sqrt{\hat{A}_c^2 + 2\sigma^2}}{\sigma^2} \right) \varepsilon\left(\hat{A}_c^2 + 2\sigma^2 \right). \tag{4.87}$$

Because $p_{\sqrt{\underline{x}}}(\sqrt{x}) = 2\sqrt{x}\,p_{\underline{x}}(x)$ (cf. Appendix), it is now easy to derive the PDF of $\hat{\underline{A}}_c = \sqrt{\hat{\underline{A}}_c^2}$:

$$p_{\hat{\underline{A}}_c}(\hat{\underline{A}}_c) = 2\hat{\underline{A}}_c\, p_{\hat{\underline{A}}_c^2}\left(\hat{\underline{A}}_c^2\right) \tag{4.88}$$

$$= \hat{A}_c\, \frac{N}{\sigma^2}\left(\frac{\hat{A}_c^2 + 2\sigma^2}{A^2}\right)^{\frac{N-1}{2}} \exp\left(-N\frac{\left(\hat{A}_c^2 + 2\sigma^2\right) + A^2}{2\sigma^2}\right)$$

$$\times I_{N-1}\left(\frac{NA\sqrt{\hat{A}_c^2 + 2\sigma^2}}{\sigma^2}\right)\varepsilon\left(\hat{A}_c^2 + 2\sigma^2\right). \tag{4.89}$$

The expectation value of $\hat{\underline{A}}_c$ is thus defined by

$$\mathbb{E}\left[\hat{\underline{A}}_c\right] = \frac{N}{\sigma^2}\int \hat{A}_c^2\left(\frac{\hat{A}_c^2 + 2\sigma^2}{A^2}\right)^{\frac{N-1}{2}} \exp\left(-N\frac{\left(\hat{A}_c^2 + 2\sigma^2\right) + A^2}{2\sigma^2}\right)$$

$$\times I_{N-1}\left(\frac{NA\sqrt{\hat{A}_c^2 + 2\sigma^2}}{\sigma^2}\right)\varepsilon\left(\hat{A}_c^2 + 2\sigma^2\right)d\hat{\underline{A}}_c. \tag{4.90}$$

In Figure 4.4, the probability that $\hat{\underline{A}}_c^2$ is negative, $\Pr[\hat{\underline{A}}_c^2 < 0]$, is plotted as a function of SNR for several values of N. From the figure one can conclude that

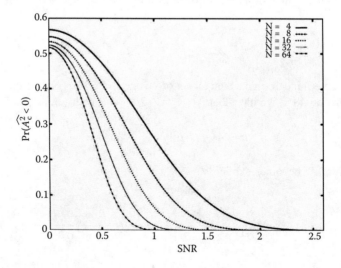

FIGURE 4.4 The probability $\Pr[\hat{A}_c^2 < 0]$ as a function of the SNR for various N. (From Sijbers, J., den Dekker, A.J., Scheunders, P., Van Dyck, D (1998). Maximum likelihood estimation of Rician distribution parameters. *IEEE Trans. Med. Imaging* 17(3): 357–361.)

for low SNR, $\hat{\underline{A}}_c$ cannot be a valid estimator of A unless a large number of data points are used for the estimation. Therefore, in practice, $\hat{\underline{A}}_c$ will only be a useful estimator if the SNR is high.

However, even if the condition of high SNR is met, the use of $\hat{\underline{A}}_c$ as an estimator of A should still not be recommended, because the results obtained are biased because of the square root operation in Equation 4.65. For high SNR, the expectation value of $\hat{\underline{A}}_c$ is approximately given by (see Appendix)

$$\mathbb{E}\left[\hat{\underline{A}}_c\right] \approx A\left(1 - \frac{\sigma^2}{2NA^2}\right). \tag{4.91}$$

Equation 4.91 is valid for high SNR. The bias appears in the second term of Equation 4.72. Note that it decreases with increasing SNR and increasing number of data points N. Furthermore, as $\mathbb{E}[\hat{\underline{A}}_c^2] = A^2$, the variance of $\hat{\underline{A}}_c$ can be easily found from Equation 4.91. It is given by

$$\mathrm{Var}\left(\hat{\underline{A}}_c\right) = \mathbb{E}\left[\hat{\underline{A}}_c^2\right] - \mathbb{E}\left[\hat{\underline{A}}_c\right]^2 \tag{4.92}$$

$$\approx \frac{\sigma^2}{N} - \frac{\sigma^4}{4N^2A^2} \tag{4.93}$$

Then, it follows that the MSE is given by

$$\mathrm{MSE} \approx \frac{\sigma^2}{N}. \tag{4.94}$$

4.4.3.1.4 ML Estimation

In what follows, we will consider the ML estimator of A from a set N Rician-distributed magnitude data points $\underline{m} = (\underline{m}_1, ..., \underline{m}_N)$.

The joint PDF $p_{\underline{m}}$ is given by

$$p_{\underline{m}} = \prod_{n=1}^{N} \frac{m_n}{\sigma^2} e^{-\frac{m_n^2 + A^2}{2\sigma^2}} I_0\left(\frac{Am_n}{\sigma^2}\right) \varepsilon(m_n), \tag{4.95}$$

where $\{m_n\}$ are the magnitude variables corresponding to the magnitude observations $\{\underline{m}_n\}$. The ML estimate of A is constructed by substituting the available observations $\{m_n\}$ in the expression for the joint PDF (Equation 4.75) and maximizing the resulting function $L(A)$, or equivalently $\ln L(A)$, with respect to A. Hence, it follows that

$$\ln L = \sum_{n=1}^{N} \ln\left(\frac{m_n}{\sigma^2}\right) - \sum_{n=1}^{N} \frac{m_n^2 + A^2}{2\sigma^2} + \sum_{n=1}^{N} \ln I_0\left(\frac{Am_n}{\sigma^2}\right), \tag{4.96}$$

or as a function of only A

$$\ln L \sim \sum_{n=1}^{N} \ln I_0 \left(\frac{Am_n}{\sigma^2} \right) - \frac{NA^2}{2\sigma^2}. \tag{4.97}$$

The ML estimate is then found from the global maximum of $\ln L$:

$$\hat{A}_{ML} = \arg \left\{ \max_A \left(\ln L \right) \right\}. \tag{4.98}$$

Note that Equation 4.98 cannot be solved analytically. Finding the maximum of the (log-)likelihood is therefore a numerical optimization problem.

4.4.3.1.5 Discussion

It is not possible to find the maximum of the $\ln L$ function directly because the parameter A enters that function in a nonlinear way. Therefore, finding the maximum of the $\ln L$ function will, in general, be an iterative numerical process.

In order to get some insight into the properties of the ML estimator, the structure of the $\ln L$ function is now studied. This structure is established by the number and nature of the stationary points of the function. Stationary points are defined as points where the gradient vanishes, i.e., where

$$\frac{\partial}{\partial A} \ln L = 0. \tag{4.99}$$

Substituting Equation 4.97 into Equation 4.99, we obtain

$$\sum_{n=1}^{N} \frac{\partial}{\partial A} \ln I_0 \left(\frac{Am_n}{\sigma^2} \right) - \frac{NA}{\sigma^2} = 0. \tag{4.100}$$

Given that

$$\frac{dI_v(z)}{dz} = I_{v+1}(z) + \frac{v}{z} I_v(z), \tag{4.101}$$

or explicitly that the derivative of $I_0(z)$, with respect to z, equals $I_1(z)$, it follows that

$$\sum_{n=1}^{N} \frac{m_n}{\sigma^2} \frac{I_1 \left(\frac{Am_n}{\sigma^2} \right)}{I_0 \left(\frac{Am_n}{\sigma^2} \right)} - \frac{NA}{\sigma^2} = 0, \tag{4.102}$$

Hence, the condition for the stationary points becomes

$$A = \frac{1}{N} \sum_{n=1}^{N} m_n \frac{I_1\left(\frac{Am_n}{\sigma^2}\right)}{I_0\left(\frac{Am_n}{\sigma^2}\right)}. \tag{4.103}$$

It follows from Equation 4.103 that $A = 0$ is a stationary point of $\ln L$, independent of the particular data set. The nature of a stationary point is determined by the sign of the second-order derivative of the function in that point. From this derivative, it follows whether a stationary point is a minimum or a maximum and whether or not it is degenerate. From Equation 4.97, the second-order derivative of the $\ln L$ function can be computed to yield

$$\frac{\partial^2 \ln L}{\partial A^2} = \sum_{n=1}^{N} \frac{m_n^2}{\sigma^4}\left[1 - \frac{\sigma^2}{Am_n}\frac{I_1\left(\frac{Am_n}{\sigma^2}\right)}{I_0\left(\frac{Am_n}{\sigma^2}\right)} - \frac{I_1^2\left(\frac{Am_n}{\sigma^2}\right)}{I_0^2\left(\frac{Am_n}{\sigma^2}\right)}\right] - \frac{N}{\sigma^2}. \tag{4.104}$$

From the knowledge that [20]

$$I_v(z) \sim \left(\frac{z}{2}\right)^v \Gamma(v+1) \quad \text{when} \quad z \to 0, \tag{4.105}$$

it is easy to verify that $A = 0$ is a minimum of $\ln L$ whenever

$$\frac{1}{N} \sum_{n=1}^{N} m_n^2 > 2\sigma^2. \tag{4.106}$$

If this condition is met, the $\ln L$ function will have two further stationary points being maxima. This can be seen by studying the possible structures of the $\ln L$ function using catastrophe theory. *Catastrophe theory* is concerned with the structural change of a parametric function under influence of its parameters [41]. It tells us that a structural change of the function is always preceded by a degeneracy of one of its stationary points. In order to analyze such a structural change, the parametric function can be replaced by a Taylor expansion of the essential variables about the latter stationary point. The essential variables correspond to the directions in which degeneracy may occur. According to the catastrophe theory, the global structure of a parametric function, with only one essential variable, is completely set by its Taylor expansion with terms up to the degree to which the coefficient cannot vanish under the influence of its parameters. The function studied here is $\ln L$ as a function of A. Its parameters are the observations. Thus, the structural change of the $\ln L$ function under the influence of the observations has to be studied.

The only essential variable is the signal parameter A. The stationary point that may become degenerate is the point $A = 0$ (degeneracy occurs whenever Equation 4.104 becomes equal to zero). If the $\ln L$ function is Taylor expanded about the stationary point $A = 0$, we yield

$$\ln L = a + \frac{b}{2!}A^2 + \frac{c}{4!}A^4 + O(A^6), \tag{4.107}$$

with

$$a = \sum_{n=1}^{N} \frac{m_n^2}{\sigma^4}\left[1 - \frac{1}{2}\frac{\sigma^2}{m_n}\right] - \frac{N}{\sigma^2}, \tag{4.108}$$

$$b = \sum_{n=1}^{N} \frac{m_n^2}{2\sigma^4} - \frac{N}{\sigma^2}, \tag{4.109}$$

$$c = -\frac{3}{8}\sum_{n=1}^{N} \frac{m_n^2}{\sigma^8}, \tag{4.110}$$

and $O(.)$ the order symbol of Landau. Notice that because the $\ln L$ function is symmetric about $A = 0$, the odd terms are absent in Equation 4.107. In order to investigate if the expansion up to the quartic term in Equation 4.107 is sufficient, it has to be determined whether the coefficients may change sign under influence of the observations. It is clear from Equation 4.109 that the coefficient b may change sign. The coefficient c, however, will always be negative, independent of the particular set of observations. This means that the expansion (Equation 4.107) is sufficient to describe the possible structures of the $\ln L$ function. Consequently, the study of the $\ln L$ as a function of the observations can be replaced by a study of the following quartic Taylor polynomial in the essential variable A:

$$\frac{b}{2!}A^2 + \frac{c}{4!}A^4, \tag{4.111}$$

where the term a has been omitted because it does not influence the structure. The polynomial in Equation 4.111 is always stationary at $A = 0$. This will be a minimum, a degenerate maximum, or a maximum when b is positive, equal to zero, or negative, respectively. It follows directly from Equation 4.111 that $\ln L$ has two additional stationary points (being maxima) if b is positive, that is, if Equation 4.106 is met. Note that the condition in (Equation 4.106) is always met for noise-free data. However, in practice, the data will be corrupted by noise, and for particular realizations of the noise, the condition in Equation 4.106 may not be met. Then $A = 0$ will be a maximum. Moreover, if the condition in Equation 4.106 is not met, b in Equation 4.111 is negative and thus $\ln L$ is convex, which means that $A = 0$ will be the only and, therefore, global maximum of the $\ln L$ function. This implies that under the influence of noise, the two maxima and one minimum will merge into one single maximum at $A = 0$. This maximum then corresponds to the ML estimate. Note that because the condition in Equation 4.106 is identical to (and therefore can be replaced by) the condition $\hat{A}_c^2 > 0$, the probability that the ML estimate is found at $A = 0$ is equal to the probability that $\hat{A}_c^2 \leq 0$. This probability can be computed from the PDF given in Equation 4.87.

It follows from these considerations that when the conventional estimator becomes invalid, the ML estimator will still yield physically relevant results.

4.4.3.2 Region of Constant Amplitude and Unknown Noise Variance

If the noise variance is unknown, the signal amplitude and the noise variance have to be estimated simultaneously (i.e., the noise variance is a nuisance parameter).

4.4.3.2.1 CRLB

The Fisher information matrix of the data with respect to the parameters (A, σ^2) is given by

$$I = -\mathbb{E}\left[\begin{pmatrix} \dfrac{\partial^2 \ln p_{\underline{m}}}{\partial A^2} & \dfrac{\partial^2 \ln p_{\underline{m}}}{\partial A \partial \sigma^2} \\[3mm] \dfrac{\partial^2 \ln p_{\underline{m}}}{\partial \sigma^2 \partial A} & \dfrac{\partial^2 \ln p_{\underline{m}}}{\partial (\sigma^2)^2} \end{pmatrix}\right]. \tag{4.112}$$

After some calculations, it can be shown that

$$I(1,1) = \frac{N}{\sigma^2}\left(Z - \frac{A^2}{\sigma^2}\right), \tag{4.113}$$

$$I(1,2) = I(2,1) = \frac{NA}{\sigma^4}\left(1 + \frac{A^2}{\sigma^2} - Z\right), \tag{4.114}$$

$$I(2,2) = \frac{N}{\sigma^4}\left(1 + \frac{A^2}{\sigma^2}(Z-1) - \frac{A^4}{\sigma^4}\right), \tag{4.115}$$

where $I(i, j)$ denotes the (i, j)th element of the matrix I, and Z is given by Equation 4.79. Finally, the CRLB for unbiased estimation of (A, σ^2) is obtained by simple inversion of the 2×2 matrix given in Equation 4.90:

$$\mathrm{CRLB} = \frac{1}{\det \boldsymbol{I}} \begin{pmatrix} I(2,2) & -I(2,1) \\ -I(1,2) & I(1,1) \end{pmatrix} \qquad (4.116)$$

4.4.3.2.2 Geometric Average

In subsection 4.4.3.1, it was shown that a simple modified root-mean-square (RMS) estimator can be used to estimate the signal amplitude. However, this estimator requires the knowledge of the noise variance. An estimator that does not require this knowledge is given by the geometric average defined as [42]

$$\widehat{\underline{A}}_{\mathrm{geom}} = \sqrt{\prod_{n=1}^{N} \sqrt[N]{\underline{m}_n^2}}. \qquad (4.117)$$

4.4.3.2.3 Discussion

The bias of this estimator is given by

$$b(\widehat{\underline{A}}_{\mathrm{geom}}) = \frac{\sigma^2}{2(N-1)A}, \qquad (4.118)$$

and the variance is given by

$$\mathrm{Var}(\widehat{\underline{A}}_{\mathrm{geom}}) = \frac{\sigma^2}{N-1}. \qquad (4.119)$$

4.4.3.2.4 ML Estimation

If a background region is not available for noise variance estimation, the signal A and variance σ^2 have to be estimated simultaneously from the N available data points by maximizing the log-likelihood function with respect to A and σ^2:

$$\left\{\widehat{A}_{\mathrm{ML}}, \widehat{\sigma^2}_{\mathrm{ML}}\right\} = \arg\left\{\max_{A,\sigma^2}(\ln L)\right\}, \qquad (4.120)$$

where $\ln L$ is given by Equation 4.96. Although optimization of a two-dimensional function is more difficult, computational requirements were observed to be limited because the likelihood function was observed to yield only one maximum.

4.4.4 DISCUSSION

In this subsection, the CRLB for unbiased estimation of the signal amplitude and the performance of the ML signal amplitude estimators, elaborated in previous subsections, are discussed.

4.4.4.1 CRLB

We will first discuss the expressions for the CRLB for unbiased estimation of the signal amplitude from complex and magnitude data separately.

> *Complex data:* Analytical expressions for the CRLBs for unbiased estimation of *A* from complex data with identical and different phase values were derived in Subsection 4.4.2. They are given by Equation 4.55 and Equation 4.68, respectively. Note that both CRLBs do not depend on the phase values.
>
> *Magnitude data:* In contrast to the CRLBs for complex data, no analytical expressions for the CRLB for unbiased estimation of the signal amplitude from magnitude MR data with known and unknown noise variance can be derived. However, the lower bounds given by Equation 4.80 and Equation 4.116 in Subsection 4.4.3 can be evaluated numerically. Thereby, the expectation values can be evaluated from Monte Carlo simulations.

Note also that all lower bounds are inversely proportional to the number of data points. Figure 4.5 shows the CRLBs as a function of the SNR, for unbiased estimation of the signal amplitude from complex as well as from magnitude data

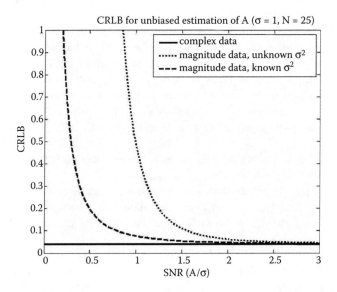

FIGURE 4.5 CRLB for unbiased estimation of A from complex and magnitude data, with known and unknown noise variance. (From Sijbers, J., den Dekker, A.J. (2004). Maximum likelihood estimation of signal amplitude and noise variance from MR data. *Magn. Reson. Med.* 51(3): 586–594.)

for known and for unknown noise variance. From the figure, one can see that the following:

- For low SNR, the CRLB for unbiased estimation of A
  - From complex data is significantly smaller than for estimation from magnitude data.
  - From magnitude data with known noise variance is significantly larger than for estimation from magnitude data with unknown noise variance (i.e., in which the noise variance is a nuisance parameter).
- Recall that knowledge of the noise variance is not required when estimating the signal amplitude from complex data.
- For increasing SNR, the CRLBs for unbiased estimation from magnitude data tend to the CRLB for unbiased estimation from complex data, in which the CRLB equals σ^2/N.

4.4.4.2 MSE

The bias, variance, and MSE of the ML estimators of A were computed, where the number of data points was set to $N = 25$ and the true variance was set to $\sigma^2 = 1$. For complex data with identical and different phase values, the bias, variance, and MSE of $\hat{A}_{ML}$ were computed from Equation 4.64 and Equation 4.65, and Equation 4.76 and Equation 4.77, respectively. On the other hand, for magnitude data with known and unknown noise variance, the bias, variance, and MSE of $\hat{A}_{ML}$ were obtained from a Monte Carlo simulation experiment with a sample size of 10^5. Thereby, $\hat{A}_{ML}$ was obtained by maximizing the log-likelihood function (Equation 4.96) with respect to A and $\{A, \sigma^2\}$ using Equation 4.98 and Equation 4.120, respectively.

The bias of $\hat{A}_{ML}$ has been plotted as a function of the SNR in Figure 4.6. In Figure 4.7, the MSE of $\hat{A}_{ML}$ has been plotted as a function of the SNR. Both figures show the results obtained for complex data with identical and different phase values as well as for magnitude data with known and unknown noise variance. From these figures, one can observe that in terms of the MSE:

- $\hat{A}_{ML}$ for complex data with identical phase values performs best, independent of the SNR.
- $\hat{A}_{ML}$ for magnitude data with known noise variance is significantly better compared with $\hat{A}_{ML}$ for magnitude data with unknown noise variance and $\hat{A}_{ML}$ for complex data with different phase values.

Also, for increasing SNR, the performance difference in terms of the MSE between the ML estimators of A based on complex and magnitude data tend to zero. As, in practice, the assumption of identical phases for complex data is generally invalid, it may be concluded that the signal amplitude is preferentially estimated from magnitude MR data for which the noise variance is known. The latter requisite is not too restrictive as often, in practice, the noise variance can be estimated with a much higher precision than the signal amplitude.

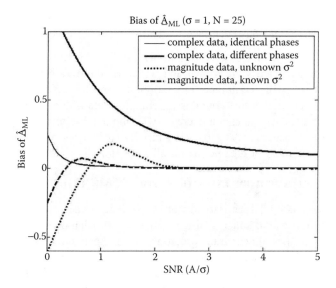

FIGURE 4.6 Bias of the ML estimator of A for complex data with identical and different phases and for magnitude data with unknown and known noise variance. (From Sijbers, J., den Dekker, A.J. (2004). Maximum likelihood estimation of signal amplitude and noise variance from MR data. *Magn. Reson. Med.* 51(3): 586–594.)

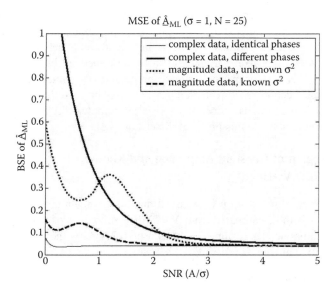

FIGURE 4.7 MSE of the ML estimator of A for complex data with identical and different phases and for magnitude data with unknown and known noise variance. (From Sijbers, J., den Dekker, A.J. (2004). Maximum likelihood estimation of signal amplitude and noise variance from MR data. *Magn. Reson. Med.* 51(3): 586–594.)

In addition, the dependence of both the bias and the variance of $\hat{\underline{A}}_{ML}$ on the number of data points has been investigated, yielding the following results:

- The bias of $\hat{\underline{A}}_{ML}$ for complex data with identical phases as well as for magnitude data with known and unknown noise variance generally decreases with the number of data points used for the estimation. On the other hand, it turns out that for complex data with different phases, the bias of $\hat{\underline{A}}_{ML}$ does not decrease with the number of data points.
- The variance, as may be expected, turns out to be inversely proportional to the number of data points for all estimators.

4.4.5 SIGNAL AMPLITUDE ESTIMATION FROM PCMR DATA

From Subsection 4.2.2.4 we learned that PCMR data are derived from the square root of the sum of the squares of a number of Gaussian-distributed variables, which is again a nonlinear transformation. It has been shown that results are biased when PCMR data are being used in quantitative analysis as an estimate of the underlying flow-related signal component magnitude [22]. The bias is due to the contributions from inherent random noise, which is not Gaussian distributed. Because the bias is not merely an additive component, it cannot be simply subtracted out. To remove the bias, knowledge of the actual shape of the data PDF becomes essential. In this section, the full knowledge of the PDF of the PCMR data is exploited for optimal estimation of the underlying signal.

Although this section focuses on complex difference processed images, the estimation techniques derived in this section can, under certain conditions, also be applied to images obtained by phase difference processing. This is because for both methods one has to estimate the underlying signal component from magnitude images for which the pixel variable m can be described by Equation 4.26. The only difference is that for phase difference processing, the dimension, or number of degrees of freedom K, directly equals the number of orthogonal Cartesian directions in which flow is encoded, whereas for complex difference processing, the dimension K is twice this physical dimension [22].

4.4.5.1 Region of Constant Amplitude and Known Noise Variance

In the following discussion, it is assumed that an unknown deterministic signal component A is to be estimated from N PCMR pixel values of a region Ω where the signal component is assumed to be constant. Thereby, the noise variance is assumed to be known.

4.4.5.1.1 CRLB

The Fisher information matrix of the data with respect to the parameter A is given by

$$I = -\mathbb{E}\left[\frac{\partial^2 \ln p_m}{\partial A^2}\right], \qquad (4.121)$$

where $p_{\underline{m}_n}$ is given by Equation 4.128. The expectation value in Equation 4.121 can be evaluated numerically. The CRLB can easily be obtained by applying the inverse operator.

4.4.5.1.2 Mean Estimator

The most intuitive way of estimating the unknown signal component is through a simple averaging of pixel values in the region Ω. Without *a priori* knowledge of the proper data PDF, this action would be justified as it is the optimal (i.e., ML) estimation procedure if the data is corrupted by Gaussian-distributed noise. This average or "mean estimator" is given by

$$\hat{\underline{A}}_{\mathrm{m}} = \frac{1}{N} \sum_{n=1}^{N} \underline{m}_n. \tag{4122}$$

The variance of this mean estimator is given by

$$\mathrm{Var}(\hat{\underline{A}}_{\mathrm{m}}) = \frac{1}{N} \left(\mathbb{E}\left[\underline{m}_n^2 \right] - E[\underline{m}_n]^2 \right). \tag{4.123}$$

However, as PCMR data are not Gaussian distributed, it is clear that a huge bias would be introduced if the signal is estimated by averaging pixel values. The bias, relative to the true signal component A, is in general defined by

$$\text{Relative bias} = \left| \frac{\mathbb{E}[\hat{A}] - A}{A} \right| \times 100\%. \tag{4.124}$$

In the definition, the absolute value was taken as to make the relative bias logarithmically plottable. For the mean estimator Equation 4.22, the expectation value $\mathbb{E}[\hat{\underline{A}}_{\mathrm{m}}]$ is given by $E[\underline{m}_n]$, because the average operator is an unbiased estimator of the expectation value. Hence, the relative bias of $\hat{A}_{\mathrm{m}}$ can be computed from the expression for the moments of the generalized Rician PDF as given in Equation 4.30 with $v = 1$. Note that it follows from Equation 4.30 and Equation 4.124 that the relative bias can be written solely in terms of the SNR and is independent of the number of averaged pixel values N. Figure 4.8 shows the relative bias of $\hat{A}_{\mathrm{m}}$ as a function of the SNR for various values of K. From the figure, it is clear that the bias increases rapidly with decreasing SNR. Also, the bias increases with increasing number of flow-encoding directions.

4.4.5.1.3 Modified RMS Estimator

An easy way to reduce the bias is by exploiting the second moment of the generalized Rician distribution, as was given in Equation 4.31. Indeed, an

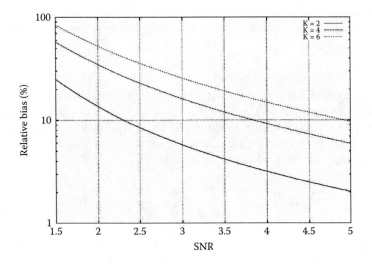

FIGURE 4.8 Relative bias of a simple spatial average estimator for $K = 2$, 4, and 6. The relative bias is independent of the number of averaged pixels N.

unbiased estimator of A^2 is given by

$$\widehat{\underline{A^2}} = \frac{1}{N} \sum_{n=1}^{N} \underline{m}_n^2 - K\sigma^2. \tag{4.125}$$

The PDF of $\widehat{\underline{A^2}}$ can be computed to yield

$$p_{\widehat{\underline{A^2}}}(\widehat{A^2}) = \frac{N}{2\sigma^2} \left(\frac{\widehat{A^2} + K\sigma^2}{A^2} \right)^{\frac{NK-2}{4}} \exp\left(-N \frac{\widehat{A^2} + K\sigma^2 + A^2}{2\sigma^2} \right)$$

$$\times I_{\frac{NK}{2}-1}\left(\frac{\sqrt{\widehat{A^2} + K\sigma^2}\, NA}{\sigma^2} \right) \varepsilon(\widehat{A^2} + K\sigma^2). \tag{4.126}$$

From the unbiased estimator of A^2, given in Equation 4.125, a modified RMS estimator of A would be given by

$$\widehat{\underline{A}}_{\mathrm{rms}} = \sqrt{\frac{1}{N} \sum_{n=1}^{N} \underline{m}_n^2 - K\sigma^2}. \tag{4.127}$$

However, root extraction is a nonlinear operation that makes the estimator $\widehat{A}_{\mathrm{rms}}$ biased. Also, the modified RMS estimator is appropriate only when the argument of the square root operator is nonnegative. A possible, at first sight quite arbitrary, solution to this problem is to artificially put the estimator $\hat{\underline{A}}_{\text{rms}}$ to zero whenever $\hat{\underline{A}}^2$ is negative:

$$\hat{\underline{A}}_{\text{rms}} = \begin{cases} \sqrt{\hat{\underline{A}}^2} & \text{if} \quad \hat{\underline{A}}^2 \geq 0 \\ 0 & \text{if} \quad \hat{\underline{A}}^2 < 0. \end{cases} \tag{4.128}$$

The PDF of $\hat{\underline{A}}_{\text{rms}}$ is then given by

$$p_{\hat{\underline{A}}_{\text{rms}}}(\hat{\underline{A}}_{\text{rms}}) = \delta(\hat{\underline{A}}_{\text{rms}}) \int_{-K\sigma^2}^{0} p_{\hat{\underline{A}}^2}(x)dx + \frac{N\hat{\underline{A}}_{\text{rms}}}{\sigma^2} \left(\frac{\hat{\underline{A}}_{\text{rms}}^2 + K\sigma^2}{A^2} \right)^{\frac{NK-2}{4}}$$

$$\times \exp\left(-N \frac{\hat{\underline{A}}_{\text{rms}}^2 + K\sigma^2 + A^2}{2\sigma^2} \right) I_{\frac{NK}{2}-1}\left(\frac{\sqrt{\hat{\underline{A}}_{\text{rms}}^2 + K\sigma^2} NA}{\sigma^2} \right) \varepsilon(\hat{\underline{A}}_{\text{rms}}),$$

$$\tag{4.129}$$

where $\delta(.)$ denotes the Dirac delta function. Notice that the first term of Equation 4.129 vanishes for high SNR. The bias of the modified estimator Equation 4.128 can be computed from

$$E[\hat{\underline{A}}_{\text{rms}}] = \int_{0}^{\infty} p_{\hat{\underline{A}}_{\text{rms}}}(x)x dx \tag{4.130}$$

Figure 4.9 shows the bias for various values of K and N. In general, the bias of the modified estimator is significantly smaller compared to the bias of the

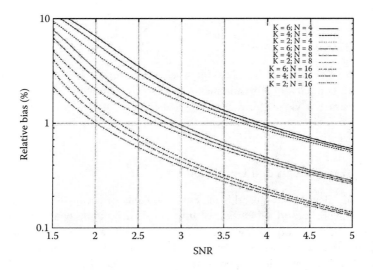

FIGURE 4.9 Relative bias of the RMS estimator of A for $K = 2$, 4, and 6 and $N = 4$, 8, and 16. The true value is $A = 100$.

mean estimator. It should be noticed that the bias of the mean estimator turns out to be positively valued, whereas the bias of the modified RMS estimator has a negative sign. That, however, cannot be observed from Figure 4.9 or Figure 4.8 because of the absolute value operator in Equation 4.124. For both estimators, the bias increases with increasing number of flow-encoding directions. However, in contrast to the mean estimator Equation 4.122, the bias of the modified estimator Equation 4.128 decreases with increasing N.

4.4.5.1.4 ML Estimator

In this subsection, the ML approach is clarified for the estimation of the unknown signal parameter A from a set of N independent magnitude PCMR data points $\underline{m} = (\underline{m}_1,...,\underline{m}_N)$. The proposed technique consists of maximizing the likelihood function of N generalized Rician-distributed data points with respect to A. The likelihood function of N independent magnitude data points $\{m_n\}$ is given by

$$L(A \mid \{m_n\}) = \prod_{n=1}^{N} p_{\underline{m}_n}(m_n \mid A). \tag{4.131}$$

where $p_{\underline{m}_n}$ is given by Equation 4.28. Then the ML estimate A of the PCMR signal is the global maximum of L, or equivalently the maximum of $\ln L$, with respect to A:

$$\hat{A}_{ML} = \arg\left\{\max_A (\ln L)\right\}. \tag{4.132}$$

Leaving in only the terms that depend on the variable A, we have explicitly

$$\ln L \sim -N\left(\frac{K}{2}-1\right)\ln A - \frac{NA^2}{2\sigma^2} + \sum_{n=1}^{N} \ln I_{\frac{K}{2}-1}\left(\frac{m_n A}{\sigma^2}\right). \tag{4.133}$$

It can be shown that $\ln L$ has only one maximum for positive values of A. Hence, the computational requirements are very low. It can also be shown that the ML estimator yields the value zero whenever $\widehat{A^2}$, given by Equation 4.125, is negative. This observation makes the modification of $\hat{\underline{A}}_{rms}$, as described by Equation 4.128, less arbitrary.

4.4.5.2 Experiments and Discussion

It is already clear from Figure 4.8 and Figure 4.9 that the accuracy of the modified RMS estimator, described earlier, is an order of magnitude better than that of the mean estimator Equation 4.122, although a significant bias still remains. To compare the modified RMS estimator with the ML estimator described in Subsection 4.4.5, a simulation experiment was set up. Thereby, K data points with deterministic signal component a_k were polluted with zero-mean Gaussian-distributed noise with equal variance, after which a magnitude data point m was

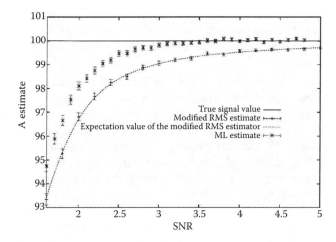

FIGURE 4.10 Comparison of the modified RMS estimator of A with the ML estimator of A for $K = 6$ and $N = 8$. Each time, the average value of 10^5 signal estimates is given along with the 95% confidence interval. The true value is $A = 100$. The expectation value of the ML estimator of A according to Equation 4.130 has also been shown.

computed according to Equation 4.26. The same procedure was repeated N times to obtain N generalized Rician-distributed magnitude data points. The deterministic signal component A of $\underline{m}$, given in Equation 4.27, was then estimated, once using the modified RMS estimator and once using the ML estimator. The data generation process and posterior signal estimation was repeated 10^5 times as a function of the SNR, after which the average value and the 95% confidence interval was computed. Figure 4.10 shows the signal estimation results for $K = 6$ and $N = 8$ as a function of the SNR. When the percentage for obtaining negative $\widehat{A^2}$ values was larger than 5%, the modified RMS estimator was regarded inappropriate. The SNR levels for this to occur were observed to be smaller than 1.5. For this reason, the modified RMS estimator was compared to the ML estimator only for SNR values higher than 1.5.

From Figure 4.10 it is clear that the ML estimator is slightly but significantly more accurate than the modified RMS estimator. A similar behavior of the performance of the estimators was observed for all combinations of $K = 2$, 4, and 6 and $N = 4$, 8, and 16. The results also show that the precision, i.e., the standard deviation, of both estimators is approximately equal. It can also be seen that at high SNR, the ML estimator cannot be distinguished from an unbiased estimator.

4.5 NOISE VARIANCE ESTIMATION

4.5.1 INTRODUCTION

Many image processing methods require the knowledge of the image noise variance. In general, these methods assume the data to be Gaussian or Poisson distributed [43–48]. As magnitude MR data are Rician distributed, these images require noise estimation methods that exploit this knowledge. In this subsection, several methods will be discussed, which can be classified as follows:

> *Single-acquisition methods:* In MRI, the image noise variance is commonly estimated from a large uniform-signal region or nonsignal regions within a single-magnitude MR image [19,39]. In this section, we consider ML estimation of the noise variance from complex (Subsection 4.5.2) and magnitude MR data (Subsection 4.5.3) from a region in which the underlying signal amplitude is nonzero but constant, as well as from a background region, i.e., a region in which the underlying signal amplitude is zero[49].
>
> *Double-acquisition methods:* Furthermore, methods were developed based on two acquisitions of the same image: the so-called double-acquisition methods [40,50,51]. At the end of Subsection 4.5.3.3, we will briefly describe a robust method based on a double-acquisition scheme[1,24].

4.5.2 Noise Variance Estimation from Complex Data

Suppose the noise variance σ^2 needs to be estimated from N complex-valued observations $\underline{c} = \{(\underline{w}_{r,n}, \underline{w}_{i,n})\}$. We will consider the case of identical underlying phase values as well as the case of different underlying phase values.

4.5.2.1 Region of Constant Amplitude and Phase

Let us first consider a region with a constant, nonzero underlying signal amplitude and identical underlying phase values.

4.5.2.1.1 CRLB
The Fisher information matrix of (A, φ, σ^2) is given by

$$
I = -\mathbb{E}
\begin{bmatrix}
\begin{pmatrix}
\dfrac{\partial^2 \ln p_{\underline{c}}}{\partial A^2} & \dfrac{\partial^2 \ln p_{\underline{c}}}{\partial A \partial \varphi} & \dfrac{\partial^2 \ln p_{\underline{c}}}{\partial A \partial \sigma^2} \\[2ex]
\dfrac{\partial^2 \ln p_{\underline{c}}}{\partial \varphi \partial A} & \dfrac{\partial^2 \ln p_{\underline{c}}}{\partial \varphi^2} & \dfrac{\partial^2 \ln p_{\underline{c}}}{\partial \varphi \partial \sigma^2} \\[2ex]
\dfrac{\partial^2 \ln p_{\underline{c}}}{\partial \sigma^2 \partial A} & \dfrac{\partial^2 \ln p_{\underline{c}}}{\partial \sigma^2 \partial \varphi} & \dfrac{\partial^2 \ln p_{\underline{c}}}{\partial (\sigma^2)^2}
\end{pmatrix}
\end{bmatrix}
=
\begin{pmatrix}
\dfrac{N}{\sigma^2} & 0 & 0 \\[2ex]
0 & \dfrac{NA^2}{\sigma^2} & 0 \\[2ex]
0 & 0 & \dfrac{N}{\sigma^4}
\end{pmatrix},
\qquad (4.134)
$$

Hence, the CRLB for unbiased estimation of (A, φ, σ^2) is given by

$$\text{CRLB} = I^{-1} = \begin{pmatrix} \dfrac{\sigma^2}{N} & 0 & 0 \\[2ex] 0 & \dfrac{\sigma^2}{NA^2} & 0 \\[2ex] 0 & 0 & \dfrac{\sigma^4}{N} \end{pmatrix}. \tag{4.135}$$

4.5.2.1.2　ML Estimation

For identical true phase values, the ML estimator of σ^2 can be shown to be given by

$$\hat{\sigma}^2_{\text{ML}} = \frac{1}{2N} \sum_{n=1}^{N} \left[\left(\hat{\underline{A}}_{\text{ML}} \cos \hat{\underline{\varphi}}_{\text{ML}} - \underline{w}_{r,n} \right)^2 + \left(\hat{\underline{A}}_{\text{ML}} \sin \hat{\underline{\varphi}}_{\text{ML}} - \underline{w}_{i,n} \right)^2 \right], \tag{4.136}$$

with $\hat{\underline{A}}_{\text{ML}}$ and $\hat{\underline{\varphi}}_{\text{ML}}$ given by Equation 4.60 and Equation 4.61, respectively. Notice that for $N = 1$, the estimator $\hat{\sigma}^2_{\text{ML}}$ will be equal to zero.

4.5.2.1.3　MSE

It can be shown that for large N the quantity $2N\hat{\sigma}^2_{\text{ML}}/\sigma^2$ is approximately distributed as χ^2_{2N-2} (i.e., chi-square distributed with $2N-2$ degrees of freedom). Because the mean and variance of a chi-squared variable with λ degrees of freedom are given by λ and 2λ, respectively, we find for the bias and variance of $\hat{\sigma}^2_{\text{ML}}$ that

$$b\left(\hat{\underline{\sigma}}^2_{\text{ML}} \right) \simeq \frac{\sigma^2}{2N} (2N - 2) - \sigma^2 = -\frac{\sigma^2}{N} \tag{4.137}$$

and

$$\text{Var}\left(\hat{\underline{\sigma}}^2_{\text{ML}} \right) \simeq \frac{\sigma^4}{N} \left(1 - \frac{1}{N} \right), \tag{4.138}$$

respectively. Then, the MSE of $\hat{\underline{\sigma}}^2_{\text{ML}}$ is given by

$$\text{MSE}\left(\hat{\underline{\sigma}}^2_{\text{ML}} \right) \simeq \frac{\sigma^4}{N}. \tag{4.139}$$

4.5.2.2　Region of Constant Amplitude and Different Phases

Next, consider a region with a constant nonzero underlying signal amplitude and different underlying phase values.

4.5.2.2.1 CRLB

It can easily be shown that the Fisher information matrix of the data with respect to the parameters $(A, \varphi_1, \ldots, \varphi_N, \sigma^2)$ is given by

$$
\boldsymbol{I} = -E
$$

$$
\times
\begin{bmatrix}
\begin{pmatrix}
\dfrac{\partial^2 \ln p_c}{\partial A^2} & \dfrac{\partial^2 \ln p_c}{\partial A \partial \varphi_1} & \cdots & \dfrac{\partial^2 \ln p_c}{\partial A \partial \varphi_N} & \dfrac{\partial^2 \ln p_c}{\partial A \partial \sigma^2} \\[2ex]
\dfrac{\partial^2 \ln p_c}{\partial \varphi_1 \partial A} & \dfrac{\partial^2 \ln p_c}{\partial \varphi_1^2} & \cdots & \dfrac{\partial^2 \ln p_c}{\partial \varphi_1 \partial \varphi_N} & \dfrac{\partial^2 \ln p_c}{\varphi_1 \partial \sigma^2} \\[2ex]
\cdots & \cdots & \cdots & \cdots & \cdots \\[2ex]
\dfrac{\partial^2 \ln p_c}{\partial \varphi_N \partial A} & \dfrac{\partial^2 \ln p_c}{\partial \varphi_N \partial \varphi_1} & \cdots & \dfrac{\partial^2 \ln p_c}{\partial \varphi_N^2} & \dfrac{\partial^2 \ln p_c}{\partial \varphi_N \partial \sigma^2} \\[2ex]
\dfrac{\partial^2 \ln p_c}{\partial \sigma^2 \partial A} & \dfrac{\partial^2 \ln p_c}{\partial \sigma^2 \partial \varphi_1} & \cdots & \dfrac{\partial^2 \ln p_c}{\partial \sigma^2 \partial \varphi_N} & \dfrac{\partial^2 \ln p_c}{\partial (\sigma^2)^2}
\end{pmatrix}
\end{bmatrix}
=
\begin{pmatrix}
\dfrac{N}{\sigma^2} & 0 & \cdots & 0 & 0 \\[2ex]
0 & \dfrac{A^2}{\sigma^2} & \cdots & 0 & 0 \\[2ex]
\cdots & \cdots & \cdots & \cdots & \cdots \\[2ex]
0 & 0 & \cdots & \dfrac{A^2}{\sigma^2} & 0 \\[2ex]
0 & 0 & \cdots & 0 & \dfrac{N}{\sigma^4}
\end{pmatrix}
$$

$$(4.140)$$

and the CRLB for unbiased estimation of $(A, \varphi_1, \ldots, \varphi_N, \sigma^2)$ is given by

$$
\mathrm{CRLB} =
\begin{pmatrix}
\dfrac{\sigma^2}{N} & 0 & \cdots & 0 & 0 \\[2ex]
0 & \dfrac{\sigma^2}{A^2} & \cdots & 0 & 0 \\[2ex]
\cdots & \cdots & \cdots & \cdots & \cdots \\[2ex]
0 & 0 & \cdots & \dfrac{\sigma^2}{A^2} & 0 \\[2ex]
0 & 0 & \cdots & 0 & \dfrac{\sigma^4}{N}
\end{pmatrix},
\qquad (4.141)
$$

4.5.2.2.2 ML Estimation

In case of different phase values, we have

$$
\hat{\sigma}^2_{\mathrm{ML}} = \frac{1}{2N} \sum_{n=1}^{N} \left[\left(\hat{A}_{\mathrm{ML}} \cos \hat{\varphi}_{n,\mathrm{ML}} - w_{r,n} \right)^2 + \left(\hat{A}_{\mathrm{ML}} \sin \hat{\varphi}_{n,\mathrm{ML}} - w_{i,n} \right)^2 \right], \quad (4.142)
$$

with $\hat{A}_{\mathrm{ML}}$ and $\hat{\varphi}_{n,\mathrm{ML}}$ given by Equation 4.73 and Equation 4.74, respectively.

4.5.2.2.3 MSE

It can be shown that, for large N, the quantity $2N\hat{\underline{\sigma}}^2_{\mathrm{ML}}/\sigma^2$ is approximately distributed as $\chi^2_{2N-(N+1)} = \chi^2_{N-1}$ [52]. This means that

$$b\left(\hat{\underline{\sigma}}^2_{\mathrm{ML}}\right) \simeq \frac{\sigma^2}{2N}(N-1) - \sigma^2 = -\frac{\sigma^2}{2}\left(1 + \frac{1}{N}\right) \qquad (4.143)$$

and

$$\mathrm{Var}\left(\hat{\underline{\sigma}}^2_{\mathrm{ML}}\right) \simeq \frac{\sigma^4}{2N}\left(1 - \frac{1}{N}\right). \qquad (4.144)$$

Hence, the MSE of $\hat{\underline{\sigma}}^2_{\mathrm{ML}}$ is given by

$$\mathrm{MSE}\left(\hat{\underline{\sigma}}^2_{\mathrm{ML}}\right) \simeq \frac{\sigma^4}{4}\left(1 + \frac{4}{N} - \frac{1}{N^2}\right). \qquad (4.145)$$

4.5.2.3 Background Region

Next, consider the case in which the noise variance is estimated from a background region (i.e., a region where A is known to be zero).

4.5.2.3.1 CRLB

It can easily be shown that the CRLB for unbiased estimation of σ^2 is given by

$$\mathrm{CRLB} = \frac{\sigma^4}{N}, \qquad (4.146)$$

independent of the underlying phase values.

4.5.2.3.2 ML Estimation

The ML estimator is given by

$$\hat{\underline{\sigma}}^2_{\mathrm{ML}} = \frac{1}{2N}\sum_{n=1}^{N}\left(\underline{w}^2_{r,n} + \underline{w}^2_{i,n}\right), \qquad (4.147)$$

independent of the underlying phase values.

4.5.2.3.3 MSE

It can easily be shown that the ML estimator Equation 4.147 is unbiased and that its variance equals the CRLB. Therefore, its MSE is simply given by

$$\mathrm{MSE}\left(\hat{\underline{\sigma}}^2_{\mathrm{ML}}\right) = \frac{\sigma^4}{N}. \qquad (4.148)$$

4.5.3 NOISE VARIANCE ESTIMATION FROM MAGNITUDE DATA

We will now describe the ML estimation of noise variance (and standard deviation) from magnitude MR data. First, we will consider ML estimation of the noise variance from a so-called background region, that is, a region in which the underlying signal is zero. The CRLB for unbiased estimation of both noise variance and standard deviation will be computed, and the ML estimators will be derived. Next, we will consider the ML estimation of noise parameters from a so-called constant region, that is, a region in which the (nonzero) signal amplitude is assumed to be constant. In this case, the noise parameters have to be estimated simultaneously with the signal amplitude.

It will be assumed that the available data is governed by a generalized Rician distribution. The methods described in the following subsections can also be applied to conventional Rician-distributed magnitude MR data, which would be the case when $K = 2$.

4.5.3.1 Background Region

Suppose that a set of N statistically independent magnitude data points $\underline{m} = \{\underline{m}_n\}$ is available from a region where the true signal value A is zero for each data point (background region). Hence, these data points are governed by a Rayleigh distribution and their joint PDF $p_{\underline{m}}$ is given by (cf. Equation 4.29)

$$p_{\underline{m}}(\{m_n\}) = \prod_{n=1}^{N} \frac{2m_n^{K-1}}{(2\sigma^2)^{K/2}\Gamma(K/2)} \exp\left(-\frac{m_n^2}{2\sigma^2}\right), \quad (4.149)$$

where $\{m_n\}$ are the magnitude variables corresponding with the magnitude observations $\{\underline{m}_n\}$.

4.5.3.1.1 CRLB (Variance)

The Fisher information matrix I with respect to σ^2 is simply given by

$$I = -\mathbb{E}\left[\frac{\partial^2 \ln p_{\underline{m}}}{(\partial \sigma^2)^2}\right] = \frac{NK}{2\sigma^4}, \quad (4.150)$$

from which the CRLB for unbiased estimation of σ^2 is easily found [6] as

$$\text{CRLB} = \frac{2\sigma^4}{NK}. \quad (4.151)$$

4.5.3.1.2 CRLB (Standard Deviation)

From the knowledge of the CRLB for unbiased estimation of σ^2, the CRLB for unbiased estimation of σ can be derived using Equation 4.48, i.e., $\left(\frac{\partial \tau}{\partial \theta}\right)I^{-1}\left(\frac{\partial \tau}{\partial \theta}\right)$, where $\tau \equiv \tau(\sigma^2) = \sigma$, and I^{-1} is given by Equation 4.151

$$\mathrm{CRLB} = \left(\frac{1}{2\sigma}\right)\frac{2\sigma^4}{NK}\left(\frac{1}{2\sigma}\right) \tag{4.152}$$

$$= \frac{\sigma^2}{2NK}. \tag{4.153}$$

4.5.3.1.3 ML Estimation

We will now describe the ML method for the estimation of the noise standard deviation or the noise variance from background magnitude MR data. Thereby, it will be assumed that the available data is governed by a generalized Rician distribution. The methods described in the following text can also be applied to conventional Rician-distributed data, which would be the case when $K = 2$.

The likelihood function is obtained by substituting the available background data points $\{m_n\}$ for the variables $\{\mathrm{m}_n\}$ in Equation 4.125. Then the log-likelihood function, only as a function of σ^2, is given by:

$$\ln L \sim -N \ln \sigma^2 - \frac{1}{K\sigma^2}\sum_{n=1}^{N} m_n^2. \tag{4.154}$$

Maximizing with respect to σ^2 yields the ML estimator of σ^2 as

$$\widehat{\sigma^2}_{\mathrm{ML}} = \frac{1}{KN}\sum_{n=1}^{N} m_n^2. \tag{4.155}$$

It can be shown that Equation 4.155 is an unbiased estimator, that is, its mean is equal to σ^2. Furthermore, the variance of the ML estimator Equation 4.155 is equal to $2\sigma^4/NK$, which equals the CRLB given by Equation 4.151 for all values of N.

One might be interested in the value of the standard deviation σ, e.g., to estimate the SNR A/σ. Simply taking the square root of the ML estimator of σ^2 in Equation 4.155 yields an estimator of σ as

$$\widehat{\sigma}_{\mathrm{ML}} = \sqrt{\frac{1}{KN}\sum_{n=1}^{N} m_n^2}, \tag{4.156}$$

This estimator is identical to the ML estimator of σ because the square root operation has a single-valued inverse (cf. Invariance property of ML estimators [53]). Its variance is approximately equal to

$$\mathrm{Var}(\widehat{\sigma}_{\mathrm{ML}}) \approx \frac{\sigma^2}{2NK}, \tag{4.157}$$

which equals the CRLB (cf. Equation 4.153). The estimator (Equation 4.156) is, however, biased because of the square root operation. Its expectation value is approximately equal to

$$\mathbb{E}[\widehat{\underline{\sigma}}_{ML}] \approx \sigma \left(1 - \frac{1}{4NK} \right). \tag{4.158}$$

This means that it is possible to apply a bias correction. This, however, would increase the variance of the estimator.

4.5.3.1.4 Conventional Estimation

Another commonly used estimator of σ can be found from the first moment of the Rayleigh PDF. Because the mean value of the generalized Rayleigh PDF is given by

$$\mathbb{E}[\underline{m}_n] = \sqrt{2}\sigma \frac{\Gamma((K+1)/2)}{\Gamma(K/2)}, \tag{4.159}$$

an unbiased estimator of σ is easily seen to be

$$\widehat{\underline{\sigma}}_c = \frac{\Gamma(K/2)}{\Gamma((K+1)/2)} \frac{1}{\sqrt{2N}} \sum_{n=1}^{N} \underline{m}_n. \tag{4.160}$$

The variance of this estimator is given by

$$\mathrm{Var}(\widehat{\underline{\sigma}}_c) = \frac{\sigma^2}{N} \left(\frac{K}{2} \left(\frac{\Gamma(K/2)}{\Gamma((K+1)/2)} \right)^2 - 1 \right), \tag{4.161}$$

which is always larger than the CRLB. Next, we can compare both estimators of σ, described in the preceding text by Equation 4.160 and Equation 4.156, in terms of the MSE. The MSE ratio, defined as

$$\mathrm{MSE\ ratio} = \frac{\mathrm{MSE}(\widehat{\underline{\sigma}}_c)}{\mathrm{MSE}(\widehat{\underline{\sigma}}_{ML})}, \tag{4.162}$$

is shown in Figure 4.11 as a function of the number of data points for $K = 2, 4$, and 6. For large N, the MSE of the common estimator (Equation 4.160) is significantly larger than that of the ML estimator (Equation 4.156). The performance of the conventional estimator, compared with the ML estimator, is not good for conventional magnitude MR images where $K = 2$.

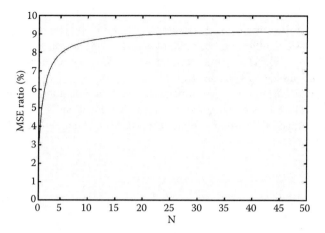

FIGURE 4.11 Performance comparison between the conventional and the ML estimator of the noise standard deviation: MSE ratio as a function of the number of data points N for $K = 2, 4$, and 6.

4.5.3.2 Region of Constant Amplitude

Suppose that a set of magnitude data points $\{\underline{m}_n\}$ is available from a region where the true signal value A is the same for each data point. Then, the CRLB for unbiased estimation of σ^2 from magnitude data of a constant region is given by Equation 4.116, in which the elements are defined by Equation 4.113 to Equation 4.115.

Furthermore, the value of σ^2 can be estimated using the ML method as follows (cf. Equation 4.120):

$$\left(\hat{A}_{ML}, \hat{\sigma}_{ML}^2\right) = \arg\left\{\max_{A,\sigma^2}\left(-N\ln\sigma^2 - \sum_{n=1}^{N}\frac{m_n^2 + A^2}{2\sigma^2} + \sum_{n=1}^{N}\ln I_0\left(\frac{Am_n}{\sigma^2}\right)\right)\right\}. \quad (4.163)$$

Note that it requires the optimization of a two-dimensional function, which cannot be solved analytically.

4.5.3.3 Double-Acquisition Method

Estimation of the noise variance from a single-magnitude image requires homogeneous regions in the image. However, large homogeneous regions are often hard to find, and therefore only a small number of data points are available for estimation. Also, background data points sometimes suffer from systematic intensity variations. To overcome these disadvantages, methods were developed based

132

Advanced Image Processing in Magnetic Resonance Imaging

on two acquisitions of the same image: the so-called double-acquisition methods. Thereby, the noise variance is, for example, computed by subtracting two acquisitions of the same object and calculating the standard deviation of the resulting image pixels [40,50,51]. Alternatively, the image noise variance can be computed from two magnitude MR images as follows.

When two conventional MR images ($K = 2$) are acquired under identical imaging conditions, one can solve σ^2 from two equations and two unknowns using the averaged (averaging is done in K-space) and single images because

$$\mathbb{E}\left[\left\langle \underline{m}_s^2 \right\rangle\right] = \frac{1}{N}\sum_{n=1}^{N} A_n^2 + 2\sigma^2, \tag{4.164}$$

$$\mathbb{E}\left[\left\langle \underline{m}_a^2 \right\rangle\right] = \frac{1}{N}\sum_{n=1}^{N} A_n^2 + 2\left(\frac{\sigma}{\sqrt{2}}\right)^2, \tag{4.165}$$

where $\langle\rangle$ denotes the spatial average of the whole image. The subscripts s and a refer to the single and averaged images, respectively. From Equation 4.165 and Equation 4.164, an unbiased estimator of the noise variance is derived as

$$\widehat{\sigma^2} = \left\langle \underline{m}_s^2 \right\rangle - \left\langle \underline{m}_a^2 \right\rangle. \tag{4.166}$$

This approach has the following advantages:

- It does not require any user interaction, as no background pixels need to be selected.
- It is insensitive to systematic errors such as ghosting, ringing, and direct current (DC) artifacts as long as these appear in both images. It is clear that if this type of error appears in only one of the two images, none of the double-acquisition methods will yield the correct result.
- The precision of the noise variance estimator (Equation 4.140) is drastically increased compared to the precision of the estimator given in Equation 4.130, as all the data points (not only those from the background region) are involved in the estimation.
- It is valid for any SNR of the image.

An obvious disadvantage is the double acquisition itself. However, in MR acquisition schemes, it is common practice to acquire two or more images for averaging. Hence, those images may be used for the proposed noise estimation procedure, without additional acquisition time. In addition, the images require proper geometrical registration, i.e., no movement of the object during acquisition is allowed.

4.5.4 Discussion

4.5.4.1 CRLB

The CRLB for unbiased estimation of the noise variance from complex data with identical and different phase values is given by Equation 4.135 and Equation 4.141, respectively. In both cases, independent of the signal amplitude of the data points, the CRLB is equal to σ^4/N.

Furthermore, the CRLB for unbiased estimation of the noise variance from magnitude data has been computed for a background region as well as for a constant region. In Figure 4.14, the CRLB is shown as a function of the SNR.

- If the noise variance is estimated from N magnitude data points of a background region, the CRLB is equal to σ^4/N (cf. Equation 4.151). Then the CRLB is the same as that for estimation from N complex data points of a background region. This might be surprising, as estimation from N complex data points actually exploits $2N$ real-valued (N real and N imaginary) observations, whereas estimation from N magnitude data points only exploits N real-valued observations. However, this is compensated by the fact that the Rayleigh PDF has a smaller standard deviation.

- If the noise variance is estimated from N magnitude data points of a constant region, the CRLB is given by Equation 4.116. It can be shown numerically that for magnitude data this CRLB tends to $2\sigma^4/N$ when the SNR increases, which is a factor 2 larger compared to estimation from complex data (cf. Figure 4.12). This is not surprising because the Rician PDF tends to a Gaussian PDF for high SNR, with the same variance as the PDF of the real or imaginary data. Hence for high SNR, the difference in CRLB between magnitude and complex data can simply be explained by the number of observations available for the estimation of the noise variance.

4.5.4.2 MSE

For complex data from a region with constant amplitude with identical and different phase values, expressions for the bias, variance, and MSE of the ML estimator $\hat{\sigma}^2_{\mathrm{ML}}$ of σ^2 were derived. The bias of $\hat{\sigma}^2_{\mathrm{ML}}$ is given by Equation 4.137 and Equation 4.143 for the two cases, respectively. From these expressions, it is clear that:

- Both noise variance estimators are biased. Also, the bias of $\hat{\sigma}^2_{\mathrm{ML}}$ is independent of the true signal amplitude.
- For identical phases, the bias of $\hat{\sigma}^2_{\mathrm{ML}}$ is inversely proportional to the number of observations (N). In contrast, the bias of $\hat{\sigma}^2_{\mathrm{ML}}$ for different phases does not decrease with increasing N; for large N, it converges to $\sigma^2/2$.

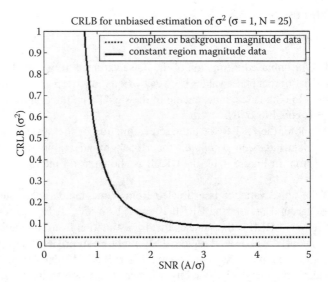

FIGURE 4.12 CRLB for unbiased estimation of the noise variance for complex data and magnitude background data (dotted line) and for magnitude data from a constant region, that is, where A is estimated simultaneously (full line). (From Sijbers, J., den Dekker, A.J. (2004). Maximum likelihood estimation of signal amplitude and noise variance from MR data. *Magn. Reson. Med.* 51(3): 586–594.)

The variance of $\hat{\underline{\sigma}}^2_{ML}$ for complex data from a region with constant amplitude for identical and different phase values is given by Equation 4.138 and Equation 4.144, respectively. Only for complex data with identical phases, the variance of $\hat{\underline{\sigma}}^2_{ML}$ asymptotically attains the CRLB. This is because for estimation from complex data with different phases the number of unknown parameters to be estimated simultaneously with σ^2 is proportional to N.

The MSE of $\hat{\underline{\sigma}}^2_{ML}$ for complex data from a region with constant amplitude for identical and different phase values is given by Equation 4.139 and Equation 4.145, respectively. Both are shown as a function of N in Figure 4.13. Moreover, the MSE of $\hat{\underline{\sigma}}^2_{ML}$ as a function of the SNR is shown in Figure 4.14.

The MSE of $\hat{\underline{\sigma}}^2_{ML}$ for magnitude data from a region with constant amplitude can be found numerically from Equation 4.163. The results for this estimator as a function of the SNR are also shown in Figure 4.14.

Finally, the MSE of $\widehat{\underline{\sigma}^2}_{ML}$ from background MR data is given by σ^4/N, for magnitude and for complex data, and is independent of the phases. From this, it is clear that the noise variance should be estimated from background data points whenever possible. If a background region is not available, a similar reasoning for the estimation of σ^2 as that for the estimation of A holds. That is, estimation of σ^2 from complex data with identical phases is then preferred to estimation from magnitude data, which, in turn, is preferred to estimation from complex data with different phases.

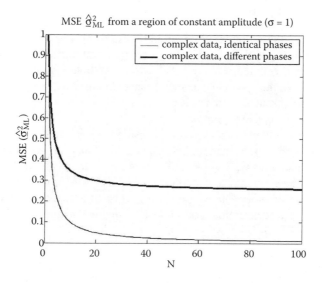

FIGURE 4.13 MSE for unbiased estimation of the noise variance from a region of constant amplitude for complex data with identical and different phases. (From Sijbers, J., den Dekker, A.J. (2004). Maximum likelihood estimation of signal amplitude and noise variance from MR data. *Magn. Reson. Med.* 51(3): 586–594.)

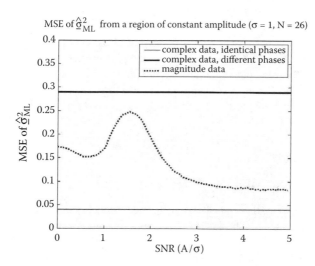

FIGURE 4.14 MSE for unbiased estimation of the noise variance from a region of constant amplitude for complex data with identical and different phases and for magnitude data. (From Sijbers, J., den Dekker, A.J. (2004). Maximum likelihood estimation of signal amplitude and noise variance from MR data. *Magn. Reson. Med.* 51(3): 586–594.)

4.6 CONCLUSIONS

In this chapter, the problem of signal and noise estimation from MR data was addressed. It was noted that original data coming from the scanner are complex and Gaussian distributed. However, because of multiple digital-data-processing steps, the PDF of the resulting data may change. In this chapter, most of the PDFs one may be confronted with when processing MRI data were discussed along with their moments and asymptotic behavior.

Furthermore, it was shown how to deal with various PDFs to optimally estimate signal and noise parameters. Conventional and ML techniques to estimate such parameters were compared. It was shown that methods based on ML estimation outperform conventional estimators. The ML signal estimator yields physically relevant solutions for the whole range of SNRs. Also, it was shown that the ML estimator, unlike conventional signal estimators, cannot be distinguished from an unbiased estimator at high SNR.

Finally, the question was addressed as to whether complex or magnitude data should be used to estimate signal or noise parameters from low SNR data when using the ML method. In summary, the following conclusions can be drawn:

- The image noise variance should preferentially be estimated from background data (i.e., from a region of interest in which the true magnitude values are zero). Thereby, it does not matter whether the noise variance is estimated from magnitude or complex data.
- On the other hand, whether or not the signal amplitude should be estimated from magnitude or complex data depends on the underlying phase values:
  - If the true phase values are known to be constant, the signal amplitude should be estimated from complex-valued data.
  - If the true phase values are unknown or if the true phase model deviates from a constant model, it is generally better, with respect to the MSE, to estimate the signal amplitude from magnitude data.

4.7 APPENDIX

4.7.1 TRANSFORMATIONS OF PDFs

In image processing, data are often transformed through various arithmetic manipulations. Here, we describe how a PDF changes as a result of such a transformation [54].

4.7.1.1 Theorem

Suppose we wish to determine the PDF, $p_{\underline{y}}(y)$, of $\underline{y}$, which is given by

$$\underline{y} = g(\underline{x}), \tag{4.167}$$

where $\underline{x}$ is a known random variable of which the PDF $p_{\underline{x}}(x)$ is known. Then to find $p_{\underline{y}}(y)$, it suffices to solve the equation $y = g(x)$. Indeed, if the real roots of Equation 4.167 are denoted by $x_1,\ldots,x_r,\ldots$, such that

$$y = g(x_1) = \cdots = g(x_r) = \cdots, \tag{4.168}$$

Then $p_{\underline{y}}(y)$ is given by

$$p_{\underline{y}}(y) = \frac{p_{\underline{x}}(x_1)}{|g'(x_1)|} + \cdots + \frac{p_{\underline{x}}(x_n)}{|g'(x_r)|} + \cdots, \tag{4.169}$$

where $g'(x)$ is the derivative of $g(x)$.

4.7.1.2 Example

Let $y = \sqrt{x}$

As an example, suppose the PDF $p_{\underline{x}}(x)$ of $\underline{x}$ is known. Then, the PDF of $\underline{y}$, given by

$$y = \sqrt{\underline{x}}, \tag{4.170}$$

is

$$p_{\underline{y}}(y) = 2yp_{\underline{x}}(y^2), \tag{4.171}$$

for $y \geq 0$.

4.7.2 GENERAL THEOREM

Given a random vector

$$\underline{x} = (\underline{x}_1,\ldots,\underline{x}_n)^T, \tag{4.172}$$

whose components $\underline{x}_i$ are random variables and given k functions

$$g_1(x),\ldots,g_k(x), \tag{4.173}$$

we form a new set of random variables:

$$\underline{y}_1 = g_1(\underline{x}),\ldots,\underline{y}_k = g_k(\underline{x}). \tag{4.174}$$

Assume $k = n$. To find the PDF $p_{\underline{y}}(y_1,\ldots,y_n)$ of the random vector $\underline{y} = (y_1,\ldots,y_n)^T$ for a specific set of numbers $y_1,\ldots,y_n$, we solve the system

$$g_1(x) = y_1,\ldots,g_n(x) = y_n, \tag{4.175}$$

If this system has no solutions, then $p_{\underline{y}}(y_1,...,y_n) = 0$. If it has a single solution $x = (x_1,...,x_n)^T$, then

$$p_{\underline{y}}(y_1,...,y_n) = \frac{p_{\underline{x}}(x_1,...,x_n)}{|J(x_1,...,x_n)|}, \tag{4.176}$$

where

$$J(x_1,...,x_n) = \begin{vmatrix} \dfrac{\partial g_1}{\partial x_1} & \cdots & \dfrac{\partial g_1}{\partial x_n} \\ \cdots & \cdots & \cdots \\ \dfrac{\partial g_n}{\partial x_1} & \cdots & \dfrac{\partial g_n}{\partial x_n} \end{vmatrix} \tag{4.177}$$

is the Jacobian of the transformation in Equation 4.175. If it has several solutions, then we add the corresponding terms as in Equation 4.169.

4.7.3 Approximation of the Mean of a Random Variable

4.7.3.1 Theorem

The mean of a function $\underline{y} = g(\underline{x})$ of a random variable $\underline{x}$ is given by

$$\mathbb{E}[g(\underline{x})] = \int_{-\infty}^{+\infty} g(x)p_{\underline{x}}(x)dx. \tag{4.178}$$

If $\underline{x}$ is concentrated near its mean μ, the mean of $g(\underline{x})$ can be approximated by a Taylor expansion about $g(\mu)$:

$$g(\underline{x}) \simeq g(\mu) + g'(\mu)(\underline{x} - \mu) + \cdots + \frac{1}{n!}g^{(n)}(\mu)(\underline{x} - \mu)^n. \tag{4.179}$$

Taking the expectation value of both sides yields

$$\mathbb{E}[g(\underline{x})] \simeq g(\mu) + \frac{1}{2}g''(\mu)\text{Var}(\underline{x}) + \cdots + \frac{1}{n!}g^{(n)}(\mu)\mu_n. \tag{4.180}$$

4.7.3.2 Example

As an example, consider the estimator of the signal amplitude given in Equation 4.65 as $\hat{\underline{A}}_c = \sqrt{\langle \underline{m}^2 \rangle - 2\sigma^2}$. To find the mean value of $\hat{\underline{A}}_c$, an expansion about the mean of the argument of the square root is employed. If we write $\underline{y} = \langle \underline{m}^2 \rangle - 2\sigma^2$, then $\mathbb{E}[\underline{y}] = A^2$. Furthermore, from Equation 4.18 and Equation 4.20, we know that $\mathrm{Var}(\underline{y}) = 4\sigma^2(A^2 + \sigma^2)/N$. Hence, from Equation 4.180 we then have for high SNR

$$\mathbb{E}[\underline{\hat{A}}_c] \simeq A + \frac{1}{2}\left(\frac{-1}{4}\right)A^{-3}\frac{4\sigma^2}{N}(A^2 + \sigma^2) \qquad (4.181)$$

$$\simeq A\left(1 - \frac{\sigma^2}{2NA^2}\right). \qquad (4.182)$$

ABBREVIATIONS

CRLB	Cramér-Rao lower bound
DC	direct current
FT	Fourier transform
ML	maximum likelihood
MRI	magnetic resonance imaging
MSE	mean squared error
PDF	probability density function
RF	radio frequent
SNR	signal-to-noise ratio

SYMBOLS

A	signal amplitude
$\hat{A}$	estimate of the signal amplitude
$\underline{\hat{A}}$	estimator of the signal amplitude
b	bias
$\underline{c}$	set of complex observations (stochastic variables)
C	covariance matrix
$\delta(.)$	delta Dirac function
$\varepsilon(.)$	unit step Heaviside function
$\mathbb{E}[.]$	expectation operator
$_1F_1(.)$	confluent hypergeometric function of the first kind
$\Gamma(.)$	gamma function
$I_0(.)$	zeroth-order modified Bessel function of the first kind
$I_1(.)$	first-order modified Bessel function of the first kind
I	Fisher information matrix
K	number of parameters
$L(.)$	likelihood function
m	magnitude observation (number)
$\underline{m}$	magnitude observation (stochastic variable)
...	magnitude variable corresponding to the observation m
μ	mean value of the Gaussian distribution
N	number of observations

$p(.)$	probability density function
φ	phase value
$\hat{\varphi}$	estimator of the phase
$\underline{\phi}$	phase variable corresponding to φ
σ	noise standard deviation
σ^2	noise variance
$\hat{\sigma}^2$	estimate of the noise variance
$\underline{\hat{\sigma}}^2$	estimator of the noise variance
$\boldsymbol{\theta}$	parameter vector
$\hat{\boldsymbol{\theta}}$	estimate of $\boldsymbol{\theta}$
$\underline{\hat{\boldsymbol{\theta}}}$	estimator of $\boldsymbol{\theta}$
τ	vector function of $\boldsymbol{\theta}$
$\underline{\hat{\tau}}$	estimator of τ
$\omega_{i,n}$	imaginary variable corresponding to the observation $\omega_{i,n}$
$\omega_{i,n}$	imaginary observation (number)
$\underline{\omega}_{i,n}$	imaginary observation (stochastic variable)
$\omega_{r,n}$	real variable corresponding to the observation $\omega_{r,n}$
$\omega_{r,n}$	real observation (number)
$\underline{\omega}_{r,n}$	real observation (stochastic variable)

REFERENCES

1. Sijbers, J., den Dekker, A.J., Verhoye, M., Van Audekerke, J., and Van Dyck, D. (1998). Estimation of noise from magnitude MR images. *Magn. Reson. Imaging.* 16(1): 87–90.
2. Sijbers, J., den Dekker, A.J., Van der Linden, A., Verhoye, M., and Van Dyck, D. (1999). Adaptive, anisotropic noise filtering for magnitude MR data. *Magn. Reson. Imaging.* 17(10): 1533–1539.
3. Nowak, R.D. (1999). Wavelet-based Rician noise removal for magnetic resonance images. *IEEE Trans. Image Processing.* 10(8): 1408–1419.
4. Wood, J.C. and Johnson, K.M. (1999). Wavelet packet denoising of magnetic resonance images: importance of Rician noise at low SNR. *Magn. Reson. Med.* 41(3): 631–635.
5. Karlsen, O.T., Verhagen, R., and Bovée, W.M. (1999). Parameter estimation from Rician-distributed data sets using maximum likelihood estimator: application to T_1 and perfusion measurements. *Magn. Reson. Med.* 41: 614–623.
6. Sijbers, J., den Dekker, A.J., Raman, E., and Van Dyck, D. (1999). Parameter estimation from magnitude MR images. *Intl. J. Imag. Syst. Tech.* 10(2): 109–114.
7. Borgia, G.C., Brown, R.J.S., and Fantazzini, P. (2000). Uniform-penalty inversion of multiexponential decay data. *J. Magn. Reson.* 147: 273–285.
8. van der Weerd, L., Vergeldt, F.J., Jager, P.A., and Van As, H., (2000). Evaluation of algorithms for analysis of NMR relaxation decay curves. *Magn. Reson. Imaging.* 18: 1151–1157.
9. Does, M.D. and Gore, J.C. (2002). Compartmental study of T_1 and T_2 in rat brain and trigeminal nerve *in vivo. Magn. Reson. Med.* 47: 274–283.

10. Sijbers, J. and Van der Linden, A. (2003). *Encyclopedia of Optical Engineering,* chap. Magnetic resonance imaging, pp. 1237–1258. Marcel Dekker, ISBN: 0-8247-4258-3.

11. Breiman, L. (1968). *Probability.* Addison-Wesley, Reading, MA.

12. Wang, Y. and Lei, T. (1994). Statistical analysis of MR imaging and its applications in image modeling. in *Proceedings of the IEEE International Conference on Image Processing and Neural Networks,* Vol. I, pp. 866–870.

13. Wang, Y., Lei, T., Sewchand, W., and Mun, S.K. (1996). MR imaging statistics and its application in image modeling. in *Proceedings of the SPIE Conference on Medical Imaging,* Newport Beach (CA).

14. Ross, S. (1976). *A first course in probability.* Collier Macmillan Publishers, New York.

15. van den Bos, A. (1989). Estimation of Fourier coefficients. *IEEE Trans. Instrum. Meas.* 38(4): 1005–1007.

16. Cunningham, I.A. and Shaw, R. (1999). Signal-to-noise optimization of medical imaging systems. *J. Opt. Soc. Am. A.* 16(3): 621–631.

17. Gradshteyn, I.S. and Ryzhik, I.M. (1965). *Table of integrals, series and products.* 4th ed., Academic Press, New York and London.

18. Rice, S.O. (1944). Mathematical analysis of random noise. *Bell Syst. Tech.* 23: 282–332.

19. Henkelman, R.M. (1985). Measurement of signal intensities in the presence of noise in MR images. *Med. Phys.* 12(2): 232–233.

20. Abramowitz, M. and Stegun, I.A. (1970). *Handbook of mathematical functions.* Dover Publications, New York.

21. Constantinides, C.D., Atalar, E., and McVeigh, E.R. (1997). Signal-to-noise measurements in magnitude images from NMR phased arrays. *Magn. Reson. Med.* 38: 852–857.

22. Andersen, A.H. and Kirsch, J.E. (1996). Analysis of noise in phase contrast MR imaging. *Med. Phys.* 23(6):857–869.

23. Pelc, N.J., Bernstein, M.A., Shimakawa, A., and Glover, G.H. (1991). Encoding strategies for three-direction phase-contrast MR imaging of flow. *J. Magn. Reson. Imaging.* 1(4): 405–413.

24. Gudbjartsson, H. and Patz, S. (1995). The Rician distribution of noisy MRI data. *Magn. Reson. Med.* 34: 910–914.

25. van den Bos, A. (1982). Parameter estimation. *Handbook of Measurement Science,* Ed. P.H. Sydenham, Vol. 1, chap. 8. Wiley, Chichester, England. pp. 331–377.

26. Zacks, S., (1971). *The theory of statistical inference.* Wiley. New York.

27. Fisher, R.A. (1922). On the mathematical foundations of theoretical statistics. *Phil. Trans. R. Soc. Lond. A,* 222: 309–368.

28. Fisher, R.A. (1925). Theory of statistical estimation. *Proc. Camb. Phil. Soc.* 22: 700–725.

29. Callaghan, P.T. (1995). *Principles of nuclear magnetic resonance microscopy.* 2nd ed. Clarendon Press, Oxford.

30. Garnier, S.J. and Bilbro, G.L. (1995). Magnetic resonance image restoration. *J. Math. Imaging Vis.* 5: 7–19.

31. Yang, G.Z., Burger, P., Firmin, D.N., and Underwood, S.R. (1995). Structure adaptive anisotropic filtering for magnetic resonance image enhancement. in *Proceedings of CAIP: Computer Analysis of Images and Patterns,* pp. 384–391.

32. Gerig, G., Kubler, O., Kikinis, R., and Jolesz, F.A. (1992). Nonlinear anisotropic filtering of MRI data. *IEEE Trans. Med. Imaging.* 11(2): 221–232.

33. Bernstein, M.A., Thomasson, D.M. and Perman, W.H. (1989). Improved detectability in low signal-to-noise ratio magnetic resonance images by means of phase-corrected real construction. *Med. Phys.* 16(5): 813–817.

34. McGibney, G. and Smith, M.R. (1993). An unbiased signal-to-noise ratio measure for magnetic resonance images. *Med. Phys.* 20(4): 1077–1078.

35. Miller, A.J. and Joseph, P.M. (1993). The use of power images to perform quantitative analysis on low SNR MR images. *Magn. Reson. Imaging,* 11: 1051–1056.

36. Sijbers, J., den Dekker, A.J., Van Dyck, D., and Raman, E. (1998). Estimation of signal and noise from Rician distributed data. in *Proceedings of the International Conference on Signal Processing and Communications,* Gran Canaria, Canary Islands, Spain.

37. Sijbers, J. and den Dekker, A.J. (2004). Maximum likelihood estimation of signal amplitude and noise variance from MR data. *Magn. Reson. Med.* 51(3): 586–594.

38. Bonny, J.M., Renou, J.P., and Zanca, M. (1996). Optimal measurement of magnitude and phase from MR data. *J. Magn. Reson.* 113: 136–144.

39. Kaufman, L., Kramer, D.M., Crooks, L.E., and Ortendahl, D.A. (1989). Measuring signal-to-noise ratios in MR imaging. *Radiology* 173: 265–267.

40. Murphy, B.W., Carson, P.L., Ellis, J.H., Zhang, Y.T., Hyde, R.J., and Chenevert, T.L. (1993). Signal-to-noise measures for magnetic resonance. *Magn. Reson. Imaging.* 11: 425–428.

41. Poston, T. and Stewart, I.N. (1980). *Catastrophe theory and its applications.* Pitman, London.

42. Pintelon, R., Schoukens, J., and Renneboog, J. (1988). The geometric mean of power (amplitude) spectra has a much smaller bias than the classical arithmetic (RMS) averaging. *IEEE Trans. Instrum. Meas.* 37(2): 213–218.

43. Olsen, S.I. (1993). Estimation of noise in images: an evaluation. *Graph. Model. Im. Proc.* 55(4): 319–323.

44. Close, R.A. and Whiting, J.S. (1996). Maximum likelihood technique for blind noise estimation. in *Proceedings of SPIE Medical Imaging,* Newport Beach, CA.

45. Lee J.S. (1981). Refined filtering of image noise using local statistics. *Comput. Vis. Graph.* 15: 380–389.

46. Mastin, G.A. (1985). Adaptive filters for digital image smoothing, an evaluation. *Comput. Vis. Graph.* 31: 103–121.

47. Lee, J.S. and Hoppel, K. (1989). Noise modeling and estimation of remotely-sensed images. in Proceedings of the IEEE International Geoscience and Remote Sensing Symposium, Vancouver.

48. Meer, P., Jolion, J., and Rosenfeld, A. (1990). A fast parallel algorithm for blind noise estimation of noise variance. *IEEE Trans. Pattern Anal. Machine Intell.* 12(2): 216–223.

49. De Wilde, J.P., Hunt, J.A., and Straughan, K. (1997). Information in magnetic resonance images: evaluation of signal, noise and contrast. *Med. Biol. Eng. Comput.* 35: 259–265.

50. Sano, R.M. (1988). *MRI: Acceptance testing and quality control-the role of the clinical medical physicist.* Medical Physics Publishing Corporation, Madison, WI.

51. Firbank, M.J., Coulthard, A., Harrison, R.M., and Williams, E.D. (1999). A comparison of two methods for measuring the signal-to-noise ratio on MR images. *Phys. Med. Biol.,* 44: 261–264.

52. Seber, G.A.F. and Wild, C.J. (1989). *Nonlinear regression.* John Wiley & Sons, New York.

53. Mood, A.M., Graybill, F.A., and Boes, D.C. (1974). *Introduction to the theory of statistics*. 3rd ed. McGraw-Hill, Tokyo.

54. Papoulis, A. (1984). *Probability, random variables and stochastic processes*. 3rd ed. McGraw-Hill, Tokyo, Japan.

55. Sijbers, J., den Dekker, A.J., Scheunders, P., and Van Dyck, D. (1998). Maximum likelihood estimation of Rician distribution parameters. *IEEE Trans. Med. Imaging.* 17(3): 357–361.

5 Retrospective Evaluation and Correction of Intensity Inhomogeneity

Martin Styner and Koen Van Leemput

CONTENTS

5.1 INTRODUCTION

Image intensity inhomogeneity, also called intensity nonuniformity (INU), in magnetic resonance imaging (MRI) is an adverse effect that affects qualitative and quantitative analysis of the images. The inhomogeneities are characterized by slowly varying intensity values of the same tissue over the image domain. The variability of the tissue intensity values with respect to image location can severely affect visual evaluation as well as the result of image processing algorithms based on absolute intensity values. The effect of the intensity inhomogeneities on the subsequent segmentation has been observed and discussed already in early applications (e.g., [1–5]; see also Figure 5.1). It has also become a highly important issue in quantitative morphometry, which is either based on volumetric measurements from segmentation or residual image differences after registration. There are numerous sources for the emergence of MRI intensity inhomogeneity. Some of them can be overcome by regular calibration, whereas many of them are different from scan to scan, for example, when inhomogeneities are due to geometry and

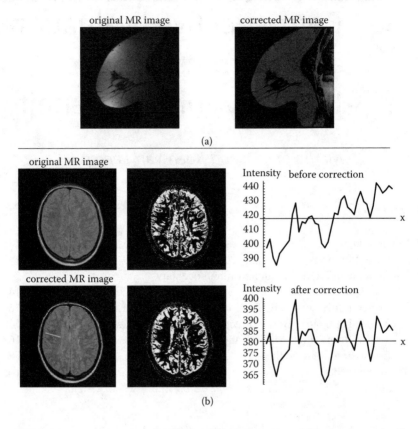

FIGURE 5.1 (a) Top row: Severe intensity distortions in breast MR mammography (from data provided by M. Brady, University of Oxford) are due to the RF surface coil and strongly hamper visual assessment. The corrected image (right) can clearly be better assessed. (b) Bottom row: The correction of subtle intensity distortions is visualized on a head MR slice from a multiple sclerosis patient (from data provided by the BIOMORPH consortium [EU BIOMED 2]). The original image and the corrected image shown in the first column are visually similar. An optimal threshold-based segmentation of the gray matter (second column) is greatly improved after correction. The clear inhomogeneity of the intensity values along a reference line that lies fully within the same tissue type (white matter) has also been removed by the correction (third column). The reference line is indicated as a yellow line in the corrected MR image.

electrical properties of the subject. Here is a list of the most common sources of MRI intensity inhomogeneity:

1. Static magnetic field $\mathbf{B}_0$ inhomogeneities are created by the technical difficulties to create a perfectly uniform field, as well as by ferromagnetic objects in the imaged object. This leads to both intensity inhomogeneities as well as spatial distortions. $\mathbf{B}_0$ inhomogeneities can be

corrected prospectively using phantoms with known reference intensities and reference points. Retrospectively, intensity inhomogeneities can be corrected with the methods presented in this chapter. It is, though, very difficult if not impossible to correct the spatial distortions retrospectively.

2. Imperfections in the gradient coils and abnormal currents through the gradient coils often lead to effects similar to those created by B_0 inhomogeneities and thus need to be corrected similarly.

3. Radio frequency (RF) coil-related effects are caused by coil imperfections, by nonuniform sensitivity of the coil (e.g., in surface coils) or by ferromagnetic objects in the imaged object (e.g., amalgam dental fillings, orthopedic implants). The main effect is a variation of the intensity values across the MR image. This type of inhomogeneity is often scan-specific and cannot be corrected prospectively using phantoms, but rather needs to be corrected retrospectively.

4. Poor RF penetration due to absorption of the RF signal leads to darker inner regions of the imaged objects compared to the outer regions. This effect can only be corrected retrospectively.

5. Minor intensity inhomogeneities of the same tissue might also be a true observation of the subject's anatomy. In general, none of the human organs or tissue types are fully homogeneous. Retrospective corrections of these subject-specific inhomogeneities can be beneficial and desired in order to enhance homogeneity within the tissue. On the other hand, the separation of inhomogeneity effects from true tissue contrast can be difficult at the border of two tissues and can lead to a subsequent incorrect segmentation[6].

6. Any of the sources listed in the preceding text can cause slowly varying intra- and interslice intensity inhomogeneities. Especially early magnetic resonance (MR) acquisition protocols, but also contemporary MR–angiography (MRA) sequences, often produce additional interslice intensity inhomogeneities that do not vary slowly, but rather in step-like fashion ("venetian blind artifact"). It is noteworthy that only a small set of "modern" methods can deal with this artifact (e.g., [7–10]).

7. Intensity variations in images of the same subject taken at different time points, as well as variations in images of different subjects, can be regarded as a special case of intensity inhomogeneity. Some of the methods presented in this chapter can correct both intrasubject as well as intersubject intensity inhomogeneity between images.

Severe intensity distortions due mainly to RF coil effects can lead to misleading visual interpretations (see Figure 5.1a). More subtle distortions that are barely visible are commonly less relevant for visual interpretation, but can still severely corrupt segmentation and registration methods (see Figure 5.1b). The task of retrospectively correcting both severe and subtle intensity inhomogeneities in MR scans has been extensively researched and discussed in the medical imaging research over the past 15 yr. This chapter gives an overview of the most important correction methods. In the next sections, we first discuss early attempts of retrospective correction, and then we present the two main classes of methods routinely used nowadays in many image processing studies.

5.2 EARLY SOLUTIONS

Starting in the mid-1980s, MRI researchers began to develop methods for the correction of intensity inhomogeneity. The need for correction was most imminent in images acquired using surface coil images with strong inhomogeneity caused by the falloff of coil sensitivity with distance from the coil center. In 1986, Haselgrove and Prammer [11] (see also Merickel et al. [3]) suggested the use of smoothing in order to reduce the inhomogeneity. They proposed that each MR slice is divided by its spatially smoothed copy. This reduces the low-frequency inhomogeneity effects of the surface coil. In order to correct for step-like interslice intensity inhomogeneities, they linearly normalized each slice to have the same average intensity values for a selected set of manually identified tissue classes visible in all slices. Smoothing was also proposed by Lim and Pfefferbaum [4] in 1989 to correct for inhomogeneities in brain MRI scans. After the manual extraction of the head, the intensity values were extended radially toward the image boundaries and smoothed with a Gaussian filter of large kernel size. They assumed that the resulting blurred image represents one homogeneous region that is only distorted by the scanner inhomogeneities. The images were corrected with this approximation of the inhomogeneity characteristics.

Many more researchers proposed different methods using smoothing filters or homomorphic unsharp masking (e.g., [12]). These early filtering methods undesirably reduce the contrast between tissues, and they often generate new artifacts in the corrected images. Homomorphic filtering also falls into the same class of correction methods [13,14]. Like smoothing methods, homomorphic filtering assumes a separation of the low-frequency inhomogeneity field from the higher frequencies of the image structures. The assumption is often valid in microscope images of small particles but often fails for the structured MR images. A scene, such as a head structure, contains a considerable amount of low-frequency components. Homomorphic filtering assumes further that the local intensity statistics are constant across the whole image. This assumption is not true for most of the inhomogeneity effects observed in MR.

In 1988, Vannier et al. [2] were one of the first to model the intensity inhomogeneity as a parametric inhomogeneity field. They used a simple linear ramp in the transverse anatomic plane as the parametric field model. This model was fitted to the linewise average intensity values of the image. Later the same year, they proposed an improved model that fitted a fourth-order polynomial to the line-by-line histogram [15]. In both approaches, only vertical distortions were taken into account. The fitted inhomogeneity field was then subtracted from the original image.

Several researchers proposed the estimation of the correction from a prior segmentation. Dawant et al. [8] proposed that users select typical samples of each tissue class in the MR image as an input to the estimation of the parametric inhomogeneity field. Tincher et al. [16] and Meyer et al. [17] present automatic techniques that fit polynomial functions to presegmented regional patches in the image. The individual fits are then combined to find an estimate for a global inhomogeneity field.

Some of the early methods assumed an additive inhomogeneity effect (e.g., [15]) whereas other methods proposed a multiplicative inhomogeneity effect (e.g., [11,16]). Nowadays, researchers agree that inhomogeneity effects in MRI are better modeled as a multiplicative effect. While a smooth multiplicative inhomogeneity effect is consistent with most characteristics of the underlying acquisition principles, it is noteworthy that minor sources of intensity inhomogeneity cannot be fully incorporated by this model [18]. Additive inhomogeneity effects are rarely observed in MRI, but can be seen in images acquired with other imaging means, such as confocal microscopy.

5.3 COMBINED SEGMENTATION AND INHOMOGENEITY CORRECTION METHODS

Whereas inhomogeneity correction methods are often needed to obtain good segmentations, the early approaches of Dawant, Tincher, and Meyer indicate that a good segmentation in turn facilitates inhomogeneity estimation. Observing this dependency between segmentation and inhomogeneity correction, the idea emerged to solve both problems simultaneously using an iterative approach in which increasingly accurate inhomogeneity corrections yield increasingly accurate segmentations and vice versa. Although other techniques exist [19,20], a landmark paper in this respect was published by Wells et al. [7], described in the following text.

Assuming a multiplicative inhomogeneity field model, Wells et al. logarithmically transformed the MR intensity data in order to make the inhomogeneity an additive artifact. Let $y_i = \ln(z_i)$ denote the log-transformed intensity at voxel i, where z_i is the voxel's measured MRI signal intensity.* Assuming that there are K different tissue types present in the image, and that the log-transformed intensity distribution of each of these tissues can be modeled by a normal distribution after taking the inhomogeneity field model into account, we have

$$p(y_i \mid l_i, \beta_i) = G_{\sigma_{l_i}}(y_i - \mu_{l_i} - \beta_i) \tag{5.1}$$

where $l_i \in \{1, 2, \ldots, K\}$ and β_i denote the tissue type and the inhomogeneity field at the ith voxel, respectively, μ_k denotes the mean intensity of tissue type k, σ_k^2

---

* For the sake of simplicity, only a single intensity value per voxel is assumed, although the technique readily applies to multispectral MRI data.

denotes the variance around that mean for tissue k, and

$$G_\sigma(x) = \frac{1}{\sqrt{2\pi\sigma^2}} exp\left(-\frac{x^2}{2\sigma^2}\right)$$

is a zero-mean Gaussian distribution with variance σ^2. It is further assumed that the prior (before the image data is seen) probability for a certain tissue type k in the ith voxel is identical for all voxels

$$p(l_i = k) = \pi_k \tag{5.2}$$

where π_k is assumed to be known a priori. Finally, knowledge about the spatial smoothness of MR inhomogeneities is incorporated by modeling the entire inhomogeneity field, denoted as $\beta = [\beta_1, \beta_2, ..., \beta_N]^T$, where N is the number of voxels in the image, by a N-dimensional zero-mean Gaussian prior probability density

$$p(\beta) = G_{\Sigma_\beta}(\beta) \tag{5.3}$$

where

$$G_\Sigma(x) = \frac{1}{\sqrt{(2\pi)^N \det(\Sigma)}} exp\left(-\frac{1}{2}x^T\Sigma^{-1}x\right).$$

Using Bayes' rule, the posterior probability of the inhomogeneity field in a given log-transformed intensity image $y = [y_1, y_2, ..., y_N]^T$ is given by the distribution

$$p(\beta|y) = \frac{p(y|\beta)p(\beta)}{\int_{\beta'} p(y|\beta')p(\beta')} \tag{5.4}$$

where $p(\beta)$ depends on the parameter Σ_β through Equation 5.3, and where

$$p(y|\beta) = \prod_i p(y_i|\beta_i)$$

with

$$p(y_i|\beta_i) = \sum_k p(y_i|l_i = k, \beta_i)p(l_i = k) \tag{5.5}$$

depends on the parameters μ_k, σ_k, and π_k through Equation 5.1 and Equation 5.2. Before Equation 5.4 can be used to assess the inhomogeneity field, appropriate values must be chosen for each of these parameters. Wells et al. determined the tissue-specific means μ_k and variances σ_k^2 by manually selecting representative voxels of each of the tissue types considered in a prototypical image and calculating sample means and variances. Once this training phase was performed, the same model parameters were used to analyze thousands of images acquired with the same MRI acquisition protocol. For the prior probabilities π_k, both uniform and nonuniform distributions were used. The exact choice of the $N \times N$ covariance matrix Σ_β is a difficult issue and will be discussed later.

Having obtained the posterior probability on the inhomogeneity field, Wells et al. used the maximum a posteriori (MAP) principle to formulate an estimate of the inhomogeneity field as the value of β having the largest posterior probability

$$\hat{\beta} = \arg\max_\beta p(\beta \,|\, y),$$

which is equivalent to

$$\hat{\beta} = \arg\max_\beta \log p(\beta \,|\, y).$$

By requiring the gradient of $\log p(\beta \,|\, y)$ with respect to β_i to be zero for all voxels, Wells et al. derived the following equations for the inhomogeneity field estimation:

$$\hat{\beta} = Hr \tag{5.6}$$

with

$$r = \begin{bmatrix} y_1 - \tilde{y}_1 \\ y_2 - \tilde{y}_2 \\ \vdots \end{bmatrix}, \quad \tilde{y}_i = \frac{\sum_k w_{ik} \mu_k}{w_i}, \quad w_i = \sum_k w_{ik}, \quad w_{ik} = p_{ik}/\sigma_k^2$$

where

$$p_{ik} = p(l_i = k \,|\, y_i, \beta_i)$$

$$= \frac{p(y_i \,|\, l_i = k, \beta_i) p(l_i = k)}{\sum_{k'} p(y_i \,|\, l_i = k', \beta_i) p(l_i = k')} \tag{5.7}$$

and H is a linear operator defined by

$$H = \left[I + W^{-1} \Sigma_\beta^{-1} \right]^{-1}, \quad W = \begin{bmatrix} w_1 & 0 & \cdots \\ 0 & w_2 & \cdots \\ \vdots & \vdots & \ddots \end{bmatrix}.$$

These equations can be interpreted as follows (see Figure 5.2). Given an estimate of the inhomogeneity field, a statistical classification of the image voxels is performed by calculating the posterior tissue-class probabilities p_{ik} (Figure 5.2b). From this classification and the Gaussian distribution parameters, a prediction $\tilde{y}$ of the MR intensities without the inhomogeneity field is reconstructed (Figure 5.2c). A residue image r is obtained by subtracting this predicted signal from the original image, giving a local estimate of the inhomogeneity field in each voxel (Figure 5.2d), along with a weight image w that reflects confidence of these local estimates (Figure 5.2e). Finally, the inhomogeneity field β is estimated by applying the linear operator H, which, as will be shown below, results in smoothing the residue image r while taking the weights w into account (Figure 5.2f).

Recall that the exact choice of the covariance matrix Σ_β, governing the prior distribution of the inhomogeneity field, has not been specified so far. In general, this matrix is impracticably large, making the evaluation of Equation 5.6 computationally intractable. In the case of equal covariances σ_k^2 for all tissue types, the confidence weights w_i are constant across the image, and Wells et al. showed that Σ_β can be chosen so that the linear operator H in Equation 5.6 simplifies to a shift-invariant linear low-pass filter. In general, however, this is not the case, and the authors heuristically use a computationally efficient approximation by low-pass-filtering the voxelwise product of the weights and the residues, and dividing the result by the low-pass-filtered version of the weights:

$$\hat{\beta}_i \simeq \frac{[FWr]_i}{[Fw]_i}$$

with F a low-pass filter.

It should be noted that the inhomogeneity field estimation involves knowledge of the tissue classification (Equation 5.6), but that this tissue classification in turn requires knowledge of the inhomogeneity field (Equation 5.7). Intuitively, both equations can be solved simultaneously by iteratively alternating between these two steps. Indeed, it can be shown that such an iterative approach is an instance of the so-called expectation-maximization (EM) algorithm [21], often used in estimation problems where some of the data is "missing." In this case, the missing data is the tissue type of each voxel; if these were known, estimation of the inhomogeneity field would be straightforward. In effect, the algorithm of Wells et al. iteratively fills in the missing tissue types based on the current inhomogeneity field estimation during the E step (Equation 5.7) and updates the inhomogeneity field accordingly during the M step (Equation 5.6); cf. Figure 5.3. Such an EM scheme guarantees increasingly better estimates of the inhomogeneity field with respect to Equation 5.4 at each iteration [22], although there is no guarantee of finding the global optimum. The iterative EM process is typically started on the E step, using a flat initial inhomogeneity field, although Wells et al. also reported results by starting on the M step using equal tissue-class probabilities. The authors reported

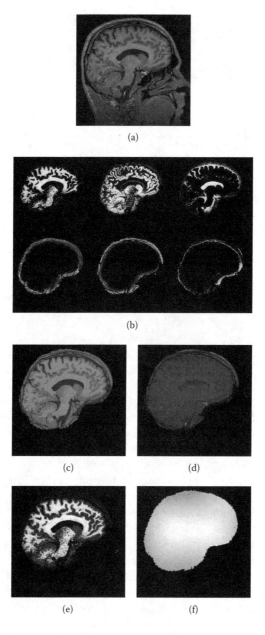

FIGURE 5.2 Illustration of the inhomogeneity field estimation component of the algorithm of Wells et al. [7]: (a) original image, (b) intermediate classification, (c) reconstructed image, (d) residue image, (e) weights, and (f) estimated inhomogeneity field. Although the calculations are performed on logarithmically transformed intensities, the reconstructed image, the residue image, and the inhomogeneity field are displayed in the original intensity domain for visualization purposes.

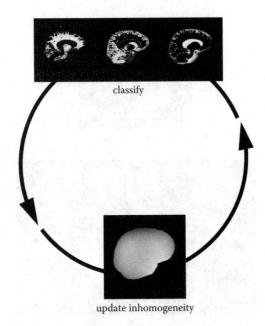

classify

update inhomogeneity

FIGURE 5.3 In the EM approach of Wells et al., inhomogeneity field estimation is repeatedly interleaved with tissue classification, yielding improved results at each iteration.

that the estimates typically stabilize in five to ten iterations, after which the algorithm is stopped. Upon convergence, an inhomogeneity field corrected image can be obtained by subtracting the estimated inhomogeneity field from the log-transformed intensities y and transforming the result back into the original MR intensity domain using the exponential transformation.

Although Wells et al. showed excellent results on MR images of the brain, a number of shortcomings were quickly identified, spawning a considerable amount of papers aiming at improving upon the original method. One line of research has concentrated on improving the segmentation results produced by the algorithm, in particular, by employing so-called Markov random field (MRF) models to minimize the effects of noise in the resulting segmentations [23–29]. While a detailed treatise of these methods is outside the scope of this text, suffice it here to say that they encourage neighboring voxels to be classified to the same tissue type, at the expense of making the E step computationally intractable, necessitating approximative solutions. Another area of improvement upon the original Wells algorithm involves the manual selection of a number of representative points for each of the tissues considered in order to estimate appropriate values of the Gaussian distribution parameters μ_k and σ_k^2. Although it is straightforward to estimate these parameters for large well-defined regions such as white and gray matter in brain MRI, other regions consist of several different types of tissue, some of which are easily overlooked during training. Guillemaud and

Brady [30] noted that tissue types for which the algorithm was not properly trained cause severe errors in the residue image and, therefore, corrupt the inhomogeneity field estimation. They proposed to account for this by replacing the collection of Gaussian distribution models for such "difficult" tissue types by one single "other" probability distribution that is uniform over the set of intensities of voxels in the image. In particular, Equation 5.5 is replaced by

$$p(y_i \mid \beta_i) = \sum_{k \text{ is Gaussian}} p(y_i \mid l_i = k, \beta_i) p(l_i = k) + \lambda \cdot p(l_i = other) \quad (5.8)$$

where λ is a small constant and $p(l_i = other)$ is the a priori probability for tissue type "other." This new class can be seen as a rejection class, gathering all the pixels with intensities far from the Gaussian distributions of well-defined tissue classes for which the model parameters can be easily estimated during the manual training phase. Deriving the EM equations for this new model yields an iterative algorithm that is identical to the original Wells algorithm, except that the inhomogeneity field is only estimated with respect to the Gaussian classes: voxels classified to the uniform rejection class have zero weight w_i in the inhomogeneity field estimation. In other words, the inhomogeneity field is estimated from regions with Gaussian tissue types and diffused through to regions of class "other."

While the approach of Guillemaud and Brady reduces the risk of mistraining the classifier by concentrating on a small number of tissue classes that are well-defined, it does not address the inherent limitations associated with such a manual training phase. Indeed, the selection of appropriate training samples is somewhat subjective, precluding full reproducibility of the results. Furthermore, the cumbersome process of training the classifier needs to be repeated for each new set of similar scans to be processed. To solve these issues, Van Leemput et al. [9] proposed to have the classifier automatically train itself on each individual scan. Rather than assuming that the Gaussian distribution parameters are known in advance, as Wells et al. do, they formulated the inhomogeneity field estimation problem as a joint inhomogeneity field and Gaussian distribution parameter estimation problem. In their approach, the smoothness of MR inhomogeneity fields is incorporated by explicitly modeling the inhomogeneity field as a linear combination of P smooth basis functions $\psi_p(x_i), p = 1, 2, \dots, P$ where x_i denotes the spatial position of the i th voxel*

$$\beta_i = \sum_p c_p \psi_p(x_i)$$

where $c_p, p = 1, 2, \dots, P$ are the (unknown) inhomogeneity field coefficients. Letting $\Phi = \{\mu_k, \sigma_k^2, c_p, k = 1, 2, \dots, K, p = 1, 2, \dots, P\}$ denote the total set of parameters

---

* Van Leemput et al. used polynomial basis functions, but the theory is valid for any kind of smooth basis functions such as splines.

in the model, Van Leemput et al. aimed at finding the MAP model parameters

$$\hat{\Phi} = \arg \max_{\Phi} p(\Phi \,|\, y).$$

In the absence of prior knowledge about the parameters Φ, a uniform prior distribution $p(\Phi)$was assumed, reducing the MAP principle to a maximum-likelihood (ML) approach

$$\hat{\Phi} = \arg \max_{\Phi} p(y \,|\, \Phi).$$

Using a zero-gradient condition on the logarithm of $p(y \,|\, \Phi)$, Van Leemput et al. derived an EM algorithm that iteratively repeats three consecutive steps. In the E step, the image voxels are classified based on the current inhomogeneity field and Gaussian distribution estimates, using Equation 5.7. The M step involves two separate steps, in which first the inhomogeneity field is updated according to the current classification and Gaussian distribution parameters, followed by an update of the Gaussian distributions. The inhomogeneity field estimation is given by

$$\begin{bmatrix} \hat{c}_1 \\ \hat{c}_2 \\ \vdots \end{bmatrix} = (A^T W A)^{-1} A^T W r \quad \text{with} \quad A = \begin{bmatrix} \psi_1(x_1) & \psi_2(x_1) & \psi_3(x_1) & \cdots \\ \psi_1(x_2) & \psi_2(x_2) & \psi_3(x_2) & \cdots \\ \vdots & \vdots & \vdots & \ddots \end{bmatrix}.$$

where the weight matrix W and the residue image r are the same as in the Wells algorithm. In other words, the inhomogeneity field is estimated as a weighed least-squares fit to the residue r (Figure 5.2d), using the confidence weights w (Figure 5.2e).* The Gaussian distribution parameters are updated according to

$$\mu_k = \frac{\sum_i p_{ik}(y_i - \beta_i)}{\sum_i p_{ik}}$$

and

$$\sigma_k^2 = \frac{\sum_i p_{ik}(y_i - \beta_i - \mu_k)^2}{\sum_i p_{ik}}$$

---

* Note that by modeling the inhomogeneity field as a linear combination of smooth basis functions, rather than by the Gaussian prior proposed by Wells et al., no approximations are required to compute the inhomogeneity estimation.

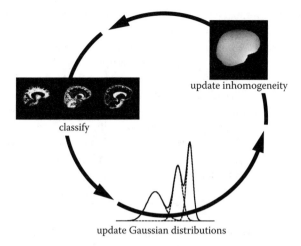

update inhomogeneity

classify

update Gaussian distributions

FIGURE 5.4 Rather than using a manually trained classifier, as Wells et al., do, Van Leemput et al., automatically train the classifier during the iterative EM procedure. The resulting three-step algorithm adapts to each individual scan to be analyzed, allowing to process images obtained with a variety of acquisition protocols without user intervention.

which simply states that the updated mean and variance estimations for class k are given by the sample mean and sample variance of the inhomogeneity-corrected intensities of voxels classified as tissue type k.

To summarize, the algorithm of Van Leemput et al. iteratively alternates between tissue classification, inhomogeneity field estimation, and retraining of the classifier (see Figure 5.4). The iterative scheme is initialized by setting the inhomogeneity field coefficients to zero (no inhomogeneity) and providing a first rough estimate of the tissue-class probabilities p_{ik}, allowing start of the iterative EM scheme with the Gaussian distribution parameter estimation step. The initial class probability estimates are given by prealigning the image under study with a so-called atlas that contains information about the expected location of the tissue types of interest in a normal population. The alignment is performed fully automatically by maximizing the mutual information [31,32] between the image under study and an anatomical template associated with the atlas, which works irrespective of the tissue characteristics in the images. Because the classifier is additionally trained automatically, images acquired with a previously unseen MR sequence can readily be analyzed without requiring user intervention. Van Leemput et al. originally applied their method to brain MRI,* but Lorenzo-Valdés et al. recently extended the technique to analyze 4-D cardiac MR images [33]. A similar algorithm for brain MRI was developed independently by Ashburner et al.** working on the original MR intensities rather than on log-transformed intensities, and using a linear

---

* The software of Van Leemput et al. is freely available under the name EMS (expectation-maximization segmentation) at http://www.medicalimagecomputing.com/EMS
** Freely available as part of the SPM99 package at http://www.fil.ion.ucl.ac.uk/spm/spm99.html.

combination of low-frequency discrete cosine transformation (DCT) basis functions as the inhomogeneity field model [34].

In parallel with EM-based approaches, a number of methods based on extensions of the K-means algorithm [35,36] were developed that also iteratively alternate between segmentation and inhomogeneity field correction. In order to explain the original K-means algorithm, let us ignore the inhomogeneity field artifact for now, and model MR images as consisting of K tissue types, each governed by a Gaussian intensity distribution with mean μ_k and a variance that is assumed equal for all tissue types. Both the tissue labels $l = [l_1, l_2, \ldots, l_N]^T$ and the tissue means $\mu = [\mu_1, \mu_2, \ldots, \mu_K]^T$ can be estimated simultaneously from an image y using the MAP principle

$$\arg\max_{\mu, l} \log p(\mu, l \mid y).$$

Assuming a uniform prior for the tissue means and for the configuration of the tissue labels in the image, this is equivalent to minimizing the objective function

$$Q_1(\mu, u \mid y) = \sum_i \sum_k u_{ik}(y_i - \mu_k)^2 \qquad (5.9)$$

where

$$u_{ik} = \begin{cases} 1 & \text{if } l_i = k \\ 0 & \text{otherwise.} \end{cases},$$

Equation 5.9 can be optimized by iteratively performing a crisp classification that assigns each voxel exclusively to the tissue type that best explains its intensity

$$u_{ik} = \begin{cases} 1 & \text{if } |y_i - \mu_k| < |y_i - \mu_j|, j = 1, \ldots, K, j \neq k \\ 0 & \text{otherwise.} \end{cases} \qquad (5.10)$$

and updating the tissue means accordingly

$$\mu_k = \frac{\sum_i u_{ik} y_i}{\sum_i u_{ik}}. \qquad (5.11)$$

A well-known generalization of the K-means algorithm is the fuzzy C-means algorithm [37,38], which aims at minimizing*

$$Q_m(\mu, u \mid y) = \sum_i \sum_k (u_{ik})^m (y_i - \mu_k)^2$$

$$\text{with} \quad m \in [1, \infty), \, u_{ik} \in [0,1], \quad \text{and} \quad \sum_k u_{ik} = 1. \qquad (5.12)$$

---

* Unequal tissue-specific covariances can also be taken into account.

For $m = 1$, the fuzzy C-means algorithm devolves to the K-means algorithm. For $m > 1$, the fuzzy C-means algorithm alternates between a fuzzy segmentation

$$u_{ik} = \left[\sum_j \left(\frac{|y_i - \mu_k|}{|y_i - \mu_j|} \right)^{2/(m-1)} \right]^{-1} \tag{5.13}$$

and an update of the tissue means

$$\mu_k = \frac{\sum_i (u_{ik})^m y_i}{\sum_i (u_{ik})^m}. \tag{5.14}$$

Whereas the classification step in the K-means algorithm uniquely assigns each voxel to one single tissue type, the fuzzy C-means algorithm calculates for each voxel fuzzy membership values in each of the tissue types. The parameter m is a weighting exponent that regulates how fuzzy those membership values are; typically, $m = 2$ is used.

In [39], Pappas extended the K-means algorithm to account for local intensity variations by letting the tissue means gradually vary over the image area.* Rather than using one single mean intensity μ_k for every tissue type throughout the image, a different mean μ_k^i is used in every voxel i, which is calculated by performing the averaging operation in Equations 5.11 over a sliding window rather than over the whole image area:

$$\mu_k^i = \frac{\sum_{i \in W_i} u_{ik} y_i}{\sum_{i \in W_i} u_{ik}}$$

with W_i, a window centered at voxel i. This approach, and variations upon it, was applied to MR images of the brain by a number of authors [40–44]. Variations include describing the spatially varying mean intensities by B-splines [41], using the fuzzy C-means algorithm instead of the K-means algorithm [42], calculating spatially varying tissue covariances as well [43], and replacing the crisp tissue classification by a partial volume estimation [44]. One notable aspect of these approaches is that the mean intensities of the tissue types are allowed to vary independently of one another, i.e., no continuity of the inhomogeneity field over the tissue boundaries is assumed. As a result, in places where the number of voxels assigned to class k is small, the mean μ_k cannot be reliably estimated, and appropriate precautions must be taken.

---

* Pappas also took an MRF model into account, but that is outside the scope of this text.

More recently, a number of authors have also used a single inhomogeneity field model that is continuous over tissue boundaries within the K-means/fuzzy C-means framework. Assuming that tissue-specific means and covariances are determined in advance, Rajapakse and Kruggel iteratively alternated between a crisp classification and an inhomogeneity field estimation step, the latter being performed by averaging local inhomogeneity field estimates with a sliding window [45]. Pham and Prince extended the fuzzy C-means algorithm with an explicit inhomogeneity field model, by optimizing the objective function

$$Q_{AFCM}(\mu,u,\beta \mid y) = \sum_i \sum_k (u_{ik})^m (y_i - \beta_i \mu_k)^2 + Q(\beta) \qquad (5.15)$$

where β is the inhomogeneity field and $Q(\beta)$ is a regularization term that ensures that the inhomogeneity field is spatially smooth and slowly varying [46]. Optimizing Equation 5.15 with respect to the membership values u, tissue means μ and inhomogeneity field β yields a 3-step algorithm that iteratively alternates between a fuzzy tissue classification, estimation of the mean intensities, and inhomogeneity field estimation. Whereas the mathematical aspects differ significantly from the EM-based method proposed by Van Leemput et al., described earlier, both techniques are conceptually very similar. A similar 3-step approach, based on an extension of the fuzzy C-means algorithm that takes interactions between neighboring voxels into account, was described by Ahmed et al. [47].

5.4 CORRECTION BASED ON EVALUATION OF THE HISTOGRAM

In contrast to the methods discussed in the previous section, the methods discussed in this section do not perform a segmentation of the image but rather base their computations mainly on the histogram of the image. All of these methods formulate the intensity inhomogeneity correction as an optimization problem based on a metric computed directly from the histogram of the corrected image (see Figure 5.5).

5.4.1 PARAMETRIC BIAS CORRECTION

We first discuss the parametric bias correction (PABIC) method [10,48] because this method can be regarded as a hybrid between the combined segmentation/correction methods and the other histogram-based methods.* While no explicit segmentation is computed, a prior parametric model for all tissue classes in the image is a necessary parameter of the histogram metric.

---

* PABIC is part of the National Library of Medicine Insight Segmentation and Registration Toolkit (ITK), an open-source software system available at http://www.itk.org.

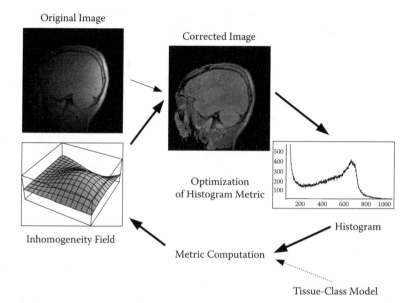

FIGURE 5.5 Schematic illustration of histogram-based inhomogeneity correction methods. In an optimization process (red arrows), the shape of the inhomogeneity field is adapted such that the chosen metric computed from the histogram of the corrected image is minimal. For the case of the PABIC method, an additional prior tissue-class model is needed for the computation of the histogram metric (dashed arrow).

PABIC assumes that images are composed of regions with piecewise constant intensities of mean values μ_k and standard deviation σ_k. Each voxel of the idealized signal corrupted by noise must take values close to one of these class means. These assumptions are violated by effects like partial voluming and the natural inhomogeneity of biological tissue. To account for these violations, PABIC incorporates a robust M-estimator [49] function f_k (16) into the histogram metric e_{tot} (17) (see Figure 5.6). Thanks to the robust estimator,

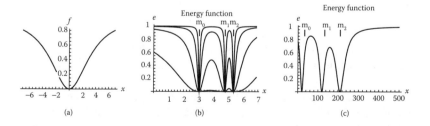

FIGURE 5.6 (a) Robust estimator function (Equation 5.16) for $\mu = 0$ and $\sigma = 2$. (b) Three-class energy function e (Equation 5.17) for $\sigma_1 = \sigma_2 = \sigma_3 = 0.03, 0.1, 0.3$ and 1.0, from top to bottom respectively. (c) Example energy function e from a head MRI correction.

most violations of the class model do not result in incorrect solutions. Failure of the class model was only observed in very low resolution MR datasets, where as much as 30% of partial volume voxels can be present. In this case, an additional preprocessing step identifies partial voluming voxels using a dilated edge filter. The identified partial volume voxels are then excluded from further computations.

$$f_k(x) = \frac{x^2}{x^2 + 3\sigma_k^2} \tag{5.16}$$

$$e_{tot} = \sum_{y_i} e(y_i) = \sum_{y_i} \prod_{k=1..K} f_k(y_i - \beta_i - \mu_k) \tag{5.17}$$

Both additive and multiplicative inhomogeneity fields are modeled as a parametric inhomogeneity field using Legendre polynomials ψ_L as basis functions in x, y, and z. In case of a multiplicative inhomogeneity field, all computations are performed in log-space. For Legendre polynomials up to the degree l, the size m of the parameter vector c_{ijk} is given by $m = (l+1)\frac{(l+2)}{2}\frac{(l+3)}{3}$. For instance, Legendre polynomials up to the third degree would therefore require 20 coefficients. The choice of the maximal degree of Legendre polynomials largely depends on prior knowledge of the coil and the expected type and smoothness of the inhomogeneity field. The inhomogeneity field estimate β_i is determined as follows:

$$\beta_i = \sum_{i=0}^{l} \sum_{j=0}^{(l-i)} \sum_{k=0}^{(l-i-j)} c_{ijk}\psi_{L,i}(x_i)\psi_{L,j}(y_i)\psi_{L,k}(z_i) \tag{5.18}$$

with $\psi_{L,i}(.)$ denoting a Legendre polynomial of degree i.

Finding the parameter vector c_{ijk} with minimum energy e_{tot} is a nonlinear optimization problem, independent of the type of inhomogeneity field and energy function. In principle, any nonlinear optimization method could be applied. PABIC uses an adapted version of the (1 + 1)-evolution strategy (ES), which belongs to the family of evolutionary algorithms (for an introduction see [50]). This method adjusts locally the search direction and step size, and provides a mechanism to step out of nonoptimal minima. Furthermore, the method is fast enough to cope with the large data sets and overcomes the problem of parameters with different scaling.

The optimal polynomial parameters correct the original MR image so that the intensity statistics of the corrected image fit the given class model optimally. This property of PABIC can also be employed for normalizing the intensity statistics for different images. A slice-by-slice intensity normalization is necessary when venetian blind artifacts are present in the image (see Section 5.1). A volume-by-volume intensity normalization is necessary when analyzing the absolute intensities of MRI time series of the same subject. For example, PABIC was used as a required preprocessing step for a method developed at ETH Zürich to segment and characterize multiple sclerosis lesions based on dynamic changes over time [51].

Figure 5.1 shows the results of a MR inhomogeneity correction using PABIC. PABIC cannot only be applied to MR images but to any kind of image data satisfying the initial assumptions of piecewise constant intensity regions. The algorithm was tested successfully on different kinds of microscopy images and biological scenes measured by a video camera.

5.4.2 Information Minimization and N3

The methods discussed in this section do not require any explicit model of the intensities or the spatial distributions of the different classes present in an MR image. This property stabilizes the correction method against pathological intensity distributions in MR data that might violate the class model. Known intensity pathologies, such as multiple sclerosis lesions in brain MR images, are handled without adaptions.

Paul Viola [52] was the first to propose in his Ph.D. thesis the use of a criterion based on information theory. He proposed to minimize the information content of the histogram. Other image processing applications such as image restoration and classification have successfully employed information minimization. The information I of an image z can be quantitatively computed using the Shannon entropy $H(z)$ as

$$I(z) = H(z) = -\sum_i p(i) \cdot \log p(i) \qquad (5.19)$$

with $p(n)$ denoting the probability that an element of image z has intensity i. The value of $H(z)$ is positive and $H(z)$ is maximal when all intensities have the same probability. Thus, the main assumption of the information minimization approach is that the histogram shows thin "spiky" unimodal intensity distributions. If this assumption is satisfied, then the information of a perfect MR image is minimal. Intensity inhomogeneities in the MR image result in spreading the class distributions, leading to a more uniform histogram with higher information content.

The information minimization idea was then picked up by several researchers such as Mangin [18], Likar et al. [53], and many more.* Mangin models a multiplicative inhomogeneity field using B-splines, whereas Likar's method incorporates both a multiplicative and an additive inhomogeneity field using orthonormal polynomial basis functions. Likar's method is currently probably the most versatile histogram-based method.

Based on similar assumptions, John Sled [54,55] proposed another histogram-sharpening approach called nonparametric, nonuniform intensity normalization

---

* The SPM2 package freely available at http://www.fil.ion.ucl.ac.uk/spm/spm2.html contains an implementation of intensity inhomogeneity correction via information minimization.

(N3).* This signal processing motivated approach takes advantage of the usually simple form of the tissue-class distributions. It searches for a smooth inhomogeneity field that maximizes the frequency content of the image intensity distribution. In order to achieve this, the method iteratively deconvolves narrow Gaussian distributions from the image intensity distribution.

The properties of both information minimization and N3 are quite similar, whereas their implementations vary quite a bit. Both methods have a minimal set of parameters and are sometimes also called "nonparametric." The remaining parameters mainly control the smoothness of the parametric inhomogeneity field and the histogram computation. The most relevant parameter of histogram-based methods is the size of the histogram bins (in the discrete case for information minimization), or the variance of the kernel (in the continuous case for N3). Choosing this parameter too low will result in a flat inhomogeneity field estimation, whereas choosing it too high will basically low-pass-filter the image to produce an inhomogeneity field estimation.

Both information minimization methods and N3 perform well on standard MR images, and have proven to be very useful in small and large MR studies. However, N3 is reported to be less suited than information minimization for images with large-scale structures [53,55].

5.5 DISCUSSION AND CONCLUSION

In this chapter, we reviewed the large body of literature concerning the retrospective evaluation and correction of intensity inhomogeneities in MRI. The proposed approaches range from early solutions, through combined segmentation and inhomogeneity correction methods, to histogram-based techniques. Whereas the early methods were instrumental in advancing the state of the art in the field, they are now largely abandoned because they typically need a high degree of user interaction (e.g., for masking or tissue labeling), and are often based on inadequate assumptions (e.g., additive inhomogeneity models or single homogeneous region assumptions after large kernel size smoothing). In contrast, both histogram-based techniques and combined segmentation and inhomogeneity correction methods are currently being used as part of everyday image processing pipelines in institutions around the world.

Whether the histogram-based approach is to be preferred over the combined segmentation and inhomogeneity correction approach or vice versa remains the subject of continuous debate. On the one hand, approaches that explicitly segment images while estimating the inhomogeneity field have the distinct advantage that inhomogeneities are estimated using extensive domain information, rather than using voxel intensities alone. Furthermore, the goal of inhomogeneity correction in image processing pipelines is typically to obtain accurate image segmentations, anyway, in which case solving the segmentation and the inhomogeneity estimation problem simultaneously makes perfect sense. However, combined segmentation and inhomogeneity correction approaches rely heavily on the availability of accurate image

---

* *N3 is freely available at http://www.bic.mni.mcgill.ca/software/distribution.

models, which restricts their applicability to the specific image processing scenarios they were designed for. For instance, the number of tissue types present in the images must be specified in advance, and appropriate intensity models have to be provided for each of the tissues considered. Histogram-based techniques, on the other hand, do not depend on such models, and can therefore be more generally applied, for instance, to correct images of different anatomical areas or images of pathological cases that might otherwise violate the model assumptions. It should be noted, however, that histogram-based techniques typically require the removal of undesired structures, such as air and nonbrain tissues in MR images of the head, prior to the inhomogeneity estimation. As the employed tissue removal steps are often tailored to specific body regions and acquisition protocols, applying histogram-based methods to different types of images does require some adaptation in practice. Furthermore, it remains to be seen how robust histogram-based techniques are with respect to intensity variations due to pathology such as brain lesions, whose intensities vary across the range of nonpathological tissue intensities. Whereas combined segmentation and inhomogeneity estimation methods can explicitly take such pathological tissues into account, either by incorporating them in their model or by detecting them as model outliers [24], histogram-based approaches treat all voxels alike, which may lead to erroneous inhomogeneity field estimations in such cases.

As a final remark, the methods presented in this chapter can be mainly applied to structural MR imaging and MR angiography. Images acquired with the more recently introduced diffusion tensor imaging (DTI) sequences remain currently uncorrected, and no methods have been proposed thus far.

ACKNOWLEDGMENTS

Both authors Koen Van Leemput and Martin Styner contributed equally to this chapter. The authors would like to thank many colleagues and friends in the medical image processing community for the great discussions and collaborations in regard to intensity inhomogeneity correction. Especially they would like to mention Guido Gerig, Christian Brechbühler, Gábor Székely, Mike Brady, Frederik Maes, Dirk Vandermeulen, Alan Colchester, and Nicholas Ayache.

REFERENCES

1. Jungke, M., von Seelen, W., Bielke, G., Meindl, S. et al. (1987). A system for the diagnostic use of tissue characterizing parameters in NMR-tomography. in *Proceedings of Information Processing in Medical Imaging,* IPMI'87, Vol. 39, pp. 471–481.

2. Vannier, M.W., Speidel, Ch.M., Rickman, D.L., Schertz, L.D. et al. (November 1988). Validation of magnetic resonance imaging (MRI) multispectral tissue classification. in *Proceedings of 9th International Conference on Pattern Recognition,* ICPR'88, pp. 1182–1186.

3. Merickel, M.B., Carman, C.S., Watterson, W.K., Brookeman, J.R., and Ayers, C.R. (November 1988). Multispectral pattern recognition of MR imagery for the noninvasive analysis of atherosclerosis. in *Proceedings of 9th International Conference on Pattern Recognition,* ICPR'88. pp. 1192–1197.

4. Lim, K.O. and Pfefferbaum, A.J. (1989). Segmentation of MR brain images into cerebrospinal fluid spaces, white and gray matter. *J. Comput. Assist. Tomogr.* 13: 588–593.

5. Kohn, M.I., Tanna, N.K., Herman, G.T., Resnick, S.M., Mozley, P.D., Gur, R.E., Alavi, A., Zimmerman, R.A., and Gur, R.C. (January 1991). Analysis of brain and cerebrospinal fluid volumes with MR imaging. Part I. methods, reliability, and validation. *Radiology.* 178: 115–122.

6. Studholme, C., Cardenas, V., Song, E., Ezekiel, F., Maudsley, A., and Weiner, M. (2004). Accurate template-based correction of brain MRI intensity distortion with application to dementia and aging. *IEEE Trans. Med. Imaging.* 23(1): 99–110.

7. Wells, W.M., III, Grimson, W.E.L., Kikinis, R., and Jolesz, F.A. (August 1996). Adaptive segmentation of MRI data. *IEEE Trans. Med. Imaging.* 15(4): 429–442.

8. Dawant, B.M., Zjidenbos, A.P., and Margolin, R.A. (1993). Correction of intensity variations in MR images for computer-aided tissue classification. *IEEE Trans. Med. Imaging.* 12(4): 770–781.

9. Van Leemput, K., Maes, F., Vandermeulen, D., and Suetens, P. (October 1999). Automated model-based bias field correction of MR images of the brain. *IEEE Transactions on Medical Imaging,* 18(10): 885–896.

10. Styner, M., Brechbuhler, C., Szekely, G., and Gerig, G. (2000). Parametric estimate of intensity inhomogeneities applied to MRI. *IEEE Trans. Med. Imaging.* 19(3): 153–165.

11. Haselgrove, J. and Prammer, M. (1986). An algorithm for compensation of surface-coil images for sensitivity of the surface coil. *Magn. Reson. Imaging.* 4: 469–472.

12. Axel, L., Constantini, J., and Listerud, J. (1987). Intensity correction in surface coil MR imaging. *Am. J. Roentgenol,* 148: 418–420. 25.

13. Gonzalez, R.C. and Woods, R.E. (1992). *Digital Image Processing.* Reading, MA: Addison-Wesley.

14. Johnston, B., Atkins, M.S., Mackiewich, B., and Anderson, M. (1996). Segmentation of multiple sclerosis lesions in intensity corrected multispectral MRI. *IEEE Trans. Med. Imaging.* 15: 154–169.

15. Vannier, M.W., Speidel, Ch.M., and Rickman, D.L. (August 1988). Magnetic resonance imaging multispectral tissue classification. *NIPS,* 3: 148–154.

16. Tincher, M., Meyer, C.R., Gupta, R., and Williams, D.M. (1993). Polynomial modeling and reduction of RF body coil spatial inhomogeneity in MRI. *IEEE Trans. Med. Imaging.* 12(2): 361–365.

17. Meyer, C.R., Bland, P.H., and Pipe. James (March 1995). Retrospective correction of intensity inhomogenities in MRI. *IEEE Trans. Med. Imaging,* 14(1): 36–41.

18. Mangin, J.F. (2000). Entropy minimization for automatic correction of intensity nonuniformity. in *Mathematical Methods in Biomedical Image Analysis,* pp. 162–169.

19. Gilles, S., Brady, M., Declerck, J., Thirion, J.-P., and Ayache, N. (1996). Bias field correction of breast MR images. in *Proceedings of Visualization in Biomedical Computing VBC'96,* Vol. 1131 of *Lecture Notes in Computer Science,* pp. 153–158.

20. Madabhushi, A., Udupa, J.K., and Souza, A. (May 2004). Generalized scale: theory, algorithms, and application to image inhomogeneity correction. in J.M. Fitzpatrick and M. Sonka, Eds. *Medical Imaging 2004: Image Processing,* Vol. 5370 of Proceedings of SPIE, pp. 765–776.

21. Dempster, A.P., Laird, N.M., and Rubin, D.B. (1977). Maximum likelihood from incomplete data via the EM algorithm. *J. R. Stat. Soc.* 39: 1–38.

22. Wu, C.F.J. (1983). On the convergence properties of the EM algorithm. *Ann. Stat.* 11(1): 95–103.

23. Van Leemput, K., Maes, F., Vandermeulen, D., and Suetens, P. (October 1999). Automated model-based tissue classification of MR images of the brain. *IEEE Trans. Med. Imaging.* 18(10): 897–908.

24. Van Leemput, K., Maes, F., Vandermeulen, D., Colchester, A., and Suetens, P. (August 2001). Automated segmentation of multiple sclerosis lesions by model outlier detection. *IEEE Trans. Med. Imaging.* 20(8): 677–688.

25. Kapur, T., Grimson, W.E.L., Kikinis, R., and Wells, W.M. (1998). Enhanced spatial priors for segmentation of magnetic resonance imaging. In *Proceedings of Medical Image Computing and Computer-Assisted Intervention — MICCAI'98,* Vol. 1496 of *Lecture Notes in Computer Science,* pp. 457–468. Springer-Verlag, New York.

26. Held, K., Kops, E.R., Krause, B.J., Wells, W.M., III, Kikinis, R., and Müller-Gärtner, H.W. (December 1997). Markov random field segmentation of brain MR images. *IEEE Trans. Med. Imaging.* 16(6): 878–886.

27. Marroquin, J.L., Vemuri, B.C., Botello, S., Calderon, F., and Fernandez-Bouzas, A. (August 2002). An accurate and efficient Bayesian method for automatic segmentation of brain MRI. *IEEE Trans. Med. Imaging.* 21(8): 934–945.

28. Zhang, Y., Brady, M., and Smith, S. (January 2001). Segmentation of brain MR images through a hidden Markov random field model and the expectation-maximization algorithm. *IEEE Trans. Med. Imaging.* 20(1): 45–57.

29. Xiao, G., Brady, M., Noble, J.A., and Zhang, Y. (January 2002). Segmentation of ultrasound B-mode images with intensity inhomogeneity correction. *IEEE Trans. Med. Imaging.* 21(1): 48–57.

30. Guillemaud, R. and Brady, M. (June 1997). Estimating the bias field of MR images. *IEEE Trans. Med. Imaging.* 16(3): 238–251.

31. Maes, F., Collignon, A., Vandermeulen, D., Marchal, G., and Suetens, P. (April 1997). Multi-modality image registration by maximization of mutual information. *IEEE Trans. Med. Imaging.* 16(2): 187–198.

32. Wells, W.M., Viola, P., Atsumi, H., Nakajima, S., and Kikinis, R. (March 1996). Multi-modal volume registration by maximization of mutual information. *Med. Image Anal.* 1(1): 35–51.

33. Lorenzo-Valdés, M., Sanchez-Ortiz, G.I., Mohiaddin, R., and Rueckert, D. (2003). Segmentation of 4D cardiac MR images using a probabilistic atlas and the EM algorithm. in R.E. Ellis and T.M. Peters, Eds. *Proceedings of MICCAI 2003,* Vol. 2878 of *Lecture Notes in Computer Science,* pp. 440–450.

34. Ashburner, J.T. (July 2000). Computational Neuroanatomy. Ph.D. thesis, University of London.

35. Tou, J.T. and Gonzalez, R.C. (1974). *Pattern Recognition Principles.* Reading, MA: Addison-Wesley.

36. Gray, R.M. and Linde, Y. (February 1982). Vector quantizers and predictive quantizers for gauss-markov sources. *IEEE Trans. Commn.* COM-30(2): 381–389.

37. Dunn, J.C. (1973). A fuzzy relative of the ISODATA process and its use in detecting compact well-separated clusters. *J. Cybern.* 3: 32–57.

38. Bezdek, J.C. (1981). *Pattern Recognition with Fuzzy Objective Function Algorithms.* Plenum Press, New York.

39. Pappas, T.N. (April 1992). An adaptive clustering algorithm for image segmentation. IEEE *Transactions on Signal Processing,* 40: 901–914.

40. Yan, M.X.H. and Karp, J.S. (1995). Segmentation of 3D brain MR using an adaptive k-means clustering algorithm. in *Proceedings of the 1994 Nuclear Science Symposium and Medical Imaging Conference*, pp. 1529–1533.
41. Yan, M.X.H. and Karp, J.S. (1995). An adaptive bayesian approach to three-dimensional MR brain segmentation. in Bizais, Y., Barillot, C., DiPaol, R., Eds. *Proceedings of Information Processing in Medical Imaging*, pp. 201–213.
42. Lee, S.K. and Vannier, M.W. (August 1996). Post-acquisition correction of MR inhomogeneities. *Magn. Reson. Med.* 36(2): 275–286.
43. Rajapakse, J.C., Giedd, J.N., and Rapoport, J.L., (April 1997). Statistical approach to segmentation of single-channel cerebral MR images. *IEEE Trans. Med. Imaging.* 16(2): 176–186.
44. Nocera, L. and Gee, J.C. (February 1997). Robust partial volume tissue classification of cerebral MRI scans. in Hanson, K.M. Ed. *Medical Imaging 1997: Image Processing*. SPIE.
45. Rajapakse, J.C. and Kruggel, F. (1998). Segmentation of MR images with intensity inhomogeneities. *Image and Vision Computing,* 16:165–180.
46. Pham, D.L. and Prince, J.L. (September 1999). Adaptive fuzzy segmentation of magnetic resonance images. *IEEE Trans. Med. Images.* 18(9): 737–752.
47. Ahmed, M.N., Yamany, S.M., Mohamed, N., Farag, A.A., and Moriarty, T. (March 2002). A modified fuzzy c-means algorithm for bias field estimation and segmentation of MRI data. *IEEE Trans. Med. Imaging.* 21(3): 193–199.
48. Brechbühler, C., Gerig, G., and Székely, G. (September 1996). Compensation of spatial inhomogeneity in MRI based on a multi-valued image model and a parametric bias estimate. in *Visualization in Biomedical Computing (VBC) '96*, pp. 141–146.
49. Huber, P. 1981. *Robust Statistics*. John Wiley & Sons, New York.
50. Schwefel, H.-P. 1995. *Evolution and Optimum Seeking*. John Wiley & Sons, New York.
51. Gerig, G., Welti, D., Guttmann, C., Colchester, A., and Székely. G. (1998). Exploring the discriminating power of the time domain for segmentation and characterization of lesions in serial MR data. in *Proceedings of Medical Image Computing and Computer-Assisted Intervention (MICCAI 98)*. pp. 469–480.
52. Viola, P. (1995). Alignment by Maximization of Mutual Information. Ph.D. thesis, Massachusetts Institute of Technology.
53. Likar, B., Viergever, M. and Pernus, F. (2001). Retrospective correction of MR intensity inhomogeneity by information minimization. *IEEE Trans. Med. Imaging.* 20(12): 1398–1410.
54. Sled, J.G., Zijdenbos, P., and Evans, A.C. (1997). A comparison of retrospective intensity nonuniformity correction methods for MRI. in *Information Processing in Medical Imaging,* Vol. 1230, pp. 459–464. Proceedings 15th Int. Conf. IMPI'97.
55. Sled, J.G., Zijdenbos P., and Evans, A.C. (February 1998). A nonparametric method for automatic correction of intensity nonuniformity in MRI data. *IEEE Trans. Med. Imaging.* 17: 87–97.

6 Noise Filtering Methods in MRI

L. Landini, M. Lombardi, and A. Benassi

CONTENTS

6.1 INTRODUCTION

It is well known that magnetic resonance imaging (MRI) methods are increasing diagnostic efficacy due to the recent progress in real-time image acquisition technology. Such advanced imaging techniques provide access to important anatomical and functional information through high-speed acquisition and high spatial resolution.

In MRI applications, there is an intrinsic trade-off between signal-to-noise ratio (SNR), contrast-to-noise ratio (CNR), and resolution. Depending on specific diagnostic tasks, high spatial resolution and high contrast may be required, whereas for image processing applications, a high SNR is usually necessary because most of the algorithms are very sensitive to noise. In fact, magnetic resonance (MR) images often require application of noise filtering techniques before visual inspection or application of noise-sensitive postprocessing methods such as segmentation algorithms [1]. As a rule, such filtering is desired to significantly decrease image noise and, simultaneously, to preserve fine image details.

Many efforts were devoted to SNR improvement, including time and spatial averaging during acquisition. Time averaging has the major advantage that the SNR increases while the spatial resolution is preserved, provided the imaging process is stationary; the disadvantage is the time required to perform the examination. Achieving a high SNR at an elevated spatial resolution may necessitate additional time averages (higher number of excitations [NEX]) and, consequently, a longer acquisition time. But a long acquisition time is undesirable because of constraints such as patient comfort, system throughput, and physical limitations arising in dynamic applications such as cardiac imaging and functional MRI. In such cases, time averaging is replaced by spatial averaging.

Spatial filtering of MR image data should ideally be fulfilled by removing the noise without loss of resolution, improving the image contrast, to obtain piecewise constant or slowly varying signals in homogeneous tissue regions, minimizing information loss by preserving detailed structures inside objects and object boundaries. Spatial filtering techniques are applied under two important assumptions: (a) the image is supposed to consist of many regions in which the signal is stationary and ergodic in the mean and variance [2] and (b) the image noise is assumed to be zero mean and Gaussian distributed. The main problem is to find these stationary regions.

The problem of finding the proper stationary area for local signal estimation is partly solved by choosing filters that are able to distinguish homogeneous regions from those with edge regions. In the literature, many approaches to improve SNR, CNR and edge blurring effects have been proposed, such as adaptive filters [3–5], wavelet filters [6–11], and anisotropic diffusion filters [12–17].

In particular, anisotropic diffusion [12] is an accepted filtering technique that is well suited for practical use because of its computational speed and algorithmic simplicity. The filter assumes image noise to be Gaussian distributed. The aniso-tropic diffusion filter has proved to be particularly effective in prefiltering of MR images before the automatic image segmentation procedure [16,18] and before MRI inhomogeneity correction [19]. Subsequently, the standard anisotropic dif-fusion method was extended by Yang [20] using both a local intensity orientation and an anisotropic measure of level contours, instead of utilizing local gradients to control the anisotropism of the filters.

When processing magnitude MR data, a Gaussian assumption for image noise is not acceptable as it can be shown to be Rice distributed, especially in regions with low SNR [3,21–23].

Not incorporating this knowledge leads inevitably to biased results, in par-ticular, when applying such filters in regions with low SNR. In order to reduce this bias, Sijbers et al. [24] proposed a modified version of the anisotropic filter suggested by Yang [20], in which the Rician nature of the data is exploited. Wavelet-based methods that explicitly account for the Rician nature of the data are described in [25–27].

How noise is spatially distributed is another issue concerning MR image noise that should be considered before applying noise filters. Examples include images multiplicatively corrected for intensity inhomogeneity [19,28], and particularly, images obtained with partially parallel imaging techniques [29–35]. Retrospective denoising with a nonlinear technique such as anisotropic diffusion filtering has been demonstrated to be an attractive option for improving the SNR of partially parallel images [36]. Topics dealing with noise in parallel MRI will be discussed in this book in another chapter.

In this chapter, we will review some of the recent methods to improve SNR and CNR in MR images, also accounting for the statistical nature of the process. Emphasis will be given to an application dealing with edge enhancement in myocardium image segmentation.

6.2 THE MR IMAGE MODEL

Usually, noise in MR images is defined simply as a deviation from the true value considered representative of a tissue category; noise is expressed by the standard deviation, whereas the true value is evaluated by the mean value of a set of pixels expected to belong to the same tissue category. It has been shown [23,37] that due to nonlinearity of the magnitude reconstruction process introduced to obtain real images, such assumptions about the image model are too weak. In fact, the first step in reconstructing MR images is to compute the inverse discrete Fourier transform of raw frequency-domain (k-space) measurements. Let y(m, n) denote a complex image with additive noise as follows:

$$y(m, n) = r(m, n) + n_r(m, n) + j[i(m, n) + n_i(m, n)] \qquad (6.1)$$

where r(m, n) and i(m, n) are real and imaginary parts of the noiseless image, with $n_r(m, n)$ and $n_i(m, n)$ representing the Gaussian noise in the real and imaginary images, respectively, with standard deviations σ_n.

By considering the magnitude reconstruction process of MR images, the magnitude of y(m, n) is given by:

$$|y(m,n)| = \sqrt{\{r(m,n) + n_r(m,n)\}^2 + \{i(m,n) + n_i(m,n)\}^2} \qquad (6.2)$$

We recall that in MR images, the signal magnitude is simply the square root of the sum of two independent Gaussian random variables, and the magnitude image data are described by a Rician distribution.

If the image intensity is much larger than the noise standard deviation, Equation 6.2 may be approximated as [3]:

$$|y(m,n)| = x(m,n) + n_z(m,n) \qquad (6.3)$$

where

$$x(m,n) = \sqrt{[r(m,n)]^2 + [i(m,n)]^2}$$

$$n_z(m,n) = \frac{r(m,n) \cdot n_r(m,n) + i(m,n) \cdot n_i(m,n)}{x(m,n)};$$

The term x(m, n) is the noiseless magnitude image and n_z(m, n) represents the noise in the magnitude image.

From Equation 6.3, the noise in the magnitude image can be given in terms of the standard deviation as follows:

$$\sigma_{|y|} = \sqrt{\frac{r(m,n)^2 \cdot \sigma_n^2 + i(m,n)^2 \cdot \sigma_n^2}{x(m,n)}} = \sigma_n \qquad (6.4)$$

Thus, the standard deviation of the noise in the magnitude image is identical to those of the real or imaginary parts and independent of the relative intensities of r(m, n) and i(m, n), assuming that x(m, n) is much larger than the noise.

Furthermore, because the noise in the magnitude image is a linear combination of the noise in the real and imaginary parts, the distribution function of the noise in the magnitude image is also Gaussian. It means that separation of signal and noise is fairly straightforward with proper filtering, so the filtered signal can be assumed representative of the physical properties of a tissue category.

In low-SNR regions, the Rician distribution equals the Rayleigh distribution. This means that the magnitude image does not equal the underlying noise-free image. Hence filtering methods yield biased results, which increase with decreasing SNR. This bias in the MR magnitude image can significantly reduce image contrast.

The noise standard deviation in the background region σ_b, where the image intensity is zero, is related to σ_n by the following relationship [38]:

$$\sigma_n = 1.526 \cdot \sigma_b \qquad (6.5)$$

It gives a practical way to estimate σ_n.

As a rule of thumb, it was found that below 5 (15 dB), the Rician distribution equals the Rayleigh distribution, whereas for SNR > 5, it approximates a Gaussian distribution. Although clinical images have overall SNR values exceeding 5:1, some important image features, particularly edge information, may have local SNRs < 5.

6.3 WAVELET-BASED FILTERING

This section considers both standard wavelet-domain filtering methods and wavelet-domain methods that account for the Rician nature of the data.

In the wavelet-domain approach, the discrete wavelet transform (DWT) tends to concentrate the energy of the desired signal into a small number of coefficients. So, the DWT of the noisy image consists of a small number of coefficients with high SNR (which should be kept), and a large number of coefficients with low SNR (which can be discarded). After discarding the noisy coefficients, the noiseless image can be reconstructed using the inverse DWT. As a result, noise is removed or filtered from the observations. Therefore, the key to effective noise removal in the wavelet domain is to determine which wavelet coefficients do not have significant signal energy and, hence, can be discarded without detrimental signal loss. Of course, a similar procedure could be carried out using Fourier-domain filtering. But the Fourier method is a spatially global operation that cannot be adjusted to local spatial variations, thus leading to uncontrolled smoothing in regions with high-frequency content, such as the edges. On the other hand, the wavelet basis functions enable DWT-based filtering procedures to adapt to such spatial variations. Only a brief algorithm outline will be given here; details can be found in [7–9].

There are three fundamental concepts in the wavelet algorithm: wavelet packet transformation, best-basis selection, and coefficient thresholding.

In wavelet packet transformation, the original 2-D images are split in terms of shifts and dilations of the low-pass scaling function and band-pass wavelet to obtain the relevant 2-D coefficients. For special choices of these functions, the shifts and dilations form an orthonormal basis. The wavelet decomposition can be implemented iteratively by successive filtering and downsampling (by a factor of 2 at each iteration) operations, using the so-called quadrature mirrors filters. At each step four 2-D images of coefficients are obtained (one low-pass subimage and three subimages corresponding to wavelet orientations that are horizontal, vertical, and diagonal).

Wavelet bases are bases of nested function spaces, which can be used to analyze signals at multiple scales. Best-basis selection consists of finding the mathematical function that not only completely represents the original signal, but that also concentrates the maximum amount of structured signal into the minimum number of coefficients. However, in many applications such as MRI, the key signal features are not well known and the optimal basis functions cannot be specified in advance. In such cases, a basis with general properties is preferable.

The thresholding operation in the wavelet domain is the operation that determines which wavelet coefficients do not have significant signal energy and hence can be discarded without detrimental signal loss. Ideally, the wavelet-domain filtering procedure should be adapted to the local SNR in each wavelet coefficient, so that wavelet coefficients with very low SNR can be suppressed. Methods for making this determination are called wavelet-domain filters.

When designing a filter in the wavelet domain, we have to account for the nature of the noise process. For example, if the SNR in a pixel is greater than 15 dB, the Rician distribution is assumed to become approximately Gaussian. If the SNR is below 15 dB, then the Rician distribution deviates from the Gaussian.

The goal of wavelet-domain filtering is to obtain a better estimate of the noise-free image wavelet coefficients by filtering the observed $d_j^o(\mathbf{k})$ coefficients at any level j, spatial position $\mathbf{k}$ and wavelet orientation o. In fact, the wavelet filter should adapt to the local SNR in each wavelet coefficient in order to suppress wavelet coefficients with very low SNR.

Several wavelet-domain filtering methods to reduce the noise power from images are reported in the literature [6–11,25–27].

In particular, Nowak [25] described a method that explicitly accounts for the Rician nature of the data. He demonstrated that the filter weight α that minimizes the mean squared error of each coefficient is given by

$$\alpha = \left(\frac{\left(d_j^o(\mathbf{k}) \right)^2 - 3 \left(\sigma_j^o(\mathbf{k}) \right)^2}{\left(d_j^o(\mathbf{k}) \right)^2} \right)_+ \qquad (6.6)$$

where (+) means that $(x)_+ = x$ if $x \geq 0$ and $(x)_+ = 0$ if $x < 0$, and $\sigma_j^o(\mathbf{k})$ is the variance of the observed coefficients. It means that the filter sets small wavelet coefficients with squared magnitude less than three times the estimated variance to zero and leaves larger coefficients approximately unaltered.

In the presence of additive white Gaussian noise with variance σ_n^2 (high SNR), the filter weight α given by Equation 6.6 can be evaluated with $\sigma_j^o(\mathbf{k}) = \sigma_n$ (assuming that each wavelet coefficient has equal variance at any spatial position $\mathbf{k}$).

A simple procedure for estimating σ_n is reported in Reference 25, and it consists of evaluating one half of the mean of the squared pixel values in the region outside the patient within the scanner.

In low-SNR situations, the mean of the magnitude image is not equal to the noise-free image (as in the Gaussian approximation) and, hence, the magnitude image is biased. Nowak demonstrated that when operating on the squared magnitude image rather than the magnitude, the wavelet coefficients are approximately Gaussian distributed. In such situations, the filter weight α can still be used. He also suggested a method for $\sigma_j^o(\mathbf{k})$ estimation and compensation. The advantage of such an algorithm is that it can be used in both high- and low-SNR imaging situations [25–27]. An advantage of the algorithm in Equation 6.6 in the high-SNR case is that its complexity is slightly lower than that in the low-SNR case.

6.4 ADAPTIVE TEMPLATE FILTERING

An interesting approach to MRI image filtering is the adaptive template filtering technique proposed by C.B. Ahn et al. [3]. Ahn proposed a local shape-adaptive template filtering for the enhancement of the SNR without resolution loss. Unlike conventional filtering, in which the template shape and coefficients are predefined and the filtered output is given as the weighted sum of the image gray levels surrounding the current pixel, in adaptive template filtering, multiple templates are defined. It prevents the edge blurring usually observed when using fixed-coefficient filters that employ spatial averaging to obtain SNR enhancement in almost constant regions. Using the proposed process, edge blurring is minimized and SNR maximized by selecting the optimally matched template.

The proposed method can be applied without any a priori evaluation of SNR. It is assumed that image gray levels are composed of 2-D locally constant or slowly varying regions separated by discontinuities called edges. In locally constant regions, local variations are mainly due to random noise, and the values of the standard deviations are relatively small. Across edges, the local standard deviations are relatively large because of the large intensity differences across image structures.

The adaptive filter should be able to distinguish between locally constant regions and edge regions: the edge regions should be excluded from the filtering process in order to prevent resolution degradation.

The algorithm proposed in [3] defines multiple templates, each template being composed of free cells and active cells, i.e., cells made available for the filter coefficients. The number NT of multiple templates is given according to the following relationship

$$NT = \sum_{k=0}^{N} C_N^k = \sum_{k=0}^{N} \frac{N!}{k!(N-k)!} \qquad (6.7)$$

where N is the template dimension minus the current pixel (for a 3×3 template, N = 8), k the number of free cells in the template, and (N − k) the number of active cells. The number of active cells defines the filter size. Equation 6.7 defines the number of templates for any filter size, i.e., the possible combinations of (N − k) active cells in a template.

In Figure 6.1, an example is reported for a 3×3 template (N = 8): when k = 0 we have NT = 1 possible combinations, whereas for k = 4, the number of templates is NT = 70. In the figure, "1" denotes active cells, "0" corresponds to free cells, and the reference cell is drawn in black.

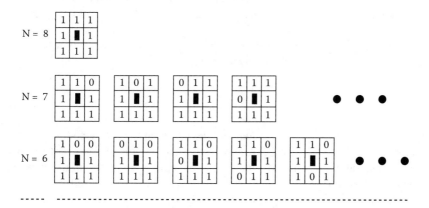

FIGURE 6.1 Example of the number of template configurations for a 3×3 template.

In order to select the best template configuration for each pixel, a quality measure must be defined. Ahn [3] proposed the use of the local standard deviation of the pixel values on the templates. For each template T_j, the local standard deviation $\sigma_j(m, n)$ of the pixel values is given by:

$$\sigma_j(m,n) = \sqrt{\frac{1}{N_j - 1} \sum_{i,j \in T_j} [x(i,j) - \bar{x}(m,n)]^2}$$

$$\bar{x}(m,n) = \frac{1}{N_j} \sum_{i,j \in T_j} x(i,j)$$

(6.8)

and x(m, n) is the input pixel value at the (m, n) coordinates.

For an implementation of the adaptive filtering procedure, first, local standard deviations for each pixel in the image and for each template must be evaluated. Next, templates are classified into two categories based on standard deviation value: templates with local standard deviation less than the threshold value (corresponding to the random noise standard deviation) and templates with local standard deviation larger than the threshold value. When more elements reflect the former condition, then the optimal template is the one having the maximum filter size. As far as the latter condition is concerned, the optimal template is the one having the minimum standard deviation

$$y(m,n) = \frac{\sigma_y^2(m,n)x(m,n) + \sigma_n^2 \bar{x}(m,n)}{\sigma_y^2(m,n) + \sigma_n^2}$$

$$\sigma_y^2(m,n) = \max\{0, \sigma_x^2(m,n) - \sigma_n^2\}$$

(6.9)

where y(m, n) is the filtered output, σ_x^2 the local variance evaluated at the (m, n) pixel, and σ_n^2 the noise variance. Local variance close to the noise variance implies almost constant regions. The relevant filtered output is a smoothed version of the input image on the template. Otherwise, i.e., in edge regions, the filtering contribution is negligible.

A key issue in the optimization of the adaptive filtering algorithm based on templates is the choice of the threshold value. A popular method is to define the threshold as

$$\tau = a\sigma_n$$

(6.10)

where σ_n is the estimated noise standard deviation on the image and a is a scale factor. Values from $1.2\sigma_n$ to $1.6\sigma_n$ may be chosen to optimize filter performance [3]. Estimation of σ_n from MR images can be done by multiplying the standard deviation of the image background σ_b by the scale factor 1.526 as in Equation 6.5.

The computational complexity of the algorithm for a brute-force search of templates is M^2NT, where M is the image dimension and NT the number of possible templates. In order to reduce the algorithm complexity, templates can be arranged according to their size and the evaluation of the standard deviation can be started from larger templates. Then, the first template having a standard deviation less than the threshold may be the optimal template. If no template is found having a standard deviation less than the threshold, the template having the minimum standard deviation is selected as the optimal template. Through this procedure, the computational time spent searching for the optimal template is substantially reduced compared to searching for the templates in arbitrary order.

The filtering algorithm can be applied iteratively so that more noise reduction can be achieved.

6.5   ANISOTROPIC DIFFUSION FILTERING

Perona and Malik [12] first proposed a nonlinear anisotropic smoothing filter for removal of background noise in images. It uses local gradients to control the anisotropy of the filter. A comprehensive review of anisotropic filter theory can be found in [17].

The smoothing operation is assumed to be a diffusive process that is suppressed or stopped at boundaries by selecting appropriate spatial diffusion strengths. In particular, depending on the values assumed by diffusion strength, the filter is able to realize intraregion smoothing in preference to smoothing across boundaries. In other words, the nonlinear anisotropic diffusion equation is:

$$\frac{\partial}{\partial t} I(\boldsymbol{x},t) = div[c(\boldsymbol{x},t) \cdot \nabla I(\boldsymbol{x},t)] \qquad (6.11)$$

The diffusion strength is controlled by $c(\boldsymbol{x}, t)$. The vector $\boldsymbol{x}$ represents the spatial coordinate, and the variable t in our discrete implementation corresponds to iteration step n. The function $I(\boldsymbol{x}, t)$ is the image intensity. The diffusion function $c(\boldsymbol{x}, t)$ assumes a constant value for linear isotropic diffusion. In that case, the diffused image is derived from isotropic application of the Laplacian operator to the image. But the price of eliminating the noise with linear diffusion is blurring of the edges. This results in their detection and localization being difficult.

In order to preserve the edges, the diffusion must be reduced or even blocked when close to a discontinuity. The diffusion function $c(\boldsymbol{x}, t)$ can be chosen to be a function of gradient magnitude evaluated on image intensity $I(\boldsymbol{x}, t)$:

$$c(\mathrm{x,t}) = e^{-\frac{|\nabla I(\mathrm{x},t)|^2}{2K^2}} \qquad (6.12)$$

Figure 6.2 shows the monotonic decrease of the diffusion coefficient $c(\boldsymbol{x}, t)$ with increasing gradient ∇I. A more effective view of the relationship between parameter K and image gradient ∇I is obtained by defining the flow function $\phi(\nabla I)$ as the product $c \cdot \nabla I$.

The parameter K is the diffusion constant, and it is chosen in order to preserve edge strength at the object boundary and to reduce the noise contribution. The

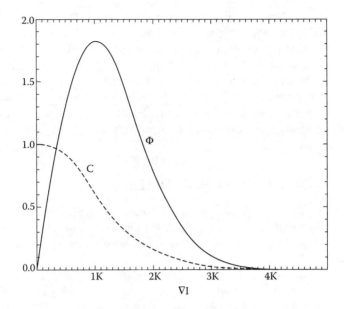

FIGURE 6.2 Monotonic decrease of the diffusion coefficient c(x, t) and flow function $\phi(\nabla I)$ as a function of gradient ∇I.

maximum flow is produced at the image location when $\nabla I = K$. When ∇I is below K, the flow function reduces to zero because in almost homogeneous regions, the flow is minimal. For ∇I larger than K, the flow function again decreases to zero, halting diffusion at locations of high gradients. A proper setting of the K parameter in the diffusion function not only preserves, but also enhances object edges. This property will be exploited in the next section in an application dealing with contour enhancement for myocardial image segmentation.

The performance of the anisotropic filter is also related to the choice of the diffusion function. An alternative choice of the diffusion function is:

$$c(\pmb{x},t) = \frac{1}{1 + (\|\nabla I(\pmb{x},t)\|/k)^{1/\alpha}} \qquad (6.13)$$

or [14]:

$$c(\pmb{x},t) = \frac{1}{2}[\tanh(\gamma(k - \|\nabla I(\pmb{x},t)\|)) + 1] \qquad (6.14)$$

Anisotropic diffusion, in its original form, is a well-accepted filtering technique because of its computational speed and algorithmic simplicity. It was applied to 2-D and 3-D MRI data by Gerig [13]. Such a filter has shown maximum performance in local filtering applications such as automatic MR cardiac image segmentation [18].

Subsequently, the standard anisotropic diffusion method was extended by Yang [20], who, instead of using local gradients to control the filter anisotropy, introduced a filter whose shape is pointwise adapted to the local structure of the image within a neighborhood, using both a local intensity orientation and an anisotropic measure of level contours. Yang demonstrated that the noise filtering efficacy of the algorithm is good both in simulated and real images, although the computational efficiency needs improvement.

These anisotropic filters applied to magnitude MR data introduce a bias in the image, because they do not account for the Rician nature of the data, which is more effective in low SNR. In order to reduce this bias, Sijbers [24] proposed a modified version of the Yang filter that introduces the Rice distribution into the maximum likelihood estimation of the filter parameters. Results obtained with simulated images and experimental magnitude MR data confirm that the differences between Gaussian- and Rician-based filters are visible in regions with low SNR.

6.6 APPLICATION OF ANISOTROPIC DIFFUSION FILTERING

The filter proposed by Perona et al. [12] is able to produce object edge enhancement if the proper choice of the diffusion constant K is made. In this section, we exploit this property to improve myocardial contours of cardiac images for a subsequent automatic segmentation operation.

In MR images of myocardium, gradient strength at the endocardium is usually different from that at the epicardium. Moreover, MRI of the myocardium is strongly influenced by gray-scale inhomogeneities that are responsible for local changes in tissue mean and variance. To account for such drawbacks, the proper diffusion parameter was determined by a simulation procedure.

We first exploited the relation between the K parameter and image gradient ∇I, using a 1-D simulation study. The error function, sampled with 16 data points, was used as an ideal model of a blurred step edge to simulate a gray-level discontinuity in the image. The sampling frequency was determined by considering that a typical MR image of 256×256 data points requires approximately 16 pixels to realize a gray-level transition at the myocardium interfaces. On the gradient profile, the slope measured at the inflection point multiplied by a constant term was used for simulation purposes. The relationship between the slope increment ΔS and K was computed as a function of the iteration step N (Figure 6.3). It demonstrates that the slope is a function of K and exhibits band-pass behavior.

Starting from the K value where ΔS is maximum, the relationship between K and ∇I was derived by simulation. In the simulation, ∇I ranged from 0 to 30 to include gray-level excursion at the endocardium and epicardium interfaces (typically ranging from 10 to 25). A linear relationship between K and ∇I was found (Figure 6.4), where $\nabla I = 2.85K$. It means that to obtain a maximum slope ΔS at any ∇I in the image, the K value should be adapted according to the data of Figure 6.4. In our implementation, we assessed the best compromise value for

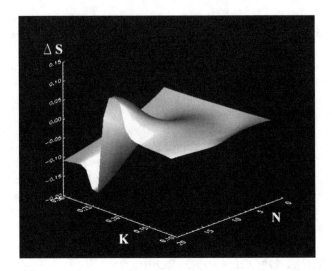

FIGURE 6.3 Relationship between slope increment ΔS and K as a function of iteration step N.

K in order to maintain ΔS at a reasonable high value (see Figure 6.3), by exploiting the band-pass shape of the ΔS − K curves (Figure 6.5). In fact, partial overlapping of ΔS − K curves corresponding to typical myocardium interface gradient value suggests that the choice of a single K value could be a reasonable compromise to enhance endocardium and epicardium edges simultaneously.

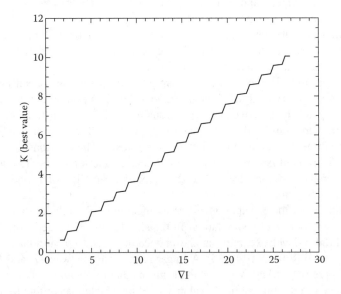

FIGURE 6.4 Relationship between K and ∇I.

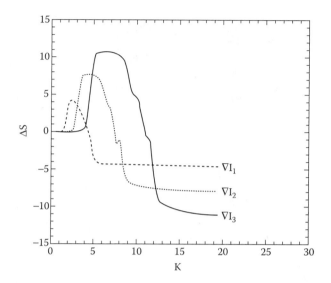

FIGURE 6.5 Plot of the slope increment (ΔS) as a function of K for $\Delta I_1 = 10$, $\Delta I_2 = 20$, and $\Delta I_3 = 30$.

The performance of anisotropic filtering in reducing noise while preserving image structure was compared to results obtained at different NEX values. Two sets of images were derived: one set obtained from a two-cylinder phantom (a 60-mm-diameter cylinder filled with olive oil and immersed in a 100-mm-diameter water-filled cylinder), and one set of images derived from a normal human heart. Cardiac images were obtained from a group of five normal volunteers.

The series of MR images was acquired with a 1.5T GE Signa CV/i scanner with a fast spoiled gradient echo (fast SPGR) sequence, using cardiovascular phase array coils. A number of 10-sections were acquired with a slice thickness of 8 mm. Three acquisitions (TR = 9.2 and TE = 1.9) were performed at 1, 2, and 4 NEX.

In Figure 6.6, phantom images at 1 NEX before (Figure 6.6a) and after (Figure 6.6c) filtering are reported, together with the 4 NEX image (Figure 6.6b).

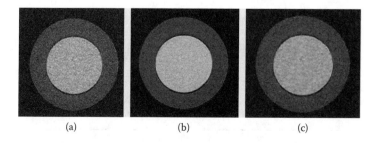

(a)   (b)   (c)

FIGURE 6.6 Phantom images: (a) 1 NEX image, (b) 4 NEX image, and (c) anisotropic-filtered 1 NEX image.

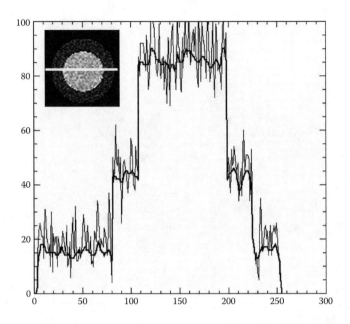

FIGURE 6.7 Phantom image profile relevant to Figure 6.6a (thin line) and to Figure 6.6c (bold line).

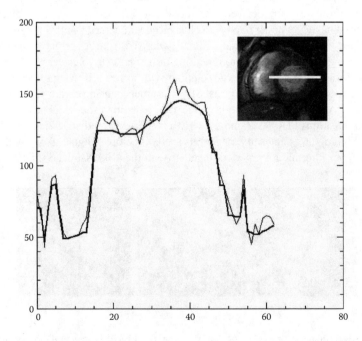

FIGURE 6.8 Left-ventricle image profile before (thin line) and after (bold line) anisotropic filtering.

Typical line profiles extracted from the phantom image before filtering (thin line) and after filtering (bold line) are shown in Figure 6.7, showing the filter's ability to preserve edges while removing noise.

In cardiac images, anisotropic filtering still preserves contrast at the myocardium interfaces: it clearly appears in 1 NEX image aquisition as shown in Figure 6.8. The relevant line profiles before (thin line) and after anisotropic filtering (bold line) are shown.

Such results are important for the later segmentation operation by an automatic procedure.

REFERENCES

1. Clarke, L.P., Velthuizen, R.P., Camacho, M.A., Heine, J.J., Vaidyanathan, M., Hall, L.O., Thatcher, R.W., and Silbiger, M.L. (1995). MRI segmentation: methods and applications. *Magn. Reson. Imaging.* 13: 343–368.
2. Wang Y. and Lei T. (1994). Statistical analysis of MR imaging and its applications in image modelling. *Proceedings of the IEEE International Conference on Image Processing and Neural Networks* I: 866–870.
3. Ahn, C.B., Song, Y.C., and Park, D.J. (1999). Adaptive template filtering for signal-to-noise ratio enhancement in magnetic resonance imaging. *IEEE Trans. Med. Imaging* 18(6): 549–556.
4. Rank, K. and Unbehauen, R. (1992). An adaptive recursive 2-D filter for removal of Gaussian noise in images. *IEEE Trans. Image Process.* 1: 431–436.
5. Westin, C.F., Wigström, L., Loock, T., Sjökvist, L., Kikinis, R., and Knutsson, H. (2001). Three-dimensional adaptive filtering in magnetic resonance angiography. *Magn. Reson. Imaging.* 14: 63–71.
6. Xu, Y., Weaver, J.B., Healy, D.M., and Lu, J. (1994). Wavelet transform domain filters: A spatially selective noise filtration technique. *IEEE Trans. Image Process.* 3, 747–757.
7. Mallat S. and Hwang, W. (1992). Singularity detection and processing with wavelets. *IEEE Trans. Inform. Theory* 38: 617–643.
8. Crouse, M.S., Novak, R.D., and Baraniuk, R.G. (1998). Wavelet-based signal processing using hidden Markov models. *IEEE Trans. On Signal Proc.* 46: 886–902.
9. Coifman, R. and Donoho, D. (1995). *Wavelets and Statistics, Lecture Notes in Statistics.* Berlin: Springer-Verlag.
10. Zaroubi, S. and Goelman, G. (2000). Complex denoising of MR data via wavelet analysis: application for functional MRI. *Magn. Reson. Imaging* 18: 59–68.
11. Placidi, G., Alecci, M., and Sotgiu, A. (2003). Post-processing noise removal algorithm for magnetic resonance imaging based on edge detection and wavelet analysis. *Phys. Med. Biol.* 48: 1987–1995.
12. Perona, P. and Malik, J. (1990). Scale space and edge detection using anisotropic diffusion. *IEEE Trans. Pattern Anal. Machine Intell.* 12(7): 629–639.
13. Gerig, G., Kubler, O., Kikins, R., and Jolesz, F.A. (1992). Nonlinear anisotropic filtering of MRI data. *IEEE Trans. Med. Imaging* 11(2): 221–232.
14. Alvarez, L., Lions, P.L., and Morel, J.M. (1992). Image selective smoothing and edge detection by nonlinear diffusion. *SIAM J. Number. Anal.* 29(3): 845–866.

15. Nordstrom, N. (1990). Biased anisotropic diffusion: a unified generalization and diffusion approach to edge detection. *Image Vision Comput.* 8(4): 318–327.

16. Atkins, M.S. and Mackiewich, B.T. (1998). Fully automatic segmentation of the brain in MRI. *IEEE Trans. Med. Imaging* 17(1): 98–107.

17. Weickert, J. (1998). *Anisotropic Diffusion in Image Processing*, ECMI Series, Teubner, Stuttgart.

18. Santarelli, M.F., Positano, V., Michelassi, C., Lombardi, M., and Landini, L. (2003). Automated cardiac MR image segmentation: theory and measurement evaluation. *Med. Eng. Phys.* 25(2): 149–59.

19. Styner, M., Brechbuhler, C., Szekely, G., and Gerig, G. (2000). Parametric estimate of intensity inhomogeneities applied to MRI. *IEEE Trans. Med. Imaging* 19(3): 153–165.

20. Yang, G.Z., Burger, P., Firmin, D.N., and Underwood, S.R. (1996). Structure filtering. *Image and Vision Computing* 14: 135–45.

21. Sijbers, J., den Dekker, A.J., Van Audekerke, J., Verhoye, M., and Van Dyck, D. (1998). Estimation of the noise in magnitude MR images. *Magn. Reson. Imaging* 16(1): 87–90.

22. Sijbers, J. and den Dekker, A.J. (2004). Maximum likelihood estimation of signal amplitude and noise variance from MR data. *Magn. Reson. Med.* 51(3): 586–594.

23. Gudbjartsson, H. and Patz, S. (1995). The Rician distribution of noisy MRI data. *Magn. Reson. Med.* 34: 910–914.

24. Sijbers, J., den Dekker, A.J., Van der Linden, Verhoye, A.M., and Van Dyck, D. (1999). Adaptive anisotropic noise filtering for magnitude MR data. *Magn. Reson. Imaging* 17(10): 1533–1539.

25. Novak, D. (1999). Wavelet-based Rician noise removal for magnetic resonance imaging. *IEEE Trans. Image Process.* 8(10): 1408–1419.

26. Wood, J.C. and Johnson, M.K. (1999). Wavelet packet denoising of magnetic resonance images: importance of Rician noise at low SNR. *Magn. Reson. Imaging* 41(3): 631–635.

27. Alexander, M.E., Baumgarter, R., Summers, A.R., Windishberger, C., Klarhoefer, M., Moser, E., and Somorjai, R.L. (2000). A wavelet-based method for improving signal-to-noise ratio and contrast in MR images. *Magn. Reson. Imaging* 18: 169–180.

28. Likar, B., Viergever, M.A., and Pernus, F. (2001). Retrospective correction of MR intensity inhomogeneity by information minimization. *IEEE Trans. Med. Imaging* 20: 1398–1410.

29. Pruessmann, K.P., Weiger, M., Scheidegger, M.B., and Boesiger, P. (1999). SENSE: sensitivity encoding for fast MRI. *Magn. Reson. Med.* 42: 952–962.

30. Sodickson, D.K. and Manning, W.J. (1997). Simultaneous acquisition of spatial harmonics (SMASH): ultra-fast imaging with radiofrequency coil arrays. *Magn. Reson. Med.* 38: 591–603.

31. Bydder, M., Larkman, D.J., and Hajnal, J.V. (2002). Generalized SMASH imaging. *Magn. Reson. Med.* 47: 160–170.

32. Griswold, M.A., Jakob, P.M., Heidemann, R.M., and Haase, A. (2000). Parallel imaging with localized sensitivities (PILS). *Magn. Reson. Med.* 44: 243–251.

33. Kyriakos, W.E., Panych, L.P., Kacher, D.F., Westin, C.F., Bao, S.M., Mulkern, R.V., and Jolesz, F.A. (2000). Sensitivity profiles from an array of coils for encoding and reconstruction in parallel (SPACE RIP). *Magn. Reson. Med.* 44: 201–208.

34. Griswold, M.A., Jakob, P.M., Heidemann, R.M., Nittka, M., Jellus, V., Wang, J., Kiefer, B., and Haase, A. (2002). Generalized autocalibrating partially parallel acquisition (GRAPPA). *Magn. Reson. Med.* 47: 1202–1210.
35. Sodickson, D.K. and McKenzie, C.A. (2001). A generalized approach to parallel magnetic resonance imaging. *Med. Phys.* 28: 1629–1643.
36. Samsonov, A.A. and Johnson, C.R. (2004). Noise-adaptive nonlinear diffusion filtering of MR images with spatially varying noise levels. *Magn. Reson. Med.* 52 (4): 798–806.
37. Macovski, A. (1996). Noise in MRI. *Magn. Reson. Med.* 36: 494–497.
38. Papoulis, A. (1965). *Probability, Random Variables And Stochastic Processes.* McGraw-Hill, New York.

Part III

Image Processing and
Quantitative Analysis

7 Image Registration Methods in MRI

Vincenzo Positano

CONTENTS

7.1   INTRODUCTION

Magnetic resonance imaging (MRI) provides information about the size, shape, and spatial relationships among anatomical structures, together with functional information with or without the use of contrast media. Combining the information provided by MRI with that provided by other acquisition modalities is an important issue in MRI-based diagnostic. For instance, computer-assisted tomography (CT) images of bony structures and ultrasound (US) views of soft tissues can improve the anatomical information provided by MRI. Positron emission tomography (PET) and single-photon computed tomography (SPECT) imaging provide quantitative information on blood flow and metabolic processes that can be combined with MRI. Other important fields of application of multimodal registration are image-guided therapy, neurosurgery, and orthopedic surgery [1]. Registration is also used in treatment planning [2], and brain atlases and mapping [2]. Developing applications include multimedia patient records, postgenomic registration to characterize gene function, registration of intra- and preoperative images in surgical interventions, and treatment monitoring.

In the MRI field, registration along MR images acquired from the same subject is also extensively used. Some examples are treatment verification by comparison of pre- and postintervention images and growth monitoring. Another important application is the spatial image realignment of each slice to its neighbors in volumetric image data and of each temporal frame in respect to the other acquired frames. In fact, acquisitions requiring multiple breath-hold and gated imaging studies can be expected to exhibit slice-to-slice and frame-by-frame misalignments due to patient motion. Last, in functional magnetic resonance imaging (fMRI) [3,4] registration is used both to correct image misalignment due to patient movement and to map the functional images on anatomical images to localize activation regions.

To join the information provided by different images, the images must be appropriately combined or fused. Before images can be fused, they must first be geometrically and temporally aligned. This alignment process is known as *registration*. The medical image registration field showed an impressive growth over the past decades. In the PubMed database, the number of publications about image registration increased from 10 in 1990 to about 140 in 2002 [5]. Among the different image modalities, MRI holds the first place with about 25% of all records. In the last decade, interest in registration in MRI is rather constant, while there is a growing interest in functional MRI.

Many criteria can be used to classify registration methods [6]. If the registration procedure involves image coming from different modalities it is defined as *multimodal registration*. When the registration involves images produced by the same modality, the registration is called *unimodal*.

Criteria can be also related to the dimensionality of the images (2-D, 3-D, or dynamic 3-D images), the nature of registration basis, the nature of the transformation (rigid, affine, projective, or curved), the interaction (manual, semiautomatic, and automatic methods), the modalities involved, and the subject (intrasubject, intersubject, or atlas).

Regarding the nature of registration basis, image-based registration can be divided into two main classes:

1. Extrinsic, i.e., based on foreign objects introduced into the imaged space designed to be well visible in the pertinent modalities. The main disadvantage of extrinsic methods is that they are invasive in nature.
2. Intrinsic methods, based only on the image data. Intrinsic methods can be based on a limited set of salient points (landmarks), on alignment of segmented structures (segmentation of features based), or on measures computed from the image gray values (voxel based). Landmarks can be manually selected by the user (anatomical landmarks) or automatically extracted (geometrical landmarks). Geometrical landmarks correspond to the optimum of some geometrical proprieties, as corners

and local curvature extrema. In the segmentation-based methods, two lines or surfaces sets (i.e., image features) are extracted from both images and used as input for the alignment procedure. The voxel-based methods operate directly on the image gray values. In principal axes and moment-based methods, the image content is reduced to a representative set of vectors, and the registration is performed using the extracted vector set. The methods using the full image content attempt to perform the registration maximizing the cross-correlation, the mutual information of some other relationship between images. Voxel-based registration does not generally require extensive preprocessing, such as segmentation or feature extraction.

About the interaction, the registration methods can be divided into:

1. Automatic, when the user only supplies the algorithm with the image data
2. Semiautomatic, when the user has to initialize the algorithm performing the segmentation or have to accept or reject suggested registrations
3. Interactive, when the user does the registration himself, helped by the software

In automatic or semiautomatic registration algorithms, there are generally three main aspects:

1. The search space is the class of potential transformations, such as rigid, affine, and elastic, used to align the images. Three-dimensional (3-D) rigid-body registration has six degrees of freedom: x, y, and z translation and rotation about x, y, and z axes. Affine transformations add shearing and scaling. The most general class of transformation, elastic, or nonlinear registration, has in theory infinite degrees of freedom.
2. The similarity metric is an indicator of how well the features or intensity values of two images match. The sum of squared intensity difference [7], generalized correlation coefficient [8], ratio image uniformity, and information theoretic measures [2,3,9] are commonly used similarity measures.
3. The search strategy optimizes the similarity metric. Examples include local or global searches, multiresolution approaches, and other optimization techniques.

In this chapter, we will first formulate the registration problem, focusing on rigid 3-D registration. Although nonlinear registration is more realistic in principle because tissues are deformable in some manner, rigid registration is often used in unimodal registrations, which is a field of particular interest in MRI. Moreover, when a large set of data is involved, as usually happens in MRI (e.g., fMRI, cardiac imaging), nonlinear registration requires excessive computational power. In the following text, similarity metrics are discussed with a focus on the mutual information measure, which is most often used in the MRI registration field. We will also discuss some optimization techniques, from the classical ones (i.e., simplex and Powell methods) to the more advanced (i.e., genetic algorithms). Because registration of image sequences is often needed in the MRI field, we cover extension of the standard registration procedures. Finally, we describe two examples of the application of registration procedures to MRI data (fMRI images and cardiac perfusion MRI images).

7.2 THE REGISTRATION PROBLEM

In the following, we assume that an image can have two or three dimensions. Let T denote the spatial transformation that maps coordinates (spatial locations) from one image or coordinate space to another image or coordinate space. Let p_A and p_B denote coordinate points (pixel locations) in images A and B, respectively. The image registration problem is to determine T so that the mapping

$$T : p_A \rightarrow p_B \Leftrightarrow T(p_A) = p_B \qquad (7.1)$$

results in the best alignment of A and B. The domain where T is defined is named the *search space* of the registration problem. The function that defines the quality factor for the alignment is named the *similarity metric* or *registration metric*. The algorithm used for the search of the function T that maximizes the chosen metric is named the *search strategy*.

In the most general case, two medical images may differ from another by any amount of rotation about an axis, by any amount of translation in any direction, may differ in scale, and nonrigid transformation can be present. Moreover, these features may vary locally throughout the volumetric extent of the images. The nature of the T transformation characterizes the search space of the registration problem, ranging from nonlinear transformation, with virtually infinite degrees of freedom, to rigid registration with six degrees (for 3-D volumes) of freedom. Intermediate cases are affine transformations, in which the images can be scaled and sheared.

It is important to note that MRI are codified in digital image format, typically the Digital Imaging and Communications in Medicine (DICOM). The DICOM format includes some information that can be useful for image registration, as the position and the orientation of the image in respect to the acquisition device and in respect to the patient (as well as the voxel size) so that a preliminary registration can be performed using this geometrical data, reducing the image misalignment. In the unimodal registration of MRI images, the absence of image scaling can be ensured by the use of the same acquisition device with the same acquisition parameters. Moreover, because the pixel dimension in both images is known from the acquisition parameters, the scaling factor can be easily computed and image scaling can be easily applied. The main sources of nonrigid distortions in MRI are due to the subject's breathing during acquisition. These kinds of distortions affect cardiovascular and abdominal MRI, whereas they are negligible in brain imaging. Moreover, because heart movement is not rigid in nature, some deformation of the heart shape can occur due to poor EEG synchronization in cardiac MRI. In practice, we can often suppose that the images to be registered will differ only for a rigid transformation.

For 3-D rigid-body registration, the mapping of coordinates $p_A = [x\ y\ z]^T$ into $p_B = [x'\ y'\ z']^T$ can be formulated as a matrix multiplication in homogeneous coordinates:

$$\begin{bmatrix} x' \\ y' \\ z' \\ 1 \end{bmatrix} = \begin{bmatrix} \cos\beta\cos\gamma & \cos\alpha\sin\gamma+\sin\alpha\sin\beta\cos\gamma & \sin\alpha\sin\gamma-\cos\alpha\sin\beta\cos\gamma & t_x \\ -\cos\beta\sin\gamma & \cos\alpha\cos\gamma-\sin\alpha\sin\beta\sin\gamma & \sin\alpha\cos\gamma+\cos\alpha\sin\beta\sin\gamma & t_y \\ \sin\beta & -\sin\alpha\cos\beta & \cos\alpha\cos\beta & t_z \\ 0 & 0 & 0 & 1 \end{bmatrix} \begin{bmatrix} x \\ y \\ z \\ 1 \end{bmatrix}$$

$$(7.2)$$

where $[t_x, t_y, t_z]$ are the translation vectors and $[\alpha, \beta, \gamma]$ are the rotation values around the three axes.

Because digital images are sampled on a discrete grid, but T generally maps to continuous values, interpolation of intensities is required. The interpolation process can affect the effectiveness of the registration, so that the choice of an appropriate interpolation algorithm plays an important role in the development of the registration procedure. This topic will be extensively covered in the following text. The general registration algorithm between two images is shown in Figure 7.1.

First, we define a reference image and a floating image, which is the image to be registered in respect to the reference one. The similarity function between the reference image and the floating one is evaluated. An optimization algorithm is used in order to estimate the best transformation function (T) that maximizes the similarity function; the estimate function is used to transform the floating image. An interpolation operation is also required. If the result is satisfactory, the procedure ends; if not, a new transformation function is evaluated, and a new loop is executed. The key issues in the registration algorithm are the choice of the similarity metric and the choice of the optimization algorithm. These issues will be described in the following subsections.

Sometimes, the images involved in registration can be preprocessed in order to improve the effectiveness of the registration algorithm. The most common preprocessing step is defining a region of interests in images to exclude structures that may negatively affect the registration process. Other preprocessing techniques consist of image filtering to remove noise, correction for intensity inhomogeneities, and image resampling to achieve the same spatial resolution in both images.

7.3 SIMILARITY METRICS

The similarity metric for image registration should satisfy some constraints. First, similarity metrics must be robust; that is, they should converge to a global maximum at the correct registration. The best registration can be in some cases

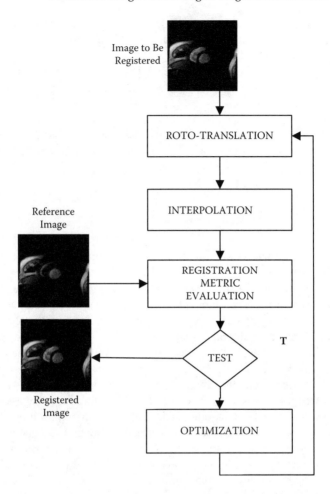

FIGURE 7.1 Flowchart of the general registration problem between two images.

a local (not global) optimum. However, this problem can be overcome by selecting an initial orientation close to the correct registration, as we have seen in the previous paragraph. We will assume in the following text that the global optimum is obtained at the correct registration transformation. Another important quality of the similarity metric is the computational complexity that affects the time required to perform the registration.

Generally, three types of similarity metrics have been proposed in image registration [10]. They are based on corresponding points, corresponding surfaces, and corresponding image intensities. We can group the first two methods in one and summarize the two main approaches to registration as follows:

1. Similarity measure by extraction of some geometrical features from the two images: The extracted features are compared, and a similarity

index is extracted. The main advantage of this approach is about the dramatic reduction of required processing time. In fact, the dimension of extracted features is usually at least a magnitude order below the dimension of images (i.e., lines vs. 2-D images, surfaces vs. 3-D images). The main disadvantages are the arbitrary choice of the feature to extract and the drawbacks in the correct extraction of the features.

2. Similarity measure by direct comparison of images to be registered: The main disadvantage of this approach is the required processing time. In fact, all image data are involved in the analysis. The main advantage is the independence from any user input. This kind of method is also known as voxel-based (3-D) or pixel-based (2-D) methods.

Both approaches were extensively used for medical image registration. An example of the first approach is the registration procedure for two MRI cardiac images. We have to extract the same geometrical feature from both reference and floating images. In the case of cardiac MR image registration, the left ventricle contour is a natural choice and can be extracted from both images with an automatic algorithm such as the one described in Reference 11. If the left ventricle contours were correctly extracted from both images, the similarity between two images can be defined as the difference between the two extracted contours, introducing a definition of the distance between two closed curves. An optimization process can be used to minimize the previously defined distance in performing the image registration. An example of the previously described procedure is the iterative closest point (ICP) algorithm introduced by Besl and McKay [12]; it is a general-purpose method for the registration of two generic point sets representations, including line segments sets, implicit curves, parametric curves, and generic surfaces. At the end of the optimization process, we have the rotation matrix and the translation vector that register the two curves with each other. The convergence theorem guarantees the achievement of a local minimum. The roto-translation matrix can be now applied to the floating image to perform the registration. The example shows both the advantages and the disadvantages of the feature-based approach: the registration operation involves 1-D data (i.e., the extracted contour) and consequently is very fast and accurate. On the other hand, the feature extraction (i.e., the localization of left ventricle contours) can be difficult and error prone. In conclusion, the use of feature-based algorithms is suggested only when fast and effective segmentation algorithms are available on the images to be registered.

The term *voxel-based methods* implies the comparison of gray levels of the images to be registered.

The simplest metric involves the use of difference or absolute difference between images (mean square difference):

$$MQ(A, B) = \sqrt{\sum_{i,j \in S} (a_i - b_j)^2} \qquad (7.3)$$

where (i, j) denotes the couple of corresponding voxels in the two images, and S is the domain where both images are defined. The introduction of the S domain means that the metric must be computed only in the geometrical region where both images are defined. This approach holds in all voxel-based metrics. This simple metrics can be effectively used when the images to be registered are sufficiently similar.

7.3.1 MUTUAL INFORMATION

Much of the current work on biomedical image registration utilizes information theoretic voxel similarity measures, in particular, mutual information (MI) based on the Shannon definition of entropy[13,14]. The MI concept comes from information theory, measuring the dependence between two variables or, in other words, the amount of information that one variable contains about the other. The MI concept measures the relationship between two random variables, i.e., intensity values in two images: if the two variables are independent, MI is equal to zero. If one variable provides some information about the second one, the MI becomes greater than zero. MI is related to the image entropy by:

$$MI(X; Y) = H(X) + H(Y) - H(X, Y) \tag{7.4}$$

where X and Y are the two images and $H(\cdot)$ is the entropy of a random variable, and is defined as:

$$H(X) = - \sum_{x_i \in \Omega_X} \log[P(X = x_i)] \cdot P(X = x_i) \tag{7.5}$$

The joint entropy of two images X and Y is:

$$H(X, Y) = - \sum_{x_i \in \Omega_X} \sum_{y_j \in \Omega_Y} \log[P(X = x_i, Y = y_j)] \cdot P(X = x_i, Y = y_j) \tag{7.6}$$

All entropies must be evaluated on the domain where both images are defined, usually as overlapping areas. Normalized mutual information (NMI), given as

$$I_N(X, Y) = \frac{H(X) + H(Y)}{2H(X, Y)} \tag{7.7}$$

is less sensitive to the size of the overlap [15] and can be used instead of MI.

The probability distribution for the evaluation of MI and NMI can be estimated with Parzen windows, histograms, or other probability density estimators. The most common method uses images histograms. Let Q and K be images with M

pixels each assuming N gray levels $g_1, g_2, \ldots, g_N$. The mutual information MI between Q and K can be defined as:

$$MI(Q, K) = H(Q) + H(K) - H(Q, K) \tag{7.8}$$

where $H(\cdot)$ is the entropy of an image. $H(Q)$ can be written as:

$$H(Q) = -\sum_{i=1}^{N} \log[P(Q = g_i)] \cdot P(Q = g_i) \tag{7.9}$$

$P(Q = g_i)$ means the probability that a pixel in Q image will assume the value g_i.

So, the image entropy can be written in terms of the image histogram His_Q:

$$H(Q) = -\frac{1}{M} \sum_{i=1}^{N} \log\left[\frac{His_Q(i)}{M}\right] \cdot His_Q(i) \tag{7.10}$$

Also, the joint entropy of two images Q and K with the same number of pixels M and the same gray-level range N can be written in terms of joint image histogram His_{QK}:

$$H(Q, K) = -\frac{1}{M^2} \sum_{i=1}^{N} \sum_{j=1}^{N} \log\left[\frac{His_{QK}(i, j)}{M^2}\right] \cdot His_{QK}(i, j) \tag{7.11}$$

$His_{QK}(i, j)$ is equal to the number of simultaneous occurrences of $Q = i$ and $K = j$.

The MI registration criterion states that the MI of the image intensity values of corresponding voxel pairs is maximal if the images are geometrically aligned. Because no assumption is made about the nature of the relation between the image intensities, this criterion is very general and powerful. MI has been shown to be robust for both multimodal and unimodal registration, and does not depend on the specific dynamic range or intensity scaling of the images. The MI as previously defined is not a negative number. Because many optimization algorithms are formulated as minimization algorithms, the negative of MI (MI) is often used as a similarity metric.

7.3.2 PHANTOM EXPERIMENTS

In order to explain the differences among similarity metrics, an experiment was performed using synthetic images. A 3-D phantom was realized, constituted by two coaxial elliptical cylinders. Three regions were defined with three different signal levels, as depicted in Figure 7.2A. Gaussian noise was added to the phantom.

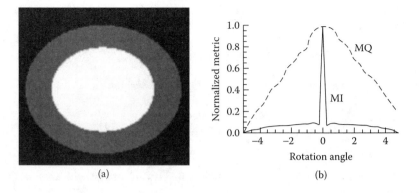

FIGURE 7.2 (A) Synthetic phantom and (B) MQ and MI metrics values for rotation angles from −5° to +5°.

The phantom was rotated from −5° to +5° in 40 steps, and the normalized values of square root metric (MQ) and MI between all the rotated images and the original image (corresponding to 20th frame) were plotted (Figure 7.2B).

From the plot, the two metrics converge to the correct alignment in a continuous manner; the MQ metric shows a "smooth" convergence around the correct alignment value, whereas MI presents a clear maximum corresponding to the 20th frame.

The same image sequence was modified including a signal value change along frames (Figure 7.3A), simulating the presence of an MR contrast medium that diffuses in blood and in muscular tissues. In this case, the shape of the MI metrics remains continuous and without local maxima, while the MQ metric shows the presence of a local maxima at the −3° location (Figure 7.3B). The proposed example shows that, in the presence of signal changes that do not modify the pattern distribution in images to be registered, metrics related to the image histogram are

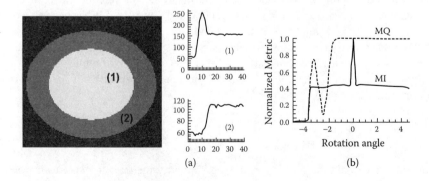

FIGURE 7.3 (A) Synthetic phantom with signal change during time and (B) the related MQ and MI metrics values for rotation angles from −5° to +5°.

more robust in respect to those directly related to voxel values. In MR image processing, signal changes in the image of the same anatomical district can often occur, so the use of information-based metrics can be useful. Signal changes can be related to the use of a contrast agent (i.e., first-pass cardiac and brain perfusion) and to different acquisition device settings (i.e., images of the same patient acquired in different times).

7.4 THE INTERPOLATION EFFECT IN THE REGISTRATION PROBLEM

When subvoxel translation or image rotation is involved, image interpolation (also defined as *resampling*) is also required to obtain the roto-translated image. In the interpolation operation, the coordinate grid is defined by the voxel locations in the reference image. The voxel values of the roto-translated floating image must be recomputed in the coordinate grid. Several methods have been proposed for interpolation of medical images; an extended review can be found in [16,17]. The main interpolation methods are truncated and windowed sinc, nearest neighbor, linear, quadratic, cubic B-spline, and Lagrange and Gaussian interpolations. Because the interpolation operation has to be repeated for each computation of the similarity function, both interpolation accuracy and computational complexity are important in the choice of the interpolation method. In the MRI field, some interpolation algorithm optimized for MR images [18] have been proposed. These algorithms use sinc-based interpolation, taking into account the bandwidth of the MR signal to find the best sinc shape. These methodologies, although more effective in respect to traditional methods, are usually too slow to be adopted in the solution of the registration problem. Instead, the interpolation required to obtain the final, registered image is performed only once, and the choice of an accurate interpolation method is appropriate.

The main trouble with interpolation operation is that it can modify the gray-level values, affecting the evaluation of the similarity function. This effect can lead to incorrect registration results if a histogram-based metric such as MI or NMI is used. Often, the image dynamic is reduced to avoid this effect.

The simplest interpolation algorithm is the nearest neighbor interpolation, in which the new voxel values are recomputed as the value of the closest neighboring voxel. This algorithm preserves the gray values of the original voxels but makes the registration metric insensitive to intravoxel misalignment because image movements less than half of the pixel size do not modify the interpolated images.

A more effective interpolation algorithm is trilinear interpolation (bilinear in 2-D images), in which the values of recomputed voxels are evaluated as the weighted sum of the neighboring voxels. This technique introduces new gray values in the interpolated image.

When the similarity function is based on joint histogram computation as in MI, it is preferable to use an interpolation technique that preserves the gray-level distribution. This technique, called *trilinear partial volume distribution* (PV)

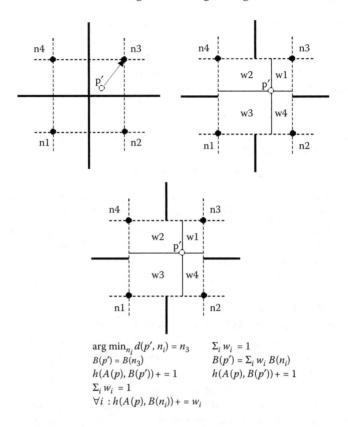

$$\arg\min_{n_i} d(p', n_i) = n_3 \qquad \Sigma_i w_i = 1$$
$$B(p') = B(n_3) \qquad\qquad B(p') = \Sigma_i w_i B(n_i)$$
$$h(A(p), B(p')) + = 1 \qquad h(A(p), B(p')) + = 1$$
$$\Sigma_i w_i = 1$$
$$\forall i : h(A(p), B(n_i)) + = w_i$$

FIGURE 7.4 Three types of interpolation for the evaluation of the joint histogram.

interpolation, updates the joint histogram for each voxel pair in the two images. Instead of interpolating new intensity values, the contribution of the image intensity of each voxel to the joint histogram is distributed over all the intensity values of the neighboring voxels, using the same weights as for trilinear interpolation.

Figure 7.4 shows the previously described interpolation algorithms. Nearest neighbor interpolation and trilinear (bilinear, in the present 2-D example) interpolation find the reference image intensity value at position p and update the corresponding joint histogram entry at p, whereas PV interpolation distributes the contribution of this sample over multiple histogram entries defined by its NN intensities, using the same weights as for bilinear interpolation.

To explain the differences among the three methods, we propose an experiment on a synthetic data set. A metric related to image voxel values (i.e., the mean square difference [MSD]) and a metric related to voxel statistic distribution, such as MI, were tested. Two different interpolation techniques were adopted, trilinear interpolation (TRI) and trilinear partial volume distribution (PV). The method of the experiment has been tested on a simulated data set that reproduces a real heart shape.

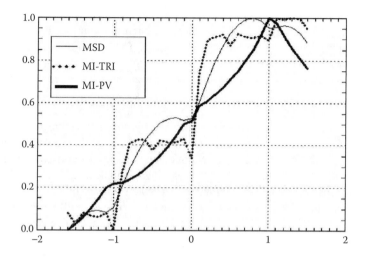

FIGURE 7.5 Registration with MSD, MI-TRI, and MI-PV approaches on simulated data.

The pixel resolution was 1.2 mm with a 5-mm slice thickness. Random 3-D roto-translation was imposed on the data set and the efficiency of the registration procedure was evaluated for the MSD metric with trilinear interpolation (MSD-TRI), MI metric with trilinear interpolation (MI-TRI), and MI metric with trilinear PV distribution interpolation (MI-PV) approaches. MI-PV performs the registration of data set best, followed by MSD and MI-TRI.

As an example, in Figure 7.5 the normalized values of MSD, MI-TRI, and MI-PV are shown vs. the translation along the z axis (i.e., the heart's longitudinal axis). The two data volumes to be registered were shifted by 1 mm along the z axis. The three approaches lead to comparable results, but MI-TRI and MSD present local maxima that can trap the optimization algorithm, leading to incorrect results.

7.5 OPTIMIZATION TECHNIQUES IN IMAGE REGISTRATION

Determination of the parameters set that maximizes a multivariable function is called the *optimization problem*. In the rigid registration problem, the optimization algorithm should find the rotation and translation parameters that will maximize the similarity function. If the registration is elastic, the number of parameters is virtually infinite and should be reduced by introducing appropriate hypotheses. A commonly used approach is to perform a rigid registration extended to the whole image, followed by a local elastic registration. The elastic registration is constrained to some mathematical model to limit the needed number of parameters.

To reduce the required processing time, a multiscale approach is often used [19,20] in which the registration is an iterative process. In the first step, low-resolution images are registered. The result of this step is used as initial estimate for the next registration that involves images at higher resolution. The process is iterated until the best available resolution is reached. The main problem encountered in the registration process is the presence of local maxima of the similarity metric. Many similarity metrics, such as MI, are irregular and rough and are often trapped in local optima. In particular, this problem affects the multiresolution approach, because the global optimum may not be present in lower resolutions.

In conclusion, the choice of the appropriate optimization technique is a compromise between the effectiveness of the method (i.e., the ability to find the global optimum) and the processing time required for the optimization process. Local methods, such as the Powell method [21], conjugate gradient [22], and the Levenberg–Marquard or simplex algorithms [23] provide good performance but do not guarantee that the global optimum of the similarity function will be reached. On the other hand, global optimization methods such as simulated annealing [24], genetic algorithms [25], tabu search [26], and particle swarm optimization [27] are generally more expensive in terms of processing time although they ensure the convergence to a global optimum under some conditions. An extensive description of the optimization is beyond the scope of this chapter. In the following text, we describe two representative methods of the two classes, the Nelder–Mead simplex algorithm and the genetic algorithm approach.

7.5.1 NELDER–MEAD SIMPLEX ALGORITHM

The Nelder–Mead algorithm is one of the most well-known optimization algorithms, also known as AMOEBA from the name of its implementation in the book *Numerical Recipes* [28]. It is often able to find reasonably good solutions quickly with only a few function evaluations per iteration. The convergence properties are less than satisfactory, so a number of variants have been proposed in the literature that attempt to address such issues [29].

The amoeba algorithm maintains at each iteration a nondegenerate simplex, a geometric figure in n dimensions (the number of dimensions is equal to the number of optimization parameters), that is, the convex hull of $n+1$ vertices, $x_0, x_1, \ldots, x_n$ in the search space, and their respective function values. In the solution of the registration problem, the function is the similarity metric. In each iteration, new points are computed, along with their function values, to form a new simplex. The algorithm terminates when the function values at the vertices of the simplex satisfy a predetermined condition. In the case of 2-D rigid registration (N = 3), one iteration of the amoeba algorithm consists of the following steps:

> *Order:* Order and relabel the four vertices as x_0, x_1, x_2, and x_3, such that
> $f(x_0)$ $f(x_1)$ $f(x_2)$ $f(x_3)$; x_0 is defined the *best vertex*, while x_3 is the
> *worst point*, and x_2 is the *next-worst point*. The algorithm computes $\bar{x}$

as the centroid of the three best points in the simplex (i.e., all vertices except x_3).

Reflect: Compute the *reflection* point x_r,

$$x_r = \bar{x} + \alpha(\bar{x} - x_3)$$

Evaluate f(x_r). If f(x_0) f(x_r) > f(x_4), accept the reflected point x_r and terminate the iteration.

Expand: If f(x_r) > f(x_0), compute the *expansion* point x_e,

$$x_e = x_r + \beta(x_r - \bar{x})$$

If f(x_e) > f (x_r), accept x_e and terminate the iteration; otherwise, accept x_r and terminate the iteration.

Contract: If f(x_r) < f($x_3 - 1$), perform a contraction between $\bar{x}$ and x_n:

$$x_e = \bar{x} + \zeta(\bar{x} - x_3)$$

If f(x_e) $\geq$ f(x_n) accept x_e and terminate the iteration.

Shrink simplex: Evaluate f() at the three new vertices:

$$x_i = x_0 + \eta(x_i - x_0)$$

The suggested values for the parameters are: $\alpha = 1$, $\beta = 1$, $\zeta = 0.5$, and $\eta = 0.5$.

The simplex algorithm is often used in the solution of the registration problem due to its small computational complexity. This allows work on large data sets to be completed in a reasonable time. Another application of the simplex algorithm is to perform a preliminary registration followed by a more effective registration executed with a more complex search strategy. In the last paragraph, an application of the Nelder–Mead algorithm, together with the Powell algorithm, will be described.

7.5.2 GENETIC ALGORITHMS

Genetic algorithms are inspired by Darwin's theory of evolution, and use an evolutionary process to find the best solution of an optimization problem. Algorithms begin with a set of individuals (represented by chromosomes) called a *population*. Each set of chromosomes represents a possible set of parameters of the function to be optimized. The value of the function (i.e., the similarity function in the registration problem) for a certain chromosome set is called *fitness of the corresponding individual*. Selecting the individuals with the best fitness, a new population is formed. This is motivated by the hope that the new population will be better than the old one. The process is repeated until some condition (for example a certain population or improvement of the best solution) is satisfied.

The technique can be summarized thus: (1) a population of individuals encoded as chromosomes propagates (2) copies of these individuals based on external fitness criteria, creating (3) a generation of new individuals by mutating chromosomes and recombining members of the population. Genetic algorithms for the solution of a registration problem are typically implemented as follows:

1. A similarity function is chosen that describes the fitness of any potential solution. The similarity function will be selected from a number of parameters, depending on the registration algorithm.
2. A population of candidate solution is initialized. Typically, each solution is described by a vector x, called a *chromosome*, with elements called *genes*. In the case of a 3-D rigid registration, the chromosome is a six-element vector.
3. Each chromosome is used to evaluate the fitness function (i.e., to perform a roto-translation of the floating image). The value of the fitness function (e.g., the mutual information value between the reference image and the roto-translated one) is used as a fitness score.
4. We assign to each chromosome a probability of reproduction, depending on the fitness score. The reproduction probability will be proportional to the fitness of the chromosome in respect to the fitness of the others in the population.
5. A new population of chromosomes is generated by taking into account the reproduction probability defined in the previous task. The new population is generated recombining the chromosomes of the existing population.
6. A random mutation of the chromosomes is introduced.
7. The process is halted if a suitable solution has been found. Otherwise, the process returns to step 3.

The key issues in the implementation of the genetic algorithm are the rules for producing the new generation and the amount of random mutation to be introduced. Regarding the latter, the mutation probability should be very small (10^4 is a typical value) to avoid damaging the good solutions that evolve during the iterative process. The role of the random mutation is to explore new search regions by avoiding premature convergence of the algorithm to suboptimal solutions.

Several rules for creation of the new generation have been proposed in literature. Usually, a percentage of old individuals (i.e., the ones with the best fitness) are saved and included in the new generation. New individuals are then created to build a population with the same size. A probability of reproduction is assigned to each individual, depending on the fitness of the individual in respect to the fitness of the others in the population. New individuals are created combining the chromosomes

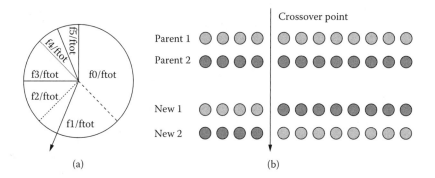

FIGURE 7.6 (A) Roulette wheel selection for genetic algorithms and (B) crossover.

of two or more parents extracted from the population. "Roulette wheel" selection is often used to perform this task. Figure 7.6A shows how roulette wheel selection works in the case of six individuals with fitness $f_0, \ldots, f_5$. A circle (the wheel) is divided into six sectors with a width proportional to the normalized fitness (*ftot* means the sum of $f_0, \ldots, f_5$). A angle is randomly extracted (the arrow in the figure), and the individual corresponding to the sector indicated by the arrow is selected for reproduction. In this way, all the individuals have a probability to reproduce, and the probability is proportional to the chromosome fitness. The process is repeated twice for each new individual to be produced in order to select the two parents involved in the reproduction process. The chromosomes of the two parents are usually combined using the random selection of a crossover point (Figure 7.6B). Genetic algorithms are generally more robust in respect to the presence of suboptimal maxima in the fitness function. The main disadvantage is the great computational complexity; in fact, the fitness function has to be evaluated for each individual in each iteration. However, genetic algorithms are inherently parallelizable, and the time required for optimization is independent of the number of parameters of the fitness function. Therefore, they can be effectively applied to registration problems involving complex transformations such as the elastic ones, and in global registration of multiple data sets.

7.6 REGISTRATION OF MULTIPLE DATA SETS

MRI data sets are often composed of a large number of parallel slices that cover a 3-D region inside the body. Slices can be acquired in different times, so patient movement, breath, or poor EEG synchronization in cardiac imaging can lead to image misalignment along the slices. Reconstruction of such data sets into 3-D volumes, via the registrations of 2-D sections, is often needed in order to perform correct 3-D visualization and morphometric analysis (e.g., surface and volume representation) of the structures of interest. Consecutive slices may differ significantly owing to the fact that they represent different anatomical locations, and the difference along slices is more pronounced when the distance between images involved in the registration is large. The purpose of the registration is to realign the slices in order to reconstruct the correct shape of the organ under examination. A typical example is in cardiovascular imaging, in which the shape of the heart or of valves should be reconstructed from parallel sections, correcting artifacts due to patient breath. In the present problem, the registration method must be robust to missing data or outliers. Registering the slices sequentially (the second with respect to the first, the third with respect to the second, etc.) sometimes leads to misregistration. In fact, if an error occurs in the registration of a slice with respect to the preceding slice, this error will propagate through the entire volume so that a global offset of the volume may be observed due to error accumulation. If all the slices are registered in respect to the one taken as reference, differences between slices can be too large to allow correct registration.

Krinidis et al. proposed a solution introducing the use of a global energy function having as variables the rigid transformation parameters of each slice [30]. The global energy function is minimized with the ICP algorithm, which is able to register multiple views of a 3-D structure. The implemented global energy function is associated with a pixel similarity metric based on the Euclidean distance transform.

Consider a set of N slices $(I_1, \ldots, I_N)$. A pixel in a slice is represented by $p = (x, y)$. The alignment of all images in the sequence can be achieved by maximizing an energy function $E(\cdot)$, which expresses the similarity between the 2-D images:

$$E(\Theta) = \sum_{i=1}^{N} \sum_{j=1}^{N} f(T_i(I_i), T_j(I_j)) \tag{7.12}$$

where $f(\cdot)$ is a similarity metric, denotes a rigid transformation matrix. The chosen metric energy function accumulates the similarity between each transformed image and all of the other already-transformed images. Assuming that the similarity function is symmetric leads to the following global maximization problem, where W_{ij}'s are appropriate weights for the similarity between each slice pair:

$$\hat{\Theta} = \arg \max_{T} \sum_{i=1}^{N} \sum_{j=i}^{N} W_{ij} f(T_i(I_i), T_j(I_{j_j})) \tag{7.13}$$

Without any additional constraints, the optimization problem has an infinite number of solutions. If the transformation applied to an arbitrary chosen image is constrained to be the identity transformation, we have $3(N-1)$ parameters to estimate in the case of 2-D rigid transformation. If the transformation is nonrigid, the number of parameters to estimate becomes virtually infinite. In this case, some constraints have to be introduced about the nature of nonrigid transformation in order to reduce the number of unknown parameters involved in the registration operation.

To simplify the optimization problem, we can suppose that $W_{ij} = 0$ for some (i, j) couples. In particular, we can state that $W_{ij} = 0$ if $|i\text{-}j| > R$. This means that the similarity function is computed only for near slices that present greater similarity. The optimization algorithm described is based on random selection of a slice I_i in the sequence followed by local registration in respect to I_i of all other slices in the neighborhood of i. The process is iterated until all slices have been processed.

The local registration is performed by extracting contours for involved images and minimizing the distance between extracted contours by an iterated conditional models (ICM) optimization algorithm. Note that image contours have to be extracted just one time at the beginning of the registration process. The described methodology clearly shows how the solution of the global registration problem will require the introduction of some hypothesis to simplify the problem, and the use of similarity metrics that can be computed in a very short time.

A 3-D data set can be also acquired several times following the evolution of a phenomenon under investigation, as happens in monitoring the perfusion of a contrast medium in tissues (e.g., the brain on cardiac perfusion) with an endogen signal change (e.g., fMRI). These data sets are usually defined as *dynamic 3-D* or *4-D acquisitions*. In these cases, misalignment in time acquisition will result in artifacts in signal monitoring. Registration of a 4-D data set is often reduced to a number of registrations of image pairs in order to reduce the required computation time and to exploit the available registration algorithms. However, global registration of multiple data sets can lead in general to better results.

In theory, it is possible to extend the concept of similarity metrics to more than two images. As an example, in square root metric, the similarity S along N images can be defined as the squared root of the sum of quadratic distances of each corresponding point in the definition field Ω, where the distance along the N corresponding points is a suitable distance metric defined in an N-dimensional space. In this approach, the similarity metric depends on all the registration parameters (e.g., 6N parameters in 3-D rigid registration) and can be optimized as previously described. Following information theory, a higher dimensional MI metric can also be defined.

In the case of three images MI becomes:

$$MI(A{:}B{:}C) = H(A) + H(B) + H(C)\ H(A,B)\ H(A,C)\ H(B,C) + H(A,B,C) \tag{7.14}$$

The MI defined in this manner may not necessarily be nonnegative, so it is not a true metric. An alternate definition is often used in medical image registration [31]:

$$MI(A{:}B{:}C) = H(A) + H(B) + H(C)\ H(A,B,C) \tag{7.15}$$

This definition is nonnegative, but is not a natural extension of the MI between two variables and requires estimation of a high-dimensional probability mass function, which is computationally very expensive when the number of involved images increases. A different approach is suggested by Zhang and Rangarajan [32]. They define a different pseudometric that for two images is:

$$R(X,Y) = H(X|Y) + H(Y|X) \qquad (7.16)$$

This metric is related to *MI* by the relationship:

$$R(X,Y) = H(X,Y)\ MI(Y,X) \qquad (7.17)$$

The main advantage of this metric is the possibility of a straightforward extension to the multidimensional case. For three images the metric becomes:

$$R(X,Y,Z) = H(X|Y,Z) + H(Y|X,Z) + H(Z|X,Y) \qquad (7.18)$$

And for N images:

$$R(X_1,X_2,...,X_N) = \sum_{i=1}^{N} H(X_i \,|\, X_1,...,X_{i-1},X_{i+1},...,X_N) \qquad (7.19)$$

To reduce the computational complexity of the metric, an upper bound of R, which is also a metric, can be used that does not require the computation of high-order joint probability. In the case of three images we have:

$$R(X,Y,Z) = H(X|Y,Z) + H(X|Y,Z) + H(Z|X,Y) \leq \frac{1}{2}(H(X|Z) + H(X|Z))$$

$$+ \frac{1}{2}(H(Y|X) + H(Y|Z)) + \frac{1}{2}(H(Z|X) + H(Z|X))$$

$$= \frac{1}{2}(R(X,Y) + R(Y,Z) + R(X,Z))$$

$$(7.20)$$

so that the metric K can be minimized instead of R:

$$K(X, Y, Z) = \frac{1}{2}(R(X, Y) + R(Y, Z) + R(X, Z)) \qquad (7.21)$$

Note that the computation of K does not require the computation of multidimensional joint histograms. The definition of upper bound can be extended to more than three images.

The use of multidimensional registration metrics is usually limited to the simultaneous registration of only three images. A common example is the contemporary registration of PD, T2, and T1 images of the same anatomical district. The two main points that discourage this technique are the difficulty in defining a suitable metric with more than two images and the high computational complexity of the related optimization process. In fact, the computational load required by optimization algorithms dramatically increases with the number of used parameters, whereas the algorithm effectiveness decreases in the same manner. Moreover, registration of only two images is a well-known problem, and many effective algorithms to solve this issue are available. Global registration methods are usually based on a series of image pair registrations.

We can define a global similarity score along a sequence of N images $I_1,\ldots,$ I_N as:

$$S_G = \sum_{i=1}^{N}\sum_{j=i}^{N} W_{ij}S(I_i,I_j) \tag{7.22}$$

The global similarity index is defined as the weighted sum of the similarity of all the possible couples of images. We suppose in this formulation, as usually happens, that the similarity between two images is a symmetric function, so $S(I_i, I_j) = S(I_j, I_i)$. Note that the computation of the global similarity function requires $N(N-1)/2$ evaluations of the metric and depends on NK parameters, where K is the parameter related to each image (3 in 2-D rigid registration, 6 in 3-D rigid registration, etc.). Because the direct research of the S_G maximum by means of an optimization algorithm is too computationally expensive, some alternative methods can be used.

In the first approach, an image in the sequence is chosen as a reference image (r), and only the weights that include the r index are different from zero. Equation 7.22 becomes:

$$S_G = \sum_{i=1,i\neq r}^{N} S(I_r,I_i) \tag{7.23}$$

and S_G can be maximized by maximizing the sum terms. This is equivalent to perform the registration of all images in the sequence respect to the referenced image.

If we define:

$$W_{ij} = \begin{cases} 1 \text{ if } j = i+1 \\ 0 \; j \neq i+1 \end{cases}$$

Equation 22 becomes:

$$S_G = \sum_{i=1}^{N-1} S(I_i, I_{i+1}) \tag{7.24}$$

This is equivalent to performing the registration of each image in the sequence in respect to the previous one. Both the methods are extensively used in image sequence registration. Regarding the second approach, it is important to note that two consecutive images in the sequence are often almost similar, so the registration algorithm can better correct the misalignment. On the other hand, an error in the registration of one image pair will affect the alignment of the temporal sequence.

A third approach often used in literature is to create a virtual reference image as the mean of all images in the sequence and to perform the registration in respect to the virtual image:

$$I_r = \frac{1}{N} \sum_{i=1}^{N} I_i \quad S_G = \sum_{i=1}^{N} S(I_r, I_i) \tag{7.25}$$

All these simplified methods go far to guarantee a global optimum in the registration. It has been empirically showed that the combination of these methods can increase registration quality [33].

In some cases, the image sequence can be divided in groups of aligned images with the presence of a misalignment along groups. As an example, in cardiac perfusion MRI a subject was asked to hold his breath during examination, so that we might have an initial image group with good alignment. When the contrast medium is injected, a first subject movement will likely happen, starting a new image group. A third group will start if the subject breaks the breath-hold state. A second example is f MRI acquisitions, in which major subject movement is related to paradigm changes.

In these cases, it may be preferable to perform a registration along quasialigned groups followed by a registration along different groups (hierarchical registration).

Figure 7.7 shows the main steps of the hierarchical registration algorithm: the value of the similarity between all image pairs is calculated, then a hierarchical tree, grouping images pairs with a high value of the similarity function. In the example, four groups (G_1, ..., G_4) are identified at the low level of the hierarchy. The algorithm performs the registration of all images inside each group. Groups G_1 and G_2 are aligned, using a representative image extracted from each group or two virtual images obtained averaging all images in each group. The resulting group G_5 is registered with the G_3 group, and the obtained G_6 group is registered with the G_4 group.

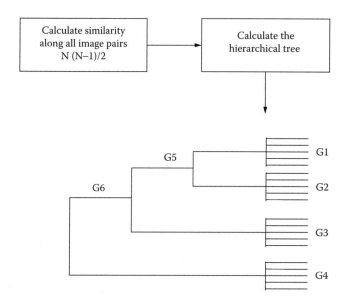

FIGURE 7.7 A hierarchical algorithm for multiple image alignment.

The main advantage of this approach is to perform a large number of registrations between image pairs that are yet almost aligned (i.e., the ones at the low hierarchical level). This reduces the processing time and increases the registration effectiveness. The small number of registrations between groups at high levels in the hierarchy can be eventually performed by sophisticated registration algorithms.

7.7 BRAIN IMAGES REGISTRATION

Image registration plays an important role in MR brain imaging, in particular, in functional image analysis. In functional magnetic resonance imaging (fMRI), in fact, the signal changes due to the hemodynamic response is small compared to signal changes produced by subject movement, heartbeat, and respiration. It should be noted that subject movement of just 100 μm can generate a change in the signal larger than the one caused by activation. Respiration causes bulk movements as well as changes in blood oxygenation that affect the BOLD signal. Subject motion and respiration in MRI scanners cannot be completely eliminated, so the functional images' alignment must be performed in the postprocessing phase. This operation usually involves rigid 3-D registration.

In a single fMRI study, the activation map obtained from a subject by processing functional images is usually superimposed on a high-resolution structural MR image (typically, a T1-weighted MRI). The registration between the two images requires a rigid registration, but in this case the gray-level distribution along the two data sets can differ due to the different acquisition techniques.

Moreover, the spatial resolution of the two sets differs as well. A local elastic transformation can also be applied to correct the geometrical distortions introduced by fMRI.

A third important application of image registration in fMRI studies is the warping of images from a number of subjects into a common standard space in order to compare the data coming from different experiments on different subjects. The most known standard data space is the Talairach atlas [34]. The warping of the images into an atlas requires some elastic transformation.

7.7.1 fMRI IMAGES REGISTRATION

Most current algorithms for motion correction in fMRI consider the head as a rigid object, so that six parameters are needed to define the head as a rigid-body transformation. The involved data sets are often anisotropic, and therefore an appropriate transformation should be defined as previously described. Because the fMRI signal is low, differences between images are small, and the sum of squared differences between the image to be registered and the reference image can be effectively used as a registration metric. This approach is used in SPM and AFNI, the most well-known realignment packages for fMRI. However, the use of more sophisticated metrics such as MI can avoid some artifacts produced by the standard approach. Freire et al. [35] compared seven different motion estimation procedures based on different choices of registration metric: two different implementations of the difference of squares measure (SPM [36] and AIR [37]); the Geman-McClure (GM) robust estimator [38], which takes into account the existence of potential outliers; the ratio image uniformity function; two symmetrical implementations of the correlation ratio (CR) [39], and MI. Results suggested that GM, CR, and MI are robust in respect to the difference of squares measure. Robust metrics such as GM or MI may be the best choices, although in many applications the computational cost of these advanced metrics may be too high in respect to the more simple algorithms required.

To improve registration quality, the registration operation is often followed by a second registration in which all images are registered in respect to the mean of all images realigned in the previous step. A variance image can be computed together with the mean difference in order to provide a better weighting for registration. In particular, the image voxels are weighted proportionally to the inverse of the variance, so that voxels with a lower variance will have more weight in the computation of the registration metric. As described in Section 7.6, more sophisticated approaches can be applied to the registration of multiple data sets, as happens in fMRI.

An important issue in fMRI experiments is the fact that subject motion is often correlated with the experimental design. As an example, if the experimental paradigm requires speech or hand motion, it is likely that head movement will be highly correlated with the paradigm. In this case, it is extremely difficult to separate the activation signals from correlated motion artifacts. In fact, methods that use some measure of correlation between signal and paradigm to remove these artifacts [40] are also likely to remove some of the true fMRI signal. Moreover, in fMRI registration some residual artifacts also can remain after an "exact" image realignment [41], due to the latency of excited tissues when they return to their original state. Both of these problems are yet unsolved.

7.7.2 MAPPING OF fMRI ON ANATOMICAL IMAGES AND BRAIN ATLAS

As previously described, the spatial normalization (i.e., the registration of the subject brain image into a standard coordinate space) is a common task in fMRI imaging. The use of this procedure can be also extended to other applications, such as those in which data coming from different subjects needs to be compared.

In the case of spatial normalization, the registration problem implies the use of an elastic transformation, whereas the registration metric can be based on a feature extraction (also known as label based) or a voxel-based (nonlabel based) approach. In the feature extraction approach, some anatomical features are extracted from the atlas. These have to be extracted from the subject image, and the registration is performed finding the best transformation that superimposes the two feature sets. Landmark points can be manually defined, but the process is error prone and time consuming, and requires the presence of an expert operator. Surfaces can be better identified in brain and can be automatically or semiautomatically extracted. Voxel-based methods use the similarity measures previously described: the sum of squared differences, correlation, and MI.

To reduce the computational complexity of the elastic registration algorithms, a multiresolution approach is often adopted in spatial normalization [42]. In this approach, only a few parameters are determined at a certain resolution. As example, the entire data volume is used to describe global frequency deformation. The volume is then split in some subvolumes and the local frequency deformations are calculated for each subvolume. The process is iterated until the needed precision is achieved. Another approach involves the reduction of the registration parameters to a small number (e.g., 9 to 12 parameters). This loss of precision in registration accuracy allows the registration procedure to be performed in a reasonable time. Moreover, in fMRI studies, different subjects can present different patterns of the brain anatomy, so high-resolution spatial normalization appears to be unnecessary in many applications. The main problem in elastic registration is the choice of constraints or priors that must drive the registration process. Priors are usually incorporated by means of some Bayesian approach, using estimators such as Maximum A Posteriori (MAP) or minimum variance estimate (MVE).

7.7.3 fMRI REGISTRATION EXPERIMENT

To show the importance of the registration operation in fMRI, the previously described procedure is applied to a data set of fMRI images acquired during a *finger–thumb tapping* experiment in which the subject is asked to touch the index finger of the right

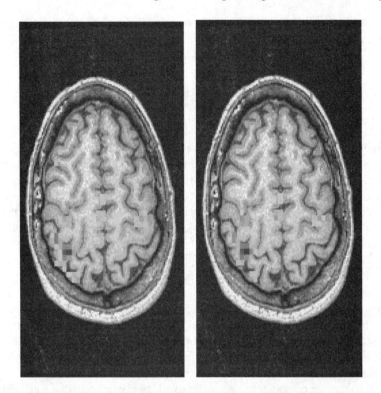

FIGURE 7.8 fMRI activation maps related to (left) not-registered and (right) registered data.

hand with the thumb. In this experiment, the motor area of the cervical cortex is activated in a well-known way. The fMRI data set is composed of 60 temporal acquisitions (acquired every 3 sec), $128 \infty 128 \infty 24$ images each, with a pixel size of $2.2 \infty 2.2$ mm and a slice thickness of 7 mm. A standard multislice EPI sequence was used. Functional images are mapped on a structural data set, composed of 8 slices, $256 \infty 256$ pixels each, with a pixel size of $1.1 \infty 1.1$ mm and a slice thickness of 7 mm. The registration operation was performed using the AFNI software, based on the AIR approach for the implementation of the registration.

Figure 7.8 shows the activation maps related to not-registered (left) and registered (right) data. It is clear from the figure that the registration is not able to correct all the activation artifacts, because of the strong correlation between subject movement and the imposed paradigm. However, registered images show a reduction of the artifacts in brain borders as well as an improved delineation of the activation zone. Figure 7.9 shows the values of the six roto-translation parameters used to perform the registration along the image sequence. The detected movements are in general very small (less than 2 mm). Continuous movements during the examination are effectively revealed, while subcutaneous movements generated by paradigm changes are often missed.

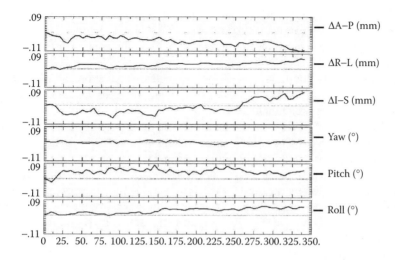

FIGURE 7.9 Values of the six roto-translation parameters used to perform the fMRI registration along image sequence.

7.8 CARDIAC IMAGES REGISTRATION

The use of contrast medium (CM) to enhance the information provided by cardiac magnetic resonance is growing. In fact, the use of contrast-enhanced images allows joining the high-resolution anatomical information provided by MRI with functional information obtained by means of the diffusion of contrast medium (typically Gadolinium) in tissues. Contrast-enhanced MR images are widely used in the study of the brain (i.e., brain perfusion imaging) as well as in medical examination of other districts, such as extremities (knees, ankles, wrists, and elbows). In this section, we refer in particular to myocardial perfusion imaging that allows assessing the extent and type of coronary artery disease (CAD).

In myocardial perfusion imaging, several slices in short-axis view are acquired over time, starting from the injection of contrast medium (Figure 7.10) in order to follow its perfusion in myocardial tissues. To avoid image misalignment, the patient under examination was asked to hold his breath during the examination. The image size is about 256 × 256 pixel with a planar resolution

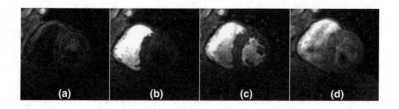

FIGURE 7.10 Example of cardiac perfusion images.

of about 1.2 mm. In each individual heartbeat (i.e., RR interval), a maximum of three to four slices can be acquired during the 300 msec interval positioned in diastole. The 3-D acquisition is repeated from 40 to 80 times, in order to follow the diffusion of the contrast agent.

Consequently, the time needed for a complete acquisition can reach 40 to 80 sec or more (40 to 80 RR intervals); in many cases the entire examination cannot be done in breath-hold state.

The quality of the myocardial perfusion can be assessed by qualitative evaluation of the signal intensity in the myocardium after the CM injection. In order to perform a quantitative analysis of the myocardial perfusion, the signal intensity changes in the acquired images must be evaluated. Therefore, intensity/time (I/T) curves are extracted by measuring the intensity value in the region of interest in the myocardium during oves time. Quantitative evaluation of the signal intensity during time provides a useful clinical index. In cardiac perfusion imaging, the maximum slope value of the I/T curve extracted from the myocardium is related to the vitality of the cardiac tissue.

Because the acquisition protocol is made to obtain spatial alignment of all frames, each pixel in an image frame should correspond to the pixels in the other frames with the same geometrical coordinates. In this case, the area of interest selection could be done on only one image in the temporal sequence, enhancing both the reliability and the performance of the analysis. Unfortunately, obtaining spatially aligned images is a difficult task in cardiac image acquisition, in which image misalignment due to patient breathing and poor ECG synchronization are commonly observed. Therefore, image registration is often needed in the postprocessing phase.

The registration methods proposed in the literature cover many of the approaches described in the present chapter: as manual registration using anatomical markers defined by an expert operator along all images in the temporal sequence [43,44] and as extractions of some geometrical features (i.e., left ventricle cavity) from each frame and image registration by registering the extracted geometrical features [45,46]. Delzescaux et al. [47] proposed a method based on the manual delineation of myocardium with right and left ventricle on one frame in the sequence. An algorithm based on template matching then performs the sequence registration. Bidaut et al. [48] proposed a method based on the minimization of intrinsic differences between each image and a reference image coupled to a 2-D (i.e., three parameters) rigid-body correction. Voxel-based methods that operate directly on the image gray values using MI as similarity metric are effective in the present problem. In fact, the pixel values can change in dependence from the transit of the contrast medium. Instead of pixel-value changes the statistics of gray-levels distribution along the images remain almost the same, leading to an MI-based registration effective in respect to other methods.

As previously described, the number of slices that can be acquired for each volume is strongly limited by the acquisition time. Typically, only three to four slices can be acquired during an RR interval. Acquiring more slices means increasing the acquisition time, making almost impossible the execution of the examination in the breath-holding state. The availability of a small number of slices implies that data are anisotropic, in the sense that the distance between two slices acquired in the z direction is large in respect to the in-plane resolution on the slice. In the 3-D registration approach, this implies that when an interpolation operation is performed to calculate the transformed volume, the interpolation in the z direction leads to large errors. For this reason, the registration problem is often reduced to the alignment of N 2-D images using a rigid transformation. In this case, the number of parameters to estimate reduces to $3(N-1)$.

In Section 7.5, the problem of global optimization of the registration function was described in detail. In this example, we first performed perfusion image registration by maximizing MI along time sequence frames in respect to a reference image for each temporal sequence. The registration is performed using the simplex optimization method. After the first step, a more accurate registration of each frame with the previous one using the Powell method was made. The user has to roughly identify the left ventricle, surrounding it with a circular mask. Without the mask, the registration algorithm may try to register structures that do not belong to the heart region. The method is consequently semiautomatic.

The method has been tested on two kinds of image data set. The first set was acquired from collaborative volunteers able to hold their breath and to reduce movements during the entire examination. The second data set was acquired from patients with suspected CAD disease scheduled for MRI examination. For each examination, a total of 120 images was acquired, consisting of 3 short-axis slices, each with 40 temporal frames acquired in diastolic phase. A total of five examinations on volunteers and five examinations on patients were used for algorithm effectiveness evaluation. Therefore, a total number of 30 temporal image sequences was used.

In order to assess the effectiveness of the automatic registration procedure, an expert user was asked to use the program with and without the use of the automatic registration algorithm. For each spatial slice, the endocardial and epicardial contours have been manually drawn. The contours were replicated along all frames, and the user was asked to manually correct the endocardial and epicardial borders. We used the overlapping area (OA) index as index for the needed correction degree. The overlapping area is the common area between the region selected in the developing image and the reference image, normalized by the reference area.

Figure 7.11 shows the average value of OA index for each frame, with and without registration, on patient images. The value of OA index on patient images is reduced by the registration procedure and becomes comparable with the index measured on volunteer images.

From the presented example, we can infer that the use of an automatic registration procedure based on maximization of the mutual information seems to be effective in order to address the requirement of fast and automatic tools for quantitative analysis of CM-enhanced MR images.

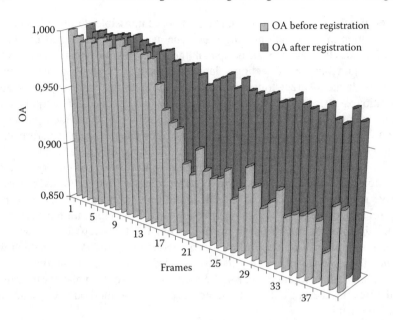

FIGURE 7.11 OA index before and after registration.

The quantitative OA index was introduced in order to measure algorithm effectiveness in a quantitative way. The results show that misalignments and artifacts introduced by patient movement during the examination are greatly reduced.

REFERENCES

1. Hajnal, J.V., Hill, D.L.G., and Hawkes, D.J. (2001). View of the future. in Hajnal, J.V., Hill, D.L.G., Hawkes, D.J., (Eds.), *Medical Image Registration*. Boca Raton, FL: CRC.
2. Hill, D.L.G., Batchelor, P.G., Holden, M., and Hawkes, D.J. (2001). Medical image registration. *Phys. Med. Biol.* 46: R1–R45.
3. Jenkinson, M. and Smith, S. (2001). The role of registration in functional magnetic resonance imaging. in Hajnal, J.V., Hill, D.L.G., and Hawkes, D.J., (Eds.), *Medical Image Registration*. Boca Raton, FL: CRC.
4. Ashburner, J. and Friston, K.J. (2000). Image registration. in *Functional MRI*, Moonen, C.T.W., Bandettini, P.A., (Eds.), Springer.
5. Pluim, J.P.W. and Fitzpatrick, J.M. (2003). Image registration. *IEEE Trans. Med. Imaging* 22(11): 1341–1343.
6. Brown, L. (1992). A survey of image registration techniques. *ACM Comput. Surv.* 24: 325–376.
7. Hill, D.L.G. and Batchelor, P. (2001). Registration methodology: concepts and algorithms. in Hajnal, J.V., Hill, D.L.G., and Hawkes, D.J., (Eds.), *Medical Image Registration*. Boca Raton, FL: CRC.

8. Roche, A., Pennec, X., Malandain, G., and Ayache, N. (2001). Rigid registration of 3-D ultrasound with MR images: A new approach combining intensity and gradient information. *IEEE Trans. Med. Imaging* 20: 1038–1049.

9. Pluim, J.P.W., Maintz, J.B.A., and Viergever, M.A. (2003). Mutual-information-based registration of medical images: a survey. *IEEE Trans. Med. Imaging.* 22(8): 986–1004.

10. Robb, R.A. (2000). *Biomedic Imaging, Visualization and Analysis.* Wiley-Liss, New York.

11. Santarelli, M.F., Positano, V., Michelassi, C., Lombardi, M., and Landini, L. (2003). Automated cardiac MR image segmentation: theory and measurement evaluation. *Medical Eng. Phys.* 25(2): 149–159.

12. Besl, P.J. and McKay, N.D. (1992). A method for registration of 3-D shapes. *IEEE Transaction on Pattern Analysis and Machine Intelligence.* 14(2): 239–256.

13. Maes, F., Collignon, A., Vandermeulen, D., Marchal, G., and Suetens, P. (1997). Multimodality image registration by maximization of mutual information. *IEEE Trans. Med. Imaging* 16: 187–198.

14. Wells, W.M., III, Viola, P., Atsumi, H., Nakajima, S., and Kikinis, R. (1996). Multi-modal volume registration by maximization of mutual information. *Med. Image Anal.* 1: 35–51.

15. Studholme, C., Hill, D.L.G., and Hawkes, D. (1999). An overlap invariant entropy measure of 3-D medical image alignment. *Pattern Recognit.* 32: 71–86.

16. Grevera, G.J. and Udupa, J.K. (1998). An objective comparison of 3-D image interpolation methods. *IEEE Trans. Med. Imaging* 17(4).

17. Lehmann, T.M., Gonner, C., and Spitzer, K. (1999). Survey: interpolation methods in medical image processing. *IEEE Trans. Med. Imaging* 18(11): 1049–1075.

18. Thacker, N.A., Jackson, A., Moriarty, D., and Vokurka, B. (1999). Improved quality of re-sliced MR images using re-normalised Sinc interpolation. *J Magn. Resonance Imaging* 10(4): 582–588.

19. Pluim, J.P.W., Maintz, J.B.A., and Viergever, M.A. (2001). Mutual information matching in multiresolution contexts. *Image Vis. Comput.* 19(1): 45–52.

20. Bernon, J.L., Boudousq, V., Rohmer, J.F., Fourcade, M., Zanca, M., Rossi, M., and Mariano-Goulart, D. (2001). A comparative study of Powell's and downhill simplex algorithms for a fast multimodal surface matching in brain imaging. *Comput. Med. Imaging Graph.* 25: 287–297.

21. Powell, M.J.D. (1964). An efficient method of finding the minimum of a function of several variables without calculating derivatives. *Comput. J.* 7: 155–163.

22. Maes, F., Vandermeulen, D., and Suetens, P. (1999). Comparative evaluation of multiresolution optimization strategies for multimodality image registration by maximization of mutual information. *Med. Image Anal.* 3: 373–386.

23. Nelder, J.A. and Mead, R. (1965). A simplex method for function minimization. *Comput. J.* 7: 308–313.

24. Matsopoulos, G.K., Mouravliansky, N.A., Delibasis, K.K., and Nikita, K.S. (1999). Automatic retinal image registration scheme using global optimization techniques. *IEEE Trans. Inform. Technol. Biomed.* 3: 47–60.

25. Fogel, D.B. (2000). *Evolutionary Computation.* IEEE Press, New York.

26. Wachowiak, M.P. and Elmaghraby, A.S. (2001). The continuous tabu search as an optimizer for 2-D-to-3-D biomedical image registration. *Lecture Notes in Computer Science,* W. Niessen and M. Viergever, Eds. New York: Springer-Verlag, Proc. MICCAI 2001–2208, 1273–1274.

27. Wachowiak, M.P., Smolikova, R., Zheng, Y., Zurada, J.M., and Elmaghraby, A.S. (2004). An approach to multimodal biomedical image registration utilizing particle swarm optimization. *IEEE Trans. on Evolutionary Computation* 8(3): 289–301.

28. Press, W.H., Flannery, B.P., Teukolsky, S.A., and Vetterling, W.T. (1988). *Numerical Recipes in C: The Art of Scientific Computing.* Cambridge University Press, Cambridge.

29. Walters, H., Parker, L.R., Morgan, S.L., and Deming, S.N. (1999). *Sequential Simplex Optimization.* CRC Press, Boca Raton, FL.

30. Krinidis, S., Nikou, C., and Pitas, I. (2003). A global energy function for the alignment of serially acquired slices. *IEEE Trans. on Information Technology in Biomedicine.* 7(2): 108–113.

31. Boes, J.L. and Meyer, C.R. (1999). Multi-variate mutual information for registration. *Medical Image Computing and Computer-Assisted Intervention.* Taylor, C., Colchester, A., Eds., *Lecture Notes in Computer Science*, Springer-Verlag. 1679: 606–612.

32. Zhang, J. and Rangarajan, A. (2004). Affine image registration using a new information metric. *Proceedings of IEEE Conference on Computer Vision and Pattern Recognition (CVPR)* 1: 848–855.

33. Positano, V., Santarelli, M.F., and Landini, L. (2003). Automatic characterization of myocardial perfusion in contrast enhanced MRI. *EURASIP J. Appl. Signal Process.* 5: 413–422.

34. Talairach, J. and Tournoux, p. (1988). Coplanar stereotaxic atlas of the human brain. Thieme Medical, New York.

35. Freire, L., Roche, A., and Mangin, J.F. (2002). What is the best similarity measure for motion correction in f MRI time series? *IEEE Trans. Med. Imaging* 21(5): 470–484.

36. Friston, K.J., Ashburner, J., Frith, C.D., Poline, J.B., Heather, J.D., and Frackowiak, R.S.J. (1995). Spatial registration and normalization of images. *Hum. Brain Mapp.* 2: 165–189.

37. Woods, R.P., Cherry, S.R., and Mazziotta, J.C. (1992). Rapid automated algorithm for aligning and reslicing PET images. *J. Comput. Assist. Tomogr.* 16: 620–633.

38. Nikou, C., Heitz, F., Armspach, J.P., Namer, I.J., and Grucker, D. (1998). Registration of MR/MR and MR/SPECT brain images by fast stochastic optimization of robust voxel similarity measures. *NeuroImage*, 8: 30–43.

39. Roche, A., Malandain, G., Pennec, X., and Ayache, N. (1998). The correlation ratio as a new similarity measure for multimodal image registration, in *Lecture Notes in Computer Science.* Berlin, Germany: Springer-Verlag, Vol. 1496, Proc. MICCAI'98, pp. 1115–1124.

40. Friston, K.J., Williams, S., Howard, R., Franckowiak, R.S.J., and Turner, R. (1996). Movement-related effects in fMRI time-series. *Magn. Reson. Med.* 35: 346–355.

41. Ashburner, J. and Friston, K.J. (2000). *Image Registration in Functional MRI*, Moonen, C.T.W., Bandettini, P.A., Eds., Springer-Verlag, New York.

42. Collins, D.L. and Neelin, P. et al. (1994). Automatic 3-D intersubject registration of mr volumetric data in standardized talairach space. *J. Comput. Assist. Tomogr.* 18(2): 192–205.

43. Schwitter, J., Nanz, D., Kneifel, S. et al. (2001). Assessment of myocardial perfusion in coronary artery disease by magnetic resonance. *Circulation* 103: 2230–2235.

44. Vallee, J.P., Sostman, H.D., MacFall, J.R., Wheeler, T., Hedlund, L.W., Spritzer, C.E., and Coleman, R.E. (1997). MRI quantitative myocardial perfusion with compartmental analysis: a rest and stress study. *Magn. Reson. Med.* 38: 981–989.

45. Yang, X., Taeymans, Y., Carlier, P., Verstraeten, P., Brunfaut, K., and Cornelis, M. (1993). Computer- aided measurement of local myocardial perfusion in MRI. *Computers in Cardiology. Proceedings.* 20: 365–368.

46. Gerig, G., Kikinis, R., Kuoni, W., von Schulthess, G.K., and Kubler, O. (1991). Semi-automated ROI analysis in dynamic MR studies. Part I: images analysis tools for automatic correction of organ displacements. *J. Comput. Assist. Tomogr.* 15: 725–732.

47. Delzescaux, T., Frouin, F., De Cesare, A., Philipp-Foliguet, S., Zeboudj, R., Janier, M., Todd-Pokropek, A., and Herment, A. (2000). Adaptive and self-evaluating registration method for myocardial perfusion assessment. *MAGMA* 13(1): 28–39.

48. Bidaut, L.M. and Vallée, J.P. (2001). Automated registration of dynamic MR images for the quantification of myocardial perfusion. *J. Magn. Reson. Imaging* 13(4): 648–55.

8 Multimodal Integration: fMRI, MRI, EEG, and MEG

Yaroslav O. Halchenko, Stephen José Hanson, and Barakk A. Pearlmutter

CONTENTS

OVERVIEW

This chapter provides a comprehensive survey of the motivations, assumptions and pitfalls associated with combining signals such as functional magnetic resonance imaging (fMRI) with electroencephalography (EEG) or magnetoencephalography (MEG). Our initial focus in the chapter concerns mathematical approaches for solving the localization problem in EEG and MEG. Next, we document the most recent and promising ways in which these signals can be combined with fMRI. Specifically, we look at correlative analysis, decomposition techniques, equivalent dipole fitting, distributed sources modeling, beamforming, and Bayesian methods. Due to difficulties in assessing the ground truth of a combined signal in any realistic experiment — a difficulty further confounded by lack of accurate biophysical models of BOLD signal — we are cautious about being optimistic in regard to multi-modal integration. Nonetheless, as we highlight and explore the technical and methodological difficulties of fusing heterogeneous signals, it seems likely that correct fusion of multimodal data will allow previously inaccessible spatiotemporal structures to be visualized and formalized and, thus, multimodal integration will eventually become a useful tool in brain-imaging research.

8.1   INTRODUCTION

Noninvasive functional brain imaging has become an important tool used by neu-rophysiologists, cognitive psychologists, cognitive scientists, and other researchers interested in brain function. In the last five decades, the technology of noninvasive functional imaging has flowered, and researchers today can choose from EEG, MEG, positron emission tomography (PET), single-photon computed tomography (SPECT), magnetic resonance imaging (MRI), and fMRI. Each method has its own strengths and weaknesses, and no single method is best suited for all experimental or clinical conditions. Because of the inadequacies of individual techniques, there is increased interest in finding ways to combine existing techniques in order to synthesize the strengths inherent in each. In this chapter, we will (a) examine specific noninvasive imaging techniques (EEG, MEG, MRI, and fMRI), (b) com-pare approaches used to analyze the data obtained from these techniques, and (c) discuss the potential for successfully combining methodologies and analyses.

Localizing neuronal activity in the brain, both in time and in space, is a central challenge to progress in understanding brain function. Localizing neural activity from EEG or MEG data is called electromagnetic source imaging (EMSI). EEG and MEG each provide data with high temporal resolution (measured in milliseconds) but limited spatial resolution. In contrast, fMRI provides good spatial but relatively poor temporal resolution. For some clinical purposes or general localization, simple techniques can be used for source imaging. However, more specific localization of the neural activity requires more sophisticated analyses; for these, researchers turned to other disciplines that face similarly difficult localization problems (seismology, remote sensing, noninvasive signal processing, and radar and sonar signal detection) for inspiration and algorithms. Because the source localization techniques used in EMSI serve as a starting point for subsequent multimodal analysis, we will discuss these methods first. We will review canonical problems of source localization and how they have been attacked by various researchers.

Following this section we discuss problems inherent in multimodal experiments and then explore how MR modalities, which have high spatial resolution, can be combined with existing EMSI techniques in order to increase localization precision (for other reviews see George et al. [1], Nunez and Silberstein [2], Salek-Haddadi et al. [3], and George et al. [4]).

Demonstrated localization accuracy remains a distant goal confounded by the lack of ground truth in any realistic experimental multimodal protocol and the lack of a complete model of the BOLD signal. Some progress on some very simple experiments in which there is a small number of isolated focal sources of activity that are consistently present in all relevant modalities gives us hope that this should be possible. We conclude that a convincing demonstration of increased accuracy for a complex protocol would constitute a major success in the field.

Throughout this chapter we provide a consistent and complete set of mathematical formulations that are stand alone; we also provide appropriate context for this notation in the existing literature (Table 8.1 presents the notation used throughout this chapter). Our conclusions and suggestions for future work make up the final section.

8.2 SOURCE LOCALIZATION IN EEG AND MEG

EEG and MEG have been widely used in research and clinical studies since the mid-20th century. Although Richard Caton (1842 to 1926) is believed to have been the first to record the spontaneous electrical activity of the brain, the term EEG first appeared in 1929 when Hans Berger, a psychiatrist working in Jena, Germany, announced to the world that "it was possible to record the feeble electric currents generated on the brain, without opening the skull, and to depict them graphically onto a strip of paper." The first SQUID-based MEG experiment with a human subject was conducted at MIT by Cohen [5] after his successful application of Zimmerman's SQUID sensors to acquire a magneto-cardiogram in 1969. EEG and MEG are closely related due to electromagnetic coupling, and we will use E/MEG to refer generically to either EEG, MEG, or both altogether. High temporal resolution (measured in milliseconds) is provided by E/MEG, but it has a major limitation: the location of neuronal activity can be hard to determine with confidence. In the following subsection we lay out the specifics of each of the E/MEG signals, the premises for conjoint E/MEG analysis, and the EMSI techniques that have been adopted for use in multimodal analysis with fMRI data.

TABLE 8.1
Notation Used throughout This Chapter

Symbol	Meaning
K	Number of simultaneously active voxels
N	Number of voxels, i.e., spatial resolution of high spatial resolution modality (fMRI)
M	Number of EEG/MEG sensors, i.e., spatial resolution of low spatial resolution modality
T	Number of time points of high temporal resolution modality (EEG, MEG)
U	Number of time points of low temporal resolution modality (fMRI)
L	Number of orthogonal axes for dipole moment components, L Å_{1, 2, 3}
$\mathbf{I}_n$	Identity matrix $(n \times n)$
0	Zero matrix of appropriate dimensionality
$\mathbf{X}$	General E/MEG data matrix; can contain EEG or MEG data or both $(M \times T)$
$\mathbf{B}$	BOLD fMRI data matrix $(N \times U)$
$\mathbf{Q}$	Dipole sources matrix
G	General E/MEG lead function, incorporating information for EEG or MEG or both
$\mathbf{G}$	General E/MEG lead matrix
$\mathbf{F}^i$	Spatial filter matrix for the ith dipole $(M \times L)$
v	Variance
$\mathbf{C}$	Covariance matrix
$\mathbf{K}$	Matrix of correlation coefficients
$\mathbf{M}_-$	Matrix transpose
$\mathbf{M}^+$	Generalized matrix inverse (pseudo-inverse)
null $\mathbf{M}$	The *null space* of $\mathbf{M}$, the set of vectors $\{\mathbf{X} \mid \mathbf{MX} = 0\}$
diag $\mathbf{M}$	The diagonal matrix with the same diagonal elements as $\mathbf{M}$

Note: We chose our notation to match the most popular conventions in the field, and at the same time minimize confusion. Regrettably, it is likely to differ from the notation used by each particular paper we reference. Following the usual conventions, we use bold uppercase symbols for matrices, bold lowercase for vectors, and nonbold symbols for scalars.

8.2.1 ASSUMPTIONS UNDERLYING INTEGRATION OF EEG AND MEG

The theory of electromagnetism and Maxwell's equations, under the assumption of quasi stationarity,* theoretically defines the relationship between observed magnetic and electric fields that are induced by the ionic currents generated inside the brain (see Malmivuo and Plonsey [7], Okada et al. [8], Murakami et al. [9], for more information about the biophysics of E/MEG signals).

---

* A signal is quasistatic if it does not change its parameters in time. The nonstationary term present in the E/MEG physical model is relatively small and can be considered zero in the range of signal frequencies that are captured by E/MEG. See [6] for a more detailed description.

The similar nature of the EEG and MEG signals means that many methods of data analysis are applicable to both E/MEG modalities. Although the SNR of E/MEG signals have improved with technological advances, and some basic analysis has been performed by experts on raw E/MEG data via visual inspection of spatial signal patterns outside of the brain, more advanced methods are required to use data efficiently. During the last two decades, many E/MEG signal analysis techniques [10] have been developed to provide insights on different levels of perceptual and cognitive processing of the human brain: event-related potential (ERP) in EEG and event-related field (ERF) in MEG, components analysis (PCA, ICA, etc.), frequency domain analysis, pattern analysis, single-trial analysis [11–13], etc. Source localization techniques were first developed for MEG because the head model required for forward modeling of the magnetic field is relatively simple. Source localization using an EEG signal has been difficult to perform because the forward propagation of the electric potentials is more complicated. However, recent advances in automatic MRI segmentation methods, together with advances in forward and inverse EEG modeling, have made EEG source localization plausible.

The theory of electromagnetism also explains why EEG and MEG signals can be considered complementary, in that they provide different views of the same physiological phenomenon [6,14,15]. On the one hand, an often-accented difference is that MEG is not capable of registering the magnetic field generated by the sources that are oriented radially to the skull surface in the case of spherical conductor geometry. On the other hand, MEG has the advantage over EEG in that the local variations in conductivity of different brain matter (e.g., white matter, gray matter) do not attenuate the MEG signal much, whereas the EEG signal is strongly influenced by the skull and different types of brain matter [8]. The orientation selectivity, combined with the higher depth precision due to homogeneity, makes MEG optimal for detecting activity in sulci (brain fissures) rather than in gyri (brain ridges). In contrast, a registered EEG signal is dominated by the gyral sources close to the skull and therefore is more radial to its surface. Yet another crucial difference is dictated by basic physics. The orthogonality of magnetic and electrical fields leads to orthogonal maps of the magnetic field and electrical potential on the scalp surface. This orthogonality means that an orthogonal localization direction is the best localization direction for both modalities [15,16]. These complementary features of the EEG and MEG signals are what make them good candidates for integration [17,18]. The conjoint E/MEG analysis has improved the fidelity of EMSI localization, but has not entirely solved the problem of source localization ambiguity. It is in the reduction of this remaining ambiguity that information from other brain imaging modalities may play a valuable role.

It is worth noting another purely technical advantage of MEG over EEG: MEG provides a reference-free recording of the actual magnetic field. Whenever EEG sensors capture scalp potentials, a reference electrode must be used as a ground to derive the signal of interest. A reference signal chosen in such a way can be arbitrarily biased relative to the EEG signal observed, even when no neuronal sources are active. The unknown in an MEG signal obtained using SQUID sensors is just a constant in time offset—the DC baseline. This baseline depends on the nearest flux quantum for which the flux-locked loop acquired a lock [19, p. 265]. Although the choice of a reference value in EEG and the DC line in MEG do not influence the analysis of potential/field topographic maps, they do impact inverse solution algorithms that assume zero net source in the head, i.e., zero baseline. In general, the simple average reference across the electrodes is used, and it has been shown to be a good approximation to the true reference signal (Reference 10, Subsection 8.2.2).

Even if the reference value (baseline) is chosen correctly, both conventional EEG and MEG face obstacles in measuring the slowly changing DC component of the signal in the low-frequency range ($f < [0.1]$Hz). In the case of EEG, the problem is due to the often-used coupling of the electrodes via capacitors, so that any DC component (slowly changing bias) of the EEG signal is filtered out. That leaves the researcher with nonzero frequency components of the signal, which often correspond to the most informative part of the signal as in the case of conventional ERP or frequency domain analysis. The DC-EEG component can be registered by using sensors with direct coupling and special scalp electrodes that are gel-filled to eliminate changes of electrical impedance at the electrode–skin interface that can cause low-frequency noise in the EEG signal. Although the MEG system does not require direct contact between sensors and skin, it is nevertheless subject to $1/f$ sensor noise, which interferes with the measurement of the neuronal DC fields. In the last decade, DC-MEG has been methodically refined by employing controlled brain-to-sensor modulation allowing the monitoring of low-frequency magnetic fields. Formalized DC-E/MEG techniques make it possible to perform E/MEG studies, which rely on the shift of DC and low-frequency components of the signal, components that occur, for example, during epileptic seizures, hyperventilation, changes in vigilance states, and cognitive or motor tasks.

8.2.2 Forward Modeling

The analysis of E/MEG signals often relies on the solution of two related problems. The forward problem concerns the calculation of scalp potentials (EEG) or magnetic fields near the scalp (MEG), given the neuronal currents in the brain, whereas the inverse problem involves estimating neuronal currents from the observed E/MEG data. The difficulty of solving the forward problem is reflected in the diversity of approaches that have been tried (see Mosher et al. [20] for an overview and unified analysis of different methods).

The basic question posed by both the inverse and forward problems is how to model any neuronal activation so that the source of the electromagnetic field can be mapped onto the observed E/MEG signal. Assuming that localized and synchronized primary currents are the generators of the observed E/MEG signals, the most successful approach is to model the ith source with a simple equivalent current dipole (ECD) $\mathbf{q}_i$ [21], uniquely defined by three factors: location represented by the vector $\mathbf{r}_i$, strength q_i, and orientation coefficients θ_i. The orientation coefficient is defined by projections of the vector $\mathbf{q}_i$ into L orthogonal Cartesian axes: $\theta_i = \mathbf{q}_i/q_i$. However, the orientation coefficient may be expressed by projections in two axes in the case of an MEG spherical model, in which the silent radial to the skull component has been removed, or even just in a single axis if normality to the cortical surface is assumed. The ECD model made it possible to derive a tractable physical model linking neuronal activation and observed E/MEG signals. In case of K simultaneously active sources at time t, the observed E/MEG signal at the sensor $\mathbf{x}_j$ positioned at $\mathbf{p}_j$ can be modeled as

$$\hat{\mathbf{x}}_j(\mathbf{r}_i,\mathbf{q}_i,t) = \sum_i^K G(\mathbf{r}_i(t),\mathbf{p}_j) \cdot \mathbf{q}_i(t) + \varepsilon \qquad (8.1)$$

where G is a lead-field function that relates the ith dipole and the potential (EEG) or magnetic field (MEG) observed at the jth sensor, and ε is the sensor noise. In the given formulation, function $G(\mathbf{r}_i(t),\mathbf{p}_j)$ returns a vector, where each element corresponds to the lead coefficient at the location $\mathbf{p}_j$ generated by a unit-strength dipole at position $\mathbf{r}_i(t)$ with the same orientation as the corresponding projection axis of θ_i. The inner product between the returned vector and dipole strength projections on the same coordinate axes yields a j-th sensor the measurement generated by the ith dipole.

The forward model (Equation 8.1) can be solved at substantial computational expense using available numerical methods [22] in combination with realistic structural information obtained from the MRI data (see Section 8.1). This high computational cost is acceptable when the forward model has to be computed once per subject and for a fixed number of dipole locations, but it can be prohibitive for dipole fitting, which requires a recomputation of the forward model for each step of nonlinear optimization. For this reason, rough approximations of the head geometry and structure are often used, e.g., the best-fit single-sphere model, which has a direct analytical solution [23], or the multiple-spheres model to accommodate the difference in conductivity parameters across different tissues. Recently proposed MEG forward modeling methods for realistic isotropic volume conductors [24,25] seem to be more accurate and faster than BEM, and hence may be useful substitutes for both crude analytical methods and computationally intensive finite-element numeric approximations. Generally, the solution of the forward problem is crucial for performing source localization using E/MEG, which is the main topic of the following subsection.

8.2.3 THE INVERSE PROBLEM

8.2.3.1 Equivalent Current Dipole Models

The E/MEG inverse problem is very challenging (see Hämäläinen et al. [6] and Baillet et al. [26] for an overview of methods.) First, it relies on the solution of the forward problem, which can be computationally expensive, especially in the case of realistic head modeling. Second, the lead-field function G from Equation 8.1 is nonlinear in $\mathbf{r}_i$, so that the forward model depends nonlinearly on the locations of activations. It is because of this nonlinearity that the inverse problem is generally treated by nonlinear optimization methods, which can lead to solutions being trapped in local minima. In the case of Gaussian sensor noise, the best estimator for the reconstruction quality of the signal is the squared error between the obtained and modeled E/MEG data:

$$E(\mathbf{r},\mathbf{q}) = \sum_{i}^{K}\sum_{t=t_1}^{t_2}\sum_{j}^{M}(\mathbf{x}_j(t) - \hat{\mathbf{x}}_j(\mathbf{r}_i,\mathbf{q}_i,t))^2 + \lambda f(\mathbf{r},\mathbf{q}) \qquad (8.2)$$

where $f(\mathbf{r}, \mathbf{q}) > 0$ is often introduced to regularize the solution, i.e., to obtain the desired features of the estimated signal (e.g., smoothness in time or in space, and lowest energy or dispersion), and $\lambda > 0$ is used to vary the trade-off between the goodness of fit and the regularization term.

This least-squares model can be applied to the individual time points ($t_1 = t_2$) ("moving dipole" model) or to a block ($t_1 < t_2$) of data points. If the sources are assumed not to change during the block (t_1, t_2), then the solution with time constant $\mathbf{q}_i(t) = \mathbf{q}_i$ is the target.

Other features derived from the data besides pure E/MEG signals as the argument $\mathbf{x}$ of Equation 8.1 and Equation 8.2 are often used, e.g., ERP/ERF waveforms that represent averaged E/MEG signals across multiple trials, mean map in the case of stable potential or field topography during some period of time, or signal frequency components to localize the sources of the oscillations of interest.

Depending on the treatment of Equation 8.2, the inverse problem can be presented in a couple of different ways. The brute-force minimization of Equation 8.2 in respect to both parameters $\mathbf{r}$ and $\mathbf{q}$ and the consideration of different K neuronal sources is generally called ECD fitting. Because of nonlinear optimization, this approach works only for cases in which there is a relatively small number of sources K, and therefore the inverse problem formulation is overdetermined, i.e., Equation 8.1 cannot be solved exactly ($E(\mathbf{r},\mathbf{q}) > 0$). If fixed time locations of the target dipoles can be assumed, the search space of nonlinear optimization is reduced, and the optimization can be split into two steps: (a) nonlinear optimization to find locations of the dipoles and then (b) analysis to determine the strength of the dipoles. This assumption constitutes the so-called spatiotemporal ECD model.

Two other frameworks have been suggested as means of avoiding the pitfalls associated with nonlinear optimization: distributed ECD (DECD) and beamforming. We discuss these two approaches in detail in the following subsections.

8.2.3.2 Linear Inverse Methods: Distributed ECD

In the case of multiple simultaneously active sources, an alternative to solving the inverse problem by ECD fitting is a distributed source model. We will use the label DECD to refer to this type of model. The DECD is based on a spatial sampling of the brain volume and distribution of the dipoles across all plausible and spatially small areas that could be a source of neuronal activation. In such cases, fixed locations ($\mathbf{r}_i$) are available for each source or dipole, removing the necessity for nonlinear optimization as in the case of the ECD fitting. The forward model (Equation 8.1) can be presented for a noiseless case in the matrix form

$$\mathbf{X} = \mathbf{GQ} \tag{8.3}$$

where $\mathbf{G}$, $M \times LN$ lead-field matrix, is assumed to be static in time. The j, i(th) entry of $\mathbf{G}$ describes how much a sensor j is influenced by a dipole i, where j varies over all sensors while i varies over every possible source, or to be more specific, every axis-aligned component of every possible source: $g_{j\bar{\imath}} = G(\mathbf{r}_i, \mathbf{p}_j)$. The vector $\bar{\imath}$ contains indices of L such projections, i.e., $\bar{\imath} = [3i, 3i+1, 3i+2]$ when $L = 3$, and $\bar{\imath} = i$ when the dipole has a fixed known orientation. Using this notation, $\mathbf{G}_{\bar{\imath}}$ corresponds to the lead matrix for a single dipole $\mathbf{q}_i$. The $M \times T$ matrix $\mathbf{X}$ holds the E/MEG data, while the $LN \times T$ matrix $\mathbf{Q}$ (note that $\mathbf{Q}_{\bar{\imath}} = \mathbf{q}_i(t)$) corresponds to the projections of the ECD's moment onto L orthogonal axes.

The solution of Equation 8.3 relies on finding an inverse $\mathbf{G}^+$ of the matrix $\mathbf{G}$ to express the estimate $\hat{\mathbf{Q}}$ in terms of $\mathbf{X}$

$$\hat{\mathbf{Q}} = \mathbf{G}^+\mathbf{X} \tag{8.4}$$

and will produce a linear map $\mathbf{X} \mapsto \hat{\mathbf{Q}}$. Other than being computationally convenient, there is not much reason to take this approach. The task is to minimize the error function (Equation 8.2), which can be generalized by weighting the data to account for the sensor noise and its covariance structure:

$$L(\mathbf{Q}) = \mathrm{tr}((\mathbf{X} - \mathbf{GQ})^\top \mathbf{W_X^{-1}}(\mathbf{X} - \mathbf{GQ})), \tag{8.5}$$

where $\mathbf{W_X^{-1}}$ is a weighting matrix in sensor space.

A zero-mean Gaussian signal can be characterized by the single covariance matrix $\mathbf{C}_\varepsilon$. In the case of a nonsingular $\mathbf{C}_\varepsilon$ we can use the simplest weighting scheme $\mathbf{W_X} = \mathbf{C}_\varepsilon$ to account for nonuniform and possibly correlated sensor noise.

Such a brute-force approach solves some problems of ECD modeling, specifically the requirement for a nonlinear optimization, but, unfortunately, it introduces another problem: the linear system (Equation 8.3) is ill-posed and underdetermined, because the number of sampled possible source locations is much higher than the dimensionality of the input data space (which cannot exceed the number of sensors), i.e., $N \gg M$. Thus, there is an infinite number of solutions for the linear system because any combination of terms from the null space of $\mathbf{G}$ will satisfy Equation 8.4 and fit the sensor noise perfectly. In other words, many different arrangements of the sources of neural activation within the brain can produce any given MEG or EEG map. To overcome this ambiguity, a regularization term is introduced into the error measure

$$L_r(\mathbf{Q}) = L(\mathbf{Q}) + \lambda f(\mathbf{Q}) \tag{8.6}$$

where $\lambda \geq 0$ controls the trade-off between the goodness of fit and the regularization term $f(\mathbf{Q})$.

Equation 8.6 can have different interpretations depending on the approach used to derive it and the meaning given to the regularization term $f(\mathbf{Q})$. All of the following methods provide the same result under specific conditions [26,27]: Bayesian methodology to maximize the posterior $p(\mathbf{Q}|\mathbf{X})$ assuming Gaussian prior on $\mathbf{Q}$ [28], Wiener estimator with proper $\mathbf{C}_\varepsilon$ and $\mathbf{C}_S$, Tikhonov regularization to trade off the goodness of fit (Equation 8.5), and the regularization term $f(\mathbf{Q}) = \operatorname{tr}(\mathbf{Q}^\top \mathbf{W}_\mathbf{Q}^{-1} \mathbf{Q})$, which attempts to find the solution with weighted by $\mathbf{W}_\mathbf{Q}^{-1}$ minimal second norm. All the frameworks lead to the solution of the next general form

$$\mathbf{G}^+ = (\mathbf{G}^\top \mathbf{W}_\mathbf{X}^{-1} \mathbf{G} + \lambda \mathbf{W}_\mathbf{Q}^{-1})^{-1} \mathbf{G}^\top \mathbf{W}_\mathbf{X}^{-1} \tag{8.7}$$

If and only if $\mathbf{W}_\mathbf{Q}$ and $\mathbf{W}_\mathbf{X}$ are positive definite [29] Equation 8.7 is equivalent to

$$\mathbf{G}^+ = \mathbf{W}_\mathbf{Q} \mathbf{G}^\top (\mathbf{G} \mathbf{W}_\mathbf{Q} \mathbf{G}^\top + \lambda \mathbf{W}_\mathbf{X})^{-1} \tag{8.8}$$

In the case when viable prior information about the source distribution is available $\mathbf{Q}_p$, it is easy to account for it by minimizing the deviation of the solution, not from 0 (which constitutes the minimal second norm solution $\mathbf{G}^+$), but from the prior $\mathbf{Q}_p$, i.e., $f(\mathbf{Q}) = \operatorname{tr}((\mathbf{Q} - \mathbf{Q}_p)^\top \mathbf{W}_\mathbf{Q}^{-1} (\mathbf{Q} - \mathbf{Q}_p))$. Then Equation 8.6 will be minimized at

$$\hat{\mathbf{Q}} = \mathbf{G}^+ \mathbf{X} + (\mathbf{I} - \mathbf{G}^+ \mathbf{G}) \mathbf{Q}_p = \mathbf{Q}_p + \mathbf{G}^+ (\mathbf{X} - \mathbf{G} \mathbf{Q}_p) \tag{8.9}$$

For the noiseless case, with a weighted L_2-norm regularizer, the Moore–Penrose pseudo-inverse gives the inverse $\mathbf{G}^+ = \mathbf{G}^\dagger$ by avoiding the null space projections of $\mathbf{G}$ in the solution, thus providing a unique solution with a minimal second norm $\mathbf{G}^\dagger = \mathbf{W}_\mathbf{Q} \mathbf{G}^\top (\mathbf{G} \mathbf{W}_\mathbf{Q} \mathbf{G}^\top)^{-1}$.

Taking $\mathbf{W}_\mathbf{Q} = \mathbf{I}_N, \mathbf{W}_\mathbf{X} = \mathbf{I}_M$ and $\mathbf{Q}_p = 0$ constitutes the simplest regularized minimum norm solution (Tikhonov regularization). Classically, λ is found using cross-validation [30] or L-curve [31] techniques, to decide how much of the noise power should be brought into the solution. Phillips et al. [32] suggested the iterative method ReML, in which the conditional expectation of the source distribution and the regularization parameters are estimated jointly. Additional constraints can be applied for greater regularization, for instance, temporal smoothness [33].

As presented in Equation 8.8, $\mathbf{G}^+$ can account for different features of the source or data space by incorporating them correspondingly into $\mathbf{W}_\mathbf{Q}$ and $\mathbf{W}_\mathbf{X}$. Next, data-driven features are commonly used in EMSI:

- $\mathbf{W}_\mathbf{X} = \mathbf{C}_\varepsilon$ accounts for any possible noise covariance structure or, if $\mathbf{C}_\varepsilon$ is diagonal, it will scale the error terms according to the noise level of each sensor.
- $\mathbf{W}_\mathbf{Q} = \mathbf{W}_{\mathbf{C}_S} = \mathbf{C}_S$ accounts for prior knowledge of the source's covariance structure.

$\mathbf{W_Q}$ can also account for different spatial features:

- $\mathbf{W_Q} = \mathbf{W_n} = (\mathrm{diag}(\mathbf{G}^\top \mathbf{G}))^{-1}$ normalizes the columns of the matrix $\mathbf{G}$ to account for deep sources by penalizing voxels too close to the sensors [34,35].
- $\mathbf{W_Q} = \mathbf{W_{gm}}$, where the ith diagonal element incorporates the gray matter content in the area of the ith dipole [36], i.e., the probability of having a large population of neurons capable of creating the detected E/MEG signal.
- $\mathbf{W_Q} = (\mathbf{W}_a{}^\top \mathbf{W}_a)^{-1}$, where rows of $\mathbf{W}_a$ represent averaging coefficients for each source [37]. So far only geometrical [38] or biophysical averaging matrices [29] have been suggested.
- $\mathbf{W_Q}$ incorporates the first-order spatial derivative of the image [39] or Laplacian form [40].

Features defined by the diagonal matrices (e.g., $\mathbf{W_n}$ and $\mathbf{W_{gm}}$) can be combined through the simple matrix product. An alternative approach is to present $\mathbf{W_Q}$ in terms of a linear basis set of the individual $\mathbf{W_Q}$ factors, i.e., $\mathbf{W_Q} = \mu_1 \mathbf{W_n} + \mu_2 \mathbf{W_{gm}} + \cdots$, with later optimization of μ_i via the EM algorithm [36].

To better condition the underdetermined linear inverse problem (Equation 8.4), Philips et al. [36] suggested to perform the inverse operation (Equation 8.4) can be performed in the space of the largest eigenvectors of the $\mathbf{W_Q}$. Such preprocessing can also be done in the temporal domain, when a similar subspace selection is performed using prior temporal covariance matrix, thus effectively selecting the frequency power spectrum of the estimated sources.

Careful selection of the described features of data and source spaces helps to improve the fidelity of the DECD solution. Nevertheless, the inherent ambiguity of the inverse solution precludes achieving a high degree of localization precision. It is for this reason that additional spatial information about the source space, readily available from other functional modalities such as fMRI and PET, can help to condition the DECD solution (Section 8.4).

8.2.3.3 Beamforming

Beamforming (sometimes called a spatial filter or a virtual sensor) is another way to solve the inverse problem, which actually does not directly minimize Equation 8.2. A beamformer attempts to find a linear combination of the input data $\hat{\mathbf{q}}_i = \mathbf{F}^i \mathbf{x}$, which represents the neuronal activity of each dipole $\mathbf{q}_i$ in the best possible way one at a given time. As in DECD methods, the search space is sampled, but, in contrast to the DECD approach, the beamformer does not try to fit all the observed data at once.

The linearly constrained minimum variance (LCMV) beamformer [41] looks for a spatial filter defined as $\mathbf{F}^i$ of size $M \times L$ minimizing the output energy $\mathbf{F}^i{}^\top \mathbf{C_X} \mathbf{F}^i$ under the constraint that only $\mathbf{q}_i$ is active at that time, i.e., that there is no attenuation of the signal of interest: $\mathbf{F}^k \mathbf{G}_{\bar{i}} = \delta_{ki} \mathbf{I}_L$, where the Kronecker delta $\delta_{ki} = 1$ only if $k = i$ and 0 otherwise. Because the beamforming filter $\mathbf{F}^i$ for the ith dipole is defined independently of the other possible dipoles, index i will be dropped from the derived results for clarity of presentation.

The constrained minimization, solved using Lagrange multipliers, yields

$$\mathbf{F} = (\mathbf{G}_{\bar{\tau}}{}^{\top}\mathbf{C}_X^{-1}\mathbf{G}_{\bar{\tau}})^{-1}\mathbf{G}_{\bar{\tau}}{}^{\top}\mathbf{C}_X^{-1} \tag{8.10}$$

This solution is equivalent to Equation 8.7, when applied to a single dipole with the regularization term omitted. Source localization is performed using Equation 8.10 to compute the variance of every dipole $\mathbf{q}$, which, in the case of uncorrelated dipole moments, is

$$v_{\mathbf{q}} = tr((\mathbf{G}_{\bar{\tau}}{}^{\top}\mathbf{C}_X^{-1}\mathbf{G}_{\bar{\tau}})^{-1}) \tag{8.11}$$

The noise sensitivity of Equation 8.11 can be reduced by using the noise variance of each dipole as normalizing factor $v_{\varepsilon} = tr((\mathbf{G}_{\bar{\tau}}{}^{\top}\mathbf{C}_{\varepsilon}^{-1}\mathbf{G}_{\bar{\tau}})^{-1})$. This produces the so-called neural activity index

$$z = \frac{v_{\mathbf{q}}}{v_{\varepsilon}} \tag{8.12}$$

An alternative beamformer, synthetic aperture magnetometry or SAM [42], is similar to the LCMV if the orientation of the dipole is defined, but it is quite different in the case of a dipole with an arbitrary orientation. We define a vector of lead coefficients $\mathbf{g}_i(\theta)$ as a function of the dipole orientation. This returns a single vector for the orientation θ of the ith dipole, as opposed to the earlier formulation in which the L columns of $\mathbf{G}_{\bar{\tau}}$ played a similar role. With this new formulation, we construct the spatial filter

$$\mathbf{f}(\theta) = \frac{1}{\mathbf{g}_i(\theta)^{\top}\mathbf{C}_X^{-1}\mathbf{g}_i(\theta)}\mathbf{g}_i(\theta)^{\top}(\mathbf{C}_X + \lambda\mathbf{C}_{\varepsilon})^{-1} \tag{8.13}$$

which, under standard assumptions, is an optimal linear estimator of the time course of the ith dipole. The variance of the dipole, accordingly, is also a function of θ, specifically $v_{\mathbf{q}}(\theta) = 1/(\mathbf{g}_i(\theta)^{\top}\mathbf{C}_X^{-1}\mathbf{g}_i(\theta))$. To compute the neuronal activity index, the original SAM formulation uses a slightly different normalization factor $v_{\varepsilon}(\theta) = \mathbf{f}(\theta)^{\top}\mathbf{C}_{\varepsilon}\mathbf{f}(\theta)$, which yields a different result if the noise variance in $\mathbf{C}_{\varepsilon}$ is not equal across the sensors.

The unknown value of θ is found via a nonlinear optimization of the neuronal activity index for the dipole:

$$\theta = \arg\max_{\vartheta} \frac{v_{\mathbf{q}}(\vartheta)}{v_{\varepsilon}(\vartheta)}$$

Despite the pitfalls of nonlinear optimization, SAM filtering provides a higher SNR to LCMV by bringing less than half of the noise power into the solution.

In addition, SAM filtering results in sharper peaks of the distribution of neuronal activity index over the volume [43].

Having computed v_q and v_ε using SAM or LCMV for the two experimental conditions, passive (p) and active (a), it is possible to compute a pseudo-t value $\hat{t}$ for each location across the two conditions

$$\hat{t} = \frac{v_q^{(a)} - v_q^{(p)}}{v_\varepsilon^{(a)} + v_\varepsilon^{(p)}} \qquad (8.14)$$

Such an approach provides the possibility of considering experimental design in the analysis of E/MEG localization.

Unlike ECD, beamforming does not require prior knowledge of the number of sources, nor does it search for a solution in an underdetermined linear system as does DECD. For these reasons, beamforming remains the favorite method of many researchers in EMSI and has been suggested for use in the integrative analysis of E/MEG and fMRI, which we cover in Section 8.5.

8.3 MULTIMODAL EXPERIMENTS

Obtaining noncorrupted simultaneous recordings of EEG and fMRI is a difficult task due to interference between the strong MR field and the EEG acquisition system. Because of this limitation, a concurrent EEG/fMRI experiment requires specialized design and preprocessing techniques to prepare the data for the analysis. The instrumental approaches described in this section are specific to collecting concurrent EEG and fMRI data. For obvious reasons MEG and fMRI data must be acquired separately in two sessions. However, even when MR and MEG are used sequentially, there is the possibility of contamination from the magnetization of a subject's metallic implants, which can potentially disturb MEG acquisition if it is performed shortly after the MR experiment.

8.3.1 MEASURING EEG DURING MRI: CHALLENGES
AND APPROACHES

Developing methods for the integrative analysis of EEG and fMRI data is difficult for several reasons, not the least of which is that the concurrent acquisition of EEG and fMRI itself has proved challenging. The nature of the problem is expressed by Faraday's law of induction: a time-varying magnetic field in a wire loop induces an electromotive force (EMF) proportional in strength to the area of the wire loop and to the rate of change of the magnetic field component orthogonal to the area. When EEG electrodes are placed in a strong ambient magnetic field resulting in the EMF effect, several undesirable complications arise:

- Rapidly changing MR gradient fields and RF pulses may induce voltages in the EEG leads placed inside the MR scanner. Introduced potentials may greatly obscure the EEG signal [44]. This kind of artifact is a real concern for concurrent EEG/MRI acquisition. Due to the deterministic nature of MR interference, hardware and algorithmic solutions may be able to unmask the EEG signal from MR disturbances. For example, Allen et al. [45] suggested an average waveform subtraction method to remove MR artifacts that seems to be effective [46]. However, it is important to note that time variations of the MR artifact waveform can reduce the success of this method [47,48]. The problem can be resolved through hardware modification that increases the precision of the synchronization of MR and EEG systems [49] or during postprocessing by using precise timings of the MR pulses during EEG waveform averaging [47]. Other techniques that have been proposed to reduce MR and ballistocardiographic artifacts include spectral domain filtering, spatial Laplacian filtering, PCA (see Figure 8.1), and ICA (see Reference 50 to Reference 53).

- Even a slight motion of the EEG electrodes within the strong static field of the magnet can induce significant EMF [54,55]. For instance, native pulsatile motion related to a heartbeat yields a ballistocardiographic artifact in the EEG that can be roughly the same magnitude as the EEG signals themselves [44–56]. Usually such artifacts are removed by the same average waveform subtraction method, in which the waveform is an averaged response to each heartbeat.
- Induced electric currents can heat up the electrode leads to painful or even potentially dangerous levels, such as to the point of burning the subject [57]. Current-limiting electric components (resistors, JFET transistors,

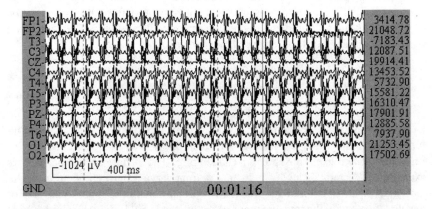

FIGURE 8.1 EEG MR artifact removal using PCA. EEG taken inside the magnet (top); EEG after PCA-based artifact removal but with ballistocardiographic artifacts present (center); and EEG with all artifacts removed (bottom). After artifact removal, it can be seen that the subject closed his eyes at time 75.9 sec. (Courtesy of M. Negishi and colleagues, Yale University School of Medicine.)

etc.) are usually necessary to prevent the development of nuisance currents that can have direct contact with subject's scalp. Simulations show the safe power range that should be used for some coil, power, or sensor configurations to comply with FDA guidelines [58].

Another concern is the impact of EEG electrodes on the quality of MR images. The introduction of EEG equipment into the scanner can potentially disturb the homogeneity of the magnetic field and distort the resulting MR images [44,59]. Recent investigations show that such artifacts can be effectively avoided [60] by using specially designed EEG equipment [56], specialized geometries, and new "MR-safe" materials (carbon fiber, plastic) for the leads. To test the influence of a given EEG system on fMRI data, a comparison of the data collected with versus without the EEG system being present, should be conducted. Analysis of such data usually demonstrates the same activation patterns in two conditions [59], although a general decrease in fMRI SNR is observed when EEG is present in the magnet. A correction to the brain matter conductivities (which are used for forward E/MEG modeling) for the Hall effect finds the following first-order correction to be negligible: $\sigma_H = 4.1 \times 10^{-8} \sigma$ for $B = [1.5]T$ [61].

8.3.2 EXPERIMENTAL DESIGN LIMITATIONS

There are two ways of avoiding the difficulties associated with collecting EEG data in the magnet: (1) collect EEG and MRI data separately or (2) use an experimental paradigm that can work around the potential contamination between the two modalities. The decision between these two alternatives will depend on the constraints associated with research goals and methodology. For example, if an experiment can be repeated more than once with a high degree of reliability of the data, separate E/MEG and fMRI acquisition may be appropriate [62–65]. In cases when simultaneous measurements are essential for the experimental objective (e.g., cognitive experiments in which a subject's state might influence the results, as in monitoring of spontaneous activity or sleep-state changes), one of the following protocols can be chosen:

- *Triggered fMRI:* Detected EEG activity of interest (epileptic discharge, etc.) triggers MRI acquisition [66–69]. Due to the slowness of the HR, relevant changes in the BOLD signal can be registered 4 to 8 sec after the event. The EEG signal can settle quickly after the end of the previous MRI block [56], so it is acquired without artifacts caused by RF pulses or gradient fields that are present only during the MRI acquisition block. Note that ballistocardiographic and motion-caused artifacts still can be present and will require postprocessing in order to be eliminated. Although this is an elegant solution and has been used with some success in the localization of epileptic seizures, this protocol does have drawbacks. Specifically, it imposes a limitation on the amount of subsequent EEG activity that can be monitored if the EEG high-pass filters do not settle down soon after the MR sequence is terminated [70]. In this case, EEG hardware that does not have a long relaxation period must be used. Another drawback with this approach is that it requires online EEG signal monitoring to trigger the fMRI acquisition in case of spontaneous activity. Often, experiments of this kind are called EEG-correlated fMRI due to the fact that offline fMRI data time analysis implicitly uses EEG triggers as the event onsets [46].

- *Interleaved EEG/fMRI:* The experiment protocol consists of time blocks and only a single modality is acquired during each time block [61–71]. This means that every stimulus has to be presented at least once per modality. To analyze ERP and fMRI activations, the triggered fMRI protocol can be used with every stimulus presentation so that EEG and MR are sequentially acquired in order to capture a clean E/MEG signal followed by the delayed HR [72].
- *Simultaneous fMRI/EEG:* Preprocessing of the EEG signal mentioned in Section 8.1 is used to remove the MR-caused artifacts and to obtain an estimate of the true EEG signal. However, neither of the existing artifact-removing methods has been proved to be general enough to work for every type of EEG experiment and analysis. It is especially difficult to use such an acquisition scheme for cognitive experiments in which the EEG-evoked responses of interest can be of small amplitude and be completely overwhelmed by the MR noise [73].

8.4 MULTIMODAL ANALYSIS

There is an increasing number of reported E/MEG/fMRI conjoint studies that attempt to gain the advantages of a multimodal analysis for experiments involving perceptual and cognitive processes: visual perception [59,72,74,75] and motor activation [59], somatosensory mapping [65,76], fMRI correlates of EEG rhythms [41,71,77–79), auditory oddball tasks [63], passive frequency oddball [80], illusory figures in visual oddball tasks [81], target detection [62,82], face perception [64], sleep [70], language tasks [74,83], and epilepsy [66,67,69,84–87].

This section begins with an explanation of the role of anatomical MRI in multimodal experiments followed by a description of multimodal analysis methods used in the above-mentioned studies or test-driven on the simulated data.

8.4.1 USING ANATOMICAL MRI

The difference in captured MRI contrasts (proton densities [PD] or T1, T2 relaxation times) for different types of organic tissue makes possible the noninvasive collection of information about the structural organization of the brain. In addition, a regular gradient or spin-echo EPI sequence is capable of detecting transient or subtle changes of the magnetic field in cortical tissue caused by neuronal activation [88,89]. However, direct application of MRI to capture functional activity remains limited due to a low signal-to-noise ratio (SNR), which is why MRI is often labeled anatomical. The next subsection briefly describes the analysis of acquired high-resolution 3-D images of the brain and how obtained structural information can be used to analyze data collected from other modalities.

8.4.1.1 Registration of EEG and MEG to MRI

If an EEG experiment is performed inside the magnet, it is possible to "mark" [90] the location of the EEG sensors to make them distinguishable on the anatomical MRI. Coordinates for these locations can then be found either manually or automatically [91] and will lie in the MRI coordinate system. In the case when MR and E/MEG data are acquired in separate sessions, spatial registration between E/MEG and MRI coordinate systems must be performed before any anatomical information can be introduced into the analysis of E/MEG data. There are two general possible ways of performing registration between MRI and E/MEG data: (a) registering a limited set of fiducial points or (b) aligning scalp surfaces obtained during MRI with a digitization of the scalp during E/MEG. Methods based on the alignment of the scalp surfaces (or point clouds) considered to perform better than those using fiducial points [92–95], but are more computationally demanding and rely on iterative optimization. In addition, it can be time consuming to obtain the dense digitization of the subject's head using a single-point 3-D digitizer. For these reasons the fiducial points approach remains the preferred E/MEG/MRI registration method (for instance, 90,96). The fiducial points method involves the alignment of a limited set of points, which have a strict known correspondence between the two spaces, so that each fiducial point in E/MEG space with coordinates ($\mathbf{x}_i^E$) has a corresponding known point ($\mathbf{x}_i^M$) in MRI space. Such coupling removes the possibility of being trapped in the local minima of the iterative surface-aligning methods and makes registration simple and fast. The precision of the derived transformation can be increased by adding more pairs of corresponding E/MEG and MRI points. A more detailed description of the registration method using fiducial points follows.

Locations of the fiducial points (e.g., anatomical points: nasion, inion, preauricular points or tragus of the left and right earlobes, vertex; MRI-visible capsules or even bite-bar points [97,98]) are captured together with the locations of E/MEG sensors using a 3-D digitizer and then matched to the locations of corresponding fiducial points obtained from the analysis of the MRI for the same subject. A 3-D rigid transformation of the points from the E/MEG ($\mathbf{x}^E$) to the MRI coordinate system ($\mathbf{x}^{E \to M}$) can be defined by the rotation matrix $\mathbf{R}$ and translation vector $\mathbf{v}$, so that $\mathbf{x}^{E \to M} = \mathbf{R}\mathbf{x}_i^E + \mathbf{v}$. Commonly, the quadratic misregistration error measure is the subject to minimization $\varepsilon(\mathbf{R}, \mathbf{v}) = \Sigma_i^P (\mathbf{x}_i^M - \mathbf{x}^{E \to M})^2$, where P is the number of the points. Solutions can be found with simplified geometrical formulations [99] or iterative search optimization using Powell's algorithm [97]. Such simplifications or complications are not necessary because the analytical form solutions have been derived in other fields

[100,101], and they are often used in the surface-matching methods discussed earlier. For instance, quaternions (vectors in L_4) can be natively used to describe a rotation in 3-D space leading to a straightforward solution of the registration problem [100]. This method is simple to implement. Its precision rapidly increases with the number of fiducial points, reaching the performance of surface-matching algorithms cheaply and efficiently.*

8.4.1.2   Segmentation and Tessellation

PD or T1/T2 3-D MR images can be used to segment different brain tissues (white matter, gray matter, cerebrospinal fluid [CSF], skull, scalp) as well as abnormal formations (tumors) [17,102]. Different kinds of MR contrasts are optimal for the segmentation of the different kinds of head and brain structures. For instance, PD-weighted MRI yields superior segmentation of the inner and outer skull surfaces, because bones have much smaller water content than brain tissue, making the skull easily distinguishable on PD images. On the other hand, exploiting T1 and T2 relaxation time differences between various sorts of brain tissue leads to higher quality segmentation of structures within the brain.

Using triangulation (tessellation) and interpolation it is possible to create fine-grained smooth mesh representations or tetrahedral assemblies of the segmented tissues [103–105]. Obtained 3-D meshes of the cortical surface alone brings valuable information to the analysis of E/MEG signals [106]: the physiology of the neuronal generators can be considered, allowing one to limit the search space for activated sources to the gray matter regions and oriented orthogonally or nearly so to the cortical surface [17,107].

Monte Carlo studies [108] tested the influence of the orientation constraint in the case of the DECD model and showed that it leads to much better conditioning

---

* To find the minimum of the error function $\varepsilon(\mathbf{R}, \mathbf{v})$, we need merely to calculate a principal eigenvector

$$\mathbf{r} = \text{max\_eigenvector} \begin{bmatrix} \text{tr}(\Sigma) & \Delta^T \\ \Delta & \Sigma + \Sigma^{\cdot} - \text{tr}(\Sigma)\mathbf{I}_3 \end{bmatrix} \tag{8.15}$$

where

$$\bar{\mathbf{x}} = \frac{1}{P}\sum_i^P \mathbf{x}_i \quad \Sigma = \frac{1}{P}\sum_i^P \left(\mathbf{x}_i^E - \bar{\mathbf{x}}^E\right)\left(\mathbf{x}_i^M - \bar{\mathbf{x}}^M\right)^{\cdot} \quad \Delta = \begin{bmatrix} (\Sigma - \Sigma^{\cdot})_{23} \\ (\Sigma - \Sigma^{\cdot})_{31} \\ (\Sigma - \Sigma^{\cdot})_{12} \end{bmatrix}$$

The eigenvector $\mathbf{r}$ can be assumed to be normalized (unit length). Regarded as a quaternion, $\mathbf{r} = [r_0, r_1, r_2, r_3]^T$ uniquely defines the rotation. This can be converted into a conventional rotation matrix

$$\mathbf{R} = \begin{bmatrix} r_0^2 + r_1^2 - r_2^2 - r_3^2 & 2(r_1 r_2 - r_0 r_3) & 2(r_1 r_3 + r_0 r_2) \\ 2(r_1 r_2 + r_0 r_3) & r_0^2 + r_2^2 - r_1^2 - r_3^2 & 2(r_2 r_3 - r_0 r_1) \\ 2(r_1 r_3 - r_0 r_2) & 2(r_2 r_3 + r_0 r_1) & r_0^2 + r_3^2 - r_1^2 - r_2^2 \end{bmatrix}.$$

The translation vector is then simply $\mathbf{v} = \bar{\mathbf{x}}^M - \mathbf{R}\bar{\mathbf{x}}^E$.

of the inverse problem while still being robust to the error of the assumed cortical surface: random deviation of the orientation in 30 range leads to just a slight increase of distortion, thus not significantly affecting the accuracy of the localization procedure. Anatomical constraints improve the localization and contrast of beam-forming imaging methods as well, but the use of anatomical constraints is found to be advantageous only in the case of good MRI/E/MEG coregistration [109].

8.4.1.3 Forward Modeling of EEG and MEG

Volumetric structures derived from the tessellation procedure are used to create a realistic geometry of the head, which is crucial for the forward modeling of E/MEG fields. Previously, rough approximations based on best-fit single or multiple sphere models were developed to overcome the burden of creating realistic head geometry, but they became less favorable as the increased availability of powerful computational resources made more realistic modeling possible. Spatial information is especially important for EEG forward modeling due to the fact that it is more strongly affected by the conductivities of the skull and the scalp than the MEG forward model. Such inhomogeneities might not affect the magnetic field at all in the case of a spherical head model, when only the inner skull surface is of the main concern for the forward modeling.

There are four numerical methods available to solve the E/MEG modeling problem, and the boundary elements method (BEM) [110] is the most commonly used when isotropy (direction independence) of the matter is assumed, so that only boundary meshes obtained by the tessellation process are required. It was shown, however, that anisotropy of the skull [111] and white matter [112] can bias EEG and MEG forward models. To solve the forward problem in the case of an anisotropic medium, the head volume is presented by a large assembly of small homogeneous tetrahedrons, and a finite elements method (FEM) [113] is used to approximate the solution. Another possible way is to use the finite difference method (FDM) on a regular computational mesh [14]. Table 8.2 lists some publicly available software that can help performs forward E/MEG modeling. Forward modeling of E/MEG signals rely on the knowledge of matter conductivities. Common values of conductivities for different tissues can be found in the literature [115], or can be estimated on a per-subject basis using electrical impedance tomography (EIT) [116] or diffusion tensor (DT) [117] MRI.

8.4.2 Forward Modeling of BOLD Signal

The successful analysis of the results of a multimodal experiment remains problematic. The main problem of multimodal analysis is the absence of a general unifying account of the BOLD fMRI signal in terms of the characteristics of a neuronal response. Various models have been suggested. On the one hand, they include naive modeling of BOLD signal in the context of a linear time invariant system (LTIS). On the other hand, there are general models of the BOLD signal in terms of detailed biophysical processes (Balloon [118] or Vein and Capillary [119] models). The naive models are not general enough to explain the variability

TABLE 8.2
Free Software Germane to Multinational Analysis of EEG/MEG/fMRI Data

Package	Forward EEG Spherical	EEG BEM	EEG FEM	MEG Spherical	MEG BEM	MEG FEM	Inverse ECD	DECD	Beamforming	MRI Brain Segmentation	Skull Segmentation	Scalp Segmentation	Tessellation	E/MEG Registration	fMRI	Matlab	POSIX‡	Mac OS X	MS Windows	Notes
Brainstorm (168)	✓	✓		✓	✓		✓		✓				⇕	✓		✓				MAP-MUSIC
NeuroFEM (169)/Pebbles	✓	✓	✓	✓	✓	✓	✓						⇕				✓	✓		
BioPSE (170)/SCIRun (171)	✓	✓	✓				✓	✓					⇕		✓	✓	✓			
Brainvisa/Anatomist (103)										✓		✓	✓		✓				✓	
FreeSurfer (172)										✓		✓	✓		✓		✓	✓		
Sure—† (173)										✓		✓	✓		✓		✓			
Brainsuite (105)										✓	✓	✓	✓						✓	
EEG/MEG/MRI tbx—† (174)				✓			✓	✓		⇕	⇕	⇕	⇕	⇕	✓	✓				
MEG tbx* (175)				✓			✓	✓		✓	✓	✓	✓	✓	✓	✓				
EEGLAB/FMRILAB (176)								✓							✓	✓				

Note: —stands for input/output facility for a feature.

An extensive MR segmentation bibliography is available online [102].

‡POSIX includes all versions of Unix and GNU/Linux. Most POSIX packages listed use X Windows for their graphical output.

*Matlab Toolbox.

of the BOLD signal, whereas complex parametric models that rely heavily on a prior knowledge of nuisance parameters (due to biophysical details), almost never do not have a reliable and straightforward means of estimation. This fact makes it unlikely that such comprehensive models will be used as reliable generative models of the BOLD signal. In the following subsections we describe modeling issues in greater detail to further underline the limited applicability of many of the multimodal analysis methods covered in Section 8.4.

8.4.2.1 Convolutional Model of BOLD Signal

Various experimenters had originally focused on simple contrast designs such as block design paradigms in order to exploit the presumed linearity between their design parameters and the HR. This assumption depends critically on the ability of the block design to amplify the SNR, and the implicit belief that the HR possesses more temporal resolution than indicated by the TR.

In order to account for the present autocorrelation of the HR caused by its temporal dispersive nature, Friston et al. [120] suggested to model HR with a LTIS. To describe the output of such a system, a convolution of an input (joint intrinsic and evoked neuronal activity $q(t)$) with a hemodynamic response function (HRF) $h(t)$ is used to model the HR

$$b(t) = [h \otimes q](t) \tag{8.16}$$

Localized neuronal activity itself is not readily available via means of non-invasive imaging, and therefore it is more appropriate to verify LTIS modeling on real data as a function of parameters of the presented stimuli (i.e., duration, contrast).

The convolutional model was used on real data to demonstrate linearity between the BOLD response and the parameters of presented stimuli [121, 122]. In fact, many experimenters have shown apparent agreement between LTIS modeling and real data. Specifically, it has been possible to model responses to longer stimuli durations by constructing them using the responses to shorter-duration stimuli, which is consistent with LTIS modeling. Because of the predictive success, its relative simplicity of application, and the resulting ignorance of biophysical details, this modeling approach became widely accepted. Unfortunately, LTIS as a modeling constraint is very weak, therefore allowing an arbitrary choice of parametric HRF based only on preference and familiarity.

Over the years, multiple models for the HRF have been suggested. The most popular and widely used up until now is a single probability density function (PDF) of gamma distribution by [123]. It was elaborated by [124] to perform the deconvolution of the HR signal, and the nuisance parameters $(n_1, t_1, n_2, t_2, a_2)$ of the next HRF were estimated for motor and auditory areas

$$h(t) = \frac{1}{c_1} t^{n_1} e^{-t/t_1} - \frac{a_2}{c_2} t^{n_2} e^{-t/t_2} \quad \text{where} \quad c_i = \max_t t^{n_i} e^{-t/t_i} = \left(\frac{e}{n_i t_i} \right)^{-n_i} \tag{8.17}$$

which can be described as the sum of two unscaled PDFs of the gamma distribution. The first term captures the positive BOLD HR, and the second term is to capture the overshoot often observed in the BOLD signal. Many other simple, as well as more sophisticated, models of HRF were suggested: Poisson PDF [120], Gaussians [125], Bayesian derivations [126–128], and others. The particular choice of any of them was primarily dictated by some motivation other than biophysics: easy Fourier transformation, presence of postresponse dip, or best-fit properties.

Since the suggestion of the convolutional model describing the BOLD response, different aspects of HR linearity became an actively debated question. If HR is linear, then with what features of the stimulus (e.g., duration, intensity) or neuronal activation (e.g., firing frequency, field potentials, frequency power) does it vary linearly? As a first approximation, it is important to define the ranges of the above-mentioned parameters in which HR was found to behave linearly. For example, early linearity tests [124] showed the difficulty in predicting long-duration stimuli based on an estimated HR from shorter-duration stimuli. Soltysik et al. [129] reviewed existing papers describing different aspects of nonlinearity in BOLD HR and attempted to determine the ranges of linearity in respect to stimuli duration in three cortical areas: motor, visual, and auditory complex. The results of these analyses have shown that although there is a strong nonlinearity observed on small stimuli durations, long stimuli durations show a higher degree of linearity.

It appears that a simple convolutional model generally is not capable of describing the BOLD responses in terms of the experimental design parameters if these vary over a wide range during the experiment. Nevertheless, LTIS might be more appropriate to model the BOLD response in terms of neuronal activation if most of the nonlinearity in the experimental design can be explained by the nonlinearity of the neuronal activation itself.

8.4.2.2 Neurophysiologic Constraints

In the previous subsection we explored the subject of linearity between the experimental design parameters and the observed BOLD signal. For the purpose of this chapter it may be more interesting to explore the relation between neuronal activity and HR.

It is a well-known fact that E/MEG signals are produced by the large-scale synchrony of neuronal activity, whereas the nature of the BOLD signal is not clearly understood. The BOLD signal does not even seem to correspond to the most energy-consuming neural activity [130], as early researchers believed. Furthermore, the transformation between the electrophysiological indicators of neuronal activity and BOLD signal cannot be linear for a whole dynamic range of signals, under all experimental conditions, and across all the brain areas. Generally, a transformation function cannot be linear as the BOLD signal is driven by a number of nuisance physiologic processes such as cerebral metabolic oxygen consumption ($CMRO^2$), cerebral blood flow (CBF), and cerebral blood volume (CBV), as suggested by the Balloon model [118], which are not generally linear.

Due to the indirect nature of the BOLD signal as a tool to measure neuronal activity, in many multimodal experiments a preliminary comparative study is done first in order to assess the localization disagreement across different modalities. Spatial displacement is often found to be very consistent across multiple runs or experiments (see Subsection 8.4.3 for an example). Specifically, observed differences can potentially be caused by the variability in the cell types and neuronal activities producing each particular signal of interest [2]. That is why it is important first to discover the types of neuronal activations that are primary sources of the BOLD signal. Some progress on this issue has been made. A series of papers generated by a project to cast light on the relationship between the BOLD signal and neurophysiology has argued that local field potentials (LFP) serve a primary role in predicting the BOLD signal ([131], and References 27, 29, 54, 55, and 81 therein). This work countered the common belief that spiking activity was the source of the BOLD signal (for example, [132]) by demonstrating a closer relation of the observed visually evoked HR to the local field potentials (LFP) of neurons than to the spiking activity. This result places most of the reported nonlinearity between experimental design and observed HR into the nonlinearity of the neural response, which would benefit a multimodal analysis.

Note that the extracellular recordings experiments described above were carried out over small ROIs, and therefore they inherit the parameters of underlying hemodynamic processes for the given limited area. Thus, even if LFP is taken as the primary electrophysiological indicator of the neuronal activity causing the BOLD signal, the relationship between the neuronal activity and the hemodynamic processes on a larger scale remains an open question.

Because near-infrared optical imaging (NIOI) is capable of capturing the individual characteristics of cerebral hemodynamics such as total, oxy-, and deoxyhemoglobin content, some researchers tried to use NIOI to reveal the nature of the BOLD signal. Rat studies using 2-D optical imaging [133] showed the nonlinear mapping between the neuronal activity and evoked hemodynamic processes. This result should be a red flag for those who try to define the general relation between neuronal activation and the BOLD signal as mostly linear. The conjoint analysis of BOLD and NIOI signals revealed the silent BOLD signal during present neural activation registered by E/MEG modalities [119]. This mismatch between E/MEG and fMRI results is known as the *sensory motor paradox* [134]. To explain this effect, the vein and capillary model was used to describe the BOLD signal in terms of hemodynamic parameters [119]. The suggested model permits the existence of silent and negative BOLD responses during positive neuronal activation. This fact, together with an increasing number of studies [135] confirming that sustained negative BOLD HR is a primary indicator of decreased neuronal activation, provide yet more evidence that the BOLD HR generally is not a simple linear function of neuronal activation but at best is a monotone function that has close-to-linear behavior in a wide range of nuisance neurophysiologic parameters. We conclude this subsection by noting that the absence of a generative model of the BOLD response prevents the development of universal methods of multimodal analysis. Nevertheless, as discussed in this subsection and shown by the results presented in the next subsection, there are specific ranges of applications in which the linearity between BOLD and neuronal activation can be assumed.

8.4.3 ANALYSIS METHODS

Whenever applicable, a simple comparative analysis of the results obtained from conventional unimodal analyses, together with findings reported elsewhere, can be considered as the first confirmatory level of a multimodal analysis. This type of analysis is very flexible, as long as the researcher knows how to interpret the results and to draw useful conclusions, especially whenever the results of comparison reveal commonalities and differences between the two [83]. On the other hand, by default, a unimodal analysis makes limited use of the data from the modalities and encourages researchers to look for analysis methods that would incorporate the advantages of each single modality. Nevertheless, simple inspection is helpful for drawing preliminary conclusions regarding the plausibility of performing a conjoint analysis using one of the methods described in this subsection, including correlative analysis, which might be considered an initial approach to try.

8.4.3.1 Correlative Analysis of EEG and MEG with fMRI

In some experiments, the E/MEG signal can serve as the detector of spontaneous neuronal activity (e.g., epileptic discharges) or changes in the processing states (e.g., vigilance states). The time onsets derived from E/MEG are alone valuable for further fMRI analysis, in which the BOLD signal often cannot provide such timing information. For instance, such use of EEG data is characteristic of the experiments performed via a triggered fMRI acquisition scheme (Subsection 8.3.2).

Correlative E/MEG/fMRI analysis becomes more intriguing if there is a stronger belief in the linear dependency between the BOLD response and features of E/MEG signal (e.g., amplitudes of ERP peaks, powers of frequency components) than between the hemodynamics of the brain and the corresponding parameter of the design (e.g., frequency of stimulus presentation or level of stimulus degradation). Then E/MEG/fMRI analysis effectively reduces the inherent bias present in the conventional fMRI analysis methods by removing the possible nonlinearity between the design parameter and the evoked neuronal response.

The correlative analysis relies on the preprocessing of E/MEG data to extract the features of interest to be compared with the fMRI time course. The obtained E/MEG features first get convolved with a hypothetical HRF (Subsection 8.4.2.1) to accommodate HR sloppiness and are then subsampled to fit the temporal resolution of fMRI. The analysis of fMRI signal correlation with amplitudes of selected peaks of ERPs revealed sets of voxels that have a close-to-linear dependency between the BOLD response and amplitude of the selected ERP peak (N170 in Horovitz et al. [64], P300 in Horovitz et al. [63], and amplitude of mismatch negativity (MMN) [80]), thus providing a strong correlation ($P < .001$ [64]). A parametric experimental design with different noise levels introduced for the stimulus degradation [64,80] or different levels of sound frequency deviant [80] helped to extend the range of detected ERP and fMRI activations, thus effectively increasing the significance of the results found. To support the suggested connection between the specific ERP peak and fMRI-activated area, the correlation of the same BOLD signal with the other ERP peaks must be lower, if there is any at all [64]. As a consequence, such analysis cannot prove that any specific peak of EEG is produced by the neurons located in the fMRI-detected areas alone, but it definitely shows that they are connected in the specific paradigm.

The search for the covariates between the BOLD signal and widespread neuronal signals, such as the alpha rhythm, remains a more difficult problem due to the ambiguity of the underlying process, as there are many possible generators of alpha rhythms corresponding to various functions [136]. As an example, Goldman et al. [77] and Laufs et al. [79] were looking for the dependency between fMRI signal and EEG alpha rhythm power during interleaved and simultaneous EEG/fMRI acquisition, correspondingly. They report similar (negative correlation in parietal and frontal cortical activity), as well as contradictory (positive correlation) findings, which can be explained by the variations in the experimental setup [137] or by the heterogeneous coupling between the alpha rhythm and the BOLD response [79]. Despite the obvious simplification of the correlative methods, they may still have a role to play in constraining and revealing the definitive forward model in multimodal applications.

8.4.3.2 Decomposition Techniques

The common drawback of the presented correlative analyses techniques is that they are based on the selection of the specific feature of the E/MEG signal to be correlated with the fMRI time trends, which are not so perfectly conditioned to be characterized primarily by the feature of interest. The variance of the background processes, which are present in the fMRI data and are possibly explained by the discarded information from the E/MEG data, can reduce the significance of the obtained correlation. That is why it was suggested [138] that the entire E/MEG signal be used, without focusing on its specific frequency band, to derive the E/MEG and fMRI signal components that have the strongest correlation among them. The introduction of decomposition techniques (such as basis pursuit, PCA, ICA, etc.) into the multimodal analysis makes this work particularly interesting.

To perform the decomposition [138], partial least-squares (PLS) regression was generalized into the tri-PLS2 model, which represents the E/MEG spectrum as a linear composition of trilinear components. Each component is the product of spatial (among E/MEG sensors), spectral and temporal factors, in which the temporal factors have to be maximally correlated with the corresponding temporal component of the similar fMRI signal decomposition into bilinear components: products of the spatial and temporal factors. Analysis using tri-PLS2 modeling on the data from Goldman et al. [77] found a decomposition into three components corresponding to alpha, theta, and gamma bands of the EEG signal. The fMRI components found had a strong correlation only in the alpha band component

(Pearson correlation .83 ($p = .005$)), although the theta component also showed a linear correlation of .56 ($p = .070$). It is interesting to note that spectral profiles of the trilinear EEG atoms received with and without fMRI influence were almost identical, which can be explained either by the noninfluential role of fMRI in tri-PLS2 decomposition of EEG or just by a good agreement between the two. On the other hand, EEG definitely guided fMRI decomposition, so that the alpha rhythm spatial fMRI component agreed very well with the previous findings [77].

8.4.3.3 Equivalent Current Dipole Models

ECD is the most elaborated and widely used technique for source localization in EMSI. It can easily account for activation areas obtained from the fMRI analysis, thus giving the necessary fine time-space resolution by minimizing the search space of nonlinear optimization to the thresholded fMRI activation map. Although very attractive, such a method has most of the problems of the ECD method mentioned in Subsection 8.2.3, and introduces another possible bias due to the belief in the strong coupling between hemodynamic and electrophysiological activities. For this reason, it needs to be approached with caution in order to carefully select the fMRI regions to be used in the ECD/fMRI combined analysis.

Although good correspondence between ECD and fMRI results is often found [139], some studies reported a significant (1 to 5 cm) displacement between locations obtained from fMRI analysis and ECD modeling [76,87,140,141]. It is interesting to note that such displacement can be very consistent across the experiments of different researchers using the same paradigm (for instance, motor activations [65,76,142]). As was already mentioned, in the first step, a simple comparison of detected activations across the two modalities can be done to increase the reliability of dipole localization alone. Further, additional weighting by the distance from the ECD to the corresponding fMRI activation foci can guide ECD optimization [143] and silent in fMRI activations can be accommodated by introducing free dipoles without the constraint on dipole location.

Auxiliary fMRI results can help to resolve the ambiguity of the inverse E/MEG problem if ECD lies in the neighborhood of multiple fMRI activations. Placing multiple ECDs inside the fMRI foci with successive optimization of ECDs orientations and magnitudes may produce more meaningful results, especially if the suggested multiple ECDs model better describes the E/MEG signal.

Due to the large number of consistent published fMRI results, it seems viable to perform a pure E/MEG experiment with consequent ECD analysis using known relevant fMRI activation areas found by the other researchers performing the same kind of experiment [144], thus providing the missing temporal explanation to the known fMRI activations.

8.4.3.4 Linear Inverse Methods

Dale and Sereno [17] formulated a simple but powerful linear framework for the integration of different imaging modalities into the inverse solution of DECD, in which the solution was presented as unregularized (just minimum norm; Equation 8.8) with $\mathbf{W_Q} = \mathbf{C}_S$ and $\lambda\mathbf{W_X} = \mathbf{C}_\varepsilon$. The simplest way to account for fMRI data is to use a thresholded fMRI activation map as the inverse solution space, but this was rejected [1] due to its incapability to account for fMRI-silent sources, which is why the idea to incorporate variance information from fMRI into $\mathbf{C}_S$ was further elaborated [108] by the introduction of relative weighting for fMRI-activated voxels via constructing a diagonal matrix $\mathbf{W_Q} = \mathbf{W}_{fMRI} = \{v_{ii}\}$, where $v_{ii} = 1$ for fMRI-activated voxels and $v_{ii} = v_0 \in [0,1]$ for voxels which are not revealed by fMRI analysis. A Monte Carlo simulation showed that $v_0 = 0.1$ (which corresponds to the 90% relative fMRI weighting) leads to a good compromise with the ability to find activation in the areas which are not found active by fMRI analysis and to detect active fMRI spots (even superficial) in the DECD inverse solution. An alternative formulation of the relative fMRI weighting in the DECD solution can be given using a subspace regularization (SSR) technique [145], in which an E/MEG source estimate is chosen from all possible solutions describing the E/MEG signal, and is such that it minimizes the distance to a subspace defined by the fMRI data (Figure 8.2). Such a formulation aids an understanding of the mechanism of fMRI influence on the inverse E/MEG solution: SSR biases underdetermined the E/MEG source locations toward the fMRI foci.

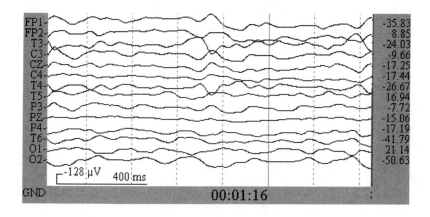

FIGURE 8.2 Geometrical interpretation of subspace regularization in the MEG/EEG source space. (A) The cerebral cortex is divided into source elements q1, q2,...,qK, each representing an ECD with a Fixed orientation. All source distributions compose a vector q in K-dimensional space. (B) The source distribution q is divided into two components $q^a \in S^a \equiv$ range(G^T), determined by the sensitivity of MEG sensors and $q^0 \in$ null G, which does not produce an MEG signal. (C) The fMRI activations define another subspace S^{fMRI}. (D) The subspace-regularized fMRI-guided solution $q^{SSR} \in M$ is closest to S^{fMRI}, minimizing the distance $\|Pq^{SSR}\|$, where P (an $N \times N$ diagonal matrix with $Pii = 1/0$ when the ith fMRI voxel is active/inactive) is the projection matrix into the orthogonal complement of S^{fMRI}. (Adapted from Figure 1 of Ahlfors, S.P. and Simpson, G.V. (2004). Geometrical interpretation of fMRI-guided MEG/EEG inverse estimates. *NeuroImage*. 22(1): 323–332. With permission.)

The relative fMRI weighting was tested [146] in an MEG experiment and conjoint fMRI/MEG analysis results were obtained similar to those reported in previous fMRI, PET, MEG, and intracranial EEG studies. F. Babiloni et al. [147] followed Dale et al. [146] in a high-resolution EEG and fMRI study to incorporate nonthresholded fMRI activation maps with other factors. First of all, $\mathbf{W_{fMRI}}$ was reformulated to $(\mathbf{W_{fMRI'}})_{ii} = v_0 + (1 - v_0)_i/_{max}$, where i corresponds to the relative change of the fMRI signal in the ith voxel, and max is the maximal detected change. This way the relative E/MEMEG/fMRI scheme is preserved and locations of stronger fMRI activations have higher prior variance. Finally, the three available weighting factors were combined: fMRI relative weighting, correlation structure obtained from fMRI described by the matrix of correlation coefficients $\mathbf{K}_S$, and the gain normalization weighting matrix $\mathbf{W_n}$ (Section 8.2.3.2): $W_Q = W_{|MR|'}^{1/2} W_n^{1/2} K_S W_n^{1/2} W_{|MR|'}^{1/2}$. Although $\mathbf{W_{fMRI'}}$ alone had improved EMSI localization, the incorporation of the $\mathbf{K}_S$ led to finer localization of neuronal activation associated with finger movement.

Although most of the previously discussed DECD methods are involved in finding minimal L_2 norm solution, the fMRI-conditioned solution with minimal L_1 - norm (regularization term in Equation 8.6, $f(\mathbf{Q}) = \mathbf{Q}_1$) is shown to provide a sparser activation map [148] with activity focalized to the seeded hotspot locations [143].

An fMRI-conditioned linear inverse is an appealing method due to its simplicity, and the rich background of DECD linear inverse methods derived for the analysis of E/MEG signals. Nonetheless, one should approach these methods with extreme caution in a domain in which nonlinear coupling between BOLD and neural activity is likely to overwhelm any linear approximation [141].

8.4.3.5 Beamforming

Lahaye et al. [149] suggest an iterative algorithm for conjoint analysis of EEG and fMRI data acquired simultaneously during an event-related experiment. Their method relies on iterated source localization by the LCMV beamformer (Equation 8.10), which makes use of both EEG and fMRI data. The covariance $\mathbf{C}_X$ used by the beamformer is calculated anew each time step, using the previously estimated sources. Although the original formulation is cumbersome, this method appears promising as (a) it makes use of both spatial and temporal information available from both modalities, and (b) it can account for silent BOLD sources using an electrometabolic coupling constant that is estimated for each dipole.

8.4.3.6 Bayesian Inference

During the last decade, Bayesian methods became dominant in probabilistic signal analysis. The idea behind them is to use Bayes' rule to derive a posterior probability of a given hypothesis having observed data D, which serves as evidence to support the hypothesis

$$p(H \mid D) = \frac{p(D \mid H)p(H)}{p(D)} \qquad (8.18)$$

where $p(H)$ and $p(D)$ are prior probabilities of the hypothesis and evidence, correspondingly, and the conditional probability $p(D|H)$ is known as a likelihood function. Thus, Equation 8.19 can be viewed as a method of combining the results of conventional likelihood analyses for multiple hypotheses into the posterior probability of the hypotheses $p(H|D)$ or some function of it, after being exposed to the data. The derived posterior probability can be used to select the most probable hypothesis, i.e., the one with the highest probability

$$\hat{H}_{|D} = \arg\max_H p(H|D) = \arg\max_H \log p(D|H) + \log p(H) \qquad (8.19)$$

leading to the maximum a posteriori (MAP) estimate, where the prior data probability $p(D)$ (often called a partition function) is omitted because the data does not depend on the choice of the hypothesis, and it does not influence the maximization over H.

For the class of problems related to signal processing, hypothesis H generally consists of a model M characterized by a set of nuisance parameters $\Theta = \{\theta_1, \theta_{2,\ldots,n}\}$. The primary goal usually is to find a MAP estimate of some quantity of interest Δ or, more generally, its posterior probability distribution $p(\Delta|D,M,\Theta)$. Δ can be an arbitrary function of the hypothesis or its components $\Delta = f(H)$ or often just a specific nuisance parameter of the model $\Delta \equiv \theta_1$. To obtain the posterior probability of the nuisance parameter, its marginal probability has to be computed by integration over the rest of the parameters of the model

$$p(\theta_1|D,M) = \int p(\theta_1, \theta_{2\ldots n}|D,M)d\theta_{2\ldots n}$$
$$= \int p(\theta_1|\theta_{2\ldots n}, D, M)p(\theta_{2\ldots n}|D,M)d\theta_{2\ldots n} \qquad (8.20)$$

Due to the integration operation involved in determination of any marginal probability, Bayesian analysis becomes very computationally intensive if an analytical integral solution does not exist. Therefore, sampling techniques (e.g., MCMC, Gibbs sampler) are often used to estimate full posterior probability $p(\Delta|D,M)$, MAP $\hat{\Delta}_{|D,M} = \mathrm{argmax}_\Delta\, p(\Delta|D,M)$, or some statistics such as an expected value $E[\Delta|D,M]$ of the quantity of interest.

The Bayesian approach sounds very appealing for the development of multimodal methods. It is inherently able to incorporate all available evidence, which is, in our case, obtained from the fMRI and E/MEG data ($D = \{\mathbf{X},\mathbf{B}\}$) to support the hypothesis on the location of neuronal activations, which in the case of DECD model is $\mathcal{H} = \{\mathbf{Q}, M\}$. However, detailed analysis of Equation 8.18 leads to necessary simplifications and assumptions of the prior probabilities in order to derive a computationally tractable formulation. Therefore, it often loses its generality. Thus, to derive a MAP estimator for $\hat{\mathbf{Q}}_{|\mathbf{X},\mathbf{B},M}$, Trujilli-Barreto et al. [150] had to condition the computation by a set of smplifying model assumptions such as: noise is formally distributed, no same parameters of forward models have inverse Gamma prior distributions, and neuronal activation is described by a linear function of hemodynamic response.

The results on simulated and experimental data from a somatosensory MEG/fMRI experiment confirmed the applicability of Bayesian formalism to the multimodal imaging even under the set of simplifying assumptions mentioned above.

Usually, model M is not explicitly mentioned in Bayesian formulations (such as Equation 8.20) because only a single model is considered. For instance, Bayesian formulation of LORETA E/MEG inverse corresponds to a DECD model, where $\Theta = \mathbf{Q}$ is constrained to be smooth (in space), and to cover a whole cortex surface. In the case of Bayesian model averaging (BMA), the analysis is carried out for different models M_i, which might have different nuisance parameters, e.g., E/MEG and BOLD signals forward models, possible spatial locations of the activations, constraints to regularize E/MEG inverse solutions. In BMA analysis, we combine results obtained using all considered models to compute the posterior distribution of the quantity of interest

$$p(\Delta \mid D) = \sum_i p(\Delta \mid D, M_i)\, p(M_i \mid D), \qquad (8.21)$$

where the posterior probability $p(M_i \mid D)$ of any given model M_i is computed via Bayes' rule using prior probabilities $p(M_i)$, $p(D)$ and the likelihood of the data given each model

$$p(D \mid M_i) = \int p(D \mid \Theta, M_i)\, p(\Theta \mid M_i)\, d\Theta. \qquad (8.22)$$

Initially, BMA was introduced into E/MEG imaging [151], in which Bayesian interpretation of Equation 8.8 was formulated to obtain $p(\mathbf{Q} \mid \mathbf{X}, \mathbf{B})$ for the case of Gaussian uncorrelated noise ($\mathbf{W_X} = \mathbf{C}_\varepsilon = v_\varepsilon \mathbf{I}$). In order to create a model, we partition the brain volume into a limited set of spatially distinct functional compartments, which are arbitrarily combined to define a M_i search space for the E/MEG inverse problem.

At the end, different models are sampled from the posterior probability $p(M_i \mid \mathbf{X})$ to get the estimate of the expected activity distribution of ECDs over all considered source models

$$E[\mathbf{Q} \mid \mathbf{X}] = \sum_i E[\mathbf{Q} \mid \mathbf{X}, M_i]\, p(M_i \mid \mathbf{X})$$

$$\mathrm{Var}[\mathbf{Q} \mid \mathbf{X}] = \sum_i \mathrm{Var}[\mathbf{Q} \mid \mathbf{X}, M_i]\, p(M_i \mid \mathbf{X})$$

where the normalized probability $p(M_i \mid \mathbf{X})$, Bayes' factor B_{i0}, and prior odds α_i, are

$$p(M_i \mid \mathbf{X}) = \frac{\alpha_i B_{i0}}{\sum_k \alpha_k B_{k0}} \qquad B_{i0} = \frac{p(\mathbf{X} \mid M_i)}{p(\mathbf{X} \mid M_0)} \qquad \alpha_i = \frac{p(M_i)}{p(M_0)}$$

In the original BMA framework for E/MEG [151] $\alpha_i = 1 \forall i$, i.e., the models had a flat prior PDF because no additional functional information was available at that point. Melie-García et al. [152] suggested to use the significance values of fMRI statistical t-maps to derive $p(M_i)$ as the mean of all such significance probabilities across the present in M_i compartments. This strategy causes the models consisting of the compartments with significantly activated voxels to get higher prior probabilities in BMA. The introduction of fMRI information prior to BMA analysis reduced the ambiguity of the inverse solution, thus leading to better localization performance. Although further analysis is necessary to define the applicability range of the BMA in E/MEG/fMRI fusion, it already looks promising because of the use of fMRI information as an additional evidence factor in E/MEG localization, rather than as a hard constraint.

Due to the flexibility of Bayesian formalism, various Bayesian methods for solving the E/MEG inverse problem already can be easily extended to partially accommodate evidence obtained from the analysis of fMRI data. For instance, correlation among different areas obtained from fMRI data analysis can be used as a prior in the Bayesian reconstruction of correlated sources [153]. The development of a neurophysiologic generative model of BOLD signal would allow many Bayesian inference methods (such as Schmidt et al. [154]) to introduce complete temporal and spatial fMRI information into the analysis of E/MEG data.

8.5 CONSIDERATIONS AND FUTURE DIRECTIONS

Although the BOLD signal is inherently nonlinear as a function of neuronal activation, there have been multiple reports of linear dependency between the observed BOLD response and the selected set of the E/MEG signal features. In general, such results are not inconsistent with the nonlinearity of BOLD, because, of course, a nonlinear function can be well approximated in the context of a specific experimental design, regions of interest, or dynamic ranges of the selected features of E/MEG signals. Besides the LFP/BOLD linearity reported by Logothetis and confirmed in the specific frequency bands of an EEG signal during a flashing checkerboard experiment [155], there have been reports of a strong correlation between the BOLD signal amplitude and other features of E/MEG responses.

In the past, DC-E/MEG signals have not been given any attention in multimodal integration, despite recent experiments showing the strong correlation between the changes of the observed DC-EEG signal and hemodynamic changes in the human brain [156]. In fact, such DC-E/MEG/BOLD coupling suggests that the integration of fMRI and DC-E/MEG might be a particularly useful way to study the nature of the time variations in the HR signal, which are usually observed during fMRI experiments but are not explicitly explained by the experimental design or the physics of the MR acquisition process.

Many EMSI methods can be naturally extended to account for fMRI data if a generative forward model of BOLD signal is available. For instance, direct universal-approximator inverse methods [157,158] have been found to be very effective (fast, robust to noise, and to complex forward models) for the E/MEG dipole localization problem and could be augmented to accept fMRI data if the generative model were augmented to produce it.

FMRI-conditioned E/MEG DECD methods have been shown to be a relatively simple and mathematically grounded for source imaging when there is good spatial agreement between E/MEG and fMRI signals. Due to the advantages of such methods, it might be valuable to consider other advanced E/MEG DECD methods such as FOCUSS [159], which is known to bring improvement of estimation of focal sources over simple linear inverse methods [160].

ICA as a signal decomposition technique has been found effective in removing artifacts in E/MEG without degrading neuronal signals [161–164], and moreover is known to be superior to PCA in the component analysis of E/MEG signals [165]. Initial research using ICA of fMRI in the spatial domain [166] was controversial; however, consecutive experiments and generalization of ICA to fMRI in the temporal domain (see Calhoun et al. [167] for an overview) has increased its normative value. The development of ICA methods for the analysis of multimodal data provides a logical extension of the decomposition techniques covered earlier in the chapter.

Because most of the multimodal methods presented in this chapter rely upon the linear dependence between signals, it is important to analyze, expand, and formalize the knowledge about the "linear" case. The formulation of a general BOLD signal model capable of describing the desired nonlinear dependency in terms of neuronal activation and nuisance physiological parameters would constitute a major step toward the development of multimodal methods with a wider range of application than in the current linear domain. Without such a model and without valid estimates of the underlying physiological parameters involved in the model, any multimodal analysis can not be considered progress.

In sum, it seems clear that fMRI should serve as a complementary evidence factor, rather than a hard constraint, in E/MEG source localization methods. The preprocessing of both fMRI and E/MEG signals should be done in order to select features of interest which had been previously reported to have good agreement between the two modalities. Any multimodal experiment should be based on the comparative study of unimodal experiments and analyses that show good agreement before performing conjoint data analysis.

REFERENCES

1. George, J.S., Aine, C.J., Mosher, J.C., Schmidt, D.M., Ranken, D.M., Schlitt, H.A., Wood, C.C., Lewine, J.D., Sanders, J.A., and Belliveau, J.W. (1995). Mapping function in the human brain with magnetoencephalography, anatomical magnetic-resonance imaging, and functional magnetic-resonance imaging. *J. Clin. Neurophysiol.* 12(5): 406–431.
2. Nunez, P.L., and Silberstein, R.B. (2000). On the relationship of synaptic activity to macroscopic measurements: does co-registration of EEG with fMRI make sense? *Brain Topogr.* 13(2): 79–96.
3. Salek-Haddadi, A., Friston, K.J., Lemieux, L., and Fish, D.R. (2003). Studying spontaneous EEG activity with fMRI. *Brain Res. Rev.* 43(1): 110–133.

4. George, J.S., Schmidt, D.M., Rector, D.M., Wood, C.C. (2002). *Functional MRI: An Introduction to Methods*, chap. 19. Dynamic functional neuroimaging intergratin multiple modalities. pp. 353–382. Oxford University Press.

5. Cohen, D. (1972). Magnetoencephalography: detection of the brain's electrical activity with a superconducting magnetometer. *Science* 175: 664–666.

6. Hämaläinen, M., Hari, R., Ilmoniemi, R.J., Knuutila, J., and Lounasmaa, O.V. (1993). Magnetoencephalography — theory, instrumentation, and applications to noninvasive studies of the working human brain. *Rev. Modern Phys.* 65(2): 413–497.

7. Malmivuo, J. and Plonsey, R. (1995). *Bioelectromagnetism—Principles and Applications of Bioelectric and Biomagnetic Fields*. Oxford University Press, New York, 1995. URL: http://butler.cc.tut._/_malmivuo/bem/bembook/index.htm.

8. Okada, Y., Lahteenmaki, A., and Xu, C. (1999). Comparison of MEG and EEG on the basis of somatic evoked responses elicited by stimulation of the snout in the juvenile swine. *Clin. Neurophysiol.* 110(2): 214–229.

9. Murakami, S., Hirose, A., and Okada, Y. (2003). Contribution of ionic currents to magnetoencephalography (MEG) and electroencephalography (EEG) signals generated by guinea-pig CA3 slices. *J. Physiol.* 553 (Pt 3): 975–985.

10. Michel, C.M., Murray, M.M., Lantz, G., Gonzalez, S., Spinelli, L., and Grave De Peralta, R. (2004). EEG source imaging. *Clin. Neurophysiol.* 115(10): 2195–2222.

11. Jung, T.-P., Makeig, S., Wester_eld, M., Townsend, J., Courchesne, E., and Sejnowski, T.J. (1999). Analyzing and visualizing single-trial event-related potentials. in *Advances in Neural Information Processing Systems 11*, pp. 118–124. MIT Press.

12. Tang, A.C., Pearlmutter, B.A., Zibulevsky, M., Hely, T.A., and Weisend, M.P. (2000). An MEG study of response latency and variability in the human visual system during a visual-motor integration task. in *Advances in Neural Information Processing Systems 12*, pp. 185–191. MIT Press.

13. Tang, A.C., and Pearlmutter, B.A. (April 2003). Independent components of magnetoencephalography: Localization and single-trial response onset detection. in Lu, Z.-L. and Kaufman, L., (Eds.). *Magnetic Source Imaging of the Human Brain*, pp. 159–201. Lawrence Erlbaum Associates, ISBN 0-8058-4511-9.

14. Wikswo, J.P., Jr., Gevins, A., and Williamson, S.J. (1993). The future of the EEG and MEG. *Electroencephalogr. Clin. Neurophysiol.* 87(1): 1–9.

15. Cohen, D. and Halgren, E. (2003). Magnetoencephalography (neuromagnetism). in *Encyclopedia of Neuroscience* 3rd ed. pp. 1–7. Amsterdam: Elsevier.

16. Malmivuo, J., Suihko, V., and Eskola, H. (1997). Sensitivity distributions of EEG and MEG measurements. *IEEE Trans. Biomed. Eng.* 44(3): 196–208.

17. Dale, A.M. and Sereno, M.I. (1993). Improved localization of cortical activity by combining EEG and MEG with MRI cortical surface reconstruction: A linear approach. *J. Cog. Neurosci.* 5(2): 162–176.

18. Baillet, S., Garnero, L., Marin, G., and Hugonin, J.P. (1999). Combined MEG and EEG source imaging by minimization of mutual information. *IEEE Trans. Biomed. Eng.* 46(5): 522–534.

19. Vrba, J. and Robinson, S.E. (2001). Signal processing in magnetoencephalography. *Methods* 25(2): 249–271.

20. Mosher, J.C., Leahy, R.M., and Lewis, P.S. (March 1999). EEG and MEG: forward solutions for inverse methods. *IEEE Trans. Biomed. Eng.* 46(3): 245–260.

21. Brazier, M.A.B. (1949). A study of the electric field at the surface of the head. *Electroencephalogr. Clin. Neurophysiol.* 2: 38–52.
22. Pruis, G.W., Gilding, B.H., and Peters, M.J. (1993). A comparison of different numerical methods for solving the forward problem in EEG and MEG. *Physiol. Meas.* (Suppl. 14). 4A: A1–9.
23. Zhang, Z. (May 1995). A fast method to compute surface potentials generated by dipoles within multilayer anisotropic spheres. *Phys. Med. Biol.* 40: 335–349.
24. Nolte, G. (November 2003). The magnetic lead field theorem in the quasi-static approximation and its use for magnetoencephalography forward calculation in realistic volume conductors. *Phys. Med. Biol.* 48: 3637–3652. URL: http://stacks.iop.org/PMB/48/3637.
25. Nolte, G. (2004). The magnetic lead field theorem in the quasi-static approximation and its use for MEG forward calculation in realistic volume conductors. *Phys. Med. Biol.* 48(22): 3637–3652.
26. Baillet, S., Mosher, J.C., and Leahy, M. (November 2001). Electromagnetic brain mapping. *IEEE Sig. Proc. Mag.*
27. Hauk, O. (2004). Keep it simple: a case for using classical minimum norm estimation in the analysis of EEG and MEG data. *NeuroImage.* 21(4): 1612–1621.
28. Baillet, S. and Garnero, L. (May 1997). A Bayesian approach to introducing anatomo-functional priors in the EEG/MEG inverse problem. *IEEE Trans. Biomed. Eng.* 44(5): 374–385.
29. Grave de Peralta Menendez, R., Murray, M.M., Michel, C.M., Martuzzi, R., and Gonzalez Andino, S.L. (2004). Electrical neuroimaging based on biophysical constraints. *NeuroImage.* 21(2): 527–539.
30. Golub, G., Heath, M., and Wahba, G. (1979). Generalized cross-validation as a method for choosing a good ridge parameter. *Technometrics* 21: 215–223.
31. Hansen, P.C. (1992). Analysis of discrete ill-posed problems by means of the L-curve. in *SIAM Review*, Vol. 34, pp. 561–580. Society for Industrial and Applied Mathematics, Philadelphia, PA, USA.
32. Phillips, C., Rugg, M.D., and Friston, K.J. (2002). Systematic regularization of linear inverse solutions of the EEG source localization problem. *NeuroImage* 17(1): 287–301.
33. Brooks, D.H., Ahmad, G.F., MacLeod, R.S., and Maratos, G.M. (1999). Inverse electrocardiography by simultaneous imposition of multiple constraints. *IEEE Trans. Biomed. Eng.* 46(1): 3–18.
34. Lawson, C.L. and Hanson, R.J. (1974). *Solving Least Squares Problems.* Series in Automatic Computation. Prentice-Hall, Englewood Cliffs, NJ 07632, USA, ISBN 0-13-822585-0.
35. Jeffs, B., Leahy, R., and Singh, M. (1987). An evaluation of methods for neuromagnetic image reconstruction. *IEEE Trans. Biomed. Eng.* 34(9): 713–723.
36. Phillips, C., Rugg, M.D., and Friston, K.J. (2002). Anatomically informed basis functions for EEG source localization: combining functional and anatomical constraints. *NeuroImage.* 16(3.1): 678–695.
37. Backus, G. and Gilbert, F. (1968). The resolving power of gross Earth data. *Geophys. J R Astron. Soc.* 16: 169–205.
38. Grave de Peralta Menendez, R., and Gonzalez Andino, S.L. (1998). A critical analysis of linear inverse solutions to the neuroelectromagnetic inverse problem. *IEEE Trans. Biomed. Eng.* pp. 440–448.

39. Wang, J., Williamson, S., and Kaufman, L. (1992). Magnetic source images determined by a lead-_eld analysis: the unique minimum-norm least-squares estimation. *IEEE T. Bio-Med. Eng.* 39(7): 665–675.
40. Pascual-Marqui, R.D., Michel, C.M., and Lehman, D. (1994). Low-resolution electromagnetic tomography: A new method for localizing electrical activity of the brain. *Int. J. Psychophysiology.* 18: 49–65.
41. Van Veen, B.D., van Drongelen, W., Yuchtman, M., and Suzuki, A. (1997). Localization of brain electrical activity via linearly constrained minimum variance spatial filtering. *IEEE Trans. Biomed. Eng.* 44(9): 867–880.
42. Robinson, S.E. and Vrba, J. (1999). Functional neuroimaging by synthetic aperture magnetometry (SAM). in Yoshimoto, T., Kotani, M., Kuriki, S., Karibe, H., and Nakasato, N., (Eds.). *Recent Advances in Biomagnetism*, pp. 302–305. Sendai, Japan. Tohoku University Press.
43. Vrba, J. and Robinson, S.E. (August 2000). Differences between synthetic aperture magnetometry (SAM) and linear beamformers. in Nenonen, J., Ilmoniemi, R., and Katila, T., (Eds.). *12th International Conference on Biomagnetism*, Helsinki, Finland, Biomag 2000. ISBN 951-22-5402-6.
44. Ives, J.R., Warach, S., Schmitt, F., Edelman, R.R., and Schomer, D.L. (1993). Monitoring the patient's EEG during echo planar MRI. *Electroencephalogr. Clin. Neurophysiol.* 87(6): 417–420.
45. Allen, P.J., Josephs, O., and Turner, R. (2000). A method for removing imaging artifact from continuous EEG recorded during functional MRI. *NeuroImage.* 12(2): 230–239.
46. Salek-Haddadi, A., Merschhemke, M., Lemieux, L., and Fish, D.R. (2002). Simultaneous EEG-correlated ictal fMRI. *NeuroImage.* 16(1): 32–40.
47. Cohen, M.S., Goldman, R.I., Stern, J., and Engel, J., Jr. (January 2001). Simultaneous EEG and fMRI made easy. *NeuroImage.* 13(6 Supp. 1).
48. Cohen, M.S. (2004). Method and Apparatus for Reducing Contamination of an Electrical Signal. United States Patent Application: 0040097802.
49. Anami, K., Mori, T., Tanaka, F., Kawagoe, Y., Okamoto, J., Yarita, M., Ohnishi, T., Yumoto, M., Matsuda, H., and Saitoh, O. (2003). Stepping stone sampling for retrieving artifact-free electroencephalogram during functional magnetic resonance imaging. *NeuroImage.* 19(2.1): 281–295.
50. Sijbers, J., Michiels, I., Verhoye, M., Van Audekerke, J., Van der Linden, A., and Van Dyck, D. (1999). Restoration of MR-induced artifacts in simultaneously recorded MR/EEG data. *Magn. Reson. Imaging.* 17(9): 1383–1391.
51. Bonmassar, G., Purdon, P.L., Jaaskelainen, I.P., Chiappa, K., Solo, V., Brown, E.N., and Belliveau, J.W. (2002). Motion and ballistocardiogram artifact removal for interleaved recording of EEG and EPs during MRI. *NeuroImage.* 16(4): 1127–1141.
52. Garreffa, G., Carni, M., Gualniera, G., Ricci, G.B., Bozzao, L., De Carli, D., Morasso, P., Pantano, P., Colonnese, C., Roma, V., and Maraviglia, B. (2003). Real-time MR artifacts filtering during continuous EEG/fMRI acquisition. *Magn. Reson. Imaging.* 21(10): 1175–1189.
53. Negishi, M., Abildgaard, M., Nixon, T., and Todd Constable, R. (2004). Removal of time-varying gradient artifacts from EEG data acquired during continuous fMRI. *Clin. Neurophysiol.* 115(9): 2181–2192.
54. Hill, R.A., Chiappa, K.H., Huang-Hellinger, F., and Jenkins, B.G. (1995). EEG during MR imaging: differentiation of movement artifact from paroxysmal cortical activity. *Neurology.* 45(10): 1942–1943.

55. Kruggel, F., Wiggins, C.J., Herrmann, C.S., and von Cramon, D.Y. (2000). Recording of the event-related potentials during functional MRI at 3.0 Tesla field strength. *Magn. Reson. Med.* 44(2): 277–282.

56. Goldman, R.I., Stern, J.M., Engel, J., Jr., and Cohen, M.S. (2000). Acquiring simultaneous EEG and functional MRI. *Clin. Neurophysiol.* 111(11): 1974–1980.

57. Lemieux, L., Allen, P.J., Franconi, F., Symms, M.R., and Fish, D.R. (1997). Recording of EEG during f MRI experiments: patient safety. *Magn. Reson. Med.* 38(6): 943–952.

58. Angelone, L.M., Potthast, A., Segonne, F., Iwaki, S., Belliveau, J.W., and Bonmassar, G. (2004). Metallic electrodes and leads in simultaneous EEG-MRI: specific absorption rate (SAR) simulation studies. *Bioelectromagnetics*. 25(4): 285–295.

59. Lazeyras, F., Zimine, I., Blanke, O., Perrig, S.H., and Seeck, M. (2001). Functional MRI with simultaneous EEG recording: feasibility and application to motor and visual activation. *J. Magn. Reson. Imaging*. 13(6): 943–948.

60. Krakow, K., Allen, P.J., Symms, M.R., Lemieux, L., Josephs, O., and Fish, D.R. (2000). EEG recording during f MRI experiments: image quality. *Hum. Brain Mapp*. 10(1): 10–15.

61. Bonmassar, G., Schwartz, D.P., Liu, A.K., Kwong, K.K., Dale, A.M., and Belliveau, J.W. (2001). Spatiotemporal brain imaging of visual-evoked activity using interleaved EEG and f MRI recordings. *NeuroImage* 13(6.1): 1035–1043.

62. Menon, V., Ford, J.M., Lim, K.O., Glover, G.H., and Pfefferbaum, A. (1997). Combined event-related f MRI and EEG evidence for temporal-parietal cortex activation during target detection. *Neuroreport*. 8(14): 3029–3037.

63. Horovitz, S.G., Skudlarski, P., and Gore, J.C. (2002). Correlations and dissociations between BOLD signal and P300 amplitude in an auditory oddball task: a parametric approach to combining f MRI and ERP. *Magn. Reson. Imaging*. 20(4): 319–325.

64. Horovitz, S.G., Rossion, B., Skudlarski, P., and Gore J.C. (2004). Parametric design and correlational analyses help integrating f MRI and electrophysiological data during face processing. *NeuroImage* 22(4): 1587–1595.

65. Schulz, M., Chau, W., Graham, S.J., McIntosh, A.R., Ross, B., Ishii, R., and Pantev, C. (2004). An integrative MEG-f MRI study of the primary somatosensory cortex using cross-modal correspondence analysis. *NeuroImage* 22(1):120–133.

66. Warach, S., Ives, J.R., Schlaug, G., Patel, M.R., Darby, D.G., Thangaraj, V., Edelman, R.R., and Schomer, D.L. (1996). EEG-triggered echo-planar functional MRI in epilepsy. *Neurology* 47(1): 89–93.

67. Seeck, M., Lazeyras, F., Michel, C.M., Blanke, O., Gericke, C.A., Ives, J., Delavelle, J., Golay, X., Haenggeli, C.A., de Tribolet, N., and Landis, T. (1998). Non-invasive epileptic focus localization using EEG-triggered functional MRI and electromagnetic tomography. *Electroencephalogr. Clin. Neurophysiol.* 106(6): 508–512.

68. Lazeyras, F., Blanke, O., Perrig, S., Zimine, I., Golay, X., Delavelle, J., Michel, C.M., de Tribolet, N., Villemure, J.G., and Seeck, M. (2000). EEG-triggered functional MRI in patients with pharmacoresistant epilepsy. *J. Magn. Reson. Imaging*. 12(1):177–185.

69. Krakow, K., Lemieux, L., Messina, D., Scott, C.A., Symms, M.R., Duncan, J.S., and Fish, D.R. (2001). Spatiotemporal imaging of focal interictal epileptiform activity using EEG-triggered functional MRI. *Epileptic Disord*. 3(2): 67–74.

70. Huang-Hellinger, F.R., Breiter, H.C., McCormack, G., Cohen, M.S., Kwong, K.K., Sutton, J.P., Savoy, R.L., Weisskoff, R.M., Davis, T.L., Baker, J.R., Belliveau, J.W., and Rosen, B.R. (1995). Simultaneous functional magnetic resonance imaging and electrophysiological recording. *Hum. Brain. Mapp.* 3: 13–25.

71. Makiranta, M.J., Ruohonen, J., Suominen, K., Sonkajarvi, E., Salomaki, T., Kiviniemi, V., Seppanen, T., Alahuhta, S., Jantti, V., and Tervonen, O. (2004). BOLD-contrast functional MRI signal changes related to intermittent rhythmic delta activity in EEG during voluntary hyperventilation-simultaneous EEG and fMRI study. *NeuroImage* 22(1):222–231.

72. Sommer, M., Meinhardt, J., and Volz, H.P. (2003). Combined measurement of event-related potentials (ERPs) and fMRI. *Acta. Neurobiol. Exp. (Wars).* 63(1): 49–53.

73. Schomer, D.L., Bonmassar, G., Lazeyras, F., Seeck, M., Blum, A., Anami, K., Schwartz, D., Belliveau, J.W., and Ives, J. (2000). EEG-linked functional magnetic resonance imaging in epilepsy and cognitive neurophysiology. *J. Clin. Neurophysiol.* 17(1): 43–58.

74. Singh, K.D., Barnes, G.R., Hillebrand, A., Forde, E.M., and Williams, A.L. (2002). Task-related changes in cortical synchronization are spatially coincident with the hemodynamic response. *NeuroImage* 16(1): 103–114.

75. Vanni, S., Warnking, J., Dojat, M., Delon-Martin, C., Bullier, J., and Segebarth, C. (2004). Sequence of pattern onset responses in the human visual areas: an fMRI constrained VEP source analysis. *NeuroImage* 21(3): 801–817.

76. Korvenoja, A., Huttunen, J., Salli, E., Pohjonen, H., Martinkauppi, S., Palva, J.M., Lauronen, L., Virtanen, J., Ilmoniemi, R.J., and Aronen, H.J. (1999). Activation of multiple cortical areas in response to somatosensory stimulation: combined magnetoencephalographic and functional magnetic resonance imaging. *Hum. Brain Mapp.* 8(1): 13–27.

77. Goldman, R.I., Stern, J.M., Engel, J., Jr., and Cohen, M.S. (2002). Simultaneous EEG and fMRI of the alpha rhythm. *Neuroreport.* 13(18): 2487–2492.

78. Moosmann, M., Ritter, P., Krastel, I., Brink, A., Thees, S., Blankenburg, F., Taskin, B., Obrig, H., and Villringer, A. (2003). Correlates of alpha rhythm in functional magnetic resonance imaging and near infrared spectroscopy. *NeuroImage* 20(1): 145–158.

79. Laufs, H., Kleinschmidt, A., Beyerle, A., Eger, E., Salek-Haddadi, A., Preibisch, C., and Krakow, K. (2003). EEG-correlated fMRI of human alpha activity. *NeuroImage* 19(4): 1463–1476.

80. Liebenthal, E., Ellingson, M.L., Spanaki, M.V., Prieto, T.E., Ropella, K.M., and Binder, J.R. (2003). Simultaneous ERP and fMRI of the auditory cortex in a passive oddball paradigm. *NeuroImage* 19(4): 1395–1404.

81. Kruggel, F., Herrmann, C.S., Wiggins, C.J., and von Cramon, D.Y. (2001). Hemodynamic and electroencephalographic responses to illusory figures: recording of the evoked potentials during functional MRI. *NeuroImage* 14(6): 1327–1336.

82. Mulert, C., Jager, L., Schmitt, R., Bussfeld, P., Pogarell, O., Moller, H.J., Juckel, G., and Hegerl, U. (2004). Integration of fMRI and simultaneous EEG: towards a comprehensive understanding of localization and time-course of brain activity in target detection. *NeuroImage* 22(1): 83–94.

83. Vitacco, D., Brandeis, D., Pascual-Marqui, R., and Martin, E. (2002). Correspondence of event-related potential tomography and functional magnetic resonance imaging during language processing. *Hum. Brain Mapp.* 17(1): 4–12.

84. Krakow, K., Wieshmann, U.C., Woermann, F.G., Symms, M.R., McLean, M.A., Lemieux, L., Allen, P.J., Barker, G.J., Fish, D.R., and Duncan, J.S. (1999). Multimodal MR imaging: functional, diffusion tensor, and chemical shift imaging in a patient with localization-related epilepsy. *Epilepsia.* 40(10): 1459–1462.

85. Krakow, K., Woermann, F.G., Symms, M.R., Allen, P.J., Lemieux, L., Barker, G.J., Duncan, J.S., and Fish, D.R. (1999). EEG-triggered functional MRI of interictal epileptiform activity in patients with partial seizures. *Brain* 122(9): 1679–1688.

86. Lantz, G., Spinelli, L., Menendez, R.G., Seeck, M., and Michel, C.M. (2001). Localization of distributed sources and comparison with functional MRI. *Epileptic. Disord.* Special Issue: 45–58.

87. Lemieux, L., Krakow, K., and Fish, D.R. (2001). Comparison of spike-triggered functional MRI BOLD activation and EEG dipole model localization. *NeuroImage* 14(5): 1097–1104.

88. Bodurka, J. and Bandettini, P.A. (2002). Toward direct mapping of neuronal activity: MRI detection of ultra-weak, transient magnetic field changes. *Magn. Reson. Med.* 47(6): 1052–1058.

89. Xiong, J., Fox, P.T., and Gao, J.H. (2003). Directly mapping magnetic field effects of neuronal activity by magnetic resonance imaging. *Hum. Brain Mapp.* 20(1): 41–49.

90. Lagerlund, T.D., Sharbrough, F.W., Jack, C.R., Jr., Erickson, B.J., Strelow, D.C., Cicora, K.M., and Busacker, N.E. (1993). Determination of 10–20 system electrode locations using magnetic resonance image scanning with markers. *Electroencephalogr. Clin. Neurophysiol.* 86(1): 7–14.

91. Sijbers, J., Vanrumste, B., Van Hoey, G., Boon, P., Verhoye, M., Van der Linden, A., and Van Dyck, D. (2000). Automatic localization of EEG electrode markers within 3-D MR data. *Magn. Reson. Imaging* 18(4): 485–488.

92. Schwartz, D.P., Poiseau, E., Lemoine, D., and Barillot, C. (Winter 1996). Registration of MEG/EEG data with 3-D MRI: Methodology and precision issues. *Brain Topogr.* pp. 101–116.

93. Huppertz, H.J., Otte, M., Grimm, C., Kristeva-Feige, R., Mergner, T., and Lucking, C.H. (1998). Estimation of the accuracy of a surface matching technique for registration of EEG and MRI data. *Electroencephalogr. Clin. Neurophysiol.* 106(5): 409–415.

94. Kozinska, D., Carducci, F., and Nowinski, K. (2001). Automatic alignment of EEG/MEG and MRI data sets. *Clin. Neurophysiol.* 112(8): 1553–1561.

95. Lamm, C., Windischberger, C., Leodolter, U., Moser, E., and Bauer, H. (2001). Co-registration of EEG and MRI data using matching of spline interpolated and MRI-segmented reconstructions of the scalp surface. *Brain Topogr.* 14(2): 93–100.

96. Towle, V.L., Bolanos, J., Suarez, D., Tan, K., Grzeszczuk, R., Levin, D.N., Cakmur, R., Frank, S.A., and Spire, J.P. (1993). The spatial location of EEG electrodes: locating the best-_tting sphere relative to cortical anatomy. *Electroencephalogr. Clin. Neurophysiol.* 86(1): 1–6.

97. Singh, K.D., Holliday, I.E., Furlong, P.L., Harding, G.F. (1997). Evaluation of MRI-MEG/EEG co-registration strategies using Monte Carlo simulation. *Electroencephalogr. Clin. Neurophysiol.* 102(2): 81–85.

98. Adjamian, P., Barnes, G.R., Hillebrand, A., Holliday, I.E., Singh, K.D., Furlong, P.L., Harrington, E., Barclay, C.W., and Route, P.J. (2004). Co-registration of magnetoencephalography with magnetic resonance imaging using bite-bar-based fiducials and surface-matching. *Clin. Neurophysiol.* 115(3): 691–698.

99. Wieringa, H.J., Peters, M.J., and Lopes da Silva, F. (1993). The estimation of a realistic localization of dipole layers within the brain based on functional (EEG, MEG) and structural (MRI) data: a preliminary note. *Brain Topogr.* 5(4): 327–330.

100. Horn, B.K.P. (April 1987). Closed-form solution of absolute orientation using unit quaternions. *J. Opt. Soc. Amer.* 4(4): 629–642.

101. Horn, B.K.P., Hilden, H., and Negahdaripour, S. (1998). Closed-form solution of absolute orientation using orthonormal matrices. *J. Opt. Soc. Amer.* 5(7).

102. Nielsen, F.A. Bibliography of segmentation in neuroimaging. URL: http://www. imm.dtu.dk/_fn/bib/ Nielsen2001BibSegmentation/.

103. Poupon, F. (December 1999), Parcellisation Systematique du Cerveau en Volumes D'internet. Le Cas Des Structures Profondes. Ph.D. thesis, INSA Lyon, Lyon, France. URL: ftp://ftp.cea.fr/pub/dsv/anatomist/ papers/fpoupon-thesis99.pdf.

104. Dale, A.M., Fischl, B., and Sereno, M.I. (1999). Cortical surface-based analysis. I. Segmentation and surface reconstruction. *NeuroImage* 9(2): 179–194.

105. Shattuck, D.W. and Leahy, R.M. BrainSuite: An automated cortical surface identification tool. *Med Im Anal.* In press.

106. Castellano, S.A. (September 1999). The Folding of the Human Brain: From Shape to Function. Ph.D. thesis, University of London, Division of Radiological Sciences and Medical Engineering, King's College, London.

107. Nunez, P.L. (1981). *Electric Fields of the Brain: The Neurophysics of EEG.* New York: Oxford University Press.

108. Liu, A.K., Belliveau, J.W., and Dale, A.M. (1998). Spatiotemporal imaging of human brain activity using functional MRI constrained magnetoencephalography data: Monte Carlo simulations. *Proc. Natl. Acad. Sci. USA* 95(15): 8945–8950.

109. Hillebrand, A. and Barnes, G.R. (2003). The use of anatomical constraints with MEG beamformers. *NeuroImage.* 20(4): 2302–2313.

110. Hamalainen, M.S. and Sarvas, J. (1989). Realistic conductivity geometry model of the human head for interpretation of neuromagnetic data. *IEEE Trans. Biomed. Eng.* 36(2): 165–171.

111. Marin, G., Guerin, C., Baillet, S., Garnero, L., and Meunier, G. (1998). In_uence of skull anisotropy for the forward and inverse problems in EEG: simulation studies using FEM on realistic head models. *Hum. Brain Mapp.* 6: 250–269.

112. Wolters, C.H., Anwander, A., Koch, M.A., Reitzinger, S., Kuhn, M., and Svens´en, M. (2001). Infiuence of head tissue conductivity anisotropy on human EEG and MEG using fast high resolution finite element modeling, based on a parallel algebraic multigrid solver. *Forschung und wissenschaftliches Rechnen.*

113. Miller, C.E. and Henriquez, C.S. (1990). Finite element analysis of bioelectric phenomena. *Crit. Rev. Biomed. Eng.* 18(3): 207–233.

114. Saleheen, H.I. and Ng, K.T. (1997). New finite difference formulations for general inhomogeneous anisotropic bioelectric problems. *IEEE Trans. Biomed. Eng.* 44(9): 800–809.

115. Geddes, L.A. and Baker, L.E. (1967). The specific resistance of biological material–a compendium of data for the biomedical engineer and physiologist. *Med. Biol. Eng.* 5(3): 271–293.

116. Goncalves, S.I., de Munck, J.C., Verbunt, J.P., Bijma, F., Heethaar, R.M., and Lopes da Silva, F. (2003). In vivo measurement of the brain and skull resistivities using an EIT-based method and realistic models for the head. *IEEE Trans. Biomed. Eng.* 50(6): 754–767.

117. Tuch, D.S., Wedeen, V.J., Dale, A.M., George, J., and Belliveau, J.W. (2001). Conductivity tensor mapping of the human brain using diffusion tensor MRI. *Proc. Natl. Acad Sci. U S A* 98(20): 11697–11701.

118. Buxton, R.B. and Frank, L.R. (1997). A model for the coupling between cerebral blood flow and oxygen metabolism during neural stimulation. *J Cereb Blood Flow Metab*, 17(1):64–72.

119. Seiyama, A., Seki, J., Tanabe, H.C., Sase, I., Takatsuki, A., Miyauchi, S., Eda, H., Hayashi, S., Imaruoka, T., Iwakura, T., and Yanagida, T. (2004). Circulatory basis of f MRI signals: relationship between changes in the hemodynamic parameters and BOLD signal intensity. *NeuroImage*, 21(4): 1204–1214.

120. Friston, K.J., Jezzard, P., and Turner, R. (1994). Analysis of functional MRI time-series. *Hum. Brain Mapp.* 1: 153–171.

121. Boynton, G.M., Engel, S.A., Glover, G.H., and Heeger, D.J. (1996). Linear systems analysis of functional magnetic resonance imaging in human V1. *J. Neurosci.* 16(13): 4207–4221.

122. Cohen, M.S. (1997). Parametric analysis of f MRI data using linear systems methods. *NeuroImage* 6(2): 93–103.

123. Lange, N. and Zeger, S.L. (1997). Non-linear fourier time series analysis for human brain mapping by functional magnetic resonance imaging. *Appl. Stat.* 46(1): 1–29.

124. Glover, G.H. (1999). Deconvolution of impulse response in event-related BOLD f MRI. *NeuroImage* 9(4): 416–429.

125. Rajapakse, J.C., Kruggel, F., Maisog, J.M., and von Cramon, D.Y. (1998). Modeling hemodynamic response for analysis of functional MRI time-series. *Hum. Brain Mapp.* 6(4): 283–300.

126. Ciuciu, P., Poline, J.B., Marrelec, G., Idier, J., Pallier, C., and Benali, H. (2003). Unsupervised robust nonparametric estimation of the hemodynamic response function for any f MRI experiment. *IEEE Trans. Med. Imaging* 22(10): 1235–1251.

127. Gitelman, D.R., Penny, W.D., Ashburner, J., and Friston, K.J. (2003). Modeling regional and psychophysiologic interactions in f MRI: the importance of hemodynamic deconvolution. *NeuroImage* 19(1): 200–207.

128. Marrelec, G., Benali, H., Ciuciu, P., Pelegrini-Issac, M., and Poline, J.B. (2003). Robust Bayesian estimation of the hemodynamic response function in event-related BOLD f MRI using basic physiological information. *Hum. Brain Mapp.* 19(1): 1–17.

129. Soltysik, D.A., Peck, K.K., White, K.D., Crosson, B., and Briggs, R.W. (2004). Comparison of hemodynamic response nonlinearity across primary cortical areas. *NeuroImage* 22(3): 1117–1127.

130. Attwell, D. and Iadecola, C. (2002). The neural basis of functional brain imaging signals. *Trends Neurosci.* 25(12): 621–625.

131. Logothetis, N.K. and Wandell, B.A. (2004). Interpreting the BOLD signal. *Annu. Rev. Physiol.* 66: 735–769.

132. Arthurs, O.J. and Boniface, S. (2002). How well do we understand the neural origins of the f MRI BOLD signal? *Trends Neurosci.* 25(1): 27–31.

133. Devor, A., Dunn, A.K., Andermann, M.L., Ulbert, I., Boas, D.A., and Dale, A.M. (2003). Coupling of total hemoglobin concentration, oxygenation, and neural activity in rat somatosensory cortex. *Neuron.* 39 (2): 353–359.

134. Paulesu, R.S., Frackowiak, R.S.J., and Bottini, G. (1997). Maps of somatosensory systems. in Frackowiak, R.S.J., (Ed.). *Human Brain Function* p. 528. Academic Press, San Diego, CA.

135. Stefanovic, B., Warnking, J.M., and Pike, G.B. (2004). Hemodynamic and metabolic responses to neuronal inhibition. *NeuroImage* 22(2): 771–778.
136. Niedermeyer, E. (1997). Alpha rhythms as physiological and abnormal phenomena. *Int. J. Psychophysiol.* 26(1–3): 31–49.
137. Laufs, H., Krakow, K., Sterzer, P., Eger, E., Beyerle, A., Salek-Haddadi, A., and Kleinschmidt, A. (2003). Electroencephalographic signatures of attentional and cognitive default modes in spontaneous brain activity fiuctuations at rest. *Proc. Natl. Acad Sci. U S A* 100(19): 11053–11058.
138. Martinez-Montes, E., Valdes-Sosa, P.A., Miwakeichi, F., Goldman, R.I., and Cohen, M.S. (2004). Concurrent EEG/fMRI analysis by multiway partial least squares. *NeuroImage* 22(3):1023–1034.
139. Ahlfors, S.P., Simpson, G.V., Dale, A.M., Belliveau, J.W., Liu, A.K., Korvenoja, A., Virtanen, J., Huotilainen, M., Tootell, R.B., Aronen, H.J., and Ilmoniemi, R.J. (1999). Spatiotemporal activity of a cortical network for processing visual motion revealed by MEG and fMRI. *J. Neurophysiol.* 82(5): 2545–2555.
140. Beisteiner, R., Erdler, M., Teichtmeister, C., Diemling, M., Moser, E., Edward, V., and Deecke, L. (1997). Magnetoencephalography may help to improve functional MRI brain mapping. *Eur. J. Neurosci.* 9(5): 1072–1077.
141. Gonzalez Andino, S.L., Blanke, O., Lantz, G., Thut, G., and Grave de Peralta Menendez, R. (2001). The use of functional constraints for the neuroelectromagnetic inverse problem: Alternatives and caveats. *Int. J. Bioelectromagnetism* 3(1).
142. Kober, H., Nimsky, C., Moller, M., Hastreiter, P., Fahlbusch, R., and Ganslandt, O. (2001). Correlation of sensorimotor activation with functional magnetic resonance imaging and magnetoencephalography in presurgical functional imaging: a spatial analysis. *NeuroImage* 14(5): 1214–1228.
143. Wagner, M. and Fuchs, M. (2001). Integration of functional MRI, structural MRI, EEG, and MEG. *Int. J. Bioelectromagnetism* 3(1).
144. Foxe, J.J., McCourt, M.E., and Javitt, D.C. Right hemisphere control of visuospatial attention: line-bisection judgments evaluated with high-density electrical mapping and source analysis. *NeuroImage* 19(3): 710–726, 2003.
145. Ahlfors, S.P. and Simpson, G.V. (2004). Geometrical interpretation of fMRI-guided MEG/EEG inverse estimates. *NeuroImage* 22(1): 323–332.
146. Dale, A.M., Liu, A.K., Fischl, B., Lewine, J.D., Buckner, R.L., Belliveau, J.W., and Halgren, E. (2000). Dynamic statistical parameter mapping: combining fMRI and MEG to produce high resolution imaging of cortical activity. *Neuron.* 26: 55–67.
147. Babiloni, F., Babiloni, C., Carducci, F., Angelone, L., Del-Gratta, C., Romani, G.L., Rossini, P.M., and Cincotti, F. (2001). Linear inverse estimation of cortical sources by using high resolution EEG and fMRI priors. *Int J. Bioelectromagnetism* 3(1).
148. Fuchs, M., Wagner, M., Kohler, T., and Wischmann, H. (1999). Linear and nonlinear current density reconstructions. *J. Clin. Neurophysiol.* 16(3): 267–295.
149. Lahaye, P.-J., Baillet, S., Poline, J.-B., and Garnero, L. (April 2004). Fusion of simultaneous fMRI/EEG data based on the electro-metabolic coupling. in *Proc. IEEE ISBI.* pp. 864–867, Arlington, VA.
150. Trujillo-Barreto, N.J., Martinez-Montes, E., Melie-García, L., and Valdés-Sosa, P.A. (2001). A symmetrical Bayesian model for fMRI and EEG/MEG neuroimage fusion. *Int J Bioelectromagnetism* 3.
151. Trujillo-Barreto, N.J., Aubert-Vazquez, E., and Valdes-Sosa, P.A. (2004). Bayesian model averaging in EEG/MEG imaging. *NeuroImage* 21(4): 1300–1319.

152. Melie-García, L., Trujillo-Barreto, N.J., Martinez-Montes, E., Koenig, T., and Valdés-Sosa, P.A. (June 2004). EEG imaging via BMA with f MRI pre-de_ned prior model probabilities. in *Hum. Brain Mapp.* Budapest, Hungary.

153. Sahani, M. and Nagarajan, S.S. (2004). Reconstructing MEG sources with unknown correlations. in *Advances in Neural Information Processing Systems 16.* MIT Press.

154. Schmidt, D.M., George, J.S., and Wood, C.C. (1999). Bayesian inference applied to the electromagnetic inverse problem. *Human Brain Mapping* 7(3): 195–212.

155. Singh, M., Kim, S., and Kim, T.S. (2003). Correlation between BOLD-f MRI and EEG signal changes in response to visual stimulus frequency in humans. *Magn. Reson. Med.* 49(1): 108–114.

156. Vanhatalo, S., Tallgren, P., Becker, C., Holmes, M.D., Miller, J.W., Kaila, K., and Voipio, J. (2003). Scalp-recorded slow EEG responses generated in response to hemodynamic changes in the human brain. *Clin. Neurophysiol.* 114(9): 1744–1754.

157. Jun, S.C., Pearlmutter, B.A., Nolte, G. (June 2003). MEG source localization using a MLP with a distributed output representation. *IEEE Trans. Biomed. Eng.* 50(6): 786–789.

158. Jun, S.C. and Pearlmutter, B.A. (2005). Fast robust subject-independent magnetoencephalographic source localization using an artificial neural network. *Hum. Brain Mapp.* 24(1): 21–34.

159. Gorodnitsky, I.F. and Rao, B.D. (1997). Sparse signal reconstruction from limited data using FOCUSS: A re-weighted minimum norm algorithm. *IEEE Trans. Sig. Proc.* 45(3): 600–616.

160. Baillet, S., Riera, J., Marin, G., Mangin, J., Aubert, J., and Garnero, L. (2001). Evaluation of inverse methods and head models for EEG source localization using a human skull phantom. *Phys. Med. Biol.* 46(1): 77–96.

161. Vigáreláki, R., Hälário, M., Sä, J., Jousmämäinen, V., and Oja, E. (2000). Independent component approach to the analysis of EEG and MEG recordings. *IEEE Trans. Biomed. Eng.* 47(5): 589–593.

162. Tang, A.C., Pearlmutter, B.A., Zibulevsky, M., and Carter, S.A. (2000). Blind separation of multichannel neuromagnetic responses. *Neurocomputing* 32–33: 1115–1120.

163. Jung, T.-P., Humphries, C., Lee, T.-W., McKeown, M.J., Iragui, V., Makeig, S., and Sejnowski, T.J. (2000). Removing electroencephalographic artifacts by blind source separation. *Psychophysiology* 37: 163–178.

164. Jung, T.-P., Makeig, S., Westerfield, M., Townsend, J., Courchesne, E., and Sejnowski, T.J. (2000). Removal of eye activity artifacts from visual event-related potentials in normal and clinical subjects. *Clin. Neurophysiol.* 111(10): 1745–1758.

165. Jung, T.-P., Humphries, C., Lee, T.-W., Makeig, S., McKeown, M.J., Iragui, V., and Sejnowski, T.J. (1999). Removing electroencephalographic artifacts: comparison between ICA and PCA. in *Neural Networks for Signal Processing VIII.* IEEE Press.

166. McKeown, M.J., Makeig, S., Brown, G.G., Jung, T.-P., Kindermann, S.S., Bell, A.J., and Sejnowski, T.J. (1998). Analysis of f MRI data by blind separation into independent spatial components. *Hum. Brain Mapp.* 6: 160–188.

167. Calhoun, V.D., Adali, T., Hansen, L.K., Larsen, J., and Pekar, J.J. (April 2003). ICA of Functional MRI Data: An Overview. pp. 281–288. URL: http://www.imm.dtu.dk/pubdb/p.php?1669. Invited Paper.

168. Leahy, R.M., Baillet, S., and Mosher, J.C. Integrated Matlab toolbox dedicated to magnetoencephalography (MEG) and electroencephalography (EEG) data visualization and processing. URL: http://neuroimage.usc.edu/brainstorm/.
169. NeuroFEM. Finite element software for fast computation of the forward solution in EEG/MEG source localisation. URL: http://www.neurofem.com/. Max Planck Institute for Human Cognitive and Brain Sciences.
170. BioPSE. (2002). Problem solving environment for modeling, simulation, and visualization of bioelectric fields, URL: http://software.sci.utah.edu/biopse.html. Scientific Computing and Imaging Institute (SCI).
171. SCIRun. (2002). SCIRun: a scientific computing problem solving environment, URL http://software.sci. utah.edu/scirun.html. Scientific Computing and Imaging Institute (SCI).
172. FreeSurfer. FreeSurfer. URL: http://surfer.nmr.mgh.harvard.edu/. CorTechs and the Athinoula A. Martinos Center for Biomedical Imaging.
173. Van Essen, D. Surface reconstruction by filtering and intensity transformations. URL: http://brainvis.wustl.edu/.
174. Weber, D. EEG and MRI Matlab toolbox. URL: http://eeg.sourceforge.net/.
175. Moran, J.E. MEG tools for Matlab software. URL: http://rambutan.phy.oakland.edu/_meg/.
176. Delorme, A. and Makeig, S. (2004). EEGLAB: an open source toolbox for analysis of single-trial EEG dynamics including independent component analysis. *J. Neurosci. Methods* 134(1): 9–21.

9 A Survey of Three-Dimensional Modeling Techniques for Quantitative Functional Analysis of Cardiac Images

*Alejandro F. Frangi, Wiro J. Niessen, M.A. Viergever, and Boodewijn P.F. Lelieveldt**

CONTENTS

* Dr. Frangi (alejandro.frangi@upf.edu) is with the Department of Technology, Pompeu Fabra University, Pg. de Circumval.lacio 8, E08003 Barcelona, Spain. *Corresponding author.*
Dr. Niessen (W.niessen@erasmosmc.nl) is with the Department of Radiology and Medical Informatics, Erasmos Medical Center Rotterdame, Room EE2151, PO Box 1738, 3000 DR Rotterdame, The Netherlands. Dr. Viergever (max@isi.uu.nl) is with the Image Sciences Institute, Utrecht University, Utrecht, Heidelberglaan 100, Room E01.335, 3584 CX Utrecht, The Netherlands.
Dr. Lelieveldt (b.lelieveldt@lumc.nl) is with the Laboratory for Clinical en Experimental Image Processing, Department of Radiology, Leiden University Medical Center, PO Box 9600, 2300 RC Leiden, The Netherlands.

267

ABSTRACT

Three-dimensional (3-D) imaging of the heart is a rapidly developing area of research in medical imaging. Advances in hardware and methods for fast spatiotemporal cardiac imaging are extending the frontiers of clinical diagnosis and research on cardiovascular diseases.

In the last few years, many approaches have been proposed to analyze images and extract parameters of cardiac shape and function from a variety of cardiac-imaging modalities. In particular, techniques based on spatiotemporal geometric models have received considerable attention. This chapter surveys the literature of two decades of research on cardiac modeling. The purpose of the chapter is threefold: (1) to serve as a tutorial on the subject for both clinicians and technologists, (2) to provide an extensive account of modeling techniques in a comprehensive and systematic manner, and (3) to critically review these approaches in terms of their performance and value in clinical evaluation with respect to the final goal of cardiac functional analysis. From this review it is concluded that whereas 3-D model-based approaches have the capability to improve the diagnostic value of cardiac images, issues such as robustness, 3-D interaction, computational complexity, and clinical validation still require significant attention.

9.1 INTRODUCTION

Cardiovascular disease (CVD) has been the leading cause of death in the U.S. since 1900 in every year but one (1918). Nearly 2600 Americans die each day of CVD, an average of one death every 34 sec [1]. CVD claims more lives each year than the next five leading causes of death combined, which are cancer, chronic lower respiratory diseases, accidents, diabetes mellitus, and influenza and pneumonia. According to the most recent computations of the Centers for Disease Control and Prevention of the National Center for Health Statistics (CDC/NCHS), if all forms of major CVD were eliminated, life expectancy would rise by almost 7 yr. If all forms of cancer were eliminated, the gain would be only 3 yr. According to the same study, the probability at birth of eventually dying from a major CVD is 47%, whereas the chance of dying from cancer is 22% [1].*

Nowadays, there is a multitude of techniques available for cardiac imaging that provide qualitative and quantitative information about the morphology and function of the heart and great vessels (Figure 9.1). Use of these technologies can help in guiding clinical diagnosis, treatment, and follow-up of cardiac diseases. Spatiotemporal imaging is a valuable research tool to understand cardiac motion and perfusion, and their relationship with different stages of disease.

Technological advances in cardiac-imaging techniques continue to provide 3-D information with increasing spatial and temporal resolution. Therefore, a single cardiac examination can result in a large amount of data (particularly in multiphase 3-D studies). These advances have led to an increasing need for efficient algorithms to plan 3-D acquisitions, automate the extraction of clinically relevant parameters, and to provide the tools for their visualization.

Segmentation of cardiac chambers is an invariable prerequisite for quantitative functional analysis. Although many clinical studies still rely on manual delineation of chamber boundaries, this procedure is time consuming and prone to intra- and

---

* The most recent European survey of CVDs is the one published by the European Society of Cardiology [2].

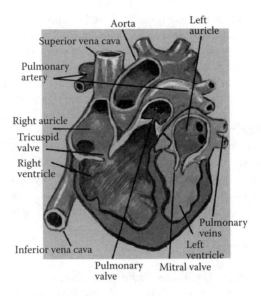

FIGURE 9.1 The heart is a complex system whose aim is to pump blood to irrigate the whole body. Assessment of cardiac performance is crucial in diagnosis and treatment of cardiovascular diseases.

interobserver variability. Therefore, many researchers have addressed the problem of automatic left (LV) and right (RV) ventricle segmentation. Because the shape of the cardiac ventricles is approximately known, it seems natural to incorporate this knowledge into the segmentation process. Such model-driven techniques have received ample attention in medical image analysis in the last decade [3,4]. Some of the advantages over model-free approaches are: (1) the model itself can constrain the segmentation process that is ill posed in nature owing to noise and image artifacts, (2) segmentation, image analysis, and shape modeling are simultaneously addressed in a common framework, (3) models can be coarse or detailed depending on the desired degree of abstraction, and (4) in some approaches, most of the chamber's shape can be explained with a few comprehensible parameters that can subsequently be used as cardiac indices (cf. References [5–9] among others).

Use of geometric models is not completely new to the analysis of cardiac images. As a matter of fact, traditional methods of obtaining parameters such as left-ventricular volume (LVV) and mass (LVM) from echo- and angiocardiography were based on (simple) geometrical models [10–13]. However, their use was mainly motivated by the need to extract 3-D parameters from 2-D images, and their accuracy was therefore limited [14].

The literature on model-driven segmentation of cardiac images has grown rapidly in the last few years, and this trend is likely to continue.* This chapter

---

* See, for instance, a recently published special issue on 3-D cardiac image analysis [15].

presents a comprehensive and critical review of the state of the art in geometric modeling of the cardiac chambers, notably the LV, and their potential for functional analysis. It is an updated version of our previous review [16]. In order to set reasonable bounds to the extent of this survey, we have confined ourselves to peer-reviewed archival publications* proposing methods for LV (or RV) segmentation, shape representation, and functional and motion analysis that fulfill the following selection criteria:

- The technique is model based.
- The reconstructed model is 3-D.**
- Illustration on cardiac images is provided.

This review is organized as follows: Section 9.2 gives a brief overview of the different acquisition modalities that have been used in imaging the heart. Section 9.3 summarizes and defines the most relevant clinical parameters that provide information on cardiac function. Section 9.4 presents a systematic classification of cardiac models by type of geometrical representation and parameterization; attention is also given to the different types of input data and features for model recovery. This section is summarized in Table 9.1. Section 9.5 discusses cardiac modeling approaches with respect to the functional parameters they provide and the degree of evaluation possible with these methods. This section leads to Table 9.2 that links the clinical target of obtaining functional information of the heart (Section 9.3) to the various technical approaches presented in Section 9.4. Finally, Section 9.6 closes the survey with conclusions and suggestions for future research.

9.2 IMAGING TECHNIQUES FOR CARDIAC EXAMINATION

The physical properties on the basis of which the imaging device reconstructs an image (e.g., radioactive emissions of an isotope) are intimately related to some specific functional aspects of the heart (e.g., its perfusion properties). Each imaging modality has advantages and limitations that influence the achievable modeling accuracy. This section briefly reviews the techniques most frequently used for 3-D clinical investigation of the heart. More extensive reviews and complementary reading can be found in References 17–24.

9.2.1 ANGIOCARDIOGRAPHY

Angiocardiography is the x-ray imaging of the heart following the injection of a radio-opaque contrast medium. Although 2-D, in principle, this technique can provide projections from two angles using a biplane system. Selective enhancement

---

* A few exceptions were made when the approaches were considered relevant and journal versions were not available.
** Even if the imaging technique is not 3-D, for instance, in the reconstruction of 3-D, models from multiple nonparallel slices or from multiple 2-D, projections.

TABLE 9.1
Overview of Cardiac Modeling Methods

	Reference	Model	Potential	Reported	Type of Input	Type of Feature
			Surface Models			
Continuous	Yettram [107], [108]	Stacked curves	BA	BA	M2DP	Manual contours
	Young [69], [109]	Bicubic Hermite patches	BA	BA	PS	Coronary bifurcation points
	Spinale [110]RV	Stacked hemiellipses	BA	BA	M2DP	Manual contours
	Pentland [111]	FE and modal analysis	NS	X	M2DP	Optic flow
	Cauvin [112]	Truncated bullet	NS	SPECT	3DV	Thresholding + Morphological skeleton
	Czegledy [113]RV	Stack of crescentic outlines	NS	CT	3DV	Linear measurements
	Gustavsson [114]	Cubic B-spline curves' mesh	US	US	MO2DS	Manual contours
	Sacks [115]RV	Biquadric surface patches	NS	MR	3DV	Manual contours
	Chen [116]	Superquadrics + spherical harmonics	NS	BA	PS	Coronary bifurcation points
	Denslow [117]RV	Ellipsoidal shell	NS	MR	3DV	Linear measurements
	Maehle [31]	Bicubic spline surface patches	NS	US	MO2DS	Edge detection + manual correction
	Chen [118]	Voxel repres./superquadric	NS	DSR	3DV	Shape and gray-level properties
	Coppini [29]	Spherical elastic surface	NS	US	MO2DS	NN edge detector
	Goshtasby [119]	Rational Gaussian surface	NS	MR	3DV	Zero-crossings Laplacian
	Matheny [120]	3-D/4-D harmonic surfaces	NS	DSR/BA	PS	Iso-surface, coronary bifurcation points
	Staib [121]	Bayesian Fourier surface	NS	MR/DSR	3DV	Gaussian gradient

	Park [6]	Superquadrics+parameter functions	MR^tag	MR^tag	PS	MR tagging-derived midwall motion field [77]
	Bardinet [9], [122]	Superquadrics + FFD	NS	DSR/SPECT	PS	Iso-surfaces
	Declerck [123]	Planispheric transformation	NS	SPECT	3DV	Normalized radial gradient
	Sato [124]	B-spline surface	BA	BA	M2DP	Apparent/occluding contours
	Sanchez-Ortiz [125]	B-spline surface + motion model	3DUS	3DUS	3DV	multiscale fuzzy clustering
	Swingen [126], [127]	B-spline surface	MR	MR	M2DP	Manual point placing
	Horkaew [128]	Hierarchical piecewise bilinear maps	CT, MR, US,	MR	3DV	Image gradient
Discrete	Geiser [129], [130]	12-sided stacked polygons	US	US	MO2DS	Manual contours
	Faber [131]	4-D discrete template	NS	MR/SPECT	3DV	Normalized radial gradient
	Gopal [27]	Polyhedral mesh	NS	US	MO2DS	Manual contours
	Friboulet [132]	Triangulated mesh	NS	MR	PS	Manual contours
	Huang [133]	Adaptive-size mesh	NS	DSR	PS	Data-to-node distance + data curvature
	Faber [134]	3-D discrete template	SPECT	SPECT	3DV	Radioactive distrib. profile
	Germano [135], [136]	Ellipsoid + local refinement	SPECT	SPECT	3DV	radioactive distrib. profile
	McInerney [137]	FE deformable balloon	NS	DSR	3DV	Gauss/Monga-Deriche grad.
	Ranganath [138]	2-D snakes + propagation	MR	MR	3DV	Intensity profile matching
	Tu [139]	Spherical template	NS	DSR	3DV	Spatiotemporal gradient
	Nastar [140]	Mass-spring mesh	NS	DSR	3DV	Edge distance map
	Rueckert [141]	Geometrically deformable template	NS	MR	3DV	Zero-crossings Laplacian
	Shi [142], [143]	Delaunay triangulation	NS	MR/DSR	PS	Bending energy

(continued)

TABLE 9.1 (Continued)
Overview of Cardiac Modeling Methods

Reference	Model	Potential	Reported	Type of Input	Type of Feature
		Surface Models			
Legget [28]	Piecewise subdivision surface	NS	US	MO2DS	Manual contours
Montagnat [144]	Simplex meshes	US	US	MO2DS	Edges in cylindrical coord.
Biedenstein [145]	Bullet-like elastic mesh	SPECT	SPECT	3DV	Radioactive distrib. profile
Gerard [146]	Simplex mesh + motion model	NS	MR, CT, US	3DV	Image gradient
Song [147]	Bayesian model of shape and features	3DUS	3DUS	3DV	Multifeature appearance model
Mäkelä [148]	Triangulated thorax model	NS	MR (+ MCG, + PET)	3DV	Deformable model [149] and image registration
Fan [150]	Superquadrics + spherical harmonics	NS	DSR	PS	Manual contours
Lorenzo-Valdes [151]	Atlas-based segmentation	MR, CT, US	MR	3DV	MRF model
van Assen [152], [153]	Fuzzy active shape model	NS	MR	3DV fuzzy inference system	
Kaus [154]	Statistically constrained deformable model	NS	MR	3DV	Spatially varying features
Wierzbicki [155]	Nonrigid template registration	MR, CT, MR	3DV	Several registration metrics	
Implicit Yezzi [156], [157]	Implicit snakes	NS	MR	3DV	Gaussian gradient

Category	Reference	Model				
Continuous	Tseng [158]	Cont. Dist. Tranf. NN	NS	US	MO2DS	Manual contours
	Niessen [159]	Implicit snakes	NS	MR/DSR	3DV	
	Lelieveldt [160]	Fuzzy implicit surfaces	CT/MR	MR	Gaussian gradient MO2DS	Air-tissue transitions
Volume Models						
	Creswell [161], [162]	Approximating	NURBS	MR	MR	PS manual contours
	Park [5], [163]	Superellipsoids + par Functions	MR^{tag}	MR^{tag}	TAG	tag line intersections + boundary points
	Haber [164]–[166]	Physics-based FE	MR^{tag}	MR^{tag}	TAG	Tag line intersections + boundary points
	Shi [167]	Biomechanical tetrahedral FE model	MR	MR	PS + 3DV	Bending energy + MR velocity image
	Hu [168]	Statistical physics-based FE	MR^{tag}	MR^{tag}	TAG	SPAMM Features + composite model
Discrete	Kuwahara [169]	Voxel representation	MR	MR	MO2DS	Manual contours
	O'Donnell [7], [8]	Hybrid volumetric ventriculoid	MR^{tag}	MR^{tag}	TAG	Tag line intersections + boundary points
	Mitchell [170]	3D AAM	MR, CT, US	MR	3DV	Texture with shape constraints
	Sermesant [171]	Deformable biomechanical model	NS	MR, SPECT	3DV	Intensity profiles
	Stegmann [172], [173]	Bitemporal AAM	MR, CT, US	MR	3DVS	Texture with shape constraints

(continued)

TABLE 9.1 (Continued)
Overview of Cardiac Modeling Methods

	Reference	Model	Potential	Reported	Type of Input	Type of Feature
			Deformation Models			
Continuous	Amini [95]	Local quadric patches	NS	DSR/MR	PS	Minimal conformal motion
	Amini [174], [175]	B-spline tag surfaces	MR^{tag}	MR^{tag}	TAG	special SPAMM protocol
	Young [77]	Bicubic Hermite FE	MR^{tag}	MR^{tag}	TAG	Tag line intersections
	Bartels [176]	Multidimensional splines	NS	Syn	3DV	Intensity conservation
	O'Dell [177]	Affine + prolate spheroidal	MR^{tag}	MR^{tag}	TAG	tag lines
	Young [81]	Bicubic Hermite FE	MR^{tag}	MR^{tag}	TAG	tag lines
	Moulton [178]	Higher-order polynomial interpolation	MR^{tag}	MR^{tag}	TAG	Tag surface intersections
	Radeva [179]	Trivariate cubic B-spline	MR^{tag}	MR^{tag}	TAG	Short-axis tag lines
	Kerwin [83]	Thin-plate splines	MR^{tag}	MR^{tag}	TAG	Tag line intersections
	Huang [181]	Quadrivariate cubic B-spline	MR^{tag}	MR^{tag}	TAG	Tag surfaces
	Young [180]	"Model tags"	MR^{tag}	MR^{tag}	TAG	Tag lines
	Chen [182]	4-D B-spline solid	MR^{tag}	MR^{tag}	TAG	Oriented filters
	Klein [183]	4-D elastic deformable model	PET	PET	3DV	Optic flow
	Masood [184]	Virtual tags	MR	MR	3DV	Myocardial velocity measurements
	Chandrashekara [185]–[187]	FFD in cylindrical coordinates	MR^{tag}	MR^{tag}	3DV	Mutual information
	Perperidis [188], [189]	Spatiotemporal FFD	MR^{tag}	MR^{tag}	3DV	Mutual information

Discrete Moore [78]	Discrete mesh	MR^{tag}	MR^{tag}	TAG	Tag line intersections
Denney [82]	Discrete grid	MR^{tag}	MR^{tag}	TAG	tag line intersections
Benayoun [99]	Adaptive-size meshes	NS	DSR	3DV	gradient
Papademetris [190], [191]	Delaunay triangulation	NS	MR/US	PS	Internal deformation energy
Shi [192]	Stochastic biomechanical model	NS	MR	PS	Displacement fields from PC MR
Lötjönen [193]	Statistical deformation model	MR, CT	MR	3DV	Normalized mutual information

Note: ^{RV} Right ventricle

Modality: BA = biplane angiocardiography; US = ultrasound; MR = magnetic resonance; DSR = dynamic spatial reconstructor; CT = computed tomography; X = transmission x-ray; SPECT = single-photon emission computed tomography; PET = positron emission tomography; MCG = magnetocardiography; Syn = synthetic images; NS = nonspecific.

Recovered from: M2DP = multiple 2-D projections; MO2DS = multiple oriented 2-D slices; 3DV = 3-D volumetric images and feature maps; 3DVS = 3-D volumetric sequences; PS = point sets; TAG = MR tag intersections, lines, or surfaces.

TABLE 9.2
Overview of Cardiac Modeling Methods: Reported Classical Functional Parameters and Their Validation

Evaluation	Reference	Modality	Parameters Global	Parameters Motion	Flexibility Complexity	Preprocessing	Automation	Ad hoc	Validation/Illustration Type	No.	Std. of Ref.
	Amini [95]	DSR, MR	—	MF	L	M	+	+	a	1	NA
	Amini [174], [175]	MR^{tag}	—	MF, SA	L	M	—	+	V	1	NA
	Bartels [176], [229]	NS	—	MF	L	N	+	+	m	1	GT
	Benayoun [99]	DSR	—	MF	L	A	+	—	a	1	NA
	Cauvin [112]	SPECT	LVV	—	C	A	+	—	P	NA	NA
	Chen [116]	BA	LVV	SA	H	M	—	—	V	1	NA
	Chen [118]	DSR	LVV	—	C	N	+	—	V	1	NA
	Chen [182]	MR^{tag}	—	SA	C	N	+	—	V	5	NA
	Fan [150]	DSR	—	MF, SA	H	M	+	—	V	1	NA
	Gustavsson [114]	US	LVV	—	H	M	=	+	V	1	NA
Qualitative or	Huang [133]	DSR	LVV	MF	L	M	+	+	a	1	NA
no evaluation	Hu [168]	MR^{tag}	—	MP, SA	L	M	+	+	V/P	1/1	NA
	Kerwin [83]	MR^{tag}	—	SA	L	A	+	—	V	1	NA
	Matheny [120]	DSR	LVV	—	G	M	+	—	a	1	NA
	Mäkelä [148]	MR(+PET, +MCG)	—	—	L	N	+	—	p	10	NA

								V/P		
Maehle [31]	US	LVV	WT	L	M	—	—	v	NA	NA
McInerney [137]	DSR	LVV	MF	L	I	—	—	v	1	NA
Niessen [159]	MR/DSR	LVV	WT	L	I	+	+	a	1/1	NA
O'Donnell [7], [8]	MR^{tag}	LVV	WT, SA	H	M	+	+	v	1	NA
Papademetris [191]	US	—	SA	L	A	+	—	a	4	NA
Papademetris [190]	MR	—	SA	L	A	+	—	v	1	NA
Park [163]	MR^{tag}	LVV	MF, SA	C	M	+	+	v	1	NA
Pentland [111]	X	LVV	MF	G	I	+	+	v	1	NA
Radeva [179]	MR^{tag}	LVV	MF, SA	L	M	—	+	v	1	NA
Rueckert [141]	MR	LVV	MF	L	I	+	+	v	1	NA
Staib [121]	DSR/MR	LVV	—	G	I	+	+	a	1/1	NA
Yezzi [156], [157]	MR	LVV	WT	L	I	+	+	v	1	NA
Young [69], [109]	BA	—	SA	L	M	+	+	a	1	NA
Young [180]	MR^{tag}	—	SA	L	I	+	—	v	1	NA
Amini [174]	MR^{tag}	LVV	MF, SA	L	M	+	+	v	1	GT
Bardinet [122]	DSR	**LVV**	MF	H	M	+	+	m	1	OB/AS
Czegledy [113]^{REV}	CT	**RVV**	—	C	M	=	—	a/m	10	AT
Denney [82], [240]	MR			L	A	+	—	p	1/1	GT/NA
Denslow [117]^{REV}	MR^{tag}	**RVV**	—	C	M	=	—	m/a	1/1	AT
Gerard [146]	US	LVV	—	L	N	+	—	p	13	AT
Germano [35], [135], [136], [204]	SPECT	**LVV, EF**	—	H	A	+	—	p	1	GT
Gopal [27]	US	**LVV**	—	L	M	+	—	p	17	AT

(continued)

TABLE 9.2 (Continued)
Overview of Cardiac Modeling Methods: Reported Classical Functional Parameters and Their Validation

Evaluation	Reference	Modality	Parameters		Flexibility Complexity	Preprocessing	Automation	Ad hoc	Validation/ Illustration		
			Global	Motion					Type	No.	Std. of Ref.
	Kerwin [83]	MR^tag	—	SA-	L	A			m	1	GT
Quantitative: synthetic, phantom and animal models	Haber [164]–[166]^RV	MR^tag	—	MF, SA	L	M	+	+	m/a	–/1	GT/NA
	Huang [181]	MR^tag	—	MF, SA	L	A	+	—			
	Legget [28], [241], [242]	US	LVV, LVM	—	L	M	=	—	p/a	6/21 +5	GT/AT
	Masood [184]	MR	—	SA	L	I	+	—	p	—	GT
	Moore [78]	MR^tag	—	SA	L	A	+	—	m	—	AS
	Moulton [178]	MR^tag	—	SA	L	M		—	m/a	–/7	NS/NS
	O'Dell [177]	MR^tag	—	SA	H	M	+	—	m	—	AS
	Papademetris [190], [243]	MR	—	SA	L	A	+	—	a/a	4/8	AT
	Sacks [115]^RV	MR	—	WT	L	M	=	+	p/a	6/1	GT/NA
	Sato [124]	BA	LVV	—	L	M	–	+	m/p	1/1	GT/AT
	Spinale [110]^RV	BA	RVV, SV	WT	L	M	=	+	p/a	22/24	AT/AT
	Shi [142]	MR/DSR	—	WT, MF	L	A	+	+	a	12	AT
	Shi [167]	MR	—	MF, SA	L	A	+	—	a	1	CL
	Shi [192]	MR	LVV	MP, SA	L	A	+	—	m/a	1	AT
	Swingen [126]	MR	LVV	—	C	N	+		p	2	GT
	Tu [139]	DSR	LVV	—	G	M	+	+	a	2	OB
	Wierzbicki [155]	CT, MR	LVV	MF	H	I	+	+	a/V	2/2	GT/OB

	Study											
Quantitative: clinical case studies without standard of reference	Yettram [107], [108]	BA	LVV	—	L	M			=	—	p	8	AT
	Young [77]	MR^tag	—	SA	L	M			+	—	m	—	AS
	Declerck [123]	SPECT	—	MF	G	A		+	+	+	V/P	3/1	NA
	Gerard [146]	US	LVV	—	L	N			+	—	V	9	NA
	Kuwahara [169], [244]	BA	LVV, EF, SV	—	L	M			=	—	P	13	NA
	Legget [28], [242]	US	LVV	—	L	M			—	—	V/P	6/2	NA
	Moore [78]	MR^tag	—	SA	L	A			+	—	V	1	NA
	O'Dell [177]	MR^tag	—	SA-	H	M			+	—	V	10	NA
	Park [5]	MR^tag	LVV,EF	MF	C	M			+	+	V/P	1/1	NA
	Park [6]	MR^tag	LVV	MF, SA	C	M			+	+	V/P	1/1	NA
	Perperidis [188], [189]	MR	LVV	—	L	I			+	+	V	7	NA
	Sermesant [171]	SPECT, MR	LVV	SA	L	I			+	·	V	3	NA
	Young [69], [109]	BA	—	SA	L	M			+	+	V	1	NA
	Young [77]	MR^tag	—	SA	L	M			+		V	1	NA
	Bardinet [122]	SPECT	LVV	WT, MF	H	M			+	—	V	1	OB
	Biedenstein [145]	SPECT	LVV	—	L	I			+	—	P	42	OB
Quantitative: clinical case studies without standard of reference	Chandrashekara [185], [186], [245]	MR^tag	—	MF	L		I	+		—	V	6	OB
	Coppini [29]	US	LVV, EF	—	L	N			+	—	V	3	OB
	Faber [131]	SPECT/MR	LVV	WT	L	I			+	+	V/P	22/16	OB
	Faber [134]	SPECT	LVV	—	L	I			+	+	P	10	OB(m)
	Germano [35], [135], [136], [246]	SPECT	LVV, EF	WT	H	A			+	—	P	144 /65	OB(m)/ AT

(continued)

TABLE 9.2 (Continued)
Overview of Cardiac Modeling Methods: Reported Classical Functional Parameters and Their Validation

Reference	Modality	Parameters Global	Parameters Motion	Flexibility Complexity	Preprocessing	Automation	Ad hoc	Validation/Illustration Type	No.	Std. of Ref.
Geiser [129], [130]	US	LVV,EF, SV,CO	WT	L	M	=	–	P	4	AT
Goshtasby [119]	MR	LVV	–	L	I	+	+	V	5	OB
Horkaew [128]	MR	LVV	–	C	I	+	+	V	160	OB
Kaus [154]	MR	LVV, RVV, BPE	–	C	N	+	+	P	121	OB
Legget [28], [241]	US	LVV, SV	–	L	M	–	–	V	5	AT
Lelieveldt [227]	MR	SP	–	L	M	–	–	V	5	AT
Lorenzo-Valdes [151], [232], [247]	MR	RVV, LVV	–	L	N	–	–	V/P	14/10	OB
Lötjönen [193]	MR	LVV,RVV, BPE	–	C	N	+	+	V	25	OB
Mitchell [170]	MR	LVV, LVM, BPE	–	C	N	+	–	V/P	38/18	OB
Ranganath [138]	MR	LVV, LVM, EF	–	G	I	+	+	V	7	OB
Sanchez-Ortiz[125]	US	LVV, EF	–	L	N	+	–	V	9	AT
Stegmann [172], [173]	MR	LVV, EF	–	C	N	+	–	V	12	OB
Song [147]	US	LVV	–	L	N	+	+	V	25	OB
Swingen [126], [127]	MR	LVV, EF, WT	–	C	N	+	–	V/P	4/1 8	OB OB

Evaluation

Tseng [224]	US	**LVV**	**WT**	—	G	I		=	—	V	1	OB
van Assen [152], [153]	MR	**LVV**		C	I	+		—	V	9	OB

Notes: NA = not available or not reported

Parameters: bold = quantitative results reported; italic = computable from the model (but quantitative results not reported). Motion parameters were classified in six categories: BPE = border-positioning error; WT = wall-thickening analysis; MF = wall/tissue motion field (not including strain analysis); MP = material parameters; SA = strain analysis; SP = scan plan.

Flexibility: C = compact model with small or medium number of degrees of freedom (DOF), G = flexible model with global-support basis function and large number of DOF, L = flexible model with local-support basis functions and large number of DOF, H = hierarchical models.

Preprocessing to initialize the model: N = none; M = manual segmentation of contours and landmarks; A = (semi) automatic delineation of contours and landmarks; I = approximated model initialization or landmark placement. Precomputation of feature images (gradient, Laplacian, etc.) was not considered as preprocessing.

Automation *after* preprocessing and selection of *ad hoc* parameters: (+) full, (−) interactive guidance may be required to correct or assist intermediate steps, (=) relying on substantial human guidance.

Ad hoc parameters: (−) none, or robustness demonstrated through sensitivity analysis; (+) yes, but no sensitivity analysis was performed.

Validation/illustration information: Type of evaluation/illustration set: m = mathematical models, p = physical phantoms (mostly balloons or heart casts), a = animal model, V = human volunteers and P = patients. Standard of reference: AS = analytic solution, AT = alternative technique, CL = comparison to literature, GT = ground truth, NS = numerical solution, OB(m) = human observer (involving multiple modalities). Papers with several evaluation studies have multiple entries. Note that only the accuracy in determining tag intersections was computed. No quantitative analysis reported on deformation field or strain analysis. Monte Carlo analysis of sensitivity for this factor is reported.

Actually, it is the conjunction of model parameterization and the recovery strategy that determines the computational load of an approach. It would have been very interesting if the computation time had been reported with each technique. Unfortunately, variability in hardware architecture over time and the variations in techniques renders any quantitative comparison unrealistic.

of the lesion to be demonstrated can be accomplished by positioning an intravascular catheter through which the contrast medium is guided and injected. Angiocardiography is usually good for anatomic delineation of lesions but is much less satisfactory for the determination of their severity and the degree of hemodynamic disturbance they have produced. This technique has been used for a long time to assess ejection fraction (EF) and ejection volumes (EV) [10] based on simplified geometric models [11–13] of the LV, but most radiologists use visual assessment based on experience [17].

9.2.2 Cardiac Ultrasound

Two-dimensional ultrasonic (US) imaging of the heart, or *echocardiography* [18,24], allows the anatomy and movements of intracardiac structures to be studied noninvasively. The application of pulsed- and continuous-wave Doppler principles to 2-D echocardiography (2DE) permits blood flow direction and magnitude to be derived and mapped onto a small region of interest of the 2DE image. In color-flow Doppler mapping (CFM), the pulsed-wave signal with respect to blood velocity and direction of flow throughout the imaging plane is color-coded and produces a color map over the 2DE image. One of the limiting factors of 2DE is the ultrasound window (presence of attenuating tissues in the interface between the US transducer and the organ of interest). To overcome this problem, *transesophageal echocardiography* can be used, which can provide high-quality color-flow images at the expense of being invasive.

Three-dimensional echocardiography (3DE)[19] is a relatively new development in the U.S. that allows 3-D quantitation of organ geometry because the complete organ structure can be imaged. This technique has been used to compute LVV and LVM [25–30] and to perform wall motion analysis [31].

9.2.3 Isotope Imaging

Isotopes have been used to study left-ventricular function and myocardial perfusion. Radionuclide techniques for monitoring global and regional ventricular function fall into two major categories: (a) first-pass studies, in which an injected bolus dose is monitored during its first passage through the heart and great vessels, and (b) gated equilibrium studies, in which the tracer mixes with the blood pool before data collection. First-pass acquisitions are typically 2-D, whereas gated equilibrium studies can be 2-D or 3-D (single-photon emission computed tomography — SPECT). Isotope imaging can be used to assess parameters like EF [32] and regional wall motion analysis [32–34]. It is also used to study myocardial perfusion [35] in cases of ischemia or myocardial infarction, and to assess myocardial viability. The overwhelming majority of radionuclide studies performed for perfusion assessment employ SPECT.

9.2.4 Cardiac Computed Tomography

Computed tomography (CT) is rapidly gaining ground as a routine cardiovascular modality. Electron beam computed tomography (EBCT)[36] was the first type of clinically applied cardiovascular CT providing 3-D data over the full cardiac cycle. Using prospective gating, a stack of slices can be acquired with a high temporal resolution (50 msec between frames) and a slice thickness of about 3–4 mm. EBCT has been mainly applied clinically for detecting coronary calcium deposits. Limitations of this system are that the spatial resolution in the transaxial direction is much less than the in-plane resolution, there is a high noise level in the images, and the geometry of the scanner is fixed, which does not allow acquisition of short-axis views.

Multislice CT (MSCT)[37,38] is less expensive and more widely available than EBCT. MSCT is characterized by a higher spatial resolution in all directions and can provide near-isotropic data sets within one breath hold, albeit at a lower temporal resolution than can EBCT (150 msec between frames). Hence, with the recent advent of multirow detectors, a faster imaging time is combined with higher resolution and lower noise levels. MSCT effectively became feasible with 4-row detectors, enabling spiral scanning of four slices at a time; nowadays 12- and 16-row detectors are commonly available, and 64-row detector gantries are under development. It is expected that the increasing number of detectors in combination with improved reconstruction algorithms will enable reduced acquisition times, and that image quality and temporal and spatial resolution will further improve in the near future. The near-isotropic resolution has two additional advantages: the data can be reformatted into any desired spatial orientation (such as the short-axis view) without interpolation, and it gives an excellent definition of the coronary vessels [39,40]. Currently, MSCT is clinically used for examining coronary anatomy, for visual coronary stenosis detection, and for assessing the amount of calcium in the coronaries. Recent studies [41,42] indicate that already a 12-row CT enables detection of coronary artery disease with 95% sensitivity and 93% specificity. However, as almost on other modality, MSCT enables the combined assessment of left-ventricular global and regional function and coronary function. It is therefore increasingly used to detect coronary artery disease in combination with left-ventricular function, and to screen asymptomatic patients with cardiovascular risk factors.

9.2.5 MAGNETIC RESONANCE IMAGING

Cardiac magnetic resonance imaging [23] (MRI) is now an established, although still rapidly advancing, technique providing information on the morphology and function of the cardiovascular system [43]. Advantages of cardiac MRI include a wide topographical field of view with visualization of the heart and its internal morphology and the surrounding mediastinal structures, multiple imaging planes, and a high soft-tissue contrast discrimination between the flowing blood and myocardium without the need for contrast media or invasive techniques. Long- and short-axis views of the heart, as used in echocardiography, can be obtained routinely because arbitrary imaging planes can be selected.

In fact, cardiac MRI can be regarded as a collection of "MR modalities," each dedicated to different aspects of cardiac function. Multislice multiphase short-axis scanning enables a detailed study of cardiac anatomy and global and regional function, both at rest and under stress. Several researchers have used

MRI to assess right- and left-ventricular parameters as represented by stroke volume (SV), EF, LVM [44–49], wall thickening (WT) [50], myocardial motion [51], and circumferential shortening of myocardial fibers [52]. Data from MRI are more accurate than those derived from left-ventricular angiocardiography, where the calculation is based on the assumption that the LV is ellipsoidal in shape. Volume measurements by MRI are independent of cavity shape, with the area from contiguous slices integrated over the chamber of interest.

Multislice multiphase imaging enables assessment of regional function in terms of the local-wall-geometry changes over the cardiac cycle. However, the complex twisting motion during cardiac contraction cannot be imaged with ordinary multislice multiphase acquisitions, and therefore the gold standard for accurate regional contractility analysis is cardiac MR tagging [53,54]. By locally changing the magnetization of the tissues, a stripe grid can be applied to the myocardium, defining a material coordinate frame. As the tagging stripes are deformed during cardiac contraction, the material coordinates inside the myocardium can be tracked over the cardiac cycle. Myocardial deformation can thus be tracked, allowing for stress and strain measurements, which are assumed to be early indicators of myocardial dysfunction. Typically, such measures are derived using an intermediary finite element (FE) continuum model, which is coupled to the tagging intersection locations. From the deformations of the continuum model, estimates for stress and strain are computed.

Apart from global function and regional myocardial contractility and motion, the perfusion of the myocardium provides important diagnostic information on coronary function [55]. The primary means to image perfusion with MR is first-pass perfusion imaging. These images monitor the arrival and subsequent distribution within the myocardium of a contrast bolus. The rate and extent of perfusion can be quantified by following the intensity profile of myocardial pixels over time. Following first-pass perfusion, delayed enhancement images are commonly acquired 15–20 min after contrast medium injection. Delayed enhancement imaging [56,57] exploits the fact that the contrast medium tends to accumulate in necrotic tissue, greatly enhancing the signal from infarcted regions, with an image resolution much higher than that seen in common nuclear scans. Quantification can be performed by measuring enhanced signal intensity within the myocardium and comparing it to nonenhanced myocardium, enabling an assessment of the extent and location of necrotic tissue. In addition, an indicator for myocardial viability can be derived from this analysis: infarct transmurality. As myocardial infarctions tend to originate from the endocardial surface, the penetration of the infarction in the myocardium toward the epicardial wall is regarded as a measure of infarct severity. Myocardial viability can be inferred from this infarct transmurality, where a higher transmurality typically signifies a decreased viability and, thus, a reduced chance of myocardial recovery after intervention.

Alternatively, global and regional function can also be quantified from phase-contrast MRI. These images depict the velocity of a material point in the scanner, where the gray values represent the velocity. By acquiring phase-contrast images perpendicular to the aorta slightly distal to the aortic root (aortic-flow images), global parameters such as SV and cardiac output (CO) can be quantified with relatively little effort. Phase-contrast imaging has also been applied to investigate regional myocardial function, in a similar way as MR tagging. By acquiring multiple short-axis images with different orthogonal phase-encoding directions, three velocity components of a material point inside the myocardium can be computed, providing a dense displacement field for the myocardium with a higher resolution than tagging data. Analogous to tagging analysis, the displacement field can serve as input to a continuum model, yielding estimates for myocardial stress and strain.

Compared to MSCT, cardiac MRI provides similar data, the additional advantages being the nonionizing nature of the modality, its versatility, and superiority in flow and perfusion imaging. MSCT, on the other hand, enables a significantly faster acquisition (± 20 sec), and thus enables a higher patient throughput and the simultaneous study of coronary and ventricular function.

9.3 CLASSICAL DESCRIPTORS OF CARDIAC FUNCTION

Development of models of the cardiac chambers has emerged from different disciplines and with various goals. Cardiac models have been used for deriving functional information, for visualization and animation, for simulation and planning of surgical interventions, and for mesh generation for FE analysis.

This survey will be confined to the application of modeling techniques for obtaining classical functional analysis. Classical functional analysis can be divided into global functional analysis (Subsection 9.3.1) and motion and deformation analysis (Subsection 9.3.2), from which the most clinically relevant parameters can be obtained.*

Model-based methods also allow one to derive new descriptors of cardiac shape and motion. Such advanced descriptors have been mainly presented in the technical literature and their clinical relevance is yet to be assessed. Without pretending to be exhaustive, Appendix A summarizes a number of nonclassical shape and motion descriptors that demonstrate the extra possibilities provided by some of the advanced methods.

9.3.1 GLOBAL FUNCTIONAL ANALYSIS

Weber and Hawthorne [58] proposed a classification of cardiac indices according to their intrinsic dimensionality: linear, surface, or volumetric descriptors. Linear parameters have been used intensively in the past because they can easily be derived from 2-D imaging techniques such as 2DE and x-ray angiocardiography.* However, these techniques assume an "idealized" geometry of the LV and strongly depend on external or internal reference and coordinate systems. Besides total ventricular wall area, other surface indices based on curvature and derived parameters have been investigated from 2-D studies [59–62]. More recently, many image processing

---

* Such parameters are, for instance, left-ventricular internal dimension (LVID), relative wall thickness (RWT), and estimates of fractional shortening of the cardiac fibers (%ΔD) and their velocity (V_{cf}). For a detailed analysis of these parameters refer to Vuille and Weyman [14].

approaches to left-ventricular modeling have suggested true 3-D global and local shape indices based on surface properties.

In practice, assessment of cardiac function still relies on simple global volumetric measures such as LVV and LVM, and EF. These and other basic parameters will be presented in the following text:

Left Ventricular Volume (LVV): This is a basic parameter required to derive other LV indices such as, e.g., EF. Angiocardiography and echocardiography have been traditionally used to assess this quantity. In the latter case, three approaches have been applied to represent the LVV: (a) as the volume of a single shape (e.g., truncated ellipse); (b) as the sum of multiple smaller volumes of similar configuration (e.g., Simpson's method), and (c) as a combination of different figures [14]. The achieved accuracy in the assessment of LVV with echocardiography varies widely with the model used to represent the LV. The best results have been obtained using Simpson's rule in which *in vitro* studies have revealed a relative error ranging from 5.9% to 26.6% depending on the particular implementation and the number of short-axis slices used in the computation [14]. It has been shown that echocardiography consistently underestimates ventricular cavity, whereas angiocardiography consistently overestimates the volumes [14]. In a recent study by Lorenz et al. [48] with a canine model and autopsy validation, it has been shown that cine MRI is a suitable and accurate method to estimate RVV and LVV. In this study, MR-based and autopsy volumes agreed to within 6 ml, yielding no statistically significant differences.

Left Ventricular Mass (LVM): Left-ventricular hypertrophy, as defined by echocardiography, is a predictor of cardiovascular risk and higher mortality [14]. Anatomically, LV hypertrophy is characterized by an increase in muscle mass or weight.

LVM is mainly determined by two factors: chamber volume and wall thickness. There are two main assumptions in the computation of LVM: (a) the interventricular septum is assumed to be part of the LV and (b) the volume, V_m, of the myocardium is equal to the total volume contained within the epicardial borders of the ventricle, $V_t(epi)$, minus the chamber volume, $V_c(endo)$; LVM is obtained by multiplying V_m by the density of the muscle tissue (1.05 g/cm^3)

$$v_m = v_t(epi) - v_c(endo) \tag{9.1}$$

$$LVM = 1.05 \times V_m \tag{9.2}$$

LVM is usually normalized to the total body surface area or weight in order to facilitate interpatient comparisons. The normal values of LVM normalized to body weight are 2.4 ± 0.3 g/kg [48].

Stroke volume (SV). This is defined as the volume ejected between the end of diastole and the end of systole.

$$SV = \text{end-diastolic volume(EDV)} - \text{end-systolic volume(ESV)} \qquad (9.3)$$

Alternatively, SV can be computed from velocity-encoded MR images of the aortic arch by integrating the flow over a complete cardiac cycle [63]. Similarly to LVM and LVV, SV can be normalized to total body surface. This corrected SV is known as *stroke volume index* (SVI). Healthy subjects have a normal SVI of 45 ± 8 ml/m^2 [48].

Ejection fraction (EF): This is a global index of left-ventricular fiber shortening and is generally considered as one of the most meaningful measures of the left-ventricular pump function. It is defined as the ratio of SV to the EDV.

$$EF = \frac{SV}{EDV} \times 100\% = \frac{EDV - ESV}{EDV} \times 100\% \qquad (9.4)$$

Lorenz et al. measured normal values of EF with MRI [48]. They found values of $67 \pm 5\%$ (57–78%) for the LV and $61 \pm 7\%$ (47–76%) for the RV. Similar values were obtained with ultrafast CT, echocardiography, and x-ray angiocardiography [48,14].

Cardiac output (CO): The role of the heart is to deliver an adequate quantity of oxygenated blood to the body. This blood flow is known as the CO and is expressed in liters per minute. Because the magnitude of the CO is proportional to body surface, one person may be compared to another by means of the cardiac index (CI), that is, the CO adjusted for body surface area. Lorenz et al. [48] reported normal CI values of 2.9 ± 0.6 l/min/m^2 and a range of 1.74–4.03 l/min/m^2.

The CO was originally assessed using Fick's method or the indicator dilution technique [64]. It is also possible to estimate this parameter as the product of the volume of blood ejected with each heartbeat (the SV) and the heart rate (HR).

$$CO = SV \times HR \qquad (9.5)$$

In patients with mitral or aortic regurgitation, a portion of the blood ejected from the LV regurgitates into the left atrium or ventricle and does not enter the systemic circulation. In these patients, the CO computed with angiocardiography exceeds the forward output. In patients with extensive wall motion abnormalities or misshapen ventricles, the determination of SV from angiocardiographic views can be erroneous. Three-dimensional imaging techniques provide a potential solution to this problem because they allow accurate estimation of the irregular left-ventricular shape.

9.3.2 MOTION AND DEFORMATION ANALYSIS

9.3.2.1 Motion Analysis

A number of techniques have been used in order to describe and quantify the motion of the heart. They can be divided into three main categories according to the method used [65]: (i) detecting endocardial motion by observing image intensity changes, (ii) determining the boundary wall of the ventricle and subsequently tracking it, and (iii) attempting to track anatomical [66–69], implanted [70–76], or induced [53,54,77–83] myocardial landmarks. There are a few problems with each of these techniques. Assumptions must be made about the motion (motion model) in the first two groups in order to obtain a unique pointwise correspondence between frames. To this end, optic flow methods [84–89]* and phase-contrast MR [91–94] have been applied for (i), and curvature-based matching [95–99] has been used to find point correspondences in (ii). Landmark-based methods [53,54,66–83] provide information on material point correspondences. However, this information is mostly sparse and, again, assumptions on the type of motion have to be made in order to regularize the problem of finding a dense displacement field. The use of implanted markers adds the extra complication of being invasive, which precludes routine use of this technique in humans. Although implanted markers are usually regarded as the gold standard, there are some concerns in the literature about their influence on image quality and about their modification of the motion patterns.

9.3.2.2 Wall Thickening

Azhari et al. [100] have compared wall thickening (WT) and wall motion in the detection of dysfunctional myocardium. From their study, it was concluded that WT is a more sensitive indicator of dysfunctional contraction [100]. This finding has prompted several researchers to define methods to quantify wall thickness. Azhari et al. [100], and Taratorin and Sideman [101] carried out a regional analysis of wall thickness by dividing the myocardium into small cuboid elements. The local wall thickness is then defined as the ratio between the volume of the particular element and the average area of its endocardial and epicardial surfaces [50].

The most widely employed method for WT computation, however, is the centerline method [102] and several improvements thereof [47,103–105]. Starting with the endo- and epicardial contours at each slice, the centerline method, in its original formulation, measures WT in chords drawn perpendicular to a line that is equidistant to both contours (the centerline). Although more accurate than methods relying on a fixed coordinate system, this method still assumes that the contours are perpendicular to the long axis of the LV. If this is not the case, the myocardial wall thickness

---

* At this point it is worth mentioning an excellent online bibliographic database maintained by the Special Interest Group on Cardiac Motion Analysis (SigCMA) that can be accessed at http://www.creatis.insa-lyon.fr/sigcma. It also provides general bibliographic information on model-based cardiac image analysis.

is overestimated, which invariably occurs, for instance, in slices that are close to the apex. Buller and coworkers [103,47] introduced an improvement on this method by estimating at each location the angle between the wall and the imaging plane. Later, Bolson and Sheehan [104,105] introduced the Center Surface method (a true 3-D extension of the centerline method), which makes use of a reference medial surface to compute the chords and subsequently wall thickness.

9.3.2.3 Strain Analysis

Strain analysis (SA) is a method to describe the internal deformation of a continuum body. It is a promising tool to study and quantify myocardial deformation. Here, we shall briefly introduce some of the concepts related to strain analysis. A comprehensive exposition of this theory can be found in Fung [106].

To describe the deformation of a body, the position of any point in the body needs to be known with respect to an initial configuration; this is called the *reference state*. Moreover, to describe the position a reference frame is needed. In the following description, a Cartesian reference frame will be assumed. It is also common to use curvilinear coordinates because some of the expressions simplify.

A myocardial point, M_r, has coordinates $\{y_i\}$ and a neighboring point, M'_r has coordinates $\{y_i + dy_i\}$. Let M_r be moved to the coordinates $\{x_i\}$ and its neighbor to $\{x_i + dx_i\}$. The deformation of the body is known completely if we know the relationship

$$x_i = x_i(y_1, y_2, y_3) \quad i = 1,2,3 \tag{9.6}$$

or its inverse,

$$y_i = y_i(x_1, x_2, x_3) \quad i = 1,2,3 \tag{9.7}$$

For every point in the body, we can write

$$x_i = y_i + u_i \quad i = 1,2,3 \tag{9.8}$$

where u_i is called the *displacement* of the particle M_r. In order to characterize the deformation of a neighborhood, the first partial derivatives of Equation 6–Equation 8 are computed. These derivatives can be arranged in matrix form to define the *deformation gradient tensor*: $F = [\partial x_i / \partial y_i]$, $(i, j = 1, 2, 3)$. The deformation gradient tensor enables estimation of the change in length between the neighboring points $\{y_i\}$ and $\{y_i + dy_i\}$, when they are deformed into $\{x_i\}$ and $\{x_i + dx_i\}$. Let $d\ell_r$ and $d\ell$ be these lengths before and after deformation. Then

$$dl^2 - dl_r^2 = 2 \sum_{i=1}^{3} \sum_{j=1}^{3} E_{ij} dy_i dy_j \tag{9.9}$$

where $E = [E_{ij}]$ is the *Green strain tensor* [106]

$$E_{ij} = \frac{1}{2}\left(\sum_{k=1}^{3} \frac{\partial x_k}{\partial y_i} \frac{\partial x_k}{\partial y_j} - \delta_{ij} \right)$$

$$= \frac{1}{2}\left(\frac{\partial u_i}{\partial y_j} + \frac{\partial u_j}{\partial y_i} - \sum_{k=1}^{3} \frac{\partial u_k}{\partial y_i} \frac{\partial u_k}{\partial y_j} \right)$$

(9.10)

where δ_{ij} is the Kronecker tensor. From the strain tensor it is possible to decompose the strains into two groups: axial and shear strains. The former correspond to the diagonal elements and represent changes in length aligned with the axes of the reference frame and the latter correspond to off-diagonal terms or deformations where the two axes are coupled.

9.4 OVERVIEW OF MODELING TECHNIQUES

Great effort has been devoted to the analysis and segmentation of cardiac images by methods guided by prior geometric knowledge. When focusing on the way models are geometrically represented, three main categories can be distinguished: surface models, volumetric models, and deformation models. In all cases both discrete and continuous models have been proposed, as well as implicitly defined surface models (Figure 9.2).

Alternatively, one may classify model-based approaches by considering the information that is used as input for model recovery. This categorization is highly influenced by the imaging modality for which the method has been developed.

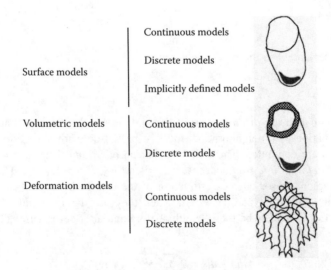

FIGURE 9.2 Proposed taxonomy of cardiac modeling approaches.

There are a variety of inputs for model recovery: (1) multiple 2-D projection images, (2) multiple oriented 2-D slices, (3) fully 3-D gray-level images, (4) 3-D point sets, (5) phase-contrast velocity fields, and (6) MR tagging information.

In this survey we will compare the different methods with respect to the type of model representation and the types of input data and features that the model is recovered from. Table 9.1, in which the different approaches are grouped according to the type of model representation, summarizes this section.

9.4.1 SURFACE MODELS

Many approaches to cardiac modeling focus on the endocardial (or epicardial) wall. Three subcategories are proposed: (a) continuous models with either global, local, or hierarchical parameterizations, (b) discrete models, and (c) implicitly defined deformable models.

9.4.1.1 Continuous Models

In the early studies of cardiac images by 2DE and angiocardiography, cardiologists used simplified models of the LV in order to compute functional parameters such as ventricular volume and mass from 2-D images. Most of the time, simple ellipsoidal models were considered (see, e.g., Vuille and Weyman [14] and Dulce et al. [46] for a comprehensive review of such models and a comparison of their accuracy). In the last decades, however, approaches have appeared that make use of 3-D acquisitions to reconstruct models varying from global parameterizations of the LV surface [5,29,112,119–121,123,194], to hierarchically parameterized models [9,114,116,136,150,160].

9.4.1.1.1 Global Approaches

In this category, we will discuss surface representations that are based on simple geometric models. In general they can provide, with a limited number of global parameters, a rough shape approximation. We also include in this category surface representations obtained as series of basis functions with global support.

Cauvin et al. [112] model the LV as a truncated bullet, a combination of an ellipsoid and a cylinder, that is fitted to the morphological skeleton of the LV. Metaxas and Terzopoulos [195] have proposed superquadrics [196] to model simple objects with a small number of parameters. Since the introduction of superquadrics, several extensions have appeared in the literature. Chen et al. [118] apply superquadrics with tapering and bending deformations to model the LV in an integrated approach for image segmentation and shape analysis. The method iterates between a region-based clusterization step [197], using statistics of image intensity and gradient, and a shape-based step that checks the consistency between the current segmentation and a superquadric model. Park, Metaxas, Young, and Axel [6] have extended the flexibility of superquadrics by introducing parameter functions: radial and longitudinal contraction, twisting, and long-axis deformation. These allow for a more detailed representation of the LV while retaining the intrinsic geometrical meaning of the superquadric parameters. LV midwall motion is recovered using preprocessed MR tagging data obtained by sampling the LV midwall surface from the 3-D FE model of Young and Axel [77].

Staib and Duncan [121] use sinusoidal basis functions to decompose the endocardial surface of the LV. The overall smoothness of the surface is controlled by decreasing the number of harmonics in the Fourier expansion. Model recovery is cast into a Bayesian framework in which prior statistics of the Fourier coefficients are used to further limit the flexibility of the model. Matheny and Goldgof [120] compare different 3-D and 4-D surface harmonic descriptions for shape recovery. Time can be incorporated in two ways in the model: (a) as hyperspherical harmonics, where an event in space–time is converted from Cartesian coordinates to hyperspherical coordinates, and (b) as "time-normal" coordinates, which are formed by including a temporal dependency in each spatial coordinate. Experiments carried out with a 3-D CT data set of a canine heart have indicated that hyperspherical harmonics can represent the beating LV with higher accuracy than direct normal extensions of spherical, prolate spheroidal, and oblate spheroidal harmonics. Coppini et al. [29] reconstruct a 3-D model of the LV based on apical views in US images. LV boundaries are obtained by grouping edges with a feedforward neural network (NN) integrating information about several edge features (position, orientation, strength, length, and acquisition angle). This allows the discarding of many edge points that are not plausible LV boundary points. The 3-D LV geometry is modeled as a spherical elastic surface under the action of radial springs (attracting the model to the edge points); a Hopfield [198] NN is used to solve the minimization problem involved in the reconstruction of this surface. Declerck, Feldmar, and Ayache [123] have introduced a spatiotemporal model to segment the LV and to analyze motion from gated SPECT sequences. The model relies on a planispheric transformation that maps endocardial points in one time frame to the corresponding material points in any other frame. First, endocardial edge points are detected in all frames using a Canny–Deriche edge detector [199] in spherical coordinates [200]. Selected points in subsequent frames are matched to the current frame using a modification of the iterative closest point (ICP) algorithm [200–202]. Based on corresponding point pairs, the parameters of a planispheric transformation are retrieved by least-squares approximation. This transformation allows the description of motion with just a few parameters that can be related to a canonical decomposition (radial motion, twisting motion around the apicobasal axis, and long-axis shortening).

9.4.1.1.2 Hierarchical Approaches

Some authors have addressed the problem of building *hierarchical representations* in which a model described with few parameters is complemented with extra deformations that capture finer details. Gustavsson et al. [114], for instance, employ a truncated ellipsoid to obtain a coarse positioning of the left-ventricular cavity from contours drawn in two short-axis and three apical echocardiographic views. Further model refinement is achieved using cubic B-spline curves approximating manually segmented contours in multiple views. Chen et al. [116], Bardinet et al. [9], and Fan et al. [150] use superquadrics [196] to coarsely describe the LV. Their approaches fundamentally differ in the representation of the additional deformation field. Chen et al. and Fan et al. use spherical harmonics in order to approximate the residual error between the superquadric estimate of the endocardial LV wall and the true wall location. Spherical harmonics have the advantage that fine-tuning can be improved *ad infinitum* with increasing number of harmonics. However, adding a new coefficient influences the shape of the model everywhere (nonlocal basis functions). Bardinet et al. [9] extend the basic superquadric deformations (tapering and bending) through the use of free-form deformations (FFD), a technique introduced in computer graphics by Sederberg and Parry [203]. The superquadric is attached to a flexible, boxlike frame, inducing a nonrigid deformation on the superquadric. Bardinet et al. use trivariate B-splines to parameterize this deformation field. In a later work, Bardinet et al. [122] apply their method to estimate left-ventricular wall motion. This is accomplished by deforming the full model (superquadric plus FFD) in the first frame, and modifying only the FFD in the subsequent frames. By tracking points with the same parametric coordinates along the cardiac cycle, a number of dynamic parameters such as wall thickening and twisting motion are computed. Germano et al. [135,136] have developed a system for automatic quantification of left-ventricular function from gated perfusion SPECT images. An iterative algorithm fits an ellipsoidal model to a semiautomatically obtained segmentation. This iterative algorithm incrementally adapts the ellipsoid's parameters and center of mass so that accurate registration of the model is obtained even in the presence of large perfusion defects. The ellipsoid defines a coordinate system that is used to refine the model. A Gaussian model of the count profiles is used to compute radial offsets corresponding to the endocardial and epicardial walls. Although simple in its formulation, this method has proved very useful in determining most of the classical cardiac functional parameters [35] from SPECT images and has been extensively validated in humans [135,136,204].

9.4.1.1.3 Local Approaches

A number of methods have been reported to provide surface reconstruction using piecewise polynomial surfaces, e.g., B-splines or bicubic Hermite surface patches. These techniques have appeared mainly in the context of surface reconstruction from multiple cross sections [31,126,169] or projections [107–110,124]. Given the ill-posed nature of this problem, most of these techniques require extensive user interaction. Usually, a set of landmarks or fiducial points are determined from each cross section or projection and, using high-level knowledge about the viewpoint and the geometry of the LV, a local surface approximation using surface patches is performed. One relevant example is the work by Sanchez-Ortiz et al. [125], in which a tensor B-spline surface model is fitted to multiple planes of a rotating US probe to recover the 3-D shape of the LV using multiscale fuzzy clustering features. A very interesting approach is the one of Horkaew and Yang [128], which uses tensor B-splines to represent a surface model of a full heart. The technique allows modeling of both the mean shape and the variability around it using landmarks obtained by optimizing a minimum description length (MDL) principle and piecewise bilinear maps (PBM).

A rather different approach is the one by Pentland and Horowitz [111], who applied modal analysis and FE to reconstruct a 3-D model of the LV from x-ray transmission data. Modal analysis offers a principled physically based strategy for reducing the number of degrees of freedom (DOF) of the model and to obtain an overconstrained problem for shape recovery. Optic flow is used to derive the deformation of the 3-D model from the 2-D views, and a Kalman filter is used for tracking the structures over time.

Instead of working with multiple cross sections or projection images, Goshtasby and Turner [119] segment left- and right-ventricular endocardial surfaces from 3-D flow-enhanced MR images. In this case, the endocardial surface is modeled as a deformable cylinder using rational Gaussian surfaces [205]. The model is deformed to fit the zero-crossings of the image Laplacian. To avoid attraction by spurious edges, prior to fitting, the feature map is masked by a rough LV region of interest obtained by intensity thresholding.

Some efforts have also been directed toward *geometric modeling of the RV*. This chamber has a more complex shape than the LV. Spinale et al. [110] fit semiellipses to model the crescentic shape of the RV from biplane ventriculograms. Czegledy and Katz [113] model the RV using a crescentic cross-sectional model composed of two intersecting circles of different radii. This 3-D model is parameterized by only a few linear dimensions that can be measured directly from CT, MR, or US images. From these dimensions, the RV volume is approximated using analytical expressions. Denslow [117] models the RV as the difference of two ellipsoids (an ellipsoidal shell model). The parameters from this shell are estimated from MR images (a long-axis and a four-chamber view) and from those, volume estimates can be derived. Sacks et al. [115] model the endocardial and epicardial walls of the RV by biquadric surface patches (contours were manually traced from MR images), and have studied surface curvature and wall thickness changes during the cardiac cycle using this representation.

9.4.1.2 Discrete Models

An alternative to continuous surface representations is the use of discrete surface models. Several methods have been reported in the literature, and they can be grouped as shown in the following subsection.

9.4.1.2.1 *Physics-Based Models*

Physics-based modeling has attracted the attention of many computer vision researchers. In this framework, surface recovery is cast into the deformation of a virtual body (the geometric model together with its material properties) under virtual external forces derived from image or point features, or user-defined constraints. In the final (deformed) state, this virtual body reaches an equilibrium between the external forces and internal (regularization) constraints. A good overview of the theory of physics-based deformable models and its applications can be found in the book by Metaxas [206] and in the survey by McInerney and Terzopoulos [3].

McInerney and Terzopoulos [137] have applied this theory to the segmentation and tracking of the LV in dynamic special reconstructor (DSR) image sequences. An FE balloon [207] deformable model is discretized using triangular elements, and deformed according to a first-order approximation of the Lagrange equations of motion. User-defined point constraints can be interactively inserted to guide the deformation of the model and to avoid local minima of the potential energy in which the model is embedded. In the Lagrangian formulation, 3-D image sequences can easily be handled by making the potential energy a function of time. Montagnat, Delingette, and Malandain [144] apply simplex meshes [208] to reconstruct the LV from multiple views of a rotating US probe. Images are acquired in cylindrical coordinates coaxial with the apicobasal axis. Accordingly, images are filtered in cylindrical coordinates. Boundary points are detected based on a combination of image gradient and intensity profiles normal to the surface. Finally, detected edge points are cast into point attraction forces deforming the model according to Newton's law of motion. Ranganath [138] reconstructs 3-D models of the LV from MRI images using multiple 2-D *snakes* [209] and by devising efficient mechanisms for interslice and interframe contour propagation. Biedenstein et al. [145] have recently published an elastic surface model and applied it to SPECT studies. The elastic surface is deformed according to a second-order partial differential equation. The external (image) forces are derived from the radioactive distribution function and push the elastic surface toward the center surface of the LV wall. Wall thickness can be then computed as the distance between the elastic surface and the mass points of the radioactivity distribution gradient. Huang and Goldgof [133] have presented an adaptive-size mesh model within a physics-based framework for shape recovery and motion tracking. The optimum mesh size is inferred from image data, growing new nodes as the surface undergoes stretching or bending, or destroying old nodes as the surface contracts or becomes less curved. The method is employed to analyze LV motion from a DSR data set. To establish point correspondences, an adaptive-size mesh is generated for the first frame to be analyzed; subsequent frames further deform this mesh while keeping its configuration fixed. Mäkelä et al. [148] propose a template-based approach for recovering a surface thorax model from MR images. To this end, the template, constructed from a normal volunteer, is adapted to fit the salient edges of the image to be segmented by optimizing the parameters of a FFD field in which the template is embedded. This technique has been successfully applied to extract a geometrical model of the torso from MR that is subsequently used to fuse 3-D functional information from position emission tomography (PET) and magnetocardiography(MCG) data.

Physics-based modeling frequently makes an assumption that can be problematic: internal constraints are usually represented in the form of controlled-continuity stabilizers [210]. It is known that, in the absence of image forces, deformable models tend to shrink. To avoid this, Rueckert and Burger [141] simultaneously model the two cardiac chambers (RV and LV) using a geometrically deformable template (GDT). The standard stabilizers on the deformed model are replaced by a stabilizer on the deformation field between a rest model and a deformed model. A GDT consists of three parts: (a) a set of vertices that defines the rest state (the template), (b) a set of vertices that defines a deformed state (an instance of the template), and (c) a penalty function that measures the amount of deformation of the template with respect to its equilibrium shape (the stabilizer). Another solution to the preceding problem was proposed by Nastar and Ayache [140], who modeled a surface as a quadrilateral or triangular mesh of virtual masses. Each mass is attached to its neighbors by perfect identical springs with predefined stiffness and natural length. The system deforms under the laws of dynamics. In addition to elastic and image forces, an "equilibrium force" determines the configuration of the mesh in the absence of external forces.

9.4.1.2.2 Spatiotemporal Models

Several researchers have developed models that explicitly incorporate both spatial and temporal variations of LV shape. Faber et al. [131] use a discrete 4-D model to segment the LV from SPECT and MR images through a relaxation labeling scheme [211]. Endo- and epicardial surfaces are modeled as a discrete template defined in a mixed spherical/cylindrical coordinate system coaxial with the LV long-axis. Each point in the template represents a radius connected to this axis. The model is spatiotemporal because the compatibility functions computed in the relaxation labeling scheme involve neighboring points both in space and time. In this way, surface smoothness and temporal coherence of motion are taken into account. Tu et al. [139] have proposed a 4-D model-based LV boundary detector for 3-D CT cardiac sequences. The method first applies a spatiotemporal gradient operator in spherical coordinates with a manually selected origin close to the center of the LV. This operator is only sensitive to moving edges and less sensitive to noise compared to a static edge detector. An iterative model-based algorithm refines the boundaries by discarding edge points that are far away from the global model. The model is parameterized by spherical harmonics, including higher-order terms, as the refinement proceeds. An interesting approach to spatiotemporal 3-D segmentation is the work by Gerard et al. [146], which is based on the concept of active objects (AO). AOs are described by means of simplex meshes [208] and embedded in a physics-based framework. Based on this approach, the authors tackle the problem of segmentation of 3-D US image sequences using a statistical motion model of heart dynamics.

9.4.1.2.3 Polygonal Models

LV *polygonal representations* have been applied by several authors [27,28,129, 132,134,142–144,] in the literature. The approaches differ either in the type of polygonal primitive (e.g., triangular or quadrilateral meshes) or the details of the shape recovery algorithm (imaging modality, input data, or recovery features). Shi et al. [142,143] use a Delaunay triangulation [212] to build a surface description from a stack of 2-D contours obtained with a combined gradient- and region-based algorithm [213]. This representation is subsequently used for motion analysis based on point correspondences. Bending energy under a local thin-plate model is used as a measure of match between models of consecutive frames.

Friboulet, Magnin, and Revel [132] have developed a polygonal model to analyze the motion of the LV from 3-D MR image sequences. LV contours are manually outlined using a trackball. After applying morphological and linear filtering to diminish quantization noise, the contours are radially resampled with constant angular step. Finally, the stack of resampled contours is fed into a triangulation procedure [214] that generates a polygonal surface with approximately equal-sized triangles. Faber et al. [134] use a combination of cylindrical and spherical coordinate systems to build a discrete model of the LV in SPECT perfusion images. A radius function defined in a discrete (orientation) space of longitudinal and circumferential coordinates describes the LV. For each orientation, the radius is determined by finding the position of maximal perfusion (which is said to occur in the center of the myocardium). After low-pass filtering to remove outlier radii, the radius function is mapped back to Cartesian space where the surface is represented using triangular or quadrilateral meshes. This approach shares some features of the work described in Faber et al. [131], but is purely static. Legget et al. [28,215] use piecewise smooth subdivision surfaces [216] to reconstruct the LV geometry from manually traced contours in 3-D US images. Some elements of the mesh can be labeled so that they allow for sharp edges (e.g., at the mitral annulus and apex) and to define regional surface descriptors. Also from 3-D US images, Gopal et al. [27] apply triangulated surfaces to reconstruct the geometry of latex balloon phantoms mimicking the LV. Three-dimensional reconstruction is directly obtained by triangulating the points of manually delineated contours from a stack of quasi-parallel slices. Song et al. [147] use a triangular surface model to represent the heart. In contrast to several other techniques, a given heart is approximated by a convex combination of shapes from a model catalog. The authors cast the surface model optimization problem in a Bayesian framework, such that the inference made about a surface model is based on the integration of both the low-level image evidence and the high-level prior shape knowledge through a pixel class prediction mechanism.

9.4.1.2.4 *Statistical Shape and Appearance Models*

These models capture the mean shape and shape variations from a training population. In 3-D cardiac modeling, these models have been developed for shape analysis and for gaining insight into commonly occurring anatomical variations. Apart from shape analysis, the learned eigenvariations can be applied to image segmentation and motion tracking by restricting the search space of an image-matching mechanism to statistically plausible directions.

In statistical shape models, a shape is expressed as a set of corresponding landmarks, which are parameterized as a coordinate vector concatenating the landmark components. These vectors are aligned using Procrustes' algorithm with respect to position, scale, and orientation, thus minimizing the sum of squared distances between the landmarks. The residual sample point distributions after alignment represent the pure shape-related differences in the population, and are modeled by computing the shape average and applying a principal component analysis (PCA) on the coordinate covariance matrix. The principal components describe the main modes of variation in the training set and the eigenvalues the amount of variance explained by each mode. A critical issue in such landmark-based models is the requirement of point correspondence: each landmark should correspond to the same anatomical location in all the training samples.

For 3-D cardiac modeling applied to MR image analysis, three classes of landmark-based approaches have been described.

Point distribution models (PDMs) represent the shape model described earlier, without image matching. Frangi et al. [217] build a PDM using nonrigid registration, where both the LV and RV are included in the model. Point correspondences are defined by nonrigidly registering the training samples (represented as labeled volumes) to a shape average computed through rigid registration. By defining a point sampling for the shape average, and inverting the nonrigid deformation, this sampling can be propagated to each individual shape. Subsequent computation of an average shape and PCA eigenvariations is identical to 2-D PDMs. Recently this technique has been applied by Ordas et al. [218] to a large database of dynamic shapes using grid computing techniques. As yet, this model has not been applied to segmentation. McLeish et al. [219] use a 3-D point distribution model to study the motion and deformation of the heart as a result of breathing. Models are constructed for a single subject, where different shape samples represent the heart shape in different inspiration levels for the same subject. Because this method tracks the heart using nonrigid registration, point correspondence is achieved by propagating a set of landmarks, similar to Frangi's approach. This yields eigenmodes per subject that characterize the motion and deformation of the heart during breathing.

Active Shape Models (ASMs) consist of a PDM, extended with a matching scheme driven by information from the target image data, enabling statistically constrained image segmentation. ASMs use a gray-level model of scan lines perpendicular to the model contour or surface to estimate new update positions for each landmark point. Alternatively, update points can be generated by an edge detector or a pixel classification approach. The differences between the cloud of candidate sample points and the model points are used for model alignment and deformation in each iteration. The model deformation is restricted to the modes of variation of the PDM. Van Assen et al. [152,153] describe an ASM built using an application-specific point correspondence based on resampling the contours of the LV to a fixed number of slices and radially spaced in-plane landmarks. The matching mechanism generates update positions using a dynamic, unsupervised tissue classification based on fuzzy clustering. Intensities are sampled for each scan line and pooled. Subsequently, the clustering distinguishes different tissues as blood, myocardium, and air, and update points are inferred on the class transitions. Model training was performed on 53 data sets, whereas the model was tested on 9 data sets. Alternatively, Kaus et al. [220,154] describe an ASM-based approach, in which the matching mechanism is embedded in the internal energy term of an elastically deformable model. Training samples are manual segmentations expressed as binary volumes, and point correspondence is achieved by fitting a template mesh with a fixed-point topology to each binary training sample. Contrary to van Assen et al. [152,153], they separately model the endo- and epicardial shape. However, a coupling is realized by integrating connecting vertices between both surfaces and adding a connection term to the internal energy. In addition, they adopt a spatially varying feature model for each landmark. Training and testing were performed on end-diastolic (ED) and end-systolic (ES) image data from 121 subjects in a leave-one-out manner, and they report an average border-positioning error of the order of 1–2 voxels. This approach has the advantage that statistical shape constraints are imposed on the allowed elastic mesh deformation, while allowing for some flexibility to deviate from the trained shapes to accommodate for untrained shape variability.

The third type of landmark-based model is the *Active Appearance Model* (AAM). AAMs are an extension of ASMs with a statistical intensity model of a complete image volume, as opposed to merely scan lines in the ASM matching. An AAM is constructed by warping the voxel volume inside the training samples to the shape average. After intensity normalization to zero mean and unit variance, the intensity average and principal components are computed. A subsequent combined PCA on the shape and intensity model parameters yields a set of components that simultaneously capture shape and texture variability. AAM matching is based on minimizing a criterion expressing the difference between model intensities and the target image. This enables a rapid search for the correct model location during the matching stage of AAMs, while utilizing precalculated derivative images for the optimizable parameters. The sum of the squares of the difference between the model-generated patch and the underlying image serves as a criterion for model convergence. Mitchell et al. [170] developed a 3-D endo- and epicardial AAM and applied it to segmentation of cardiac MR studies. They applied an application-specific point correspondence identical to Van Assen et al. [152,153]. The model was trained and tested on 55 subjects, and border position errors from 2–3 mm were reported. Stegmann [172] further expanded the AAM to three dimensions and time; an AAM is described, in which 3-D LV models in ED and ES are coupled, enabling simultaneous detection in both frames. He also reports additional improvements: the integration of "whiskers" — surface scan lines pointing outward, where the intensity is included in the intensity model, greatly extending the lock-in range of the AAM. In addition, they have developed an automated correction for respiration-induced slice shifts, which corrupt the deformation statistics. This method is also 2-D AAM based. Correction for these slice shifts during training and matching yielded considerable improvements. Using a training and testing set of 12 subjects, they report on highly accurate estimates of ventricular volume and EF in a leave-one-out validation.

9.4.1.3 Implicitly Defined Deformable Models

Either in continuous or discrete form, the two preceding models were characterized by having an explicit surface parameterization. A surface model can also be defined by means of an implicit function. For instance, in the level-set approach, a model is obtained as the zero level set of a higher-dimensional embedding function. This technique, sometimes referred as *geodesic deformable models*, has been introduced independently by Caselles et al. [221] and Malladi et al. [222] based on the work by Osher and Sethian [223]. Geodesic deformable models have been applied by Yezzi et al. [156,157] to the segmentation of MR cardiac images. Later, Niessen et al. [159] extended the method to treat multiple objects and applied it to the segmentation of 3-D cardiac CT and MR images. Although geodesic models have the ability to handle changes in topology, unwanted and uncontrollable topological changes can occur in images with low-contrast edges or with boundary gaps because this is a purely data-driven approach.

There are other types of implicit models not related to level sets. Tseng, Hwang, and Sheehan [224], for instance, use an NN to define a continuous distance transform (CDT) to the LV boundary. A feedforward NN is trained to learn the distance function to the endocardial and epicardial contours using a few hand-segmented image slices. The surface of the LV is then represented as the zeroes of the distance function. The NN can generalize the boundaries of the LV in the slices not included in the training set, thus serving as an aid to segment a 3-D image for which the user has to provide the segmentation of a few slices only. Under an affine deformation model, the distance transform is used to match different temporal frames and to derive motion parameters. Wall thickness is computed by the centerline method [102] using two CDT NNs for describing the endo- and epicardial surfaces.

A third approach to implicit modeling is the use of surface primitives that are defined in implicit form. Lelieveldt et al. [160] segment thoracic 3-D MR images using hierarchical blending of hyperquadrics [225] and concepts of constructive solid geometry (CSG)[226]. The method provides an automatic, coarse segmentation of a multiple-object scene with little sensitivity to its initial placement. The most representative organs in the torso (lungs, heart, liver, spleen, and cardiac ventricles) are incorporated in the model, which can be hierarchically registered to the scanner coordinate system using only a few coronal, sagittal, and transversal survey slices. Owing to the contextual information present in the model, this sparse information has successfully been used to estimate the orientation of the long axis of the LV. This allows observer-independent planning of 3-D, long-axis acquisitions in patients [227]. This technique was not designed to estimate accurate cardiac functional parameters but can be used to generate a first initialization for more accurate algorithms.

9.4.2 VOLUMETRIC MODELS

As opposed to the plethora of surface representations, the use of volumetric models in the analysis and segmentation of cardiac images received little attention in the early years. However, several techniques have appeared in the literature in the last few years that specifically model the myocardium.

O'Donnell et al. [7,8] were the first to suggest a volumetric model to recover myocardial motion from MR tagging. The model, termed *hybrid volumetric ventriculoid*, can be decomposed into three parts: (a) a thick-walled superquadric, (b) a local offset either in nonparametric [7] or parametric [8] form, and (c) a local deformation in the form of a polyhedrization. The thick-walled superquadric represents a high-level abstraction model of the myocardium that is further refined by the local offsets. Altogether, these two parts constitute the rest model of the myocardium that is rigidly scaled to the dimensions of a new data set. The local deformation field is responsible for capturing the detailed shape variability of different data sets. Park et al. [5] have extended their LV surface model [6] to a superellipsoid model with parameter functions. The model is fitted to tagged MR images, providing a compact and comprehensive description of motion. Radial and longitudinal contraction, twisting, long-axis deformation, and global translation and rotation are readily available from the parameter functions. Alternatively, standard strain analysis can be carried out. It is also possible to estimate other volumetric parameters such as SV, CO, LVV, and LVM. In order to fit the model, a set of boundary points is manually delineated and a set of tags semiautomatically tracked along the cardiac cycle using the algorithm of Young et al. [81]. Therefore, the accuracy of all volumetric measurements depends on the manual outlining.

Haber, Metaxas, and Axel [164] have developed a model of biventricular geometry using FEs in a physics-based modeling context. The 3-D motion of the RV is analyzed by defining external forces derived from spatial modeling of magnetization (SPAMM) MR tagging data [165,166]. Recently, Hu, Metaxas, and Axel [168] have developed a biomechanically based image analysis framework for the estimation of the in vivo material properties and the stress and strain distributions in both ventricles. A similar aim has motivated Shi et al. [192] to develop a stochastic FE framework for the simultaneous estimation of kinematic and material parameters from the heart in vivo. Creswell et al. [161] and Pirolo et al. [162] describe a mathematical (biventricular) model of the heart built from 3-D MR scans of a canine specimen. Manual contour delineation of the epicardial, and LV and RV endocardial boundaries provides a set of points that is approximated with cubic nonuniform rational B-splines (NURBS) [228]. From this representation, a hexahedral FE model is built in order to generate a realistic geometric model for biomechanical analysis.

Shi et al. [167] have introduced an integrated framework for volumetric motion analysis. This work extends the surface model of Shi et al. [142] by combining surface motion extracted from MR magnitude images, and motion cues derived from MR phase-contrast (velocity) images. The latter provide motion information inside the myocardial wall but are known to be less accurate at the boundaries [94]. The two sources of motion evidence (boundary and midwall motion) are fused by solving the discretized material constitutive law of the myocardium, assuming a linear isotropic elastic material. In this framework, the measured boundary and midwall motion estimates at two consecutive frames are used as the boundary and the initial conditions of an FE element formulation. An advantage of this method with respect to physically based techniques is that material properties can be set based on experimental knowledge about myocardial mechanical properties, and not on a virtual mechanical analog, which usually leads to *ad hoc* parameter settings.

9.4.3 DEFORMATION MODELS

Hitherto, we have focused on representing either the endocardial (or epicardial) surface or the volume of the myocardial muscle. Tissue deformation, however, can be modeled without necessarily modeling the ventricular boundaries. To this end, material point correspondences in different temporal frames are required. These correspondences can be obtained by matching certain geometric properties over time (general techniques). If images are acquired using MR tagging technology, several other approaches can be applied that exploit the explicit correspondences inferable from tag displacements (MR tagging-based techniques).

9.4.3.1 General Techniques

Several techniques have been proposed in the literature for deformation recovery based on shape properties only. These methods are attractive because of their generality. On the other hand, one must be sure of the validity of the underlying assumptions or motion models before they are applied to analyze image sequences corresponding to normal and pathological myocardial motion patterns.

9.4.3.1.1 Continuous Models

Amini and Duncan [95] have developed a surface model based on the assumption of conformal motion, in which the angles between curves are preserved but not the distances between points. The LV surface is divided into locally quadric patches from which differential properties can be computed. Interframe patch correspondences are obtained using a metric that is minimal for conformal motion. An assumption made in this model is that the subdivision into surface patches and the number of neighboring patches visited during the matching process are sufficient to accommodate the largest stretching that can occur between frames. Bartels et al. [176,229] model material deformations with multidimensional splines. The method shares the properties of optical flow techniques to estimate motion fields. However, these approaches do not return an explicit model of the deformations (only displacements at discrete positions are provided). The main assumption of this technique is that, for a given material point, luminance is a conserved quantity. As in optic flow techniques, with only this assumption the solution remains underconstrained and, therefore, a regularization term must be added. Illustrations of the method applied to 2-D cardiac x-ray sequences are provided and the formulation is readily extended to 3-D sequences. However, it is questionable whether luminance conservation can provide a reliable cue for deformation recovery in regions with homogeneous intensity or in the presence of imaging artifacts and noise. For MR tagging, in particular, the approach must be adapted because luminance is not conserved owing to the physics of the imaging process [84].

Rueckert et al. [230] originally proposed the statistical deformation models (SDMs) with application to brain modeling, and they have since been applied to 3-D cardiac modeling as well. A model is constructed by registering several training sets using multilevel free-form deformations. Because registration is based on normalized mutual information, no expert segmentations in the form of labeled volumes, as in Frangi [217], are required. These free-form deformations are parameterized using a control point grid, and statistical analysis using PCA is performed on the control point sets, yielding an average deformation and principal components. Lötjönen et al. [193,231] constructed a four-chamber model of the heart using SDMs. To more accurately represent the basal and apical level, they generated training samples by nonrigidly registering a triangulated surface template to short- and long-axis image volumes in an alternating manner. Prior to training, the long- and short-axis data are corrected for patient motion by sequentially shifting each slice while maximizing the normalized mutual information between the data. Subsequently the models from the different subjects are registered, and an SDM is computed from the registration control points. In addition, an average intensity model is generated. This enables application to segmentation by nonrigid registration of the intensity template to the image data, while constraining the deformations to statistically trained limits. Model training and segmentation tests were performed on MR data from 25 subjects in a leave-one-out manner. In the same paper, two other types of models are compared on segmentation performance: a landmark probability distribution model and a probabilistic surface atlas. These models are applied to segmentation by adding a model term to the normalized mutual information measure, and performing nonrigid registration by maximizing the combined measure. This comparison indicates that the probability-based models work better than the SDMs, mainly because of overconstraining of the statistical models with a limited training set.

Lorenzo-Valdes et al. [151,232] present a probabilistic approach to cardiac modeling. A probabilistic cardiac atlas of the left and right ventricle is constructed from a set of manual segmentations as follows: First, the cardiac cycle is phase-normalized to a fixed number of frames. Subsequently, the manual segmentations are rigidly registered to one reference subject. Probabilistic maps are generated by blurring the segmented structure for each image and averaging over all subjects. The model is applied to segmentation using expectation maximization. They evaluated the model on 14 normal subjects and 10 patients with LV hypertrophy, and demonstrated that by blurring the normal-trained model, it can be generalized to accommodate for the pathological shape variations in patients with LV hypertrophy.

Perperidis et al. [188,189] proposed a registration-based approach to recover cardiac deformation. They use a 4-D FFD, which couples space and time. In this way, they are able to correct for differences in heart rate between a reference subject and the subject under analysis, or for differences in acquisition parameters.

9.4.3.1.2 Discrete Models

Benayoun and Ayache [99] propose an adaptive mesh model to estimate nonrigid motion in 3-D image sequences. The size of the mesh is locally adapted to the magnitude of the gradient where the most relevant information is supposed to appear (e.g., cardiac walls). Mesh adaptation is carried out at the first frame only; subsequent frames only deform the mesh to recover motion. The underlying hypothesis is that the deformation is small. Meshes at two time instants are registered through an energy-minimizing approach matching differential image properties (curvature and gradient). Papademetris et al. [190,191] have proposed a deformation model inspired by continuum mechanics. The method recovers a dense deformation field using point correspondences obtained with the point-tracking algorithm of Shi et al. [142]. Regularization is accomplished by measuring the internal energy of the myocardial tissue assuming a linear elastic body model. This is equivalent to a regularization term on the strain tensor space and not on the displacement field.*Anisotropy of the fibrous structure of the LV is accounted for in the internal energy by making the model stiffer in the fiber direction [233].

Recently, Sermesant et al. [171] presented a technique for the integration of information from multiple modalities into a biomechanical model of the heart. Their representation is based on a tetrahedral mesh for the myocardium of both left and RVs. The method registers the model to multimodal image data by using a hierarchical registration technique based on a modification of the ICP algorithm to intensity and gradient features. The model incorporates both anatomical and functional information that is inherited from the imaging sources: fiber orientation from diffusion tensor imaging and anatomical labels from the visible human project. This model has been applied for the segmentation of SPECT and MR image sequences. More recently the authors incorporated electrical information into the template to generate an electromechanically coupled model [234].

Finally, Klein and Huesman [183] developed a 4-D deformable model for motion compensation in dynamic cardiac PET images. The technique uses temporal continuity and a consistency constraint to ensure that the motion between two distant frames is consistent with that of two consecutive frames. The method also uses a nonuniform elastic material model to obtain better motion estimates.

9.4.3.2 MR Tagging-Based Techniques

The introduction of MR tagging has stimulated researchers to develop models of cardiac tissue deformation. Compared to motion recovery based on point correspondences or optic flow, MR tagging has the advantage that, in principle, material point correspondences can be estimated from tag information. In this subsection, different approaches for modeling the deformation fields are reviewed. Accurate tag localization is a prerequisite for subsequent deformation recovery and, therefore, it is closely related to deformation models. A brief overview of tag-tracking techniques is given in Appendix B.

9.4.3.2.1 Continuous Models

Several approaches have been proposed in which the parameterization of the deformation field is a continuous function. The availability of continuous deformation maps allows computation of local strains. Young et al., for instance,

---

*For a survey of optic flow methods in computer vision, see Beauchemin and Barron [90].

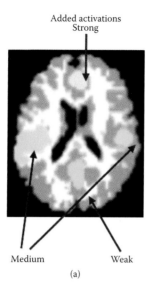

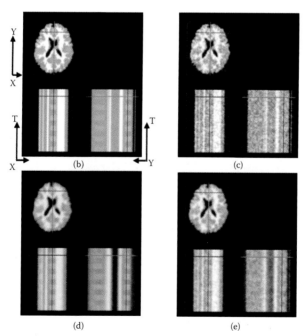

COLOR FIGURE 18.4 Simulated 2-dimensional time series (2-D + T data), visualized with 3 orthogonal slices (spatial axes: X, Y; time axis: T). (a) Added activations on a 2-D brain slice; (b) Ground truth data; (c) Simulated noisy data, with noise level $N(0, 30^2)$; (d) Restored data by the ST-SVR (*W-model* = 1); (e) Gaussian-smoothed data with Gaussian standard deviation 0.5 (s.t.d. = 0.5).

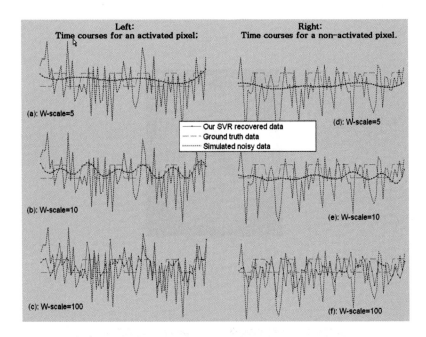

COLOR FIGURE 18.5 Effects on time course with varying *W-scale* for an activated pixel and a nonactivated pixel in the ST-SVR approach (*W-model* = 0).

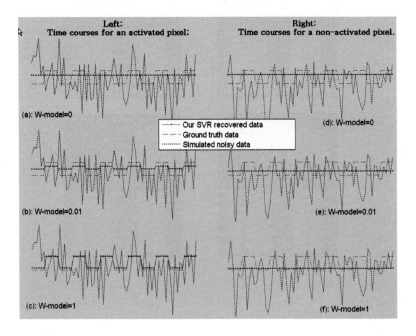

COLOR FIGURE 18.6 Effects on time course with varying *W-model* for an activated pixel and a nonactivated pixel in the ST-SVR approach (*W-scale* = 0).

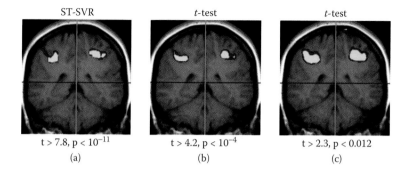

ST-SVR   *t*-test   *t*-test

t > 7.8, p < 10⁻¹¹   t > 4.2, p < 10⁻⁴   t > 2.3, p < 0.012

(a)   (b)   (c)

COLOR FIGURE 18.7 Comparison of ST-SVR and *t*-test for real fMRI data from a visuospatial task (color activation maps).

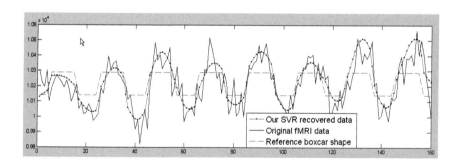

COLOR FIGURE 18.8 Time courses of an activated voxel for the real fMRI data in Figure 18.7. (Horizontal axis: temporal frame index; vertical axis: fMRI data intensity).

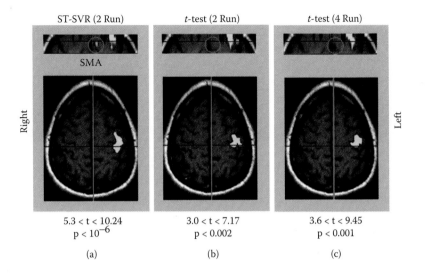

ST-SVR (2 Run) t-test (2 Run) t-test (4 Run)

SMA

Right   Left

$5.3 < t < 10.24$ $3.0 < t < 7.17$ $3.6 < t < 9.45$

$p < 10^{-6}$ $p < 0.002$ $p < 0.001$

(a)   (b)   (c)

COLOR FIGURE 18.9 Multirun result comparison for a motor experiment of real fMRI data. Top: coronal slices; bottom: axial slices (color activation maps).

developed a model-based approach for tracking tag intersections [77] and tag stripes [81]* that has been validated using silicone gel phantoms [237]. A deformation field that maps the first (undeformed) frame to a subsequent (deformed) frame is modeled through a piecewise polynomial function. Two fitting steps are involved in this method. First, the material points (tag intersections or stripes) in each deformed frame, $t > 0$, are reconstructed in the coordinate system of the undeformed state, $t = 0$ (reconstruction fit). In the latter frame, tag surfaces are arranged in true planes because no motion has occurred as yet. In the second step, the material points for $t > 0$, expressed in the reference frame ($t = 0$), are used to reconstruct a displacement field relative to $t = 0$ (*deformation fit*).**

A similar approach is followed by O'Dell et al. [177]. One-dimensional displacements are obtained by three independent sets of tag lines: one in the cardiac long axis and two orthogonal sets in the short-axis view. Reconstruction of the deformation field is performed in two interpolation steps. The first step assumes a global affine transformation between two time frames. This is done to eliminate global bulk motion, and linear stretches and shear. In the second step, the residual deformation is interpolated using a prolate spheroidal decomposition to describe the curvilinear deformations expected in the heart.

Both Young et al. [77,81] and O'Dell et al. [177] assume that the reference frame to which the strain analysis is related is the undeformed state. This is normally the first frame in the sequence (planar tag surfaces). Although this simplifies the problem by allowing decoupling of the motion component normal to the tagging plane, these methods cannot be used to compute strains between two arbitrary frames. The latter can be useful in order to retrospectively select the reference frame to coincide precisely with the diastole or systole, or to compute strains over a subset of the cardiac cycle. To circumvent this limitation, Moulton et al. [178] have proposed a Lagrangian approach that explicitly computes the intersection of the tag surfaces in two arbitrary frames. Tag surfaces are obtained by interpolating the tag curves that are stacked in different imaging planes. Surface intersections define a set of material lines for each time frame. These points were used to perform strain calculations employing a p-version of FE basis functions.

Radeva, Amini, and Huang [179] use two coupled volumetric models: a tissue deformation field and a model describing the LV geometry. The first model is represented by a cubic trivariate B-spline (termed B-solid by the authors); the second model is represented by two coupled surfaces (endocardium and epicardium) fitted to boundary points. It is assumed that the boundaries are either manually delineated or (semi)automatically detected from the tagged images. The B-solid is deformed under thin-plate internal constraints and under two external forces.

---

* Related regularization schemes are the global and body-smoothing terms described in Young and Axel [77] that act on the deformation gradient tensor. However, they are not directly interpretable as an internal deformation energy.

** Amini et al. [235] have compared landmark-based (tag intersections) with curved-based tag (stripes) tracking based on the simulator of Waks et al. [236]. They concluded that as the number of stripes or landmarks increases, the two methods give similar performances. Under large deformations, the degradation of the curve-based techniques is more graceful compared to landmark-based methods.

The first corresponds to tagging information: the isoparametric curves of the model are deformed to align with the tag strips. Simultaneously, the B-solid is attracted toward the LV boundaries by integrating a distance function to edge points on the epicardial and endocardial surfaces. Therefore, in this method, boundary and tag information are incorporated in a unified approach. Because this method has been applied in combination with short-axis tagged images only, it yields in-plane 2-D displacements. Later, Huang et al. [181] extended the method to analyze true 3-D deformations using a spatiotemporal model. The method differs from the one of Radeva et al. in that no boundary information is now incorporated. On the other hand, a spatiotemporal B-solid is constructed through a 4-D tensor product spline (three dimensions and time). The fitting process to SPAMM data is governed by a normal constraint that ensures that attraction produced by each tag plane is in its normal direction. Because multiple orthogonal tag planes are available, this allows a full 3-D reconstruction of the deformation field. A related work is that of Amini et al. [174], in which the same parametric B-splines are used to reconstruct the tag planes and track myocardial beads in three dimensions. More recently, Chen and Amini [182] have proposed a maximum *a posteriori* framework for tag line detection using oriented filters, which is used to recover the parameters of 3-D and 3-D + t deformable solids.

Kerwin and Prince [83] have developed an alternative projection technique to accurately estimate the 3-D, location of the intersection points of the tag grid. The deformation field between two frames is recovered using thin-plate spline interpolation. Myocardial points are distinguished from those in static tissues by checking whether they pass across the imaging plane over time. For points that do not fulfill the preceding criterion, a test is performed to check their inclusion within the outlined myocardial borders prior to rejection from the analysis. Such a rejection scheme is important for proper visualization and analysis of myocardial motion.

Young [180] introduced the concept of *model tags*, which represent the material surfaces within the heart tissue that are tagged with magnetic saturation. Model tags are "attached" to the heart and deform with it. They are embedded within a 3-D FE model describing the geometry of the LV; this model is linear in the transmural direction and employs bicubic Hermite interpolation in the circumferential and longitudinal directions. Instead of finding the 3-D location of the tag plane intersections, this approach finds the intersections of the model tags with the imaging planes (model tag intersections or MTIs). The FE model is subsequently deformed so that the MTIs match the tag stripes in each image plane. Matching is carried out by a local search algorithm guided by an orientation filter. Additional mechanisms are incorporated to allow efficient user interaction and to correct for erroneous MTI matches.

Wierzbicki et al. [155] evaluated the application of nonrigid registration for tracking cardiac structures over the cardiac cycle. A manual segmentation of the epicardium, LV and RV, and atria in the end-diastotic (ED) frame is nonrigidly registered to the remainder of the sequence, and several registration similarity measures are evaluated in animal experiments and healthy volunteers. The mean-square difference similarity measure showed the best performance in conventional 4-D data sets, enabling a tracking error of about 1 mm.

Rao et al. [187] describe the construction of an average atlas for cardiac motion using SDMs. Motion fields are modeled from tagged MR data as follows: The ED phase is selected as a reference frame for each subject, and each subsequent time frame is registered to this patient-specific reference frame. Subsequently, the anatomical short-axis acquisitions are used to nonrigidly register each ED reference frame. Using these intersubject registrations, the temporal deformation fields for a patient can all be transformed in the same reference frame, effectively normalizing motion for patient-specific geometry. The model is based on data from nine healthy subjects. Chandrashekara et al. [186] further extended this model by applying a PCA on the deformation fields computing two models: one model sequence with separate models for each cardiac phase; this requires a phase normalization for the cardiac cycle. The second model pools all deformation fields from all phases. They apply the cardiac deformation model to tracking the myocardium in a tagged MR sequence using nonrigid registration, while constraining the deformations to the statistically trained deformations. Both model types performed comparably in a limited validation: training on nine healthy and testing on eight normal subjects.

A work related to MR tagging is the virtual tagging technique of Masood et al. [184]. Actually, this technique does not rely on tagged MR images but on phase-contrast velocity maps. However, the authors propose a technique by which they overlay artificial grids, whose control points are then deformed such that the difference between the induced deformation velocity and that of the actually measured MR data is minimum. Therefore, this technique allows for a regularized quantification of phase-contrast images using arbitrary virtual tag patterns.

9.4.3.2.2 Discrete Models

Moore et al. [78] use MR tagging to reconstruct the location of material points through the cardiac cycle by interpolating the positions of the tags from short- and long-axis image planes using an iterative point-tracking algorithm. Discrete tag locations are arranged in cuboid volume elements that are identified in the deformed and reference frames. For each element, a 3-D strain tensor is calculated using the generalized inverse method [238]. Because the strain analysis is performed on a coarse discrete grid, only average strains can be retrieved. The tag-tracking procedure of this method compensates for through-plane motion. An important conclusion from this work is that strain analysis can be largely influenced by through-plane motion if it is not corrected for.

Denney and Prince [82] employ a multidimensional stochastic approach to obtain a dense discrete model of the displacement field from a sparse set of noisy measurements (tag displacements). The displacement field is constrained to be smooth and incompressible (isochoric deformation). This formulation leads to a partial stochastic model of the deformation field that can be solved using Fisher's estimation framework [239].

9.5 DISCUSSION

Comparison of the performance of different techniques is a difficult task due to the diversity of approaches, the different or complementary information obtained from them, the varying imaging modalities and image acquisition protocols, and, last but not least, the lack of a standard way to report performance. In order to draw some comparative conclusions, we have classified the existing methodologies according to the degree of their validation (Subsection 9.5.1). At the same time, we have introduced a number of performance criteria (Subsection 9.5.2). In this comparison we have focused on techniques leading to traditional cardiac indices, *viz*, global (Subsection 9.3.1) and motion parameters (Subsection 9.3.2). Table 9.2 summarizes this discussion.

9.5.1 VALIDATION

Three main groups of papers can be distinguished: (1) those with no evaluation or only qualitative illustrations, (2) those with quantitative evaluation on nonhuman data sets, and (3) those with quantitative evaluation on human data sets. This classification has been used in constructing Table 9.2.

Although there are always exceptions confirming the rule, Table 9.2 indicates several trends. Most papers in the first category correspond to articles presenting technical or methodological aspects of advanced modeling techniques. The results sections in these papers are restricted to either technical aspects or proof-of-concept illustration on realistic images, indicating the potential of the technique. Only a few of them have had follow-up articles confirming those hypotheses in large studies. Further evaluation of these techniques is required in order to determine their usefulness in clinical tasks.

Approaches in the second category are numerous. Methodologies in this category have been evaluated on simulated images or in phantom experiments. These have the advantage of providing the ground truth to assess the accuracy and reproducibility of the techniques. Owing to the use of idealized geometries and measurement conditions, extrapolation of the results to *in vivo* human studies remains to be demonstrated. Some papers in this second category have evaluated their techniques on *ex vivo* or *in vivo* animal models. Several researchers have reported experiments with dogs [99,110,115,117,120–122,137,140,142,159,162], swine [110,117,121,178], or calfs [113,241,242].* Only a few studies have compared measurements, obtained from *ex vivo* [241,242] or *in vivo* [110,142,243, 248] animal studies, with other standard-of-reference techniques.

---

* Both fitting steps handle sparse data and, therefore, regularization is needed. Regularization, however, is known to introduce artifactual strains. The effect of three regularization terms has been studied in [77]: (i) a thin-plate spline stabilizer, (ii) a global smoothing regularizer minimizing the deformation gradient tensor, F, and (iii) a local body regularizer minimizing the deformation gradient tensor expressed in some natural local coordinate system (e.g., aligned in circumferential, longitudinal, and radial directions). Based on simulations of an axis-symmetric deformation of a thick-walled incompressible cylinder, it was shown that all three constraints yield similar results in the strain analysis.

MR tagging techniques for reconstruction of myocardial motion or tissue deformation deserve separate attention. Most *in vivo* animal and human studies have reported on the Monte Carlo analysis of sensitivity to errors in tag localization and tracking, and on the ability to recover the location of tags in different frames [78,83,177,178,240].* Several models have been used in the literature to benchmark the accuracy of motion and deformation recovery. These evaluations were based, for instance, on spherical and cylindrical models of cardiac motion [69,78,82,177] FE solutions with realistic geometries [178], artificially generated motion trajectories [122], or synthetic images using the cardiac motion simulator [83,181,250] developed by Waks, Prince, and Douglas [236] that builds upon the kinematic model of Arts et al. [251]. A study was carried out by Declerck et al. [252] that thoroughly compared four techniques [82,177,253,254] for motion tracking from tagged MR. This paper provides results on normal and pathological subjects. Although the general trends of motion were captured correctly by all methods, this study shows that there are noticeable differences in the displacement and strain computations provided by each technique.

Finally, the third category includes studies that reported application on human volunteers and patients, including studies that provided quantitative results in terms of cardiac functional parameters. The size of the populations in most of these studies was small: with only three exceptions, all studies were conducted on less than a dozen volunteers or patients.

9.5.2 PERFORMANCE CRITERIA

In the following subsections we elaborate on the criteria that we have used to compare the different methods.

9.5.2.1 Model Complexity or Flexibility

The complexity or flexibility of a technique has been categorized in four groups according to the number of DoF or parameters involved.** The four categories are: (1) Compact models with only a few parameters (on the order of a dozen) of which prototypical examples are superquadrics, (2) flexible models with large number of DOFs and parameterized with global-support basis functions, of which representative examples are harmonic parameterizations of several types, (3) flexible models with large number of DOFs and parameterized with local-support basis functions, members of which family are B-spline and polyhedral models,

---

* Remarkably, a large number of evaluations involving canine models have been acquired with the dynamic spatial reconstructor. However, the reduced clinical availability of this technique and its specific image properties make it difficult to extrapolate the results of the evaluation to other clinical imaging techniques.

** Validation MR tagging itself for describing tissue deformation has been addressed by Young et al. [249] using a silicone gel phantom. Strains derived from MR tagging were compared to the analytic equilibrium strains under a Mooney-Rivlin material law.

and (4) flexible hierarchical models encompassing a reduced set of DOFs coarsely describing shape, plus an extended set of DOFs giving extra flexibility to the model. Representative of this family are superquadrics with free-form deformations. Complexity is, to some extent, related to the computational demands of an algorithm. Highly flexible algorithms are usually related to higher computation time for deforming them to a given image data set.* On the other hand, it is also a measure of the ability of a modeling technique to accommodate for fine shape details.

Although idealized models of ventricular geometry (mainly ellipsoids or ellipsoidal shells) are appealing for their parsimony and for historical reasons, Table 9.1 shows that no study has quantitatively demonstrated their accuracy in computing simple measurements such as LVV and EF. Compact models have developed in two different directions. On the one hand, in particular for the RV, some researchers have evaluated combinations of simple models that roughly derive RVV from a small number of linear measurements [113,117]. The models, however, remain highly constrained and have been tested on *ex vivo* casts experiments only. A second direction has been to trade off the compactness of the superquadric models and their flexibility without the need of hierarchical decompositions [5,6]. In this manner, flexibility is added in an elegant way by which each parameter function has an interpretation in terms of local and global shape changes. Park et al. [163] have demonstrated the application of this technique with a cascade of SPAMM sequences that allow for motion analysis in the whole cardiac sequence, overcoming some of the limitations of tag fading in MR.

Most approaches that reached the stage of quantitative evaluation are based on flexible or hierarchical representations. Both present challenges and advantages. Flexible representations (e.g., polyhedral meshes or harmonic decompositions) are highly versatile and can accommodate detailed shape variations. Most of the quantitative evaluation studies have been reported on local flexible models, most of which are able to cope even with complex topologies. On the other hand, restricting the space of possible shapes is usually difficult or requires substantial manual intervention or guidance [107,108,110,169]. Hierarchical or top-down approaches aim at a reduction in computational time and at improving robustness by incrementally unconstraining the space of allowed shape variation [7–9,35, 114,116]. One weak point of hierarchical approaches is the need for *ad hoc* scheduling mechanisms to determine when one level in the representation hierarchy should be fixed and a new level added, and up to which level the model should be refined. Furthermore, optimization procedures involved in the recovery of hierarchical models have to be designed with particular care. It is unclear how it can be ensured that a succession of optimizations at different modeling levels actually leads to optimum global deformation. Also, the questions arise regarding how to link different levels of model detail with the resolution of the underlying

---

* Here we disregard the obvious rigid transformation parameters to instantiate the model in world coordinates.

image data, and how to interact with the models if, after all, manual editing is required. Still, hierarchical model representation is an active and challenging field in 3-D medical image segmentation research in which several investigators have presented encouraging results in cardiac [7–9,116,122,135,136] and thorax modeling [148,149,160,231].

9.5.2.2 Robustness and Effective Automation

Processing prior to model recovery, automation of the recovery algorithm itself, and the presence of *ad hoc* parameters are factors that determine the robustness of a technique and its *effective automation*. By effective automation we refer to the automation of the overall approach, from raw images until the presentation of the functional parameters.

Before a given model can be fitted or deformed to a data set, almost every technique requires some type of preprocessing to convert the raw gray-level images into a representation suitable for shape recovery. Section 9.4 has suggested a classification of types of input data. For the sake of simplicity, Table 9.2 only indicates the degree of manual involvement required to obtain the corresponding input data. Four categories were considered: no preprocessing required (N), manual initialization of landmarks or models (I), (semi) automated initialization of landmarks and models integrated into the technique (A), and fully manual segmentation of landmarks and contours (M). Although variability inherent to the preprocessing can have a marked effect on the overall performance of a technique, this factor is usually disregarded in the evaluation of algorithms. A remarkable exception is the evaluation of MR tag-tracking algorithms using Monte Carlo analysis to assess the influence of erroneous tag localization in the recovery of tissue deformation [78,83,177,178,240]. Model initialization is also related to the issue of preprocessing. Although a few techniques make explicit mention of the procedure required to initialize the model (cf., e.g., [5,6,81,145, 159,160]), model initialization in a 3-D environment can be nontrivial or may require expert guidance.

Another factor undermining the robustness and reliability of a technique is the presence of *ad hoc* parameters that have to be set by the user. This can be particularly problematic when such parameters are highly dependent on a given data set. This is a known problem, for instance, of many physics-based deformable models for which several weights must be tuned to balance the smoothing constraints to the external energy terms. However, in the literature, analysis of sensitivity of the result to the weighting parameters is mostly missing. In Table 9.1, we have classified the different techniques into two categories according to the presence of user-defined *ad hoc* parameters: no parameters or parameters with corresponding analysis of sensitivity (−) and parameters for which no sensitivity analysis was performed (+). The fact that several methods do not present *ad hoc* parameters (−) should not be confounded with overall robustness. Even within the approaches with quantitative evaluation, many papers in the (−) category either require substantial preprocessing [9,27,28,77,177,178,190] or human guidance [113,107,108,117,

129,130,224]. Both factors influence the robustness and reproducibility of the derived functional information.

Finally, Table 9.2 also indicates the degree of user guidance (automation) of the fitting procedures for any given input data (preprocessing) and set of *ad hoc* parameters. Three degrees of automation were used to classify the approaches: relying on substantial human guidance (=), manual interaction can be necessary for guiding or correcting the deformation (−), and fully automated (+). In general terms, the larger the need for human intervention during the fitting procedure, the less robust a technique, and the more prone it is to inter- and intraobserver variability of the final results.

9.6 CONCLUSIONS AND SUGGESTIONS FOR FUTURE RESEARCH

In this chapter we have reviewed the techniques for 3-D geometric modeling and analysis of cardiac images. In particular, we have focused on those techniques leading to traditional indices of cardiac function. We have proposed a systematic classification of the approaches based on the type of representation of the geometric model, and the type of input data required for model recovery (Table 9.1). Furthermore, we have given a critical assessment of these approaches according to the type of functional parameters that they provide, their degree of evaluation, and the performance achieved in terms of modeling flexibility, complexity, and effective automation (Table 9.2).

From the surveyed literature, four main lines of future efforts can be distinguished:

1. Research on modeling and model deformation techniques: The last two decades have witnessed an enormous amount of work on 3-D models of the LV and RV. This holds true for all imaging modalities (cf. Table 9.2). In spite of the large number of attempts, no approach has simultaneously achieved robustness, automation, model flexibility, and computational speed. Manual outlining and analysis of cardiac images is still the most popular technique in clinical environments.

   Several issues will require more attention in order to integrate the advances of modeling techniques into clinical practice. Accurate 3-D modeling techniques are, in general, computationally intensive. Exploration of flexible modeling techniques that make efficient use of their DOF will be worthy of further research. So far the main flow of efforts has been focused on adopting generic geometrical representations to build cardiac shape models (e.g., superquadrics, B-splines, polyhedral meshes, Fourier descriptors, etc.). As a consequence, in generating a realistic LV shape, the representations are either too restrictive or require a considerable number of parameters. The question arises as to how to infer a compact representation giving rise to realistic shapes, possibly learned from examples. In this area, several recent efforts have shown promising results [146,152,153,172, 185,186,193,217,231] and new techniques are likely to continue to appear in the near future. These modeling approaches go from shape examples to a specific shape representation; they can reduce computational demands and improve robustness. A small number of efficiently selected model parameters reduces the dimensionality of the model recovery problem, and naturally constrains its results owing to model specificity.

Further investigation of suitable image features will be needed to improve shape recovery. In particular, incorporation of domain knowledge about the type of image modality (and acquisition protocols) can play an important role in increasing the accuracy of shape recovery techniques. In this area the use of image registration techniques is assuming increasing importance in cardiac image analysis [255], given that it facilitates the fusion of information from multiple modalities into a single model reference frame [171].

Most of the initial modeling techniques presented in this review were either purely geometric or inspired by a virtual physical analog (physics-based approaches). More recently, several papers have introduced the known biomechanical properties of the heart in the formulation of models that analyze cardiac images [167,171,190,191,243]. Further development of such approaches, and their application to segmentation tasks, can be a natural way of extending the ideas of physics-based methods and of relating some of the *ad hoc* parameters with the experimental evidence provided by biomechanics. Combination of other physical phenomena such as electromechanical coupling into image-based analysis has also been explored by some authors [234].

2. Research on interactive model-based segmentation: Table 9.2 supports the idea that model-based cardiac segmentation has not reached the status of being effectively automated because current techniques either require substantial expert guidance, *ad hoc* parameter fine-tuning, or nontrivial preprocessing. Although full automation is a desirable end goal, its difficulty has been acknowledged many times in the literature. There is a growing consensus that user interaction is, to some extent, unavoidable, and that it has to be considered as an integral part of the segmentation procedure. Therefore, development of efficient tools for 3-D interaction will play an important role in the near future. Being efficient entails the operator keeping control over the segmentation process to correct it or overrule its results where it has failed, with minimal and intuitive user interaction, and guiding the algorithm in abnormal situations (e.g., in dealing with a pathological case). Of course, the issue of reproducibility in cases of human intervention needs attention. Where well-defined repetitive tasks are recognized, or where a local user interaction can be extrapolated to a broader area,

the process should be automated, thus improving segmentation throughput and repeatability. How to devise such efficient and intuitive mechanisms for 3-D manipulation of models and volumetric data, and how to integrate them into the deformation of the models, remain topics for future research.

3. Research on functional cardiac descriptors: There are many shape and motion parameters other than traditional indices (cf. Appendix A). Unfortunately, although these new indices seem to provide either richer information or a more detailed analysis of cardiac function, their clinical evaluation has been very limited. As a consequence, it is difficult to determine their clinical relevance and the extra information provided with respect to traditional indices such as LVV, EF, etc. The lack of clinical evaluations may be related to the fact that advanced 3-D modeling techniques, from which these parameters can be derived, are computationally expensive and require considerable user intervention. The need of considerable pre- and postprocessing procedures, *ad hoc* parameter settings, and technical understanding of the modeling technique itself may explain why most of the described approaches are not available as stand-alone prototypes on which clinical studies can be carried out routinely.

   There is certainly room for development of novel shape and motion descriptors. However, there is an even larger need for evaluation of already-existing indices on reference data sets and large-scale clinical studies. It is remarkable that this lack of large-scale evaluation studies is present even in the case of many techniques aiming at the extraction of traditional functional parameters (Table 9.2).

   It is unrealistic to expect that every new technique proposed in the future will go through the process of a thorough clinical evaluation study. Unfortunately, many research institutes working on geometric modeling and shape analysis are not located in a clinical environment. Access to state-of-the-art image material and derived parameters for testing and benchmarking purposes is, therefore, difficult. In this respect, a public, common database of a representative set of images from different modalities would be highly beneficial. This database should establish a few standard data sets (both synthetic and clinical study cases) with as many independent measurements as possible of mass, SV, etc. Given the current speed of development in imaging modalities, such a database should be updated regularly to be representative of the state-of-the-art imaging technology.

4. Research into image-based in silico cardiac modeling and simulation: In recent years there has been a growing interest in research bridging the gap across different structural and functional levels in computational modeling and simulation of the cardiac system [256–260]. This field requires integration of research findings from traditional

physical and medical sciences together with the development in novel imaging systems, experimental and computational techniques, image and data analysis methods, and visualization tools that have emerged or matured more recently [261–264]. In our opinion, although some attempts have been made in this area, we will see in the near future more and more efforts directed toward integrating structural and functional information across scales and imaging modalities [265]. We believe that image-based cardiac modeling can play a fundamental role in this arena by providing the anatomical and functional information that is essential in personalized computational models for *in vivo* applications.

5. Multidisciplinary approaches: When imaging and modeling techniques get more complex, the interaction of clinicians, medical physicists, and technologists in a common environment becomes increasingly important. Several issues have to be addressed in a cooperative fashion: the interrelationship between image acquisition and cardiac modeling, the development of effective visualization techniques of 4-D data sets, realization of intuitive interfaces to interact with geometric models at the various stages of initialization, deformation, and eventual correction of results, and the concise transfer of clinical information from images and models to cardiologists.

It is to be expected that approval by clinicians of a model-based technique that provides functional parameters will depend on close collaboration between the technicians involved in image acquisition, the computer scientists devoted to the development of efficient modeling and model recovery techniques, and the cardiologists providing feedback about the desired information and display methods, the validity of the assumptions, and the design of evaluation studies.

NOMENCLATURE

n-D	*n*-dimensional, $n \in \{2, 3, 4\}$.
*n*DE	*n*-dimensional echocardiography
BA	Biplane angiography
c	Curvedness
CDT	Continuous distance transform
CFM	Color-flow (Doppler) mapping
CI	Cardiac index
CO	Cardiac output
CSG	Constructive solid geometry
CT	Computed tomography
CVD	Cardiovascular disease

DOF	Degrees of freedom
DSR	Dynamic spatial reconstructor
E	Green's strain tensor
EBCT	Electron beam computed tomography
EDV	End-diastolic volume
EF	Ejection fraction
ESV	End-systolic volume
F	Deformation gradient tensor
FE	Finite element
FFD	Free-form deformation
$\gamma(h, t)$	Shape spectrum.
GCG	Geometric cardiogram
GDT	Geometrically deformable template
H	Mean curvature
HARP	Harmonic phase
HR	Heart rate
ICP	Iterative closest point (algorithm)
K	Gaussian curvature
k_i, k_2	Principal curvatures
KLT	Karhunen-Loeve transform.
LV	Left ventricle
LVM	Left-ventricular mass
LVV	Left-ventricular volume
MF	Wall/tissue motion field
MRI	Magnetic resonance imaging
MSCT	Multi-slice computed tomography
MTI	Model tag intersections
NN	Neural network
NURBS	Non uniform rational B-spline
PBM	Piecewise bilinear maps
RV	Right ventricle
RVV	Right-ventricular volume
s	Shape index
SA	Strain analysis
SPAMM	Spatial modulation of magnetization
SPECT	Single-photon emission computed tomography
SSPs	Similar shape patches
SV	Stroke volume
SVI	Stroke volume index
τ	Local stretching factor
US	Ultrasound (imaging)
WT	Wall thickening

APPENDIX A

NONTRADITIONAL SHAPE AND MOTION DESCRIPTORS

Three-dimensional model-based analysis of left-ventricular shape and motion has the potential of providing rich morphological and functional information. Current clinical assessment of cardiac function is based mainly on global parameters such as LVV and EF. However, several researchers have demonstrated in the past the importance of local functional indices such as WT and segmental motion analysis [102,266–268], and local curvature and shape [59–62] as potential cardiac indices. Unfortunately, most of these studies were based on 2-D imaging techniques. Although they can indicate major trends about cardiac shape, a 3-D analysis would be able to better account for the true cardiac geometry. In this section, we briefly summarize several new indices proposed in the literature for the description of shape and motion. Some of them have been presented as a by-product of a specific modeling technique, whereas others are easily computable from any model representation. Therefore, this distinction seems a natural classification.

GENERIC DESCRIPTORS

Mean and Gaussian Curvature

The principal curvatures (k_1 and k_2, respectively) measure the maximum and minimum bending of a regular surface. Rather than using principal curvatures, it is more common to use two derived quantities known as Gaussian ($K = k_1 k_2$) and mean ($H = (k_1 + k_2)/2$) curvatures. By analyzing the signs of the pair (K, H), it is possible to locally distinguish between eight surface types [269].

Friboulet et al. [97] have studied the distribution of the Gaussian curvature in the LV at different phases of the cardiac cycle. From this study it was concluded that this distribution remains structurally stable over time. Whereas the LV free wall provides rich and dense curvature information, the curvature at the septal wall is less suitable for establishing point correspondences. Similar findings were made by Sacks et al. [115] with respect to the RV free wall: the RV free wall has a relatively uniform distribution of principal curvatures, and the surface geometry of the RV free wall does not change significantly from end diastole to end systole.

Shape Index and Shape Spectrum

Although mean and Gaussian curvatures are related to the concept of curvedness, there still remains scale information in these shape descriptors. To overcome this problem, Clarysse et al. [270] have used the *shape index* (s) and *curvedness* (c), two parameters that were introduced by Koenderink and van Doorn [271] and are defined as follows:

$$s = \frac{2}{\pi} \tan^{-1} \frac{k_2 + k_1}{k_2 - k_1} \qquad (9.11)$$

$$c = \left(\frac{k_1^2 + k_2^2}{2} \right)^{\frac{1}{2}} \qquad (9.12)$$

whereas c is inversely proportional to the object size, s defines a continuous distribution of surface types ranging from cuplike umbilic ($s = -1$) to peaklike umbilic ($s = 1$) points. It can be shown that whereas the shape index is invariant by homothecy, the curvedness is not. In this way, shape information and size can be easily decoupled.

The shape spectrum [272], $\gamma (h, t)$, is a global shape index defined as the fractional area of the LV with shape index value h at time t

$$\gamma(h, t) = \frac{1}{A} \iint_s \delta(s(\mathbf{x}) - h) ds \qquad (9.13)$$

where $A = \iint_s dS$ is the total area of the surface S, dS is a small region around the point $\mathbf{x}$, and $\delta(\cdot)$ is the 1-D Dirac delta function. Cardiac deformation can be analyzed by tracking the shape index and curvedness of similar shape patches (SSPs) over time. SSPs are connected surface patches whose points have similar shape indices, i.e., the shape index falls within a given range $s \pm \Delta s$. Clarysse et al. have shown the potential applicability of these indices by analyzing phantoms of normal and diseased LVs. An LV model of dilated cardiomyopathy, and a model of an ischemic LV (both akinetic and hypokinetic in the left anterior coronary territory) were generated using 4-D spherical harmonics. The curvedness spectrum was significantly altered by both pathologies, even when they were localized (as in the ischemic models). Reduction of the global function in the dilated myocardium had no significant repercussion on the shape index spectra. This could be an indicator that this pathology mostly affects the magnitude of motion only. An alternative to global analysis is to track the curvature parameters in predetermined regions. Clarysse et al. tracked three reference points over time: the apex, a point in the anterior wall, and a point in the cup of the pillar anchor. Using the local temporal variation of the curvedness and shape index, it was possible to distinguish between the normal and the diseased model. A potential problem with this technique is the reliable tracking of SSPs. If local deformations are too large, the trace of points might be lost.

LOCAL STRETCHING

Mishra et al. [96] have presented a computational scheme to derive local epicardial stretching under conformal motion. In conformal motion, it is assumed that motion can be described by a spatially variant but locally isotropic strething factor. In particular, for any two corresponding patches before and after motion, P and $\overline{P}$, the local stretching factor, τ, can be computed from the change in

Gaussian curvature and a polynomial stretching model by means of the relationship

$$\overline{K} = \frac{K}{\tau^2} + \frac{(E, F, G, \tau, \tau_u, \tau_v, \tau_{uu}, \tau_{vv})}{\tau^2} \qquad (9.14)$$

where $f(\cdot)$ is the polynomial stretching model (linear or quadratic) [96], E, F, and G are the coefficients of the first fundamental form [273], and (u, v) are coordinates of a local parameterization of the surface patch. Mishra et al. [96] present a method to solve for τ in Equation 14 and show that the local epicardial stretching factors computed over the cardiac cycle follow a similar evolution to the temporal variation of the principal strains obtained by Young et al. [109] using strain analysis techniques.

MODEL-SPECIFIC SHAPE DESCRIPTORS

Geometrical Cardiogram (GCG)

Azhari et al. [274] describe a method for classification of normal and abnormal LV geometries by defining a geometrical cardiogram (GCG), a helical sampling of the LV geometry from apex to base [275]. The GCG at end systole and at end diastole are subsequently analyzed via a Karhunen–Loeve transform (KLT) to compress their information. A truncated set of the KLT basis vectors is used to project the GCG of individual patients into a lower-dimensional space, and the mean-square error between the projected and the original GCG is used to discriminate between a normal and an abnormal LV [276]. From this vectorial representation, LVV, EF [275], and WT [100] can also be computed.

Deformable Superquadric and Related Models

One of the first 3-D primitives used to model the LV was the superquadric. It is a natural extension of the simplified geometric models originally used in 2DE [14] and angiocardiography [10–13]. Along with three main axes indicating principal dimensions, the superquadric models can be provided with additional parametric deformations such as linear tapering and bending [9,118], free-form deformations [122], displacement fields [7,8], or parametric functions providing information about radial and longitudinal contraction, twisting motion, and deformation of the LV long axis [5,6] and wall thickness [6]. In particular, Park et al. [5,6] suggest resolution of deformation and motion into a few parametric functions that can be presented to the clinician in the form of simple plots. All these functions are either independent of the total LV volume (e.g., twisting) or can be normalized with respect to the dimensions of the LV (e.g., radial and longitudinal contraction). This allows interpatient comparisons of contraction and shape change.

Global Motion Analysis Based On Departure from an Affine Model

Friboulet et al. [132] modeled the LV using a polyhedral mesh at each frame of the cardiac cycle. The state of the LV was characterized by the center of gravity and the moments of inertia of the polyhedral mesh. The deformation between two frames was hypothesized to follow an affine model. By defining a metric to compare two different polyhedral representations, the authors were able to quantify the difference between the actual interframe deformation and the corresponding deformation derived from an affine motion model. Several parameters of global motion are then derived: the temporal variation of the longitudinal and transversal moments of inertia, and the proportion of total motion explained by the affine model. By means of case studies, it was demonstrated that these global indices are able to discriminate between normal (EF = 0.71) and highly diseased (EF = 0.1) LVs. On the other hand, the global nature of these indices precludes the quantification of localized, inhomogeneous dysfunction of the LV.

Motion Decomposition through Planispheric Transformation

Declerck et al. [123] have proposed a canonical decomposition of cardiac motion into three components: radial motion, twisting motion around the apicobasal axis, and long-axis shortening. This decomposition is achieved through a transformation of the Cartesian coordinates of the LV wall to a planispheric space. In this space, a 4-D transformation is defined that regularly and smoothly parameterizes the spatiotemporal variation of the LV wall. Because the canonical decomposition of motion can be directly obtained in the planispheric space, these descriptors also vary smoothly along the cardiac cycle. Finally, by tracking the position of material points over time in the planispheric space and subsequently mapping to Cartesian coordinates, it is possible to reconstruct their 3-D trajectories.

Modal Analysis: Deformation Spectrum

Nastar and Ayache have introduced the concept of the deformation spectrum [140], which can be applied within the framework of modal analysis [194]. The deformation spectrum is the graph representing the value of the modal amplitudes as a function of mode index. The deformation spectrum corresponding to the deformation between two image frames describes which modes are excited in order to deform one object into another. It also gives an indication of the strain energy [140] of the deformation. As a consequence, a pure rigid deformation has zero strain energy. Two deformations are said to be similar when the corresponding deformation fields are equivalent up to a rigid transformation. In order to measure the dissimilarity of two deformation fields, the lower-order modes related to rigid transformation are discarded. The difference of the deformation spectra so computed can be used to define a metric between shapes (e.g., the LV in two phases of the cardiac cycle) that can be applied to classify them into specific categories (e.g., normal or abnormal motion patterns). Finally, the amplitude of the different modes can be tracked over time. Using Fourier spectral analysis, Nastar and Ayache have shown that these modes are concentrated in a few low-frequency coefficients.

APPENDIX B

MR Tag Localization Techniques

Early attempts to model myocardial tissue deformation tracked tag grid intersections manually over time [77]. Other researchers [78, 82, 83, 165, 177, 240] have used semiautomatic tools [277–279], based on snakes, to locate and track tag intersections and to define myocardial contours. Although they still require user interaction, these tools can speed up the manual procedure while reducing inter-observer variability [280].

Young et al. [81] propose an interactive scheme for tag tracking. The 2-D tag grid is modeled as a whole (active carpet). Separate manual segmentation of the LV boundaries is required to compute myocardial strains only. Tag tracking is performed using a modified snake [209] algorithm. Because tags show up in these images as dark lines (intensity valleys), the image intensity is used as external energy. Additionally interactive guidance is supported by introducing user-defined constraints. Only points in the myocardium mask are tracked in each frame, whereas carpet points outside the myocardium (inactive points) provide a weak form of continuity. Kraitchman et al. [237] have introduced an interactive method for tracking tag intersections. The method shares some of the features of the active carpet model of Young et al. [81]. The carpet of tag intersections is modeled as a mass-spring mesh of triangles. Tag intersections are tracked by means of correlation-based external energy and, eventually, by adding interactive constraints. Finally, this technique allows computation of average strains on the triangular patches. Another method for automatic tracking of the SPAMM grid has been presented by Kumar and Goldgof [79]. In the first frame, template matching is applied to provide an initial position for the tag grid. In this frame, the tag grid has a high contrast and a regular arrangement. In the subsequent frames, each line of the tag grid is independently tracked using a discrete thick snake with a width of two pixels (the typical tag width). The product of the image intensity in the two pixels is used as external energy to attract the snakes to the tag lines. Although these methods for extracting tag intersections can be useful for 3-D deformation analysis, in the original formulations, the methods proposed by Kumar and Goldgof [79], Young et al. [81], and Kraitchman et al. [237] have all been applied to 2-D strain analysis.

There are other approaches not based on snakes. Zhang et al. [281] decouple horizontal and vertical tag tracking via Fourier decomposition and spectral masking. In order to compensate for spectral cross-modulation from perpendicular lines, local histogram equalization is needed prior to spectral analysis.

Detection of tag lines is simplified in the preprocessed images, and a simple local search can then be used to track local intensity minima (tag lines) over time. Kerwin and Prince [250] have developed a method to simultaneously detect and track tag surfaces without the need for prior 2-D tag tracking. Tag surfaces are modeled using a kriging update model [282,283]. This model parameterizes tag surfaces and a global quadratic surface plus a local stochastic displacement. A recursive spatiotemporal scheme is developed that updates the kriging model. Measurements to update the model are obtained through a local search for tag lines. In this search a matched filter is employed, modeling the intensity profile across a tag line. Osman et al. [284,285] have introduced and evaluated a method for cardiac-motion tracking based on the concept of harmonic phase (HARP). The method uses isolated spectral peaks in the Fourier domain of MR tagged images as a cue for tag tracking. The inverse Fourier transform of a spectral peak is a complex image whose computed angle is called harmonic phase image. Osman et al. [286,284] show how this angle can be treated as a material property that can be related to myocardial strain. This technique has the advantage that it is fast, fully automatic, and provides dense material properties. So far the method has been applied to 2-D images and, thus, only provides information about "apparent motion." Osman and Prince [287] present several visualization techniques that can be used to display the information provided by HARP images.

REFERENCES

1. American Heart Association. (2003). *American Heart Association 2004 Heart and Stroke Statistical Update*. Dallas, Tex. http://www.americanheart.org.
2. European Society of Cardiology. (2002). *Cardiovascular Diseases in Europe 2002*. Dallas, Tex. *http://www.escardio.org*.
3. McInerney, T. and Terzopoulos, D. (1996). Deformable models in medical image analysis: a survey. *Med. Image. Anal.*, 1(2): 91–108.
4. Singh, A., Goldgof, D., and Terzopoulos, D. (Eds.). (1998). *Deformable Models in Medical Image Analysis*. Los Alamitos, CA: IEEE Computer Society Press.
5. Park, J., Metaxas, D., and Axel, L. (1996). Analysis of left ventricular wall motion based on volumetric deformable models and MRI-SPAMM. *Med. Image. Anal.* 1(1): 53–72.
6. Park, J., Metaxas, D., Young, A., and Axel, L. (1996). Deformable models with parameter functions for cardiac motion analysis from tagged MRI data. *IEEE Trans. Med. Imaging.* 15(3): 278–289.
7. O'Donnell, T., Gupta, A., and Boult, T. (1995). The hybrid volumetric ventriculoid: new model for MR-SPAMM 3-D analysis. in *IEEE Comput. in Cardiol.* 5–8. Vienna, Austria: IEEE Computer Society Press.
8. O'Donnell, T., Boult, T., and Gupta, A. (1996). Global models with parametric offsets as applied to cardiac motion recovery. in *Comput. Vis. Patt. Recogn.* 293–299. San Francisco, USA.
9. Bardinet, E., Cohen, L., and Ayache, N. (1998). A parametric deformable model to fit unstructured 3-D data. *Comput. Vis. Image Underst.* 71(1): 39–54.

10. Kennedy, J., Baxley, W., Figley, M., Dodge, H., and Blackmon, J. (1966). Quantitative angiocardiography: I. The normal left ventricle in man. *Circulation,* 34: 272–8.

11. Hermann, H. and Bartle, S. (1968). Left ventricular volumes by angiocardiography: comparison of methods and simplification of techniques. *Cardiovasc. Res.* 2(4): 404–14.

12. Davila, J. and Sanmarco, M. (1966). An analysis of the fit of mathematical models applicable to the measurement of left ventricular volume. *Am. J. Cardiol.* 18: 31–42.

13. Dodge, H., Sandler, H., Baxley, W., and Hawley, R. (1966). Usefulness and limitations of radiographic methods for determining left ventricular volume. *Am. J. Cardiol.* 18: 10–24.

14. Vuille, C. and Weyman, A. (1994). Left ventricle I: general considerations, assessment of chamber size and function. in A.E. Weyman, Ed. *Principles and Practice of Echocardiography,* 2nd ed. Lea and Febiger.

15. Frangi, A., Rueckert, D., and Duncan, J. (2002). Editorial on the special issue on three-dimensional cardiovascular image analysis. *IEEE Trans. Med. Imaging.* 21(9): 1005–10.

16. Frangi, A., Niessen, W., and Viergever, M. (2001). Three-dimensional modeling for functional analysis of cardiac images: a review. *IEEE Trans. Med. Imaging.* 20(1): 2–25.

17. Raphael, M. and Donaldson, R. (1998). The normal heart: methods of examination. in D. Sutton, Ed., *Textbook of Radiology and Imaging.* 6th ed., 541–65. Churchill Livingstone.

18. McCann, H., Sharp, J., Kinterr, T.M., McEwaninter, C., Barillot, C., and and Greenleaf, J. (1988). Multidimensional ultrasonic imaging for cardiology. *Proc. IEEE.* 76: 1063–1071.

19. Nelson, T., Downey, D., Pretorius, D., and Fenster, A. (1999). *Three-dimensional Ultrasound.* Philadelphia: Lippincott Williams and Wilkins.

20. Robb, R., Hoffman, E., Sinak, L., Harris, L., and Ritman, E. (1983). High-speed three-dimensional X-ray computed tomography: the dynamic spatial reconstructor. *Proc. IEEE.* 71: 308–319.

21. Boyd, D. and Lipton, M. (1983). Cardiac computed tomography. *Proc. IEEE.* 71: 298–307.

22. Higgins, C. (1986). Overview of MR of the heart–1986. *Am. J. Roentgenol.* 146(5): 907–918.

23. Wright, G. (1997). Magnetic resonance imaging. *IEEE Sign. Proc. Mag.* 56–66.

24. Quistgaard, J. (1997). Signal acquisition and processing in medical diagnostic ultrasound. *IEEE Sign. Proc. Mag.* 67–74.

25. Ariet, M., Geiser, E., Lupkiewicz, S., Conetta, D., and Conti, C. (1984). Evaluation of a three-dimensional reconstruction to compute left ventricular volume and mass. *Am. J. Cardiol.,* 54(3): 415–420.

26. Treece, G., Prager, R., Gee, A. V., and Berman, L. (1999). Volume measurements in sequential freehand 3-D ultrasound. in Kuba, A., Sámal, M., Todd-Pokropek, A. (Eds.). *IPMI,* Vol. 1613 of *Lecture Notes in Computer Science.* 70–83. Springer-Verlag.

27. Gopal, A., King, D., Katz, J., Boxt, L., King, D., and Shao, M. (1992). Three-dimensional echocardiographic volume computation by polyhedral surface reconstruction: *in vitro* validation and comparison to magnetic resonance imaging. *J. Am. Soc. Echocardiogr.* 5(2): 115–124.

28. Legget, M., Leotta, D., Bolson, E., McDonald, J., Martin, R., Li, X., Otto, C., and Sheehan, F. (1998). System for quantitative three-dimensional echocardiography of the left ventricle based on magnetic field position and orientation sensing system. *IEEE Trans. Biomed. Eng.* 45(4): 494–504.

29. Coppini, G., Poli, R., and Valli, G. (1995). Recovery of the 3-D shape of the left ventricle from echocardiographic images. *IEEE Trans. Med. Imaging.* 14(2): 301–317.

30. Stetten, G. and Pizer, S. (1999). Medial-node models to identify and measure objects in real-time 3-D echocardiography. *IEEE Trans. Med. Imaging.* 18(10): 1025–1034.

31. Maehle, J., Bjoernstad, K., Aakhus, S., Torp, H., and Angelsen, B. (1994). Three-dimensional echocardiography for quantitative left ventricular wall motion analysis: a method for reconstruction of endocardial surface and evaluation of regional dysfunction. *Echocardiogr.* 11(4): 397–408.

32. Germano, G., Erel, J., Kiat, H., Kavanagh, P., and Berman, D. (1997). Quantitative LVEF and qualitative regional function from gated 201-T1 perfusion SPECT. *J. Nucl. Med.* 38(5): 749–754.

33. Faber, T., Stokely, E., Templeton, G., Akers, M., Parkey, R., and Corbett, J. (1989). Quantification of three-dimensional left ventricular segmental wall motion and volumes from gated tomographic radionuclide ventriculograms. *J. Nucl. Med.* 30(5): 638–649.

34. Faber, T., Akers, M., Peshock, R., and Corbett, J. (1991). Three-dimensional motion and perfusion quantification in gated single-photon emission computed tomograms. *J. Nucl. Med.* 32(12): 2311–2317.

35. Germano, G., Kavanagh, P., and Berman, D. (1997). An automatic approach to the analysis, quantitation and review of perfusion and function from myocardial perfusion SPECT images. *Intl. J. Cardiac Imaging* 13(4): 337–346.

36. Boyd, D. and Haugland, C. (1993). Recent progress in electron beam tomography. *Med. Imag. Tech.* 11: 578–85.

37. Ohnesorge, B., Flohr, T., Becker, C., Knez, A., Kopp, A., K. F., and Reiser, M. (2000). Cardiac imaging with rapid, retrospective ECG synchronized multilevel spiral CT. *Der. Radiologe.* 40(2): 111–117. In German.

38. Klingenbeck-Regn, K., Schaller, S., Flohr, T., Ohnesorge, B., Kopp, A., and Baum, U. (1999). Subsecond multi-slice computed tomography: basics and applications. *Eur. J. Radiol.* 31(2): 110–124.

39. Nieman, K., van Geuns, R., Wielopolski, P., Pattynama, P., and de Feyter, P. (2002). Noninvasive coronary imaging in the new millennium: a comparison of computed tomography and magnetic resonance techniques. *Rev. Cardiovasc. Med.* 3(2): 77–84.

40. de Feyter, P., Mollet, N., Cadermartiri, F., Nieman, K., and Pattynama, P. (2003). MS-CT coronary imaging. *J. Interv. Cardiol.* 16(6): 465–468.

41. Ropers, D., Baum, U., Pohle, K., Ulzheimer, S., Ohnesorge, B., Schlundt, C., Bautz, W., Daniel, W., and Achenbach, S. (2003). Detection of coronary artery stenoses with thin multi-detector row spiral computed tomography and multiplanar reconstruction. *Circulation.* 107(5): 664–666.

42. Nieman, K., Cademartiri, F., Lemas, P., Raaijmakers, R., Pattynama, P., and de Feyter, P. (2002). Reliable noninvasive coronary angiography with fast submillimeter multislice spiral computed tomography. *Circulation.* 106(16): 2051–2054.

43. Higgins, C., Ed. (1999). Special issue: cardiovascular MRI, *J. Magn. Reson. Imaging.* Vol. 10(5).

44. Keller, A., Peshock, R., Malloy, C., Buja, L., Nunnally, R., Parkey, R., and Willerson, J. (1986). *In vivo* measurement of myocardial mass using nuclear magnetic resonance imaging. *J. Am. Coll. Cardiol.* 8(1): 113–117.

45. Cranney, G., Lotan, C., Dean, L., Baxley, W., Bouchard, A., and Pohost, G. (1990). Left ventricular volume measurement using cardiac axis nuclear magnetic resonance imaging. *Circulation.* 82(1): 154–163.

46. Dulce, M., Mostbeck, G., Friese, K., Caputo, G., and Higgings, C. (1993). Quantification of the left ventricular volumes and function with cine MR imaging: comparison of geometric models with three-dimensional data. *Radiology.* 188(2): 371–376.

47. van der Geest, R., de Roos, A., van der Wall, E., and Reiber, J. (1997). Quantitative analysis of cardiovascular MR images. *Intl. J. Cardiac Imaging.* 13(3): 247–258.

48. Lorenz, C., Walker, E., Morgan, V., Klein, S., and Graham, T. (1999). Normal human right and left ventricular mass, systolic function, and gender differences by cine magnetic resonance imaging. *J Cardiovasc Magn Res,* 1(1):7–21.

49. Marcus, J., Götte, M., de Waal, L., Stam, M., van der Geest, R., Heethaar, R., and van Rossum, A. (1999). The influence of through-plane motion on left ventricular volumes measured by magnetic resonance imaging: implications for image acquisition and analysis. *J. Cardiovasc. Magn. Res.* 1(1): 1–6.

50. Beyar, R., Shapiro, E., Graves, W., Rogers, W., Guier, W.H., Carey, G., Soulen, R., Zerhouni, E., Weisfeldt, M., and Weiss, J. (1990). Quantification and validation of left ventricular wall thickening by a three-dimensional volume element magnetic resonance imaging approach. *Circulation.* 81(1): 297–307.

51. McVeigh, E. (1996). MRI of myocardial function: motion tracking techniques. *Magn. Reson. Imaging.* 14(2): 137–150.

52. Clark, N., Reichek, N., Bergey, P., Hoffman, E., Brownson, D., Palmon, L., and Axel, L. (1991). Circumferential myocardial shortening in the normal human left ventricle: assessment by magnetic resonance imaging using spatial modulation of magnetization. *Circulation.* 84(1): 67–74.

53. Axel, L. and Dougherty, L. (1989). Heartwall motion: Improved method of spatial modulation of magnetization for MR imaging. *Radiology.* 172(2): 349–350.

54. Zerhouni, E., Parish, D., Rogers, W., Yang, A., and Shapiro, E. (1988). Human heart: tagging with MR imaging—a method for noninvasive assessment of myocardial motion. *Radiology.* 169(1): 59–63.

55. Edelman, R. (2004). Contrast-enhanced MR imaging of the heart: overview of the literature. *Radiology* 232(3): 653–668.

56. Kim, R., Wu, E., Rafael, A., Chen, E., Parker, M., Simonetti, O., Klocke, F., Bonow, R., and Judd, R. (2000). The use of contrast-enhanced magnetic resonance imaging to identify reversible myocardial dysfunction. *N. Engl. J. Med.* 343(20): 1445–1453.

57. Choi, K., Kim, R., Gubernikoff, G., Vargas, J., Parker, M., and Judd, R. (2001). Transmural extent of acute myocardial infarction predicts long-term improvement in contractile function. *Circulation.* 104(10): 1101–1107.

58. Weber, K., and Hawthorne, E. (1981). Descriptors and determinants of cardiac shape: an overview. *Fed. Proc.* 40(7): 2005–2010.

59. Mancini, G., De Boe, S., Anselmo, E., Simon, S., Le Free, M., and Vogel, R. (1987). Quantitative regional curvature analysis: an application of shape determination for the assessment of segmental left ventricular function in man. *Am. Heart. J.* 113(2 Pt 1): 326–334.

60. Mancini, G., DeBoe, S., McGillem, M., and Bates, E. (1988). Quantitative regional curvature analysis: a prospective evaluation of ventricular shape and wall motion measurements. *Am. Heart. J.* 116(6 Pt 1): 1611–1621.

61. Kass, D., Traill, T., Keating, M., Altieri, P., and Maughan, W. (1988). Abnormalities of dynamic ventricular shape change in patients with aortic and mitral valvular regurgitation: assessment by Fourier shape analysis and global geometric indexes. *Circ. Res.* 62(1): 127–38.

62. Barletta, G., Baroni, M., del Bene, R., Toso, A., and Fantini, F. (1998). Regional and temporal nonuniformity of shape and wall movement in the normal left ventricle. *Cardiology,* 90(3): 195–201.

63. Kondo, C., Caputo, G., Smelka, R., Foster, E., Shimakawa, A., and Higgins, C. (1991). Right and left ventricular stroke volume measurements with velocity-encoded cine MR imaging: *in vitro* and *in vivo* validation. *Am. J. Roentgenol.* 157(1): 9–16.

64. Hillis, L., Firth, B., and Winniford, M. (1986). Analysis of factors affecting the variability of Fick vs. indicator dilution measurements of cardiac output. *Am. J. Cardiol.* 56(12): 764–768.

65. McEachen, J.C., Neohorai, A., and Duncan, J.S. (1994). A recursive filter for temporal analysis of cardiac motion. in *Math. Meth. Biomed. Imag. Anal.* 124–133. Seattle, USA.

66. Potel, M., Rubin, J., MacKay, S., Aisen, A., Al-Sadir, J., and Sayre, R. (1983). Methods for evaluating cardiac wall motion in three dimensions using bifurcation points of the coronary arterial tree. *Inv. Rad.* 18(1): 47–57.

67. Kim, H., Min, B., Lee, M., Seo, J., Lee, Y., and Han, M. (1985). Estimation of local cardiac wall deformation and regional wall stress from biplane coronary cineangiograms. *IEEE Trans. Biomed. Eng.* 32(7): 503–512.

68. Chen, C. and Huang, T. (1990). Epicardial motion and deformation estimation from coronary artery bifurcation points. in *Intl. Conf. Comput. Vision.* 456–459. Osaka, Japan.

69. Young, A., Hunter, P., and Smaill, B. (1992). Estimation of epicardial strain using the motions of coronary bifurcations in biplane cineangiography. *IEEE Trans. Biomed. Eng.* 39(5): 526–531.

70. Harrison, D., Goldblatt, A., Braunwald, E., Glick, G., and Mason, D.T. (1963). Studies on cardiac dimensions in intact unanesthetized man. I. description of techniques and their validation. *Circ. Res.* 13: 448–455.

71. Rankin, J., McHale, P., Artentzen, C., Ling, D., Greenfield, J., and Anderson, R. (1976). The three-dimensional dynamic geometry of the left ventricle in the conscious dog. *Circ. Res.* 39(3): 304–313.

72. Brower, R., ten Katen, H., and Meester, G. (1978). Direct method for determining regional myocardial shortening after bypass surgery from radiopaque markers in man. *Am. J. Cardiol.* 41(7): 1222–1229.

73. Ingels, N., Daughters, G., Stinson, E., Alderman, E. (1980). Evaluation of methods for quantitating left ventricular segmental wall motion in man using myocardial markers as a standard. *Circulation.* 61(5): 966–972.

74. Meier, G., Bove, A., Santamore, W., and Lynch, P. (1980). Contractile function in canine right ventricle. *Am. J. Physiol.* 239(6): H794–804.

75. Arts, T., Hunter, W., Douglas, A., Muijtjens, A., Corsel, J., and Reneman, R. (1993). Macroscopic three-dimensional motion patterns of the left ventricle. *Adv. Exp. Med. Biol.* 346: 383–392.

76. Villarreal, F.L., Waldman, L.K., and Lew, W.Y.W. (1988). Technique for measuring regional two-dimensional finite strains in canine left ventricle. *Circ. Res.* 62(4): 711–721.

77. Young, A. and Axel, L. (1992). Three-dimensional motion and deformation of the heart wall: estimation with spatial modulation of magnetization — a model-based approach. *Radiology.* 185(1): 241–247.

78. Moore, C., O'Dell, W., McVeigh, E., and Zerhouni, E. (1992). Calculation of three-dimensional left ventricular strains form biplanar tagged MR images. *J. Magn. Reson. Imaging.* 2(2): 165–175.

79. Kumar, S., and Goldgof, D. (1994). Automatic tracking of SPAMM grid and the estimation of deformation parameters from cardiac MR images. *IEEE Trans. Med. Imaging.* 13(1): 122–132.

80. Fischer, S.E., McKinnon, G.C., Scheidegger, M. B., Prins, W., Meier, D., and Boesiger, P. (1994). True myocardial motion tracking. *Magn. Res. Med.* 31(4): 401–413.

81. Young, A., Kraitchman, D., Dougherty, L., and Axel, L. (1995). Tracking and finite element analysis of stripe deformation in magnetic resonance tagging. *IEEE Trans. Med. Imaging.* 14(3): 413–421.

82. Denney, T. and Prince, J. (1995). Reconstruction of 3-D left ventricular motion from planar tagged cardiac MR images: an estimation theoretic approach. *IEEE Trans. Med. Imaging.* 14(4): 1–11.

83. Kerwin, W. and Prince, J. (1998). Cardiac material markers from tagged MR images. *Med. Image. Anal.* 2(4): 339–353.

84. Prince, J. and McVeigh, E. (1992). Motion estimation from tagged MR image sequences. *IEEE Trans. Med. Imaging.* 11(2): 238–249.

85. Amartur, S. and Vesselle, H. (1993). A new approach to study cardiac motion: the optical flow of cine MR images. *Magn. Reson. Med.* 29(1): 59–67.

86. Tistarelli, M. and Marcennaro, G. (1994). Using optical flow to analyze the motion of human body organs from bioimages. in *Math. Meth. Biomed. Image. Anal.* 100–9. Seattle, USA.

87. Denney, T. and Prince J. (1994). Optimal brightness functions for optical flow estimation of deformable motion. *IEEE Trans. Image. Process.* 3(2): 178–191.

88. Denney, T. and Prince, J. (1995). A frequency domain performance analysis of Horn and Schunck's optical flow algorithm for deformable motion. *IEEE Trans. Image. Process.* 4(9): 1324–1327.

89. Dougherty, L., Asmuth, J., Blom, A., Axel, L., and Kumar, R. (1999). Validation of an optical flow method for tag displacement estimation. *IEEE Trans. Med. Imaging.* 18(4): 359–363.

90. Beauchemin, S. and Barron, J. (1995). The computation of optical flow. *ACM Comput. Surveys.* 27(3): 444–467.

91. Song, S. and Leahy, R. (1991). Computation of 3-D velocity fields from 3-D cine CT images of a human heart. *IEEE Trans. Med. Imaging.* 10(3): 295–306.

92. Song, S., Leahy, R., Boyd, D., Brundage, B., and Napel, S. (1994). Determining cardiac velocity fields and intraventricular pressure distribution from a sequence of ultrafast CT cardiac images. *IEEE Trans. Med. Imaging.* 14(2): 386–397.

93. Gorce, J.M., Friboulet, D., and Magnin, I. (1996). Estimation of three-dimensional cardiac velocity fields: assessment of a differential method and application to three-dimensional CT data. *Med. Image. Anal.* 1(3): 245–261.

94. Meyer, F., Constable, R., Sinusas, A., and Duncan, J. (1996). Tracking myocardial deformation using phase contrast MR velocity fields: a stochastic approach. *IEEE Trans. Med. Imaging.* 15(4): 453–465.

95. Amini, A. and Duncan, J. (1992). Bending and stretching models for LV wall motion analysis from curves and surfaces. *Image. Vis. Comp.* 10(6): 418–430.

96. Mishra, S., Goldgof, D., Huang, T., and Kambhamettu, C. (1992). Curvature-based non-rigid motion analysis from 3-D point correspondences. *Intl. J. Image Syst. Tech.* 4: 214–25.

97. Friboulet, D., Magnin, I., Mathieu, C., Pommert, A., and Hoehne, K. (1993). Assessment and visualization of the curvature of the left ventricle from 3-D medical images. *Comput. Med. Image and Graph.* 17(4–5):257–62.

98. Kambhamettu, C. and Goldgof, D. (1994). Curvature-based approach to point correspondence recovery in conformal non-rigid motion. *Comput. Vis. Graph. and Image Proc.* 60(1): 26–43.

99. Benayoun, S. and Ayache, N. (1998). Dense and non-rigid motion estimation in sequences of medical images using differential constraints. *Intl. J. Comput. Vision.* 26(1): 25–40.

100. Azhari, H., Sideman, S., Weiss, J., Shapiro, E., Weisfeldt, M., Graves, W., Rogers, W., and Beyar, R. (1990). Three-dimensional mapping of acute ischemic regions using MRI: wall thickening vs. motion analysis. *Am. J. Physiol.* 259(5 Pt 2): H1492–503.

101. Taratorin, A. and Sideman, S. (1995). 3-D functional mapping of left ventricular dynamics. *Comput. Med. Image and Graph.* 19(1): 113–129.

102. Sheehan, F., Bolson, E., Dodge, H., Mathey, D., Schofer, J., and Woo, H. (1986). Advantages and applications of the centerline method for characterizing regional ventricular function. *Circulation.* 74(2): 293–305.

103. Buller, V., van der Geest, R., Kool, M., and Reiber, J. (1994). Accurate three-dimensional wall thickness measurement from multi-slice short-axis MR imaging. in *IEEE Comput. in Cardiol.* 245–248. Maryland, USA: IEEE Computer Society Press.

104. Bolson, E. and Sheehan, F. (1993). Centersurface model for 3-D analysis of regional left biventricular function. In *IEEE Comput. in Cardiol.* 735–738. London, U.K.: IEEE Computer Society Press.

105. Bolson, E., Sheehan, F., Legget, M.E., Jin, H., McDonald, J.A., Sampson, P., Martin, R., Bashein, G., and Otto, C. (1995). Applying the centersurface model to 3-D reconstructions of the left ventricle for regional functional analysis. in *IEEE Comput. in. Cardiol.* 63–7. Vienna, Austria: IEEE Computer Society Press.

106. Fung, Y. (1990). *Biomechanics: Motion, Flow, Stress and Growth.* New York: Springer-Verlag.

107. Yettram, A. and Vinson, C. (1979). Geometric modeling of the human left ventricle. *J. Biomech.* 101: 221–223.

108. Yettram, A., Vinson, C., and Gibson, D. (1982). Computer modeling of the human left ventricle. *J. Biomech.* 104(2): 148–152.

109. Young, A., Hunter, P., and Smaill, B. (1989). Epicardial surface estimation from coronary angiograms. *Comput. Vis. Graph. and Image Process.* 47(1): 111–127.

110. Spinale, F., Carabello, B., and Crawford, F. (1990). Right ventricular function and three-dimensional modeling using computer-aided design. *J. Appl. Physiol.* 68(4): 1707–1716.

111. Pentland, A. and Horowitz, B. (1991). Recovery of nonrigid motion and structure. *IEEE Trans. Pattern Anal. Machine Intell.* 13(7): 730–742.

112. Cauvin, J., Boire, J., Zanca, M., Bonny, J., Maublant, J., and Veyre, A. (1993). 3-D modeling in myocardial 201TL SPECT. *Comput. Med. Image and Graph.* 17(4–5): 345–350.

113. Czegledy, F. and Katz, J. (1993). A new geometric description of the right ventricle. *J. Biomed. Eng.* 15(5): 387–391.

114. Gustavsson, T., Pascher, R., and Caidahl, K. (1993). Model-based dynamic 3-D reconstruction and display of the left ventricle from 2-D cross-sectional echocardiograms. *Comput. Med. Image and Graph.* 17(4–5): 273–278.

115. Sacks, M., Chuong, C., Templeton, G., and Peshock, R. (1993). Newblock *in vivo* 3-D reconstruction and geometric characterization of the right ventricular free wall. *Ann. Biomed. Eng.* 21(3): 263–275.

116. Chen, C., Huang, T., and Arrott, M. (1994). Modeling, analysis, and visualization of left ventricle shape, and motion by hierarchical decomposition. *IEEE Trans. Pattern Anal. Machine Intell.* 16(4): 342–356.

117. Denslow, S. (1994). An ellipsoidal shell model for volume estimation of the right ventricle from magnetic resonance images. *Acad. Radiol.* 1(4):345–51.

118. Chen, C. W., Luo, J., Parker, K., and Huang, T. (1995). CT volumetric data-based left ventricle motion estimation: an integrated approach. *Comput. Med. Image and Graph.* 19(1): 85–100.

119. Goshtasby, A. and Turner, D. (1995). Segmentation of cardiac cine MR images for extraction of right and left ventricular chambers. *IEEE Trans. Med. Imaging.* 14(1): 56–64.

120. Matheny, A. and Goldgof, D. (1995). The use of three- and four-dimensional surface harmonics for rigid and nonrigid shape recovery and representation. *IEEE Trans. Pattern Anal. Machine Intell.* 17(10): 967–981.

121. Staib, L. H. and Duncan, J. S. (1996). Model-based deformable surface finding for medical images. *IEEE Trans. Med. Imaging.* 15(5): 720–731.

122. Bardinet, E., Ayache, N., and Cohen, L. (1996). Tracking and motion analysis of the left ventricle with deformable superquadrics. *Med. Image Anal.* 1(2): 129–150.

123. Declerck, J., Feldmar, J., and Ayache, N. (1998). Definition of a four-dimensional continuous planisferic transformation for the tracking and the analysis of the left-ventricle motion. *Med. Image Anal.* 2(2): 197–213.

124. Sato, Y., Moriyama, M., Hanayama, M., Naito, H., and Tamura, S. (1997). Acquiring 3-D models of non-rigid moving objects from time and viewpoint varying image sequences: A step toward left ventricle recovery. *IEEE Trans. Pattern Anal. Machine Intell.* 19(3): 253–259.

125. Sanchez-Ortiz, G., Wright, G., Clarke, N., Declerck, J., Banning, A., and Noble, J. (2002). Automated 3-D echocardiography analysis compared with manual delineations and SPECT MUGA. *IEEE Trans. Med. Imaging,* 21(9): 1069–1076.

126. Swingen, C., Seethamraju, R., and Jerosch-Herold, M. (2003). Feedback-assisted three-dimensional reconstruction of the left ventricle with MRI. *J. Magn. Reson. Imaging.* 17(5): 528–537.

127. Swingen, C., Seethamraju, R., and Jerosch-Herold, M. (2003). An approach to the three-dimensional display of left ventricular function and viability using MRI. *Int. J. Card. Imag.* 19(4):325–36.

128. Horkaew, P. and Yang, G. (2003). Optimal deformable surface models for 3-D medical image analysis. in *IPMI,* Vol. 2732 of *Lecture Notes in Computer Science* 13–24.

129. Geiser, E., Lupkiewicz, S., Christie, L., Ariet, M., Conetta, D., and Conti, C. (1980). A framework for three-dimensional time-varying reconstruction of the human left ventricle: sources of error and estimation of their magnitude. *Comput. Biomed. Res.* 13(3): 225–241.

130. Geiser, E., Ariet, M., Conetta, D., Lupkiewicz, S., Christie, L., and Conti, C. (1982). Dynamic three-dimensional echocardiographic reconstruction of the intact human left ventricle: technique and initial observations in patients. *Am. Heart. J.* 103(6): 1056–1065.

131. Faber, T., Stokely, E., Peshock, R., and Corbett, J. (1991). A model-based four-dimensional left ventricular surface detector. *IEEE Trans. Med. Imaging.* 10(3): 321–329.

132. Friboulet, D., Magnin, I., and Revel, D. (1992). Assessment of a model for overall left ventricular three-dimensional motion from MRI data. *Intl. J. Cardiac Imaging.* 8(3): 175–190.

133. Huang, W.C. and Goldgof, D. (1993). Adaptive-size meshes for rigid and non-rigid analysis and synthesis. *IEEE Trans. Pattern Anal. Machine Intell.* 15(6): 611–616.

134. Faber, T., Cooke, C., Peifer, J., Pettigrew, R., Vansant, J., Leyendecker, J., and García, E. (1995). Three-dimensional displays of left ventricular epicardial surface from standard cardiac SPECT perfusion quantification techniques. *J. Nucl. Med.* 36(4): 697–703.

135. Germano, G., Kiat, H., Kavanagh, P., Moriel, M., Mazzanti, M., Su, H.T., van Train, K., and Berman, D. (1995). Automatic quantification of ejection fraction from gated myocardial perfusion SPECT. *J. Nucl. Med.* 36(11): 2138–2147.

136. Germano, G., Kavanagh, P., Chen, J., Waechter, P., Su, H.T., Kiat, H., and Berman, D.S. (1995). Operator-less processing of myocardial perfusion SPECT studies. *J. Nucl. Med.* 36(6): 2127–2132.

137. McInerney, T. and Terzopoulos, D. (1995). A dynamic finite element surface model for segmentation and tracking in multidimensional medical images with application to 4-D image analysis. *Comput. Med. Image and Graph.* 19(1): 69–83.

138. Ranganath, S. (1995). Contour extraction from cardiac MRI studies using snakes. *IEEE Trans. Med. Imaging.* 14(2): 56–64.

139. Tu, H., Matheny, A., Goldgof, D., and Bunke, H. (1995). Left ventricular boundary detection from spatio-temporal volumetric computed tomography images. *Comput. Med. Image and Graph.* 19(1): 27–46.

140. Nastar, C. and Ayache, N. (1996). Frequency-based nonrigid motion analysis: application to four dimensional medical images. *IEEE Trans. Pattern Anal. Mach. Intell.* 18(11): 1067–1079.

141. Rueckert, D., and Burger, P. (1997). Geometrically deformable templates for shape-based segmentation and tracking in cardiac MR image. in *Energy Minim. Meth. Comput. Vision and Patt. Recog.* Vol. 1223 of *Lecture Notes in Computer Science.* Venice, Italy: Springer-Verlag.

142. Shi, P., Sinusas, A., Constable, R., Ritman, E., and Duncan, J. (2000). Point-tracked quantitative analysis of left ventricular surface motion from 3-D image sequences: algorithms and validation. *IEEE Trans. Med. Imaging*, 19(1): 36–50.

143. Duncan, J., Shi, P., Constable, T., and Sinusas, A. (1998). Physical and geometrical modeling for image-based recovery of left ventricular deformation. *Prog. Biophys. Mol. Biol.* 69(1–2): 333–351.

144. Montagnat, J., Delingette, H., and Malandain, G. (1999). Cylindrical echocardio-graphic image segmentation based on 3-D deformable models. in *MICCAI, Lecture Notes Comp. Science*, 168–175.

145. Biedenstein, S., Schäfers, M., Stegger, L., Kuwert, T., and Schober, O. (1999). Three-dimensional contour detection of left ventricular myocardium using elastic surfaces. *Eur. J. Nucl. Med.* 26(3): 201–207.

146. Gerard, O., Billon, A., Rouet, J., Jacob, M., Fradkin, M., and Allouche, C. (2002). Efficient model-based quantification of left ventricular function in 3-D echocar-diography. *IEEE Trans. Med. Imaging*. 21(9): 1059–1068.

147. Song, M., Haralick, R., Sheehan, F., and Johnson, R. (2002). Integrated surface model optimization for freehand three-dimensional echocardiography. *IEEE Trans. Med. Imaging*. 21(9): 1077–1090.

148. Makela, T., Pham, Q., Clarysse, P., Nenonen, J., Lotjonen, J., Sipila, O., Hanninen, H., Lauerma, K., Knuuti, J., Katila, T., and Magnin, I. (2003). A 3-D model-based registration approach for the PET, MR and MCG cardiac data fusion. *Med. Image Anal.* 7(3): 377–389.

149. Lötjönen, J., Reissman, P.J., Magnin, I., and Katila, T. (1999). Model extraction from magnetic resonance volume data using a deformable pyramid. *Med. Image Anal.* 3(4): 387–406.

150. Fan, L., Tamez-Pena, J., and Chen, C. (2003). Local force model for cardiac dynamics analysis from volumetric image sequences. *Comput. Med. Image and Graph.* 27(6): 437–446.

151. Lorenzo-Valdes, M., Sanchez-Ortiz, G., Elkington, A., Mohiaddin, R., and Rueck-ert, D. (2004). Segmentation of 4-D cardiac MR images using a probabilistic atlas and the EM algorithm. *Med. Image Anal.* 8(3): 255–265.

152. van Assen, H., Danilouchkine, M., Behloul, F., Lamb, H., van der Geest, R., Reiber J., and Lelieveldt, B. (2003). Cardiac LV segmentation using a 3-D active shape model driven by fuzzy inference. in *MICCAI*, Vol. 2878 of *Lecture Notes in Computer Science*, 535–540.

153. van Assen, H., Danilouchkine, M., Dirksen, M., Lamb, H., van der Geest, R., Reiber, J., and Lelieveldt, B. (2004). A 3D-active shape model driven by fuzzy inference: application to cardiac CT and MR.

154. Kaus, M., von Berg, J., Weese, J., Niessen, W., and Pekar, V. (2004). Automated segmentation of the left ventricle in cardiac MRI. *Med. Image Anal.* in press.

155. Wierzbicki, M., Drangova, M., Guiraudon, G., and Peters, T. (2004). Validation of dynamic heart models obtained using non-linear registration for virtual reality training, planning, and guidance of minimally invasive cardiac surgeries. *Med. Image Anal.* 8(3): 245–254.

156. Yezzi, A., Tannenbaum, A., Kichenassamy, S., and Olver, P. (1996). A gradient surface approach to 3-D segmentation. in *Proc IS&T 49th Ann. Conf.* Minneapolis, USA.

157. Yezzi, A., Kichenassamy, S., Kumar, A., Olver, P., and Tannenbaum, A. (1997). A geometric snake model for segmentation of medical imagery. *IEEE Trans. Med. Imaging.* 16(2): 199–209.

158. Tseng, Y.H., Hwang, J.N., and Sheehan, F. (1997). Three-dimensional object representation and invariant recognition using continuous distance transform neural networks. *IEEE Trans. Neural Networks.* 8(1): 141–147.

159. Niessen, W., ter Haar Romeny, B.M., and Viergever, M. (1998). Geodesic deformable models for medical image analysis. *IEEE Trans. Med. Imaging.* 17(4): 634–641.

160. Lelieveldt, B., van der Geest, R., Rezaee, M.R., Bosch, J., and Reiber, J. (1999). Anatomical model matching with fuzzy implicit surfaces for segmentation of thoracic volume scan. *IEEE Trans. Med. Imaging.* 18(3): 231–238.

161. Creswell, L., Wyers, S., Pirolo, J., Perman, W., Vannier, M., and Pasque, M. (1992). Mathematical modeling of the heart using magnetic resonance imaging. *IEEE Trans. Med. Imaging.* 11(4): 581–589.

162. Pirolo, J., Bresina, S., Creswell, L., Myers, K., Szabó, B., Vannier, M., and Pasque, M. (1993). Mathematical three-dimensional solid modeling of biventricular geometry. *Ann. Biomed. Eng.* 21(3): 199–219.

163. Park, J., Metaxas, D., Axel, L., Yuan, Q., and Blom, A. (1999). Cascaded MRI-SPAMM for LV motion analysis during a whole cardiac cycle. *Intl. J. Med. Inf.* 55(2): 117–126.

164. Haber, E., Metaxas, D., and Axel, L. (1997). Three-dimensional geometric modeling of cardiac right and left ventricles. in *Biomed. Eng. Soc. Annu. Meeting.* San Diego, USA.

165. Haber, E., Metaxas, D., and Axel, L. (1998). Motion analysis of the right ventricle from MRI images. in *MICCAI, Lecture Notes in Computer Science,* 177–188.

166. Haber, I., Metaxas, D., and Axel, L. (2000). Three-dimensional motion reconstruction and analysis of the right ventricle using tagged MRI. *Med. Image. Anal.* 4(4): 335–355.

167. Shi, P., Sinusas, A., Constable, R., and Duncan, J. (1999). Volumetric deformation analysis using mechanics-based data fusion: application in cardiac motion recovery. *Intl. J. Comput. Vision.* 35(1): 87–107.

168. Hu, Z., Metaxas, D., and Axel, L. (2003). In vivo strain and stress estimation of the heart left and right ventricles from MRI images. *Med. Image Anal.* 7(4): 435–444.

169. Kuwahara, M. and Eiho, S. (1991). 3-D heart image reconstructed from MRI data. *Comput. Med. Image and Graph,* 15(4): 241–246.

170. Mitchell, S., Bosch, J., Lelieveldt, B., van der Geest, R., Reiber, J., and Sonka, M. (2002). 3-D active appearance models: segmentation of cardiac MR and ultrasound images. *IEEE Trans. Med. Imaging* 21(9): 1167–1178.

171. Sermesant, M., Forest, C., Pennec, X., Delingette, H., and Ayache, N. (2003). Deformable biomechanical models: application to 4D cardiac image analysis. *Med. Image. Anal.* 7(4): 475–488.

172. Stegmann, M. (2004). Generative Interpretation of Medical Images. Ph.D. thesis, Informatics and Mathematical Modeling Institute, Technical University of Denmark.

173. Stegmann, M., Ersboll, B., and Larsen, R. (2003). FAME–a flexible appearance modeling environment. *IEEE Trans. Med. Imaging.* 22(10): 1319–1331.

174. Amini, A., Chen, Y., Elayyadi, M., and Radeva, P. (2001). Tag surface reconstruction and tracking of myocardial beads from SPAMM-MRI with parametric B-spline surfaces. *IEEE Trans. Med. Imaging.* 20(2): 94–103.

175. Tustison, N., Davila-Roman, V., and Amini, A. (2003). Myocardial kinematics from tagged MRI based on a 4-D B-spline mode. *IEEE Trans. Biomed. Eng.* 50(8): 1038–1040.

176. Bartels, K., Bovik, A., Aggarwal, S., and Diller, K. (1993). The analysis of biological shape change from multi-dimensional dynamic images. *Comput. Med. Image and Graph.* 17(2): 89–99.

177. O'Dell, W., Moore, C., Hunter, W., Zerhouni, E., and McVeigh, E. (1995). Three-dimensional myocardial deformations: Calculation with displacement field fitting to tagged MR images. *Radiology* 195(3): 829–835.

178. Moulton, M., Creswell, L., Downing, S., Actis, R., Szabo, B., Vannier, M., and Pasque, M. (1996). Spline surface interpolation for calculating 3-D ventricular strains from MRI tissue tagging. *Am. J. Physiol.* 270(1 Pt 2): H281–97.

179. Radeva, P., Amini, A., and Huang, J. (1997). Deformable B-solids and implicit snakes for 3-D localization and tracking of SPAMM MRI data. *Comput. Vis. Image Underst.* 66(2): 163–78.

180. Young, A. (1999). Model tags: direct 3-D tracking of heart wall motion from tagged MR images. *Med. Image Anal.* 3(4): 361–372.

181. Huang, J., Abendschein, D., Davila, V., and Amini, A. (1999). Four-dimensional LV tissue tracking from tagged MRI with a 4-D B-spline model. *IEEE Trans. Med. Imaging.* 18(10): 957–972.

182. Chen, Y. and Amini, A. (2002). A MAP framework for tag line detection in SPAMM data using Markov random fields on the B-spline solid. *IEEE Trans. Med. Imaging.* 21(9): 1110–1122.

183. Klein, G. and Huesman, R. (2002). Four-dimensional processing of deformable cardiac PET data. *Med. Image. Anal.* 6(1): 29–46.

184. Masood, S., Gao, J., and Yang, G. (2002). Virtual tagging: numerical considerations and phantom validation. *IEEE Trans. Med. Imaging.* 21(9): 1123–1131.

185. Chandrashekara, R., Mohiaddin, R., and Rueckert, D. (2003). Analysis of myocardial motion and strain patterns using a cylindrical B-spline transformation model. in *Surgery, Simulation and Soft Tissue Modeling.* Vol. 2673 of *Lecture Notes in Computer Science.* 88–99.

186. Chandrashekara, R., Rao, A., Sanchez-Ortiz, G., Mohiaddin, R., and Rueckert, D. (2003). Construction of a statistical model for cardiac motion analysis using nonrigid image registration. in *IPMI,* Vol. 2732 of *Lecture Notes in Computer Science,* 599–610.

187. Rao, A. Sanchez-Ortiz, G.I., Chandrashekara, R., Lorenzo-Valdes, M., Mohiaddin, R., and Rueckert, D. (2003). Construction of a cardiac motion atlas from MR using non-rigid registration. in *Functional Imaging and Modelling of the Heart,* no. 2674 in *Lecture Notes in Computer Science,* 141–150.

188. Perperidis, D., Rao, A., Lorenzo-Valdes, M., Mohiaddin, R., and Rueckert, D. (2003). Spatio-temporal alignment of 4-D cardiac MR images. in *Functional Imaging and Modeling of the Heart.* Vol. 2674 of *Lecture Notes in Computer Science.* 205–12.

189. Perperidis, D., Rao, A., Mohiaddin, R., and Rueckert, D. (2003). Non-rigid spatio-temporal alignment of 4-D cardiac MR images. in *Biomedical Image Registration,* Vol. 2717 of *Lecture Notes Computer Science,* 191–200.

190. Papademetris, X., Sinusas, A., Dione, D., Constable, R., and Duncan, J. (2002). Estimation of 3-D left ventricular deformation from medical images using bio-mechanical models. *IEEE Trans. Med. Imaging.* 21(7): 786–800.

191. Papademetris, X., Sinusas, A., Dione, D., and Duncan, J. (2001). Estimation of 3-D left ventricular deformation from echocardiography. *Med. Image Anal.* 5(1): 17–28.

192. Shi, P., and Liu, H. (2003). Stochastic finite element framework for simultaneous estimation of cardiac kinematic functions and material parameters. *Med. Image. Anal.* 7(4): 445–464.

193. Lötjönen, J., Kivistö, S., Koikkalainen, J., Smutek, D., and Lauerma, K. (2004). Statistical shape model of atria, ventricles and epicardium from short- and long-axis MR images. *Med. Image Anal.* 8(3) 371–386.

194. Pentland, A. and Sclaroff, S. (1991). Closed-form solutions for physically based shape modeling and recognition. *IEEE Trans. Pattern Anal. Machine Intell.* 13(7): 703–714.

195. Metaxas, D. and Terzopoulos, D. (1993). Shape and nonrigid motion estimation through physics-based synthesis. *IEEE Trans. Pattern Anal. Machine Intell.* 15(6): 580–591.

196. Barr, A. (1981). Superquadrics and angle-preserving deformations. *IEEE Comput. Graph. and Appl.* 1(1): 11–23.

197. Chen, C., Luo, J., and Parker, K. (1998). Image segmentation via adaptive K-mean clustering and knowledge-based morphological operations with biomedical applications. *IEEE Trans. Image Process.* 7(12): 1673–1683.

198. Hopfield, J. (1984). Neurons with graded responses have collective computational properties like those of two-state neurons. *Proc. Nat. Acad. Sci.* 81: 3088–3092.

199. Monga, O., Deriche, R., and Rocchisani, J.M. (1991). 3-D edge detection using recursive filtering: application to scanner images. *Comput. Vis. Graph and Image Process: Image Underst.* 53(1): 76–87.

200. Declerck, J., Feldmar, J., Goris, M., and Betting, F. (1997). Automatic registration and alignment on a template of cardiac stress and rest SPECT images. *IEEE Trans. Med. Imaging,* 16(6):727–37.

201. Besl, P. and McKay, N. (1992). A method for registration of 3-D shapes. *IEEE Trans. Pattern Anal. Machine Intell.* 14(2): 239–255.

202. Feldmar, J., and Ayache, N. (1996). Rigid, affine and locally affine registration of free-form surfaces. *Intl. J. Comput. Vision.* 18(2): 99–119.

203. Sederberg, T.W. and Parry, S.R. (1986). Free-form deformation of solid geometric models. in *SIGGRAPH'86,* Vol. 20, 151–160. Dallas, USA: ACM.

204. Germano, G. and Berman, D. (1999). *Clinical Gated Cardiac SPECT,* chap. Quantitative gated perfusion SPECT, 115–146. Armonk, NY: Futura Publishing Co.

205. Goshtasby, A. (1993). Design and recovery of 2-D and 3-D shapes using rational Gaussian curves and surfaces. *Intl. J. Comput. Vision.* 10(3): 233–256.

206. Metaxas, D.N. (1996). Physics-based deformable models: applications to computer vision, graphics and medical imaging. Cambridge: Kluwer Academic Publisher.

207. Cohen, L. (1991). On active contour models and balloons. *Comput. Vis. Graph. and Image Process: Image Underst.* 53(2): 211–218.

208. Delingette, H. (1999). General object reconstruction based on simplex meshes. *Intl. J. Comput. Vision.* 32(2): 111–146.

209. Kass, M., Witkin A., and Terzopoulos, D. (1988). Snakes: active contour models. *Intl. J. Comput. Vision.* 1(4): 321–331.

210. Terzopoulos, D. (1986). Regularization of inverse visual problems involving discontinuities. *IEEE Trans. Pattern Anal. Machine Intell.* 8(4): 413–424.

211. Kittleri, J. and Illingworth, J. (1985). Relaxation labeling algorithms – a review. *Image Vis. Comput.* 3(4): 206–216.

212. Watson, D.F. (1981). Computing the n-dimensional Delaunay tessellation with application to voronoi polytopes. *Comput. J.* 24(2): 167–172.

213. Chakraborty, A., Staib, L., and Duncan, J. (1996). Deformable boundary finding in medical images by integrating gradient and region information. *IEEE Trans. Med. Imaging,* 15(6): 859–870.

214. Ekoulé, E., Peyrin, F., and Odet, C. (1987). Description d'une procédure de triangulation entièrament automatique. in *Proc COGNITIVA 87,* 88–95. Paris, France.

215. Sheehan, F., Bolson, E., Martin, R., Bashein, G., and McDonald, J. (1998). Quantitative three-dimensional echocardiography: methodology, validation and clinical applications. in *MICCAI, Lecture Notes in Computer Science.* 102–109.

216. Hoppe, H. (1994). Surface Reconstruction from Unorganized Points. Ph.D. thesis. University of Washington. Seattle.

217. Frangi, A., Rueckert, D., Schnabel, J., and Niessen, W. (2002). Automatic construction of multiple-object three-dimensional statistical shape models: application to cardiac modeling. *IEEE Trans. Med. Imaging* 21(9): 1151–1166.

218. Ordas, S., Boisrobert, L., Bossa, M., Laucelli, M., Huguet, M., and Frangi, A.F. (2004). GRID, enabled automatic construction of a two-chamber cardiac PDM from a large database of dynamic 3-D shapes. in *IEEE Intl. Symposium of Biomedical Imaging.*

219. McLeish, K., Hill, D., Atkinson, D., Blackall, J., and Razavi, R. (2002). A study of the motion and deformation of the heart due to respiration. *IEEE Trans. Med. Imaging.* 21(9): 1142–1150.

220. Kaus, M., Pekar, V., Lorenz, C., Truyen, R., Lobregt, S., and Weese, J. (2003). Automated 3-D PDM construction from segmented images using deformable models. *IEEE Trans. Med. Imaging.* 22(8): 1005–1013.

221. Caselles, V., Catté, F., Coll, T., and Dibos, F. (1993). A geometric model for active contours in image processing. *Num Mathematik,* 66: 1–31.

222. Malladi, R., Sethian, J., and Vemuri, B. (1995). Shape modeling with front propagation: a level set approach. *IEEE Trans. Pattern. Anal. Machine Intell.* 17(2):158–74.

223. Osher, S. and Sethian, S. (1988). Fronts propagating with curvature dependent speed: algorithms based on the Hamilton-Jacobi formalism. *J. Comput. Phys.* 79: 12–49.

224. Tseng, Y.H., Hwang, J.N., and Sheehan, F. (1998). 3-D heart modeling and motion estimation based on continuous distance transform neural networks and affine transform. *J. VLSI Sign Process. Syst. for Sign, Image and Video Technol.* 18(3): 207–218.

225. Hanson, A. (1988). Hyperquadrics: smoothly deformable shapes with convex polyhedral bounds. *Comput. Vis. Graph and Image Process* 44(2): 191–210.

226. Requicha, A.A. and Voelcker, H.B. (1982). Solid modeling: a historical summary and contemporary assessment. *IEEE Comput. Graph and Appl.* 2(2): 9–24.

227. Lelieveldt, B., van der Geest, R., Lamb, H., Kayser, H., and Reiber, J. (2001). Automated observer-independent acquisition of cardiac short-axis MR images: a pilot study. *Radiology* 221(2): 537–542.

228. Piegl, L. and Tiller, W. (1996). *The NURBS Book*. Monographs in Visual Communication, 2nd ed. Berlin: Springer-Verlag.

229. Bartels, K., Bovik, A., and Griffin, C. (1994). Spatio-temporal tracking of material shape change via multi-dimensional splines. in *Math. Meth. Biomed. Image Anal.* 110–116. Seattle, USA.

230. Rueckert, D., Frangi, A., and Schnabel, J. (2003). Automatic construction of 3-D statistical deformation models of the brain using non-rigid registration. *IEEE Trans. Med. Imaging.* 22(8): 1014–1025.

231. Lötjönen, J. (2003). Construction of patient-specific surface models from MR images: application to bioelectromagnetism. *Comput. Methods Programs Biomed.* 72(2): 167–178.

232. Lorenzo-Valdes, M., Sanchez-Ortiz, G., Mohiaddin, R., and Rueckert, D. (2003). Segmentation of 4-D cardiac MR images using a probabilistic atlas and the EM algorithm. in *MICCAI*, Vol. 2878 of *Lecture Notes in Computer Science*. 440–450.

233. Guccione, J. and McCulloch, A. (1991). Finite element modeling of ventricular mechanics. in Hunter, P., McCulloch, A., Nielsen, P., Eds. *Theory of Heart*, 122–144. Berlin: Springer-Verlag.

234. Sermesant, M., Faris, O., Evans, F., McVeigh, E., Coudiere, Y., Delingette, H., and Ayache, N. (2003). Preliminary validation using in vivo measures of a macroscopic electrical model of the heart. in *Intl. Symposium on Surgery Simulation and Soft Tissue Modeling, Lecture Notes in Computer Science*, 230–243.

235. Amini, A., Chen, Y., and Abendschein, D. (1999). Comparison of landmark-based and curved-based thin-plate warps for analysis of left-ventricular motion from tagged MRI. in *MICCAI, Lecture Notes in Computer Science*, 498–507.

236. Waks, E., Prince, J., and Douglas, A. (1996). Cardiac motion simulator for tagged MRI. in *Math. Meth. Biomed. Imag. Anal.* 182–191.

237. Kraitchman, D., Young, A., Chang, C.N., and Axel, L. (1995). Semiautomatic tracking of myocardial motion in MR tagged images. *IEEE Trans. Med. Imaging.* 14(3): 422–433.

238. Douglas, A., Hunter, W., and Wiseman, M. (1990). Inhomogeneous deformation as a source of error in strain measurements derived from implanted markers in the canine left ventricle. *J. Biomech.* 23(4): 331–341.

239. Schweppe, F. (1973). *Uncertain Dynamic Systems*. Englewood Cliffs, NJ: Prentice-Hall.

240. Denney, T. and McVeigh, E. (1997). Model-free reconstruction of three-dimensional myocardial strain from planar tagged MR images. *J. Magn. Reson. Imaging.* 7(5): 799–810.

241. Leotta, D., Munt, B., Bolson, E., Kraft, C., Martin, R.W., Otto, C., and Sheehan, F. (1997). Quantitative three-dimensional echocardiography by rapid imaging from multiple transthoracic windows: *in vitro* validation and initial *in vivo* studies. *J. Am. Soc. Echocardiogr.* 10(8): 830–839.

242. Munt, B., Leotta, D., Bolson, E., Coady, K., Martin, R., Otto, C., and Sheehan, F. (1998). Left ventricular shape analysis from three-dimensional echocardiograms. *J. Am. Soc. Echocardiogr.*, 11(8): 761–769.

243. Sinusas, A., Papademetris, X., Constable, R., Dione, D., Slade, M., Shi, P., and Duncan, J. (2001). Quantification of 3-D regional myocardial deformation: shape-based analysis of magnetic resonance images. *Am. J. Physiol. Heart. Circ. Physiol.* 281(2): H698–714.

244. Sasayama, S., Nonogi, H., Fujita, M., Sakurai, T., Wakabayashi, A., Kawai, C., Eiho, S., and Kuwahara, M. (1984). Three-dimensional analysis of regional myocardial function in response to nitroglycerin in patient with coronary artery disease. *J. Am. Coll. Cardiol.* 3(5): 1187–1196.

245. Rao, A., Sanchez-Ortiz, G., Chandrashekara, R., Lorenzo-Valdes, M., Mohiaddin, R., and Rueckert, D. (2002). Comparison of cardiac motion across subjects using non-rigid registration. in *MICCAI*, Vol. 2488, 722–729.

246. Germano, G., Erel, J., Lewin, H., Kavanagh, P., and Berman, D. (1997). Automatic quantitation of regional myocardial wall motion and thickening from gated technetium-99m sestamibi myocardial perfusion single-photon emission computed tomography. *J. Am. Coll. Cardiol.* 30(5): 1360–1367.

247. Lorenzo-Valdes, M., Sanchez-Ortiz, G., Mohiaddin, R., and Rueckert, D. (2002). Atlas-based segmentation and tracking of 3-D cardiac MR images using non-rigid registration. in *MICCAI*, Vol. 2488 of *Lecture Notes in Computer Science*. 642–650.

248. Papademetris, X., Shi, P., Dione, D., Sinusas, A., Constable, R., and Duncan, J. (1999). Recovery of soft tissue object deformation from 3-D image sequences using biomechanical models. in Kuba, A., Sámal, M., Todd-Pokropek, A., Eds., *IPMI*, Vol. 1613 of *Lecture Notes in Computer Science* 352–357. Springer-Verlag.

249. Young, A., Axel, L., Dougherty, L., Bogen, D., and Parenteau, C. (1993). Validation of tagging with MR imaging to estimate material deformation. *Radiology,* 188(1): 101–108.

250. Kerwin, W. Prince, J. (1999). Tracking MR tag surfaces using a spatiotemporal filter and interpolator. *Intl. J. Image Syst. Tech.* 10: 128–142.

251. Arts, T., Hunter, W., Douglas, A., Muijtjens, M., and Reneman, R. (1992). Description of the deformation of the left ventricle by a kinematic model. *J. Biomech.* 25(10): 1119–1127.

252. Declerk, J., Denney, T., Öztürk, C., O'Dell, W., and McVeigh, E. (2000). Left ventricular motion reconstruction from planar tagged MR images: a comparison. *Phys. Med. Biol.* 45(6): 1611–1632.

253. Declerck, J., Ayache, N., and McVeigh, E. (1998). Use of a 4-D planispheric transformation for the tracking and the analysis of LV motion with tagged MR images. Technical Report 3535, INRIA.

254. Öztürk, C. and McVeigh, E. (2000). Four-dimensional B-spline-based motion analysis of tagged cardiac MR images: introduction and *in vivo* validation. *Phys. Med. Biol.* 45(6): 1683–1702.

255. Makela, T., Clarysse, P., Sipila, O., Pauna, N., Pham, Q., Katila, T., and Magnin, I. (2002). A review of cardiac image registration method. *IEEE Trans. Med. Imaging.* 21(9): 1011–1021.

256. McCulloch, A. (2000). Modeling the human cardiome in silico. *J. Nucl. Cardiol.* 7(5): 469–496.

257. Crampin, E., Halstead, M., Hunter, P., Nielsen, P., Noble, D., Smith, N., and Tawhai, M. (2004). Computational physiology and the Physiome Project. *Exp. Physiol.* 89(1): 1–26.

258. Noble, D. (2002). Modelling the heart: insights, failures and progress. *Bioessays,* 24(12):1155–63.

259. Pullan, A., Buist, M., Sands, G., Cheng, L., and Smith, N. (2003). Cardiac electrical activity–from heart to body surface and back again. *J Electrocardiol.* 36 (suppl.): 63–67.

260. Bassingthwaighte, J. and Vinnakota, K. (2004). The computational integrated myocyte: a view into the virtual heart. *Ann. NY Acad. Sci.* 1015: 391–404.

261. Winslow, R., Helm, P., Baumgartner, W., Peddi, S., Ratnanather, T., McVeigh, E., and Miller, M. (2002). Imaging-based integrative models of the heart: closing the loop between experiment and simulation. in Angel [262], 129–41.

262. Angel, K., Ed. (2002). *In Silico Simulation of Biological Processes.* No. 247 in Novartis Foundation Symposium. Wiley.

263. Taylor, C. and Draney, M. (2004). Experimental and computational methods in cardiovascular fluid mechanics. *Annu. Rec. Fluid Mech.* 36: 197–231.

264. Smye, S. and Clayton, R. (2002). Mathematical modelling for the new millennium: medicine by numbers. *Med. Eng. Phys.* 24(9): 565–574.

265. McCulloch, A.D., Huber, G. (2002). Integrative biological modelling in silico. in Angel [262], 4–19.

266. Greenbaum, R. and Gibson, D. (1981). Regional non-uniformity of left ventricular wall movement. *Br. Heart J.* 45(1): 29–34.

267. Shapiro, E., Marier, D.L., St. John Sutton, M.G., and Gibson, D.G. (1981). Regional non-uniformity of wall dynamics in normal left ventricle. *Br. Heart J.* 45(3): 264–270.

268. Pandian, N., Skorton, D., Collins, S., Falsetti, H., Burke, E., and Kerber, R. (1983). Heterogeneity of left ventricular segmental wall thickening and excursion in 2-dimensional echocardiograms of normal human subjects. *Am. J. Cardiol.* 51(10): 1667–1673.

269. Besl, P. and Jain, R. (1986). Invariant surface characteristics for 3-D object recognition in range images. *Comput. Vis. Graph and Image Process.* 33(1): 33–80.

270. Clarysse, P., Friboulet, D., and Magnin, I. (1997). Tracking geometrical descriptors on 3-D deformable surfaces: application to the left-ventricular surface of the heart. *IEEE Trans. Med. Imaging.* 16(4): 392–404.

271. Koenderink, J. and van Doorn, A. (1992). Surface shape and curvature scales. *Image Vis Comput.* 10(8): 557–565.

272. Dorai, C. and Jain, A.K. (1997). Shape spectrum-based view grouping and matching of 3-D free-form objects. *IEEE Trans. Pattern Anal. Machine Intell.* 19(10):1139–46.

273. Do Carmo, M.P. (1976). *Differential Geometry of Curves and Surfaces.* New Jersey: Prentice-Hall.

274. Azhari, H., Grenadier, E., Dinnar, U., Beyar, R., Adam, D., Marcus, M., and Sideman, S. (1989). Quantitative characterization and sorting of three-dimensional geometries: Application to left ventricles in vivo. *IEEE Trans. Med. Imaging.* 36(3): 322–332.

275. Azhari, H., Sideman, S., Beyar, R., Grenadier, E., and Dinnar, U. (1987). An analytical descriptor of three-dimensional geometry: application to the analysis of the left ventricle shape and contraction. *IEEE Trans. Biomed. Eng.* 34(5): 345–355.

276. Azhari, H., Gath, I., Beyar, R., Marcus, M., and Sideman, S. (1991). Discrimination between healthy and diseased hearts by spectral decomposition of their left ventricular 3-D geometry. *IEEE Trans. Med. Imaging.* 10(2): 207–215.

277. Guttman, M., Prince, J., and McVeigh, E. (1994). Tag and contour detection in tagged MR images of the left ventricle. *IEEE Trans. Med. Imaging.* 13(1): 74–88.

278. Guttman, M., Zerhouni, E., and McVeigh, E. (1997). Analysis and visualization of cardiac function from MR images. *IEEE Comput. Graph and Appl.,* 17(1): 55–63.

279. Axel, L., Bloomgarden, D., Chang, C., Kraitchman, D., and Young, A. (1993). SPAMMVU: a program for the analysis of dynamic tagged MRI. in *Proc ISMRM,* 724.

280. Bazille, A., Guttman, M., McVeigh, E., and Zerhouni, E. (1994). Impact of semi-automated vs. manual image segmentation errors on myocardial strain calculation by MR tagging. *Inv. Rad.* 29(4): 427–433.

281. Zhang, S., Douglas, M.A., Yaroslavsky, L., Summers, R.M., Dilsizian, V., Fanan-apazir, L., and Bacharach, S.L. (1996). A Fourier-based algorithm for tracking SPAMMM tags in gated magnetic resonance cardiac images. *Med. Phys.* 23(8): 1359–1369.

282. Kerwin, W. and Prince, J. (1999). The Kriging update model and recursive space-time function estimation. *IEEE Trans. Sig. Proc.* 47(11): 2942–2952.

283. David, M. (1988). *Handbook of Applied Advanced Geostatistical Ore Reserve Estimation.* Amsterdam, The Netherlands: Elsevier Science Publishers B.V.

284. Osman, N., Kerwin, W., McVeigh, E., and Prince, J. (1999). Cardiac motion tracking using CINE harmonic phase (HARP) magnetic resonance imaging. *Magn. Res. Med.* 42(6): 1048–1060.

285. Garot, J., Bluemke, D., Osman, N., Rochitte, C., McVeigh, E., Zerhouni, E., and Prince, J. (2000). Fast determination of regional myocardial strain fields from tagged cardiac images using harmonic-phase MRI. *Circulation.* 101(9): 981–988.

286. Osman, N., and Prince, J. (1998). Motion estimation from tagged MR images using angle images. in *Intl. Conf. Image Process.* Vol. 1, 704–708. Chicago: Comp. Soc. Press.

287. Osman, N. and Prince, J. (2000). Visualizing myocardial function using HARP MRI. *Phys. Med. Biol.* 45(6): 1665–1682.

10 MRI Measurement of Heart Wall Motion

Alistair A. Young

CONTENTS

10.1 INTRODUCTION

Impaired cardiac function in a region of the heart is associated with abnormal local cardiac wall motion and deformation [1]. A quantitative description of regional heart wall motion is required for the assessment of cardiac performance, both in the diagnosis of disease and the evaluation of treatment effect. In addition, information on myocardial tissue kinematics is required to evaluate mathematical models of cardiac mechanics, which can be used to gain an understanding of how tissue properties such as myocyte contraction, electrical activation, and extracellular coupling combine to effect whole-organ function [2,3].

In order to measure tissue motion and strain, researchers have used implanted radiopaque beads [4], ultrasonic markers [5], or natural landmarks such as the coronary arteries [6]. These provide useful information, but are too invasive or too sparse for clinical use. Recently, there has been a lot of clinical interest in echocardiographic tissue Doppler imaging (TDI), which allows measurement of tissue velocity along the direction of the ultrasound beam [7]. Spatial derivatives of the velocity field give rise to strain rates, which can then be integrated through time to give myocardial strain [8]. Although limited in signal-to-noise ratio (SNR), regions that can be imaged, and components of the deformation available, this technique is proving useful in the clinical evaluation of tissue function.

MRI also provides quantitative information on motion and deformation in the heart, and offers the potential of precise, noninvasive assessment of all components of deformation, in all regions of the heart, throughout the cardiac cycle, at acceptable spatial and temporal resolution. Many of these techniques have been available for over a decade for research purposes, but the translation into the clinical domain has been hindered by the complex and time-consuming nature of the image postprocessing. Recently, however, there have been rapid developments in the field, including harmonic phase (HARP) and displacement encoding with stimulated echoes (DENSE) acquisition techniques. These have the potential for higher spatial and temporal resolution, and faster evaluation procedures. This chapter will review the current state of the art, with a view to demonstrating the common aspects of the different techniques. This approach will highlight a convergence of techniques for magnetic resonance imaging (MRI) assessment of cardiac tissue kinematics. An overview of image analysis and methods for reconstruction of 2-D and 3-D motion and strain is also given, with some discussion on how these methods can be applied to the different imaging procedures. For a more comprehensive review of MRI cardiac deformation analysis techniques, see Reference 9.

10.2 TAGGING

10.2.1 SELECTIVE TAGGING

Selective saturation pulses have been employed for many years to label tissue and blood [10] and thereby obtain measures of motion and blood flow [11,12]. The basic spin preparation pulse sequence involves a selective (soft) radio frequency (RF) 90° pulse combined with a slice-select gradient (typically oriented orthogonal to the imaging plane), followed by a spoiler gradient designed to dephase the transverse magnetization in the tag plane. This gives rise to a signal void in the tagged slice, which can be subsequently tracked or used to label blood flow by imaging the remaining longitudinal magnetization. Although this technique can place magnetic tag planes at any position and orientation in the

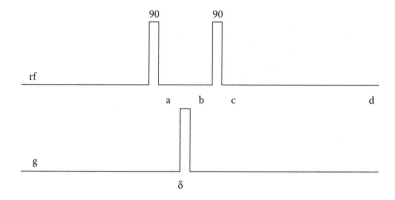

FIGURE 10.1 The 1-1 SPAMM spin preparation sequence. Labels a, b, c, and d refer to various time points, shown in Figure 10.2.

object, the number of tag planes is limited by the time required for each selective tagging pulse.

10.2.2 SPATIAL MODULATION OF MAGNETIZATION

In 1989, Axel and Dougherty developed an efficient nonselective magnetization preparation pulse sequence called spatial modulation of magnetization (SPAMM) [13], which produces a cosine modulation of the longitudinal magnetization. The basic pulse sequence is shown in Figure 10.1.

The first 90° pulse rotates the longitudinal magnetization into the transverse plane. In the standard reference coordinate system (rotating at the Lamor frequency), the transverse magnetization profile along the x direction at time point a is shown in Figure 10.2a. The gradient **g** wraps the phase of the magnetization along the gradient direction (the x direction in Figure 10.2b) by an amount $\phi = \gamma \delta \mathbf{g} \cdot \mathbf{X}$, where **X** is spatial position, γ is the gyromagnetic ratio and δ is the pulse duration (assuming an ideal gradient pulse). The second 90° pulse rotates the magnetization back to the longitudinal (z) direction as in Figure 10.2c. After a period of time in which the transverse component has been removed, due to T2 relaxation or a crusher gradient, the longitudinal magnetization profile (ignoring T1 relaxation) is shown in Figure 10.2d.

This simple tag sequence is known as a 1-1 sequence, referring to the relative magnitudes of the two RF pulses. The sequence is typically applied on detection of the R-wave ECG trigger at end-diastole (ED) and is followed by an imaging sequence to acquire cine images regularly spaced through the cardiac cycle. Variations on the imaging sequence include fast low-angle shot (FLASH), blipped echo-planar imaging (EPI), and steady-state free precession (SSFP). The tag stripe contrast fades with tissue T1 (about 800 msec for myocardium), and is also influenced by the imaging procedure (e.g., EPI offers greater tag persistence than FLASH due to reduced RF excitation).

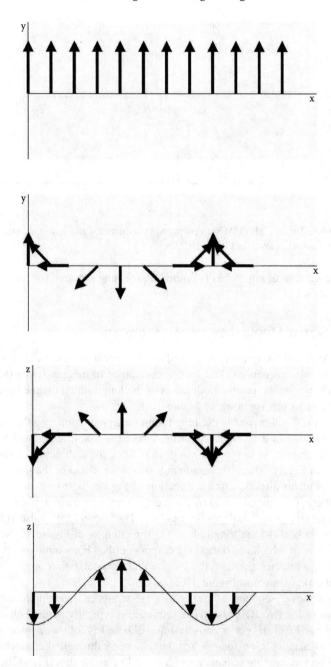

FIGURE 10.2 Magnetization profiles at various times in the spin preparation sequence as labeled in Figure 10.1.

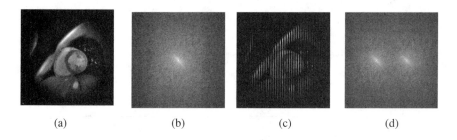

(a)   (b)   (c)   (d)

FIGURE 10.3 (a) Untagged image (multiplied by a windowing function to avoid edge effects). (b) FT of untagged image. (c) Image modulated by cosine function. (d) FT of the modulated image, showing two peaks.

The MR signal is usually acquired in k-space (spatial frequency space), with the spatial image given by a 2-D Fourier transform (FT) of the k-space signal. It is useful to examine the effect of tagging on the k-space data. The amplitude modulation of the spatial image with a cosine is equivalent to a convolution of the k-space data with the frequency space representation of a cosine, which consists of two peaks at either side of the origin (Figure 10.3).

Multiple RF pulses separated by gradients were then designed [14] to make the saturation profile thinner and more square (as in the delays alternating with nutations for tailored excitation [DANTE] sequence [15]). This results in a graphic visualization of tagged-spin motion as dark stripes in the image (Figure 10.4).

10.2.3 COMPLEMENTARY SPAMM

An interesting variant of the SPAMM sequence was introduced by Fischer et al. [16], in which two spin preparation pulses and resulting images are acquired in series. The first acquires the 1-1 SPAMM image and the second acquires an inverted 1-1 SPAMM image, by changing the sign of the last 90° nonselective RF pulse (Figure 10.5).

These two images can then be subtracted to enhance the tagged signal, and eliminate the signal due to T1 relaxation. Initially proposed to improve tag contrast,

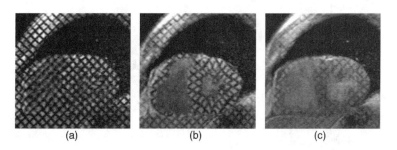

(a)   (b)   (c)

FIGURE 10.4 Five-pulse SPAMM-tagging sequence. (a) ED, (b) end-systole, and (c) late diastole.

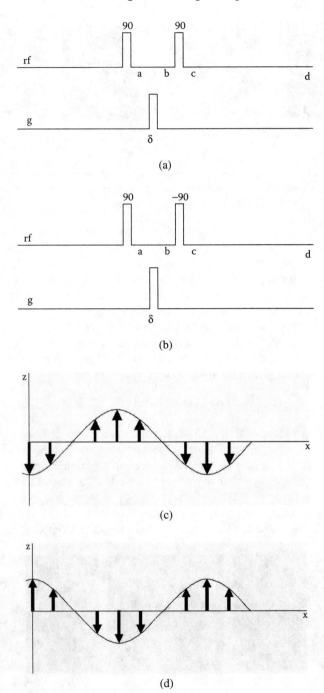

FIGURE 10.5 CSPAMM pulse sequences (a, b) and the complementary tag modulations that result (c, d).

this complementary SPAMM (CSPAMM) technique has subsequently proved useful in eliminating artifacts due to T1 relaxation in the HARP and DENSE techniques, as described in the following sections.

10.3 SPAMM IMAGE ANALYSIS

10.3.1 STRIPE TRACKING

The image stripes visualized in SPAMM images can be tracked in the image plane and used to calculate 2-D motion and deformation (ignoring for the moment the effects of through-plane motion). Kraitchman et al. [17] described a semiautomatic, interactive procedure for tracking tag intersection points. A template matching procedure was used to filter the tagged images to estimate an image intersection likelihood function. Adapting the active contour ("snakes") formulation of Kass et al. [18], a 2-D active contour mesh model comprising interconnected tag intersection points was then used to optimally find the best location of the stripe intersections, subject to constraints on the relative displacement between neighboring tag intersection points. This method was extended to track the entire stripe (including points between stripe intersections) using a 2-D active "carpet" model of interconnected snakes [19]. Each stripe was modeled as a thin, flexible beam with resistance to stretching and bending. Stripes were connected at the intersection points (in the case of grid-tagged images), and the entire mesh deformed to maximize the image-derived feature values (Figure 10.6). Similar stripe tracking procedures have been developed by Guttman et al. [20] and others [9]. User interaction is essential to these tracking processes in order to correct errors due to image contamination and resolve correspondence between frames in cases where the tissue motion was greater than stripe spacing. Image contamination arises due to respiration and gating artifacts, stripe T1 relaxation, spin dephasing due to susceptibility, and poor image SNR. It is, therefore, imperative that the user interaction is efficient, intuitive, and minimal.

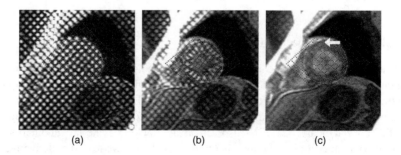

(a)   (b)   (c)

FIGURE 10.6 Grid-tagged image stripes tracked with an active carpet model. (a) end-diastole, (b) end-systole, and (c) late diastole. Note tracking error (arrow) due to spin dephasing at myocardium/fat boundary in late diastole.

10.3.2 Homogeneous Strain

Given tracked landmarks in the form of grid tag intersection points, 2-D motion and strain can be calculated from a triangulation of points within the heart wall [21]. Assuming a homogeneous strain state within the marker triangle, vectors denoting line segments between the triangle vertices ($\mathbf{dX}$ in the undeformed state and $\mathbf{dx}$ in the deformed state) are related by the deformation gradient tensor F [22]:

$$\mathbf{dx} = F \, \mathbf{dX} \tag{10.1}$$

The undeformed state has typically been ED (coinciding with the tag generation pulse or the first image, thereafter), the deformed any subsequent frame, in particular end-systole (ES). Given two or more noncollinear line segments arranged in matrices $A = [\mathbf{dX}_1 \, \mathbf{dX}_2 \, ...]$ and $B = [\mathbf{dx}_1 \, \mathbf{dx}_2 \, ...]$, F can be estimated by linear least squares as

$$F = BA^T(AA^T)^{-1} \tag{10.2}$$

By polar decomposition, the deformation gradient tensor may be separated into an orthogonal unitary rotation tensor R (which describes rotation about the triangle centroid) and a positive symmetric stretch tensor U (which describes strain):

$$F = R \, U \tag{10.3}$$

The Lagrangian (or Green's) strain tensor E can be calculated as [23]

$$E = 0.5 \, (U^2 - I) = 0.5 \, (F^T F - I) \tag{10.4}$$

This strain tensor is related to the change in length of small line segments that are initially aligned with the material coordinate system axes in the undeformed state. These homogeneous strain methods extend naturally to three dimensions [24], in which case four or more points are needed to estimate the homogeneous approximation of F (more points gain more robustness against noise, at the expense of a larger region in which the strain is assumed homogeneous).

10.3.3 Nonhomogeneous Strain

The assumption of homogeneous strain within a region can lead to errors in strain calculation if the region is too large. Typically, strain in the heart varies rapidly in the transmural direction (from outer to inner wall surfaces), and variation also occurs in the circumferential and longitudinal directions (although less rapidly in the normal heart) [21,25]. Also, errors in tracking material landmarks are amplified in the strain calculation (strain can be viewed as the derivative of displacement and as such suffers from the ill-posed nature of numerical differentiation). In order to reduce these errors, a continuous displacement field may be fitted to the tracked points [19]. The displacement field can be modeled using high-order finite elements, with C^1 (value and slope) continuity in displacement between elements, giving C^0 (value) continuity in strain between elements [19]. The finite element method can then be used to estimate kinematics, stress, and activation within the heart wall [2].

With regard to kinematics, the displacement field **u** within each finite element is given by

$$\mathbf{u}(\boldsymbol{\xi}) = \sum_{n=1}^{N} \Psi_n(\boldsymbol{\xi})\mathbf{u}_n \tag{10.5}$$

where **u** is the displacement of the model at element coordinate, $\mathbf{u}_n$ are element parameters, and Ψ_n are associated basis functions. Thus, the interpolated field within each element is a weighted average of the nodal values $\mathbf{u}_n$, with the weights given by the basis functions Ψ_n evaluated at the element coordinates. The displacement field can be fitted to displacements of the tracked points by linear least squares, minimizing the following objective function with respect to the nodal values $\mathbf{u}_n$

$$\varepsilon = S(\mathbf{u}) + \sum_{i} \| \mathbf{u}(\boldsymbol{\xi}_i) - \mathbf{d}_i \|^2 \tag{10.6}$$

where $\mathbf{d}_i$ is the displacement of the ith tracked point and $\mathbf{u}(_i)$ is the corresponding model displacement. $S(\mathbf{u})$ is a smoothing term included to regularize the problem in regions not adequately constrained by the data [19].

Given an undeformed position **X** and deformed position **x** such that $x_i = X_i + u_i$, the nonhomogeneous deformation gradient tensor can be calculated at any point in the model as

$$F_{ij} = \frac{\partial x_i}{\partial X_j} = \sum_{k} \frac{\partial x_i}{\partial \xi_k} \frac{\partial \xi_k}{\partial X_j} \tag{10.7}$$

The R, U, and E tensors are then calculated as earlier. In the following, it is useful to consider the mapping between undeformed (reference or material) position and deformed position $\mathbf{x}(\mathbf{X})$, as well as the reverse mapping $\mathbf{X}(\mathbf{x})$.

10.3.4 RECONSTRUCTION OF 3-D KINEMATICS

Given a set of tagged image slices (typically in standard short- and long-axis orientations), in which at least three linearly independent tag gradient directions have been employed, the 3-D displacement field can be reconstructed using a field fitting approach. The main problem in the reconstruction of 3-D motion and deformation from tagged MR images is the fact that image stripe points tracked through time do not represent tracked 3-D material points. Those material points imaged at ED (shortly after tag creation) move off the imaging plane at subsequent times due to through-plane components of heart motion, so their displacements are not imaged. However, material points imaged at ES (or any other time after ED) are located on the same tag saturation plane as their corresponding tracked image stripe points at ED.

Assuming the ED tag planes (shortly after tag creation) are approximately flat and orthogonal to the image plane, the component of displacement normal to the tag plane can be calculated as the distance from the material point position at ES to its projection on the original tag plane:

$$d_\perp = \mathbf{n} \cdot (\mathbf{x}_{ES} - \mathbf{x}_{ED}) \qquad (10.8)$$

where $\mathbf{n}$ is the normal to the original tag plane (i.e., a unit vector in the direction of the tagging gradient $\mathbf{g}$) and $\mathbf{x}_{ES}$ and $\mathbf{x}_{ED}$ are the positions of the tracked stripe point at ES and ED respectively. All three components of the displacement field from ES to ED can, therefore, be recovered by fitting displacement components from all images simultaneously, minimizing

$$\varepsilon = S(\mathbf{u}) + \sum_i (\mathbf{n}_i \cdot (\mathbf{u}(\boldsymbol{\xi}_i) - \mathbf{u}_i))^2 \qquad (10.9)$$

where $\mathbf{u}_i$ is the 3-D displacement of the tracked stripe point $(\mathbf{x}_{ES} - \mathbf{x}_{ED})$ and $\mathbf{u}(\boldsymbol{\xi}_i)$ is the corresponding model 3-D displacement.

The previous procedure can be used to reconstruct the 3-D displacements of all image stripe points in all desired frames. In order to calculate Lagrangian strain in each frame referred to the ED state, a single ED model must be deformed to each subsequent frame. This can be done using a least-squares fit of the ED model to each subsequent frame, minimizing the objective function defined in Equation 6. Strain can then be calculated at each frame using Equation 7 and Equation 4.

Validation experiments with a deformable MR phantom were performed to determine the magnitude of errors in motion and strain reconstruction using this method [19]. A silicone gel in the shape of a cylindrical annulus was deformed by rotating the inner cylinder with respect to the outer cylinder. This resulted in a well-controlled nonhomogeneous deformation field which could be calculated analytically using a universal solution of the finite elasticity equations. The displacement field for this problem is independent of the material stiffness of the gel. The model reconstructed the displacement field to less than 0.5 mm. The average root-mean-square (RMS) errors in strain were 6% in shear and 16% in the radial axial strain.

Subsequently, this method was extended to interactively reconstruct the 3-D motion and strain directly from the images, without the need for stripe tracking [26]. A set of "model tags" was embedded into a finite element model of the left ventricular geometry at ED. This geometry can be interactively determined in 3-D using guide-point modeling [27]. The intersections of the model tags could then be superimposed on the images (giving a set of "model stripes"), allowing direct comparison between image stripes and model stripes. Image-derived forces were then calculated to pull the model stripes toward the image stripes. A Levenburg–Marquardt nonlinear least-squares algorithm was used to minimize Equation 9, updating the material (element) positions of the image stripe points at each iteration. By interactively modeling the deformation in all slices simultaneously, without the need to stripe tracking on each image, the time required for image analysis was decreased by a factor of 10.

10.3.5 SLICE FOLLOWING

CSPAMM can also be used in conjunction with slice following [28] to give tagged images in which the effects of through-plane motion are eliminated. By making the first RF pulse slice selective, only a thin slice of tissue is excited and tagged by the SPAMM sequence. A thick slice encompassing the tagged slice (and the range of possible through-plane motions) is then imaged by the subsequent imaging pulse sequence. Subtraction of the complementary tagged images cancels the signal in the thick slice which is not tagged, leaving an image of the tagged myocardium only. This simplifies the motion tracking and strain analysis procedure, in that material points imaged at ED are also imaged in each subsequent frame. Note that slice following is not necessary for the evaluation of 2-D or 3-D motion and strain, but does allow a simplified 3-D analysis.

10.4 HARP

10.4.1 THEORY

In 1998, Osman et al. [29] introduced a fast analysis for MR tagged images using harmonic phase, or HARP. Noting that the tagged image is spatially modulated by a cosine (in the case of a 1-1 tag pulse sequence; a sum of cosines for a DANTE type tag pulse sequence), at point c in Figure 10.1 the longitudinal component of the magnetization is

$$M_z(\mathbf{X}) = M_0(\mathbf{X})\cos(\mathbf{k}_e \cdot \mathbf{X}) \tag{10.10}$$

where $\mathbf{k}_e$ is given by the strength and duration of the tagging gradient and $\mathbf{X}$ is the initial position of the material point at the time point c. The Euler equations give the exponential form:

$$M_z(\mathbf{X}) = \frac{M_0(\mathbf{X})}{2}(\exp(j\mathbf{k}_e \cdot \mathbf{X}) + \exp(-j\mathbf{k}_e \cdot \mathbf{X})) \tag{10.11}$$

This expresses the cosine modulation as the sum of two complex phasors rotating in opposite directions, as in Figure 10.7.

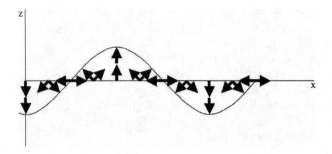

FIGURE 10.7 A cosine can be expressed as the sum of two complex phasors rotating in opposite directions.

At some later time t, the material point initially at **X** has moved to **x**, where **x** = **X** + **u**. Let **X**(**x**) denote a reference map giving the material point position **X** as a function of deformed position **x**. Ignoring T1 effects, the longitudinal magnetization at time t is

$$M_z(\mathbf{x}) = \frac{M_0(\mathbf{X}(\mathbf{x}))}{2}(\exp(j\mathbf{k}_e \cdot \mathbf{X}(\mathbf{x})) + \exp(-j\mathbf{k}_e \cdot \mathbf{X}(\mathbf{x}))) \quad (10.12)$$

Thus, the phase of each of the complex phasors encodes the original position **X**. The HARP technique applies a bandpass filter centered on one of the harmonic peaks in k-space created by the cosine modulation (Figure 10.8). The filter is designed to pass frequency and phase components associated with the harmonic peak, and remove all signal from the other harmonics. The FT of this filtered signal is called the harmonic image. The phase of this complex image, expressed as an angle in the range $[-\pi, \pi)$, is called the harmonic phase. This phase can be thought of as being fixed with respect to material coordinates; as the heart deforms, the harmonic phase of a material point is constant.

(a)   (b)   (c)   (d)

FIGURE 10.8 (a) SPAMM-tagged image at ES. (b) F T showing k-space peaks. (c) Filter to isolate one peak in k-space. (d) Phase of the HARP image, masked to the LV.

Note that we can shift the filtered k-space harmonic peak back to the origin of k-space using a shift operator (i.e., a convolution in k-space with $\delta(\mathbf{f}-\mathbf{k}_e)$: a delta function at $\mathbf{k}_e$). This has the effect of multiplying the harmonic image by $\exp(j\mathbf{k}_e \cdot \mathbf{x})$, giving

$$H(\mathbf{x}) = \frac{M_0(\mathbf{X}(\mathbf{x}))}{2}(\exp(j\mathbf{k}_e \cdot (\mathbf{X} - \mathbf{x}))) \tag{10.13}$$

This is an image whose phase is proportional to displacement $\mathbf{u}$ wrapped into the range $[-\pi, \pi)$.

The spatial resolution of the harmonic phase is determined by the size of the filter used to isolate the spectral peak. If only 32×32 pixels are included in the k-space filter, then the resolution of the harmonic image is approximately 32×32 [30].

10.4.2 KINEMATICS

For small motions (less than π/k_e), the component of displacement $\mathbf{u}$ in the tag encoding direction at a spatial point $\mathbf{x}$ can be calculated as the difference in harmonic phase at $\mathbf{x}$ between the deformed and undeformed time points. For larger displacements, aliasing occurs, and a phase unwrapping procedure must be employed.

Because the harmonic phase is linearly related to material coordinate $\mathbf{X}$, the spatial derivative of the harmonic phase is simply related to the inverse of the deformation gradient tensor F:

$$[F_{ij}]^{-1} = \frac{\partial X_i}{\partial x_j} \tag{10.14}$$

The Eulerian (or Almansi's) strain tensor e is given by

$$e = \frac{1}{2}(I - F^{-T}F^{-1}) \tag{10.15}$$

where F^{-T} is $(F^{-1})^T$. This strain tensor is related to the change in length of line segments that are currently aligned with the spatial coordinate system axes at the deformed time t. Its relationship to the Lagrangian strain tensor E is given by

$$e = \frac{1}{2}(I - R(2E + I)^{-1}R^T) \tag{10.16}$$

A 3-D analysis can be performed using the procedure outlined in Subsection 10.2.5, where $\mathbf{n}$ is a unit vector in the direction of the tagging gradient $\mathbf{g}$ and the displacements $\mathbf{u}$ are directly obtained from the (unwrapped) HARP phase offset (Equation 13).

10.4.3 CSPAMM HARP

An improvement to the original HARP technique was proposed by Kuijer et al. [31], who noted that the HARP angle is corrupted by a signal from other spectral peaks, particularly the signal arising from T1 relaxation. The effect of T1 relaxation is to modify Equation 12 to

$$M_z(\mathbf{x}, t) = (M_0 \cos(\mathbf{k}_e \cdot \mathbf{X}) - M_0) \exp(-t/T1) + M_0 \qquad (10.17)$$

This gives rise to a peak at the center of k-space, which grows with time, as the harmonic peaks decrease with time. To avoid corruption of the HARP image by this signal, the k-space filter is typically kept small (e.g., 32 × 32 pixels), resulting in a low-resolution displacement map. By employing a 1−1 CSPAMM tagging procedure, subtraction of the complementary signals cancels the central peak and only two peaks remain, either of which can be used to construct the HARP image. The size of the filter used to extract the peak can then be enlarged, resulting in a higher resolution displacement map.

10.5 PHASE-CONTRAST VELOCITY

Although only magnitude images have been considered in the preceding section, the phase of the MR image is also typically available, because the MR signal acquisition is usually done in quadrature. This phase can be made sensitive to motion using velocity-encoding gradients. In the classical pulsed gradient spin echo experiment [32], two velocity-encoding gradients are used for this purpose (Figure 10.9).

Spins that do not move during the time interval Δ refocus with no phase offset. For spins that coherently move an amount $\mathbf{u}$ during Δ, the phase shift is $\phi = \gamma \delta \mathbf{g} \cdot \mathbf{u}$ (assuming an ideal gradient profile). If the time interval Δ is kept small, there is little time for T2 decay, and the displacement over Δ can be used as an estimate of velocity.

The gradient echo version of this sequence is shown in Figure 10.10.

Here a bipolar gradient allows encoding of displacement over shorter time intervals, at the cost of T2* decay. In practice, two scans must be acquired with

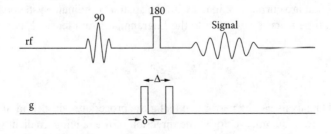

FIGURE 10.9 Pulse gradient spin echo pulse sequence in which displacement between the two gradients is encoded in the phase of the MR signal.

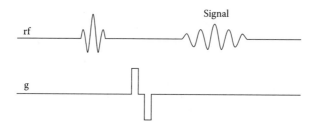

FIGURE 10.10 Gradient echo version of the velocity-encoding sequence.

different velocity-encoding gradient strengths, to allow subtraction of phase off-sets arising from sources other than motion (e.g., eddy currents). The velocity is typically calculated from the phase difference as

$$v = (v_{enc}/\pi)\, \Delta\phi \tag{10.18}$$

where v_{enc} is the velocity that produces a phase shift of π radians. Note that the act of encoding the velocity as a phase angle implies that velocities outside a certain range are aliased into a 2π range, so velocities greater than or equal to v_{enc} or less than $-v_{enc}$ are wrapped to the range $[-v_{enc}, v_{enc})$. Separate acquisitions with velocity-encoding gradients in the x, y, and z directions allow the calculation of all components of the velocity vector.

An advantage of velocity-encoded images (compared with the displacement encoding of tagging techniques such as SPAMM) is improved spatial resolution, because the velocity is measured at every pixel in the image (~1 mm, compared with a typical stripe spacing of 5–8 mm). Also, velocity images allow the simple calculation of instantaneous strain rate. The Cartesian rate of deformation tensor V is derived from the spatial derivative of velocity:

$$V = \frac{1}{2}(D + D^T); \quad D = \frac{\partial \mathbf{v}}{\partial \mathbf{x}} \tag{10.19}$$

The rate of deformation tensor is related to the material derivative of E by [23, p. 446]

$$\frac{DE}{Dt} = F^T V F \tag{10.20}$$

A disadvantage of velocity encoding is that motion through the cardiac cycle must be reconstructed by the integration of myocardial velocities over time. A nonrigid motion tracking procedure is described by Zhu et al. [33]. Due to through-plane motion, material points are not imaged through time, leading to difficulties in the calculation of 3-D displacement.

A stimulated echo variant of the velocity-encoding pulse sequence was proposed by Wedeen et al. [34] to give velocity-encoded images in which the effects of through-plane motion are eliminated. Unlike the slice following CSPAMM

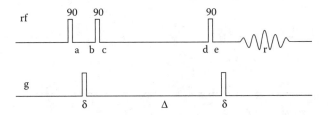

FIGURE 10.11 Stimulated echo version of the pulsed gradient spin echo. Displacement during the interval Δ is encoded in the phase of the signal measured at time point f.

technique (Subsection 10.3.5), which applied an encoding sequence at a fixed point in time and imaged at multiple subsequent times, this novel technique encoded velocity at multiple points in time but imaged at a single fixed time. This results in a series of functional images of a single plane of myocardium which describe velocities and strain rates throughout the cardiac cycle, referenced to the imaged frame.

10.6 DENSE

10.6.1 THEORY

In order to avoid the tracking errors arising from integrating velocity measurements over time, displacement information over long time periods (potentially up to the period of the cardiac cycle) can be encoded in the phase of the MR signal using stimulated echoes [32,35]. The 180° pulse of the pulse gradient spin echo (PGSE) experiment (Figure 10.9) is split into two 90° pulses separated by a time interval which is typically substantially longer than T2 but shorter than T1 (Figure 10.11). The velocity-encoding gradients of the PGSE now encode displacement over the longer "mixing time." Thus, the phase of the image is proportional to the displacement occurring over Δ; however, the signal-to-noise ratio is halved [32,35].

Due to the longer mixing time, the displacement-encoded signal fades over time, and a nonencoded signal gains in strength due to T1 relaxation. The relaxed, non-encoded signal corrupts the displacement map. Traditionally, the size of the encode and decode gradients in the slice direction were set to a large value in order to crush the relaxed component signal. The decode gradient was thus used to both refocus the encoded signal and defocus the relaxed component. However, a common problem with stimulated echo imaging techniques in the heart is that myocardial strain or rotation leads to nonhomogeneous displacement over a voxel. In this case, the phase of the transverse magnetization within the voxel will not completely refocus, leading to signal loss. The size of the signal loss depends on the magnitude of the decode gradient and the amount of strain or rotation occurring in the myocardium. Fischer et al. [35] examined this problem and concluded that stimulated echo techniques had limited application in the heart if the displacement decode gradient is also used as a crusher gradient to destroy signals arising due to T1 relaxation.

Aletras et al. [36] used this method to measure myocardial displacement, and coined the term: displacement encoding with stimulated echoes (DENSE). Although displacement could be encoded at each pixel in the image, displacement over a period of only ~100 msec was able to be imaged, due to the use of crusher gradients in the slice-select direction to destroy the relaxed component. Later [37], an inversion recovery sequence was employed to ameliorate signal due to T1 relaxation. However, due to the specific TI (inversion time), displacement to only one time in the cardiac cycle could be encoded for each imaging sequence.

10.6.2 CSPAMM DENSE (CINE-DENSE)

Kim et al. [38] exploited the similarity between DENSE and SPAMM to develop an imaging sequence which acquires displacement over all frames in the cardiac cycle (cine-DENSE). They noted that the initial displacement-encoding pulse sequence is the same as the tagging pulse sequence applied in 1-1 SPAMM. This is followed (at time d in Figure 10.11) by an imaging sequence, which includes a displacement-decoding gradient of the same magnitude as the encoding gradient. A complementary tag pattern can be created by inverting the second $90°$ RF (as in Figure 10.5), allowing subtraction of the two complementary images to eliminate the signal due to T1 relaxation.

As in CSPAMM HARP, by time d in Figure 10.11, the signal consists of a tag-modulated component together with a component due to longitudinal (T1) relaxation (Equation 17). Between d and f the longitudinal magnetization is transferred into the transverse plane, and a decoding pulse of the same duration as the encoding pulse is applied. This adds a phase to the signal proportional to the current position $\mathbf{x}$:

$$M_z(\mathbf{x},t) = \{(M_0 \cos(\mathbf{k}_e \cdot \mathbf{X}) - M_0)\exp(-t/T1) + M_0\}\exp(-j\mathbf{k}_e \cdot \mathbf{x}) \quad (10.21)$$

According to the shift theorem of the FT, the effect of the decoding gradient is to shift the spectral peaks by $\mathbf{k}_e$ in k-space. Employing the complex exponential form for the cosine and rearranging terms, the signal can be seen to arise from three components:

$$M_{xy}(\mathbf{x},t) = \frac{M_0}{2}\exp(-t/T1)\exp(-j\mathbf{k}_e \cdot (\mathbf{x} - \mathbf{X}))$$
$$+ M_0(1 - \exp(-t/T1))\exp(-j\mathbf{k}_e \cdot \mathbf{x}) \quad (10.22)$$
$$+ \frac{M_0}{2}\exp(-t/T1)\exp(-j\mathbf{k}_e \cdot (\mathbf{x} + \mathbf{X}))$$

The first component is the desired DENSE echo, whose phase is directly proportional to the tissue displacement $\mathbf{u}$. The second term arises out of the T1 relaxation and has a phase modulated by the spatial position $\mathbf{x}$. The third component has a phase modulated by $\mathbf{x} + \mathbf{X}$, i.e., a complex sinusoid that varies approximately twice as fast as the initial tag frequency. These three signal components give rise to spectral peaks in k-space at 0, $\mathbf{k}_e$ and $2\mathbf{k}_e$, respectively.

As in previous DENSE implementations [36,37], the tag gradient $\mathbf{k}_e$ can be chosen high enough (greater than 0.25 cycles per pixel) so that the third spectral peak is shifted above the maximum frequency sampled (Figure 10.12). (Note that the spatial frequencies above twice the sampling frequency are removed with an analog low-pass filter before the signal is sampled, so that the higher spatial frequencies will not be aliased into the digitized signal.)

The second term, arising due to T1 relaxation, has a phase modulated by $\mathbf{k}_e \cdot \mathbf{x}$. This term cannot be pushed out of the readout window because large values of $\mathbf{k}_e$ result in unacceptable signal loss due to myocardial strain. Instead, the CSPAMM technique can be used to subtract out the relaxed component. As in CSPAMM HARP, two data sets are acquired, in which the second RF pulse in the 1-1 SPAMM encoding is $+90$ in the first set and -90 in the second set. Subtraction of these two images reinforces the first term and cancels out the second term [38].

10.6.3 KINEMATICS

Analysis of DENSE images can be performed in the same way as for HARP (Subsection 10.4.2). Spatial derivatives of phase can be used to calculate the Eulerian strain at each time frame. For reconstruction of displacements greater than $\pi/\mathbf{k}_e$ pixels, a phase unwrapping procedure must be used. A 3-D analysis can be performed using the procedure outlined in Subsection 10.2.5, in which $\mathbf{n}$ is a unit vector in the direction of the encode gradient $\mathbf{g}$ and the displacements $\mathbf{u}$ are directly obtained from the (unwrapped) DENSE phase.

10.7 DIFFUSION

The MRI signal can also be made sensitive to diffusion, primarily due to the incoherent motion of water within the tissue [39]. In the heart, diffusion tensor imaging can be used to determine the direction of maximum diffusivity, which corresponds to the eigenvector of the maximum eigenvalue of the diffusion tensor [40]. This is typically aligned in the direction of the muscle cells [41]. The second and third eigenvectors can give information on the layered structure of the myocytes [42]. It is interesting to note that the pulse sequence for diffusion-weighted imaging is the same as shown in Figure 10.9. In the case of diffusion imaging, any incoherent motion between the two gradients results in a diminished echo magnitude (due to partial dephasing within a voxel that is not refocused by the decode gradient). The stimulated echo version is shown in Figure 10.11. This sequence has been used to measure diffusion in the beating heart [42]. Thus, DENSE imaging is intrinsically diffusion weighted. As shown in the preceding text, DENSE is also strain weighted, because any strain within the tissue will also cause loss in signal. This was exploited by Osman et al. [43], who used the

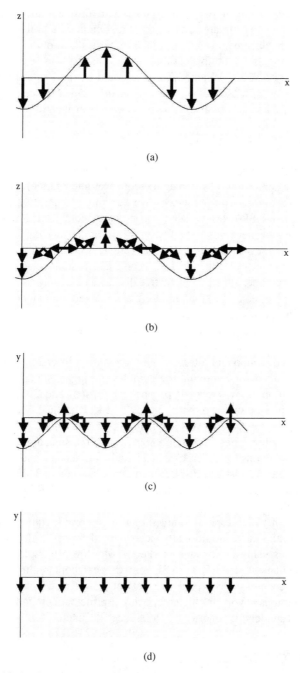

FIGURE 10.12 (a) Longitudinal magnetization after the 1-1 SPAMM preparation. (b) Representation as the sum of two complex phasors. (c) Transverse magnetization after application of the decoding gradient. (d) Sampled signal after application of the analog low-pass filter to remove the signal oscillating at $\sim 2\mathbf{k}_e$.

signal difference between two acquisitions with different through-plane tag-encoding gradients to estimate strain in the direction normal to the slice. Relative through-plane displacement was determined from the phase of the DENSE image. However, the effects of motion parallel to the image plane (e.g., shears) were ignored. In theory, strain-weighted imaging could be used to estimate all components of the strain tensor, using at least six different displacement- encoding directions, in a manner similar to MR diffusion tensor imaging.

10.8 FUTURE DIRECTIONS

The field of MRI cardiac motion estimation is developing rapidly. True fast imaging with steady-state precision (FISP) imaging sequences allow faster image acquisition of tagged images [44]. With phase-sensitive quadrature acquisitions, the phase of both peaks in a 1-1 SPAMM-encoded image can be combined to reduce artifacts due to field inhomogeneity, susceptibility, etc. [45]. Parallel imaging techniques using 8 or more element cardiac coils allow 3-D cardiac acquisitions in a single breath hold, or in real time [46]. Eulerian HARP or DENSE strain analysis can also be performed in real time, allowing interactive display as the images are acquired.

An alternative technique for removing the relaxed component of SPAMM-tagged images, proposed by Aletras et al. [47], involves inversion RF pulses regularly spaced in the imaging sequence (alternating inversion recovery [AIR] SPAMM). This effectively keeps the relaxed component alternating about zero, while the tagged component alternates in sign at each inversion.

Another method for removing the complex conjugate signal (the third term at ~$2\mathbf{k}_e$ in Equation 22) was recently proposed by Epstein and Gilson [48]. Instead of increasing the size of $\mathbf{k}_e$ so that the peak at $2\ \mathbf{k}_e$ is pushed out of the readout window, two additional CSPAMM acquisitions are performed with $\sin(\mathbf{k}_e \cdot \mathbf{X})$ and $-\sin(\mathbf{k}_e \cdot \mathbf{X})$ encoding. These can be combined with the standard $\cos(\mathbf{k}_e \cdot \mathbf{X})$ and $-\cos(\mathbf{k}_e \cdot \mathbf{X})$ acquisitions in order to cancel both the second and third terms in Equation 22, leaving just the displacement-encoded signal. This technique can be used to reduce the size of $\mathbf{k}_e$ (reducing the amount of SNR loss due to strain) or to encode displacement perpendicular to the slice.

This issue of which type of strain should be reported requires further study. Lagrangian strains have traditionally been quoted for SPAMM-tagged studies [25] and studies using radiopaque markers [24]. However, Eulerian strains are easier to calculate in HARP, DENSE and tissue Doppler ultrasound studies. Statistical procedures to compare one patient group with another need to be devised. Preliminary work in this area, using a principal component analysis of finite element models, is reported in Reference 9, Chapter 3.

10.9 CONCLUSIONS

It can be seen that the field is in a stage of rapid development, and any review of current techniques is almost immediately outdated. The techniques of SPAMM, HARP, and DENSE are closely related and can be thought of as variants on the same technique. Both HARP and DENSE are methods designed to select one spectral peak of the SPAMM-tagged signal. In HARP, the peak is selected using a software filter and can be shifted to the center of k-space to give an image whose phase is proportional to displacement. In DENSE, the spectral peak is selected using hardware filters (i.e., by manipulating the signal acquired by the scanner) and is shifted to the center of k-space using the decoding gradient. Because a k-space filter is not required for DENSE, the resolution of the displacement map is greater. This increased spatial resolution comes at the cost of reduced SNR and intravoxel dephasing at high strains or rotations.

REFERENCES

1. Herman, M.V., Heinle, R.A., and Klein, M.D., (1967). Localized disorders in myocardial contraction: Asynergy and its role in congestive heart failure. *New Eng. J. Med.* 277: 222–232.
2. Hunter, P.J. and Smaill, B.H. (1988). The analysis of cardiac function: a continuum approach. *Prog. Biophys. Mol. Biol.* 52: 101–164.
3. http://www.bioeng.auckland.ac.nz/physiome/physiome.php (accessed July 30, 2004).
4. Rodriguez, F., Tibayan, F.A., Glasson, J.R., Liang, D., Daughters, G.T., Ingels, N.B., Jr., and Miller, D.C. (2004). Fixed-apex mitral annular descent correlates better with left ventricular systolic function than does free-apex left ventricular long-axis shortening. *J. Am. Soc. Echocardiogr.* 17(2): 101–107.
5. Feng, Q., Song, W., Lu, X., Hamilton, J.A., Lei, M., Peng, T., and Yee, S.P. (2002). Development of heart failure and congenital septal defects in mice lacking endothelial nitric oxide synthase. *Circulation* 106(7): 873–879.
6. Young, A.A. (1991). Epicardial deformation from coronary cineangiograms. in Glass, L., Hunter, P.J., McCulloch, A.D., (Eds.). *Theory of Heart*, Springer-Verlag, pp. 175–208.
7. Miyatake, K., Yamagishi, M., Tanaka, N., Uematsu, M., Yamazaki, N., Mine, Y., Sano, A., and Hirama, M. (1995) New method for evaluating left ventricular wall motion by color-coded tissue Doppler imaging: in vitro and in vivo studies. *J. Am. Coll. Cardiol.* 25: 717–724.
8. D'Hooge, J., Heimdal, A., Jamal, F., Kukulski, T., Bijnens, B., Rademakers, F., Hatle, L., Suetens, P., and Sutherland, G.R. (2000). Regional strain and strain rate measurements by cardiac ultrasound: principles, implementation and limitations. *Eur. J. Echocardiogr.* 1: 154–170.
9. Amini, A.A. and Prince, J.L. (2001). *Measurement of Cardiac Deformations from MRI: Physical and Mathematical Models.* Dordrecht: Kluwer Academic Publishers.
10. Morse, O.C. and Singer, J.R. (1970). Blood velocity measurement in intact subjects. *Science* 170: 440–441.
11. Axel, L., Shimakawa, A., and MacFall, J. (1986). A time-of-flight method of measuring flow velocity by magnetic resonance imaging. *Magn. Res. Imaging* 4(3): 199–205.
12. Zerhouni, E.A., Parish, D.M., Rogers, W.J., Yang, A., and Shapiro, E.P. (1988). Human heart: tagging with MR imaging — a method for noninvasive assessment of myocardial motion. *Radiology.* 169(1): 59–63.

13. Axel, L. and Dougherty, L. (1989). MR imaging of motion with spatial modulation of magnetization. *Radiology.* 171(3): 841–845.

14. Axel, L. and Dougherty, L. (1989). Heart wall motion: improved method of spatial modulation of magnetization for MR imaging. *Radiology.* 172(2): 349–350.

15. Mosher, T.J. and Smith, M.B. (1990). A DANTE tagging sequence for the evaluation of translational sample motion. *Magn. Reson. Med.* 15(2): 334–339.

16. Fischer, S.E., McKinnon, G.C., Maier, S.E., and Boesiger, P. (1993). Improved myocardial tagging contrast. *Magn. Reson. Med.* 30(2): 191–200.

17. Kraitchman, D.L., Young, A.A., Chang, C.-N., and Axel, L. (1995). Semi-automatic tracking of myocardial motion in MR tagged images. *IEEE Trans. Med. Imaging.* 14: 422–433.

18. Kass, M., Witkin, A., and Terzopoulos, D. (1987). Snakes: Active contour models. *Int. J. Comput. Vision.* 1: 321–331.

19. Young, A.A., Kraitchman, D.L., Dougherty, L., and Axel, L. (1995). Tracking and finite element analysis of stripe deformation in magnetic resonance tagging. *IEEE Trans. Med. Imaging.* 14: 413–421.

20. Guttman, M.A., Prince, J.L., and McVeigh, E.R. (1994). Tag and contour detection in tagged MR images of the left ventricle. *IEEE Trans. Med. Imaging.* 13: 764–788.

21. Young, A.A., Imai, H., Chang, C.-N., and Axel, L. (1993). Two-dimensional left ventricle motion during systole using MRI with SPAMM. *Circulation.* 89: 740–752.

22. Meier, G.D., Ziskin, M.C., Santamore, W.P., and Bove, A.A. (1980). Kinematics of the beating heart. *IEEE Trans. Biomedical. Eng.* 27: 319–329.

23. Fung, Y.C. (1965). *Foundations of Solid Mechanics.* United Kingdom: Prentice-Hall.

24. Waldman, L.K., Fung, Y.C., and Covell, J.W. (1985). Transmural myocardial deformation in the canine left ventricle. Normal in vivo three-dimensional finite strains. *Circ. Res.* 57(1): 152–163.

25. Young, A.A., Kramer, C.M., Ferrari, V.A., Axel, L., and Reichek, N. (1994). Three-dimensional left ventricular deformation in hypertrophic cardiomyopathy. *Circulation.* 90: 854–867.

26. Young, A.A. (1999). Model Tags: Direct 3-D tracking of heart wall motion from tagged magnetic resonance images. *Med. Image Anal.* 3: 361–372.

27. Young, A.A., Cowan, B.R., Thrupp, S.F., Hedley, W.J., and Dell'Italia, L.J. (2000). Left ventricular mass and volume: Fast calculation with guide-point modeling on MR images. *Radiology.* 216(2): 597–602.

28. Fischer, S.E., McKinnon, G.C., Scheidegger, M.B., Prins, W., Meier, D. and Boesiger, P. (1994). True myocardial motion tracking. *Magn. Reson. Med.* 31(4): 401–413.

29. Osman, N.F., McVeigh, E.R., and Prince, J.L. (2000). Imaging heart motion using harmonic phase MRI. *IEEE Trans. Med. Imaging.* 19(3): 186–202.

30. Parthasarathy, V. and Prince, J.L. (2004). Strain resolution from HARP-MRI. *Proc. Int. Soc. Magn. Reson. Med.* 11: 1797. Kyoto, May 16–20.

31. Kuijer, J.P., Jansen, E., Marcus, J.T., van Rossum, A.C., and Heethaar, R.M. (2001). Improved harmonic phase myocardial strain maps. *Magn. Reson. Med.* 46(5): 993–999.

32. Callaghan, P.T. (1993). *Principles of Nuclear Magnetic Resonance Microscopy.* New York: Oxford University Press. p. 163.

33. Zhu, Y. and Pelc, N.J. (1999). A spatiotemporal model of cyclic kinematics and its application to analyzing nonrigid motion with MR velocity images. *IEEE Trans. Med. Imaging.* 18(7): 557–569.

34. Wedeen, V.J., Weisskoff, R.M., Reese, T.G., Beache, G.M., Poncelet, B.P., Rosen, B.R., and Dinsmore, R.E. (1995). Motionless movies of myocardial strain-rates using stimulated echoes. *Magn. Reson. Med.* 33(3): 401–408.

35. Fischer, S.E., Stuber, M., Scheidegger, M.B., and Boesiger, P. (1995). Limitations of stimulated echo acquisition mode (STEAM) techniques in cardiac applications. *Magn. Reson. Med.* 34(1): 80–91.

36. Aletras, A.H., Ding, S., Balaban, R.S., and Wen, H. (1999). DENSE: displacement encoding with stimulated echoes in cardiac functional MRI. *J. Magn. Reson.* 137(1): 247–252.

37. Aletras, A.H. and Wen, H. (2001). Mixed echo train acquisition displacement encoding with stimulated echoes: an optimized DENSE method for in vivo functional imaging of the human heart. *Magn. Reson. Med.* 46(3): 523–534.

38. Kim, D., Gilson, W.D., Kramer, C.M., and Epstein, F.H. (2004). Myocardial tissue tracking with two-dimensional cine displacement-encoded MR imaging: development and initial evaluation. *Radiology.* 230(3): 862–871.

39. Le Bihan, D., Breton, E., Lallemand, D., Grenier, P., Cabanis, E., and Laval-Jeantet, M. (1986). MR imaging of intravoxel incoherent motions: application to diffusion and perfusion in neurologic disorders. *Radiology.* 161(2): 401–407.

40. Basser, P.J., Mattiello, J., and LeBihan, D. (1994). MR diffusion tensor spectroscopy and imaging. *Biophys. J.* 66(1): 259–267.

41. Hsu, E.W., Muzikant, A.L., Matulevicius, S.A., Penland, R.C., and Henriquez, C.S. (1998). Magnetic resonance myocardial fiber-orientation mapping with direct histological correlation. *Am. J. Phys.* 274: H1627–1634.

42. Dou, J., Tseng, W.Y., Reese, T.G., and Wedeen, V.J. (2003). Combined diffusion and strain MRI reveals structure and function of human myocardial laminar sheets in vivo. *Magn. Reson. Med.* 50(1): 107–113.

43. Osman, N.F., Sampath, S., Atalar, E., and Prince, J.L. (2001). Imaging longitudinal cardiac strain on short-axis images using strain-encoded MRI. *Magn. Reson. Med.* 46(2): 324–334.

44. Zwanenburg, J.J., Kuijer, J.P., Marcus, J.T., and Heethaar, R.M. (2003). Steady-state free precession with myocardial tagging: CSPAMM in a single breathhold. *Magn. Reson. Med.* 49(4): 722–730.

45. Ryf, S., Schwitter, J., Tsao, J., Stuessi, A., and Boesiger, P. (2004). Peak-combination HARP for increased reproducibility of tagging analysis. *Proc. Int. Soc. Magn. Reson. in Med.* p. 1784.

46. Kellman, P., Larson, A.C., Zhang, Q., Simonetti, O.P., Arai, A.E., and McVeigh E.R. (2004). Cardiac cine 3d trueFISP parallel imaging using auto-calibrating 2d-TSENSE. *Proc. Int. Soc. Magn. Reson. Med.* p. 2120.

47. Aletras, A.H., Freidlin, R.Z., Navon, G., and Arai, A.E. (2004). AIR-SPAMM: alternative inversion recovery spatial modulation of magnetization for myocardial tagging. *J. Magn. Reson.* 166(2): 236–245.

48. Epstein, F.H. and Gilson, W.D. (2003). Displacement-encoded MRI of the heart using cosine and sine modulation to eliminate (CANSEL) artifact-generating echoes. *Proc. Int. Soc. Magn. Reson. Med.* p. 1860.

Part IV

Spectroscopy, Diffusion, Elasticity: From Modeling to Parametric Image Generation

11 Principles of MR Spectroscopy and Chemical Shift Imaging

Uwe Klose and Filip Jiru

CONTENTS

11.1 INTRODUCTION

Most of the clinical magnetic resonance (MR) scanners being used mainly for MR imaging allow additionally the acquisition of MR spectra. In MR spectroscopy, information about the distribution of chemical compounds in a chosen volume of interest can be obtained, and signals from various nuclei present in the compounds can be observed. Feasible nuclei for *in vivo* measurements on patients are hydrogen-1, phosphorus-31, carbon-13, and fluorine-19. The most frequently used nucleus for *in vivo* MR spectroscopy is hydrogen-1. This nucleus consists only of one proton, and this measurement technique is therefore often

369

termed *proton spectroscopy* (H-MRS). If a spectroscopic examination is performed under ideal conditions with *in vitro* samples of human tissue, a great variety of signals can be observed in H-MRS. Under the limitations of *in vivo* examinations, signals from only a few metabolites can be clearly identified in the spectrum. The metabolites with the signals that are easiest to evaluate are creatine, choline, and *N*-acetyl aspartate (NAA), which is a neurotransmitter observed predominantly in examinations of brain and spinal cord. Signals from other metabolites such as glutamate, glutamine, or citrate can be studied, if advanced measurement and evaluation techniques are applied, but many other molecules remain invisible in spectroscopic examinations due to their low concentration within the tissue, or due to short relaxation times or strong coupling effects. The challenge of *in vivo* spectroscopy is, therefore, the interpretation of signals of those few molecules that can be identified within the spectrum and that might give important additional diagnostic information.

In the case of H-MRS, no additional hardware is necessary to perform spectroscopic measurements at an MR scanner. The available field gradients are used for volume selection, and the radio frequency (RF) coils developed for MR imaging can be used to apply RF pulses and to acquire the spectra. Only special measurement sequences and evaluation procedures are necessary. In this chapter, the most common measurement techniques are described, whereas appropriate evaluation techniques are the topic of the following chapter. Because H-RMS is the most frequently used type of MR spectroscopy in a clinical environment, it is the chief topic of this chapter. Two possible measurement techniques for H-RMS are described in two sections: single-voxel spectroscopy (SVS), which can be used to obtain spectroscopic information from one specific selected voxel, and chemical shift imaging (CSI), in which spectroscopic information is acquired for several different locations with a single measurement.

11.2 SVS

Spectroscopic measurements were the first type of measurement to use the magnetic resonance principle in the 1950s, and they were performed to obtain information about the compounds of a chemical specimen in a probe within a test tube. In these measurements, the spatial origin of the measured signals was defined by the sensitivity of the used RF coil. In clinical patient examinations, such a rough localization is insufficient. Spectroscopic measurements should provide additional information about specific tissues within the brain that are identified in conventional MR imaging. Therefore, a spatial selectivity with an accuracy of at least 1 cm is necessary for a spectroscopic measurement.

In SVS, this selectivity is realized by a combination of slice-selective excitations. Each of these excitations works as in MR imaging: an RF pulse with a specific frequency bandwidth is applied while a field gradient is switched on. Thus, the excitation of nuclei is restricted to a selected slice. By modifying the strength of the field gradient and the center of the RF band, the thickness of the slice and the distance of the center of the slice from that of the magnet can be modified. The orientation of the selected slice can also be chosen freely by an appropriate weighted combination of the three available field gradients within the magnet. To select a volume instead of a slice, three slice selections with different orientations are necessary. The selected volume (volume of interest) is then built by those spatial points that are part of all three selected slices, i.e., the intersection of the three selected slices.

The combination of different slice selections can be realized within a single measurement sequence in such a way that only signals from the selected volume of interest are obtained (single-shot measurement). Two types of sequences use this principle: the PRESS sequence (point-resolved spectroscopy (1), with a 90° excitation and 2 180° refocusing pulses and the STEAM sequence (stimulated echo acquisition mode (2), in which 3 90° pulses are used. These single-shot sequences are mostly used in H-MRS, but they require a certain time difference between the first excitation and the beginning of the data acquisition (echo time). For measurements with other nuclei, e.g., phosphorus, it might be necessary to realize short echo times. This is possible if three measurements with different magnetization preparations are performed, and the obtained signals are combined in an appropriate way. This measurement technique is called ISIS (3), and is described at the end of this chapter.

11.2.1 PRESS AND STEAM

The scheme of the PRESS sequence is shown in Figure 11.1. In the upper line, the three RF pulses can be seen. Each of them is applied simultaneously with a

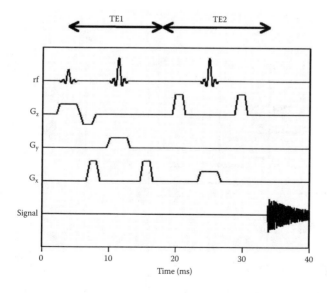

FIGURE 11.1 PRESS sequence with three RF pulses applied simultaneously with field gradients along the main axes of the magnet. Only the first part of the data acquisition time is shown.

different field gradient. After the excitation by the first 90° pulse, transversal magnetization is produced within a slice perpendicular to the z axis in this case. This magnetization starts to dephase and, after a time TE1/2, the magnetization of a part of the slice is refocused by the first 180° pulse in the same way as in conventional spin-echo sequences. The spin echo occurring at the time TE1 is not evaluated, the magnetization dephases again, and the second 180° pulse is applied at the time TE1 + TE2/2 to give an echo at the time TE1 + TE2, where the data acquisition time starts. In contrast to imaging sequences, in which the spin echo occurs usually in the center of the data acquisition time, the acquisition of data in spectroscopy sequences usually starts at the center of the spin echo and lasts for several 100 msec. This is necessary to be able to analyze small frequency differences of a few hertz in the chemical shift of the observed molecules.

The use of the double spin echo is necessary for the realization of the desired spatial selectivity. The steps of volume excitation are depicted in Figure 11.2. After selective excitation of a slice in a xy plane by the first 90° pulse, the refocusing effect of the first 180° pulse is restricted to a slice in a xz plane. Therefore, the first spin echo occurs only for spins within a bar that includes those spatial positions that are part of both these slices. At all other parts of the first excited slice, the magnetization at the echo time TE1 is dephased due to the missing refocusing effect of the 180° pulse. The second 180° pulse again affects a different slice, which is in a yz plane in the example of Figure 11.2. Only those spins that are located in the crossing volume of the previously selected bar and

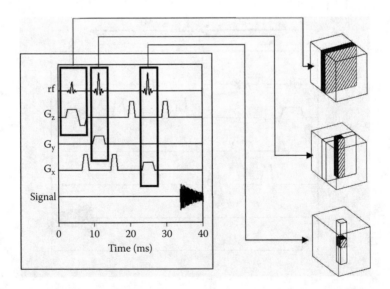

FIGURE 11.2 Selection of a cube with a PRESS sequence. The three RF pulses within the sequence are marked, and the selected regions after each pulse are shown for a cubic object.

the new slice are affected by this refocusing. This crossing volume builds a cube with an edge length, which is the slice thickness of the three RF pulses. Only in this cube does a magnetization excited by the 90° pulse and fully rephased by both 180° pulses exist.

The restriction of the acquired signal to this selected volume of interest assumes, however, that all signals from other parts of the examined subject are cancelled out. Usually, the homogeneity of the magnetic field in an examined region of the body is very high and, therefore, the dephasing due to field inhomogeneities of the static field within the patient is not strong enough to destroy the visible magnetization of the regions outside of the selected voxel of interest (VOI). To improve the suppression of signals from outside the VOI, additional field gradients (so-called spoiler gradients) are applied within the PRESS sequence, which are shown in Figure 11.2. The effect of these spoiler gradients vanishes within the VOI due to the effect of the refocusing 180° pulses, but outside the VOI they intensify the dephasing effect on the undesirably excited magnetization.

The shape and the orientation of the VOI shown in Figure 11.2 is only an example of the possibilities. In clinical examinations, the slice thickness of all three RF pulses can be chosen independently and, also, oblique slices, realized by combining of two or three of the x, y and z gradients, can be selected, resulting in a tilted cuboid VOI. In the given example, the three selected slices are perpendicular to each other. This is the most usual situation, but it is also possible to use other angles between the slices, resulting in a parallelepiped as the shape of the VOI. Curved and concave surfaces cannot be realized with this measurement technique. Several modifications of this conventional PRESS technique for special demands are described in the literature. Some of them are described in Subsection 11.2.4.

The second frequently used sequence for volume-selective spectroscopy is the STEAM sequence. This technique uses the effect of a stimulated echo occurring after the application of three successive pulses. The most intense signal strength of a stimulated echo can be obtained if all three pulses are 90° pulses (Figure 11.3).

The principle of volume selection is the same as with the double spin-echo sequence in Figure 11.2. The difference is the process of signal production. The first 90° pulse is used to produce transversal magnetization within a selected slice, identical to the 90° pulse within the PRESS sequence (Figure 11.4a, Figure 11.4b). Although most of the spins are in phase immediately after the excitation, they begin to dephase with time under the influence of local inhomogeneities of the static magnetic field and applied field gradients (Figure 11.4c). After a time TE/2, the second 90° pulse rotates the dephased magnetization within the xy plane into the zy plane (Figure 11.4d). The axes x and y are here used within a coordinate system that rotates with the resonance frequency around the z axis. Those components of the magnetization vectors that remain in the transverse plane after the effect of the second RF pulse will experience a dephasing due to inhomogeneities and spoiler gradients and, after some time, these components will cancel each other out (Figure 11.4f). For the longitudinal magnetization, however, no dephasing occurs and, therefore, the z components of the magnetization vectors are

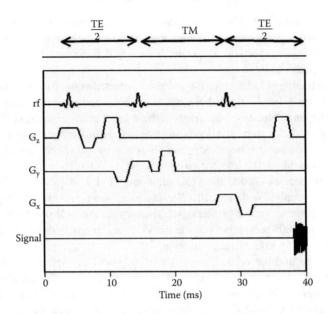

FIGURE 11.3 Scheme for the STEAM sequence. The refocusing gradients have to be positioned before the second and after the third RF pulse.

almost unchanged at the time when the third 90° pulse is applied (Figure 11.4e). The z components of the magnetization experience only T1 relaxation, which is small in the usual time interval of a few milliseconds of the so-called TM interval. After the last 90° pulse, the remaining part of the magnetization is again flipped

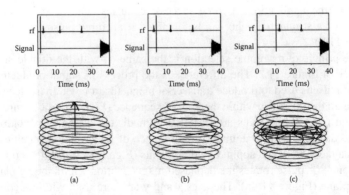

FIGURE 11.4 The orientation of the magnetization vectors of different parts within a selected volume during a STEAM sequence at different times within the sequence. In (e) – (i), the magnetization vectors with lower and larger resonance frequencies than the adjusted frequency are shown in separate spheres. The last figure shows the net magnetization in bold and the modified hypothetical length of the magnetization vector for the assumed T2 decay only (thin line).

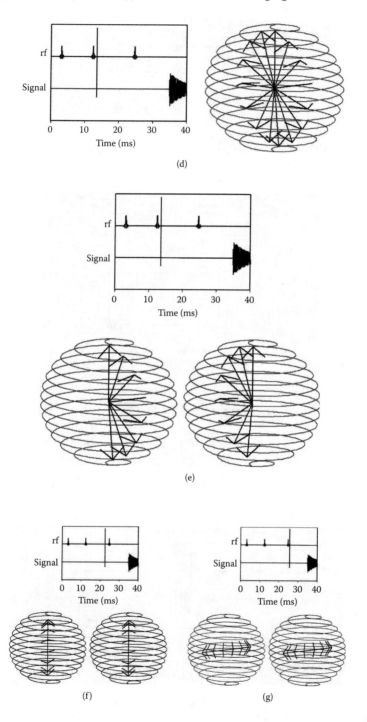

FIGURE 11.4 (Continued).

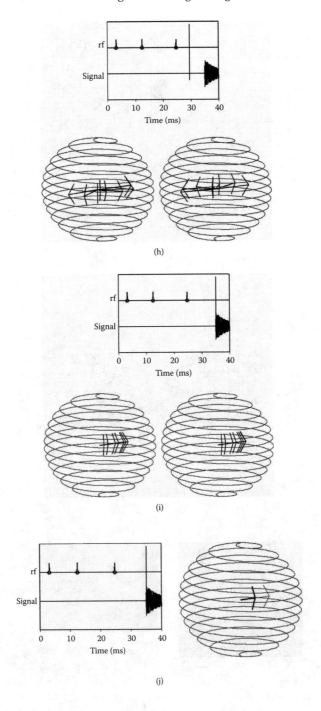

FIGURE 11.4 (Continued).

into the transverse plane (Figure 11.4g). The net magnetization after this pulse is zero because the direction of the magnetization vectors varies between the +x and the −x direction and, therefore, the magnetization contributions are cancelled out. After the third pulse, a rephasing occurs similar to conventional spin-echo sequences (Figure 11.4h) and, at the time TE + TM, all spins are again in the same phase (Figure 11.4i) and build a measurable net magnetization vector (Figure 11.4j). Due to the missing influence of that magnetization, which has been dephased within the TM interval between the second and third pulses, the amplitude of the sum vector at the beginning of the data acquisition is only half of the amplitude that can be obtained in a spin-echo sequence with the same echo time (4). The main advantage of the STEAM sequence compared to PRESS is a reduced minimal echo time. The time between the second and the third RF pulses is part of the echo time in spin-echo sequences. In STEAM sequences, however, the relevant part of the magnetization vectors have only longitudinal magnetization in this time and experience only T1 relaxation, but no T2 relaxation. Because in human tissue T1 is usually much longer than T2, the signal loss in the time between the second and third pulses is much less in STEAM sequences. This sequence is therefore used if very short echo times should be realized. Another advantage of the STEAM sequence is the avoidance of 180° pulses and, therefore, the increased difficulties with the higher pulse amplitude in the center of the pulse and with the nonideal slice profile can be avoided (5).

11.2.2 ARTIFACTS IN SVS

One of the most important quality parameters in SVS is the volume selectivity of the signal. The measurement sequence must avoid any signal from the regions outside the volume of interest. This can be realized by the careful spoiling of unwanted signals. Unwanted signals can originate from all positions in which transverse magnetization is produced during the measurement sequence. Transverse magnetization is produced at first within the whole slice excited by the first 90° pulse, but only the small part of the slice within the volume of interest should contribute to the measured signal. The two following pulses are again excitation pulses in STEAM sequences, and they lead to transverse magnetization throughout the excited slices. The entire volume with unwanted transverse magnetization in a STEAM measurement consists of three orthogonal slices and is shown in Figure 11.5, whereas the volume of interest (from which the signal should originate) is the intersection of the three slices. In order to obtain good spatial selectivity for the acquired MRS signal, any signal contribution from the unwanted transverse magnetization must be strongly reduced. This can be realized by appropriate spoiling, which leads to strong dephasing of the magnetization. Spoiling can be achieved by the application of strong field gradients without RF pulses. The effect of a spoiler gradient is a large distribution of phase angles of the transverse magnetization within even small structures. This phase distribution leads to a cancellation of the signal contributions from these structures. In the volume of interest, which is also affected by the applied spoiler gradients, this signal cancellation can be avoided by the

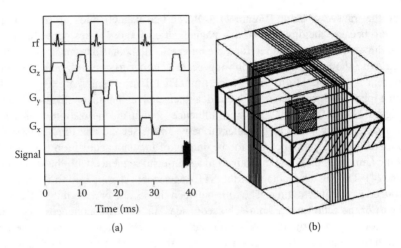

(a)   (b)

FIGURE 11.5 Visualization of the volume excited by the three RF pulses within a STEAM sequence (three orthogonal slices) in comparison to the selected cube of interest.

application of a second field gradient, which has the opposite dephasing effect. The position of this second spoiler gradient within the sequence scheme has to be chosen in such a way that only the desired transverse magnetization within the region of interest is affected, but not the unwanted magnetization within other parts of the examined subject. Strong spoiling is especially necessary if small structures with relatively high signal intensities occur in the examination volume. In H-MRS examinations of the brain, such a critical structure is the subcutaneous fat along the skull. The fat layer is partly very thin, but gives a very strong signal compared to the metabolite signals that should be examined. The adequacy of the spoiling within a sequence for brain H-MRS can, therefore, be checked by the strength of unwanted fat signals (6).

The production of unwanted transverse magnetization is not limited to the excitation pulses of the STEAM sequence. This problem arises also in the PRESS sequence, which uses refocusing 180° pulses. The central parts of the slices selected by these pulses, in which a 180° flip angle is realized, do not experience any additional magnetization. The transverse magnetization produced by the first pulse remains, and in other parts of the examined subject, the 180° pulse produces only negative longitudinal magnetization, which does not lead to any signal. Due to the unavoidable imperfect slice profile of the refocusing pulses, however, there are zones of 90° excitation at the borders of the slices, selected by the 180° pulses. The spins within these zones have transverse magnetization after the occurrence of the 180° pulse, and they make undesirable contributions to the signal if sufficient spoiling is not applied. If this second gradient pulse is applied prior to the third pulse, it compensates for the spoiling effects of the magnetization from the volume of interest, but not for the unwanted signals outside this volume (Figure 11.6).

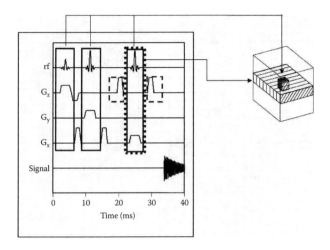

FIGURE 11.6 The different parts within the PRESS sequence. The combination of the three RF pulses (denoted by bold lines) leads to the excitation of the selected voxel. The unwanted signal contributions originating from the last pulse (denoted by dotted lines), which excites the shown slice, can be strongly reduced by additional spoiler gradients (denoted by dashed lines). The signal from the region of interest is not affected by the spoiler gradients due to the effect of the refocusing 180° pulse.

Because there are several crusher gradients in the sequences and various unwanted coherences are dephased differently, the order of the slice selection gradients has a pronounced effect on the overall performance of the sequence and, hence, on the achieved spectral quality (7).

Another problem in volume-selective spectroscopy is the chemical shift displacement. The origin of this artifact is the same as the chemical shift artifact in MR imaging: The resonance frequency of protons in different molecular surroundings varies and, therefore, the exact localization of the selected slice depends on the resonance frequency of the protons. The spatial difference between excitation profiles for protons with a difference $\Delta\omega$ in the resonance frequency is

$$\Delta x = \Delta\omega / \gamma \, G$$

where γ is the gyromagnetic ratio and G the strength of the field gradient applied simultaneously with the RF pulse (Figure 11.7a and Figure 11.7b). This effect occurs with all three excitations of a PRESS or STEAM sequence and leads to a diagonal shift of the voxel position that is dependent on the resonance frequency difference of the protons (Figure 11.7c and Figure 11.7d). Because in H-MRS it is not the water signal that is of interest but the signal of the metabolites, the resonance frequency of the MR system is often adjusted to the resonance frequency of one of the major metabolites (NAA in brain measurements). Nevertheless, it should be noted that the exact localization of the selected region of interest is different for different metabolites

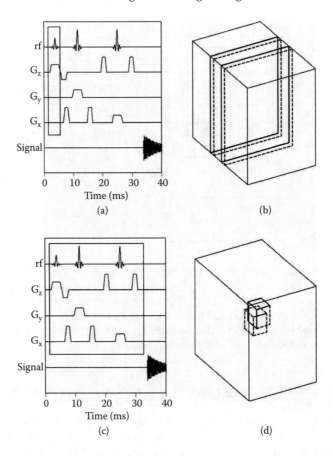

FIGURE 11.7 The effect of two different resonance frequencies (denoted by bold and dashed lines) on the position of the excited volume for one — (a) and (b) — and three RF pulses — (c) and (d) — of a PRESS sequence.

as well as for water and fat signals. Especially for the selected voxel in the proximity of the skull, strong fat signals might occur if the voxel position for the resonance frequency of fat is shifted toward the skull. The extent of the chemical shift displacement can be reduced by the use of strong field gradients for the slice selection.

11.2.3 WATER SUPPRESSION

As in MR imaging techniques, in H-MRS measurements, only signals from hydrogen nuclei are acquired. In contrast to MRI, however, the signals of interest in H-MRS do not originate from water, but from certain metabolites within the body. The concentration of these metabolites is much lower than that of water and, therefore, the intensity of the signal contributions of metabolites is much lower than that of water. The metabolite signals are separable from the water signals because the chemical shift of these signals are specific and different from that of water. This separation can fail, however, if contributions of the water signal get superimposed on the signals from metabolites. Although water has a specific chemical shift value, a typical water signal from an *in vivo* measurement has a certain range. The "foot" of the signal especially can have a considerable width and can, therefore, get superimposed on the signals of metabolites. The width of the water signal is usually described by the width of the peak at the half of the maximum (usually referred to as full width at half maximum, FWHM) and depends on the range of frequencies within the selected volume element, which can be strongly reduced by shimming. However, the difference in the signal intensities between water and the metabolites is in the range of 1000:1 and, therefore, even with a very good shim, a superposition of signals from the foot of the water signal on the metabolite signals will occur. This effect can be reduced if a water suppression technique is used. The simplest technique is the application of a frequency selective RF pulse prior to the first excitation of the volume selection part within a sequence (Figure 11.8). This technique was first used for

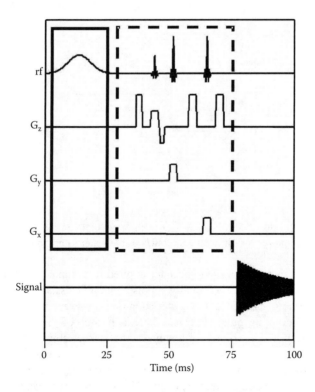

FIGURE 11.8 The implementation of a water suppression pulse within a PRESS sequence. The additional pulse (denoted by the bold line) is applied with a large pulse duration and without a gradient prior to the volume selection part (denoted by the dashed line) of the sequence.

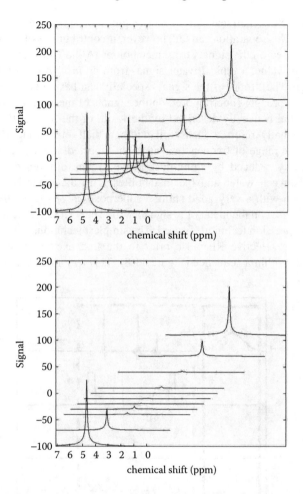

FIGURE 11.9 Results of water suppression with (a) a single pulse and (b) with a combination of three pulses for different amplitudes of the water saturation pulse.

frequency selective imaging (8) and later applied to volume-selective spectroscopy, e.g., by Sijens et. al (9). Usually, a pulse with a Gaussian shape is used for water suppression. The success of water suppression is very sensitive to the flip angle of the saturating pulse. Because the exact value of this flip angle varies within the examination volume of a subject, the correct amplitude of the water suppression pulse is often adjusted prior to each volume-selective measurement. In a series of measurements with varying transmitter values, the one with an optimal water saturation is selected (Figure 11.9a). Improved water suppression can be obtained if a combination of pulses is used instead of only one pulse. The result of a transmitter adjustment for a three-pulse combination is shown in Figure 11.9b. In this case, the relation between transmitter values was 89:83:161 with a time delay of 60 msec between the pulses. This pulse combination was

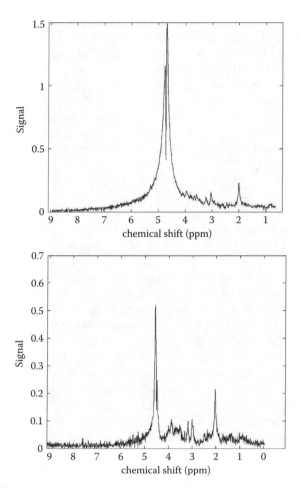

FIGURE 11.10 The effect of optimal water suppression (a) with a single pulse and (b) with the three-pulse combination. The residual contribution of the water signal was significantly lower in the measurement using the pulse combination.

suggested by Ogg et al. (10) for an optimal water suppression if different B1 and T1 values of the water signal occur in the examined object. A strong saturation of the water peak is obtained in a larger range of transmitter values in this case, and the extent of water suppression for the optimal transmitter value is larger with the three-pulse combination than with the one-pulse saturation (Figure 11.10): The reason for the incomplete saturation with one pulse is the composite nature of the water signal. It consists of several compartments with different relaxation times. The result of the optimal water suppression obtained with one pulse shows a superposition of these compartments. In Figure 11.10a, a part with a shorter T2-relaxation time and, therefore, a broader peak shape, shows a residual signal, whereas a component with a longer T2-relaxation time and a narrower line shape already has a negative longitudinal magnetization. Using the three-pulse combination, a more complete water saturation can be obtained (Figure 11.10b).

The optimization of the homogeneity of the local magnetic field within the volume of interest and the evaluation of the optimal RF pulse amplitudes for the water suppression are the two adjustment procedures that are usually performed prior to each single-voxel measurement. In the current generation of MR scanners, these adjustments are performed automatically and take less time than manual optimization. However, the time required for the adjustments has to be taken into account, especially if localized spectroscopy measurements from several positions should be performed.

11.2.4   COUPLING EFFECTS IN SVS

Most of the phenomena in MRI, including the principles of volume selection in MRS, can be explained by the simple model of magnetization vectors that are aligned to the direction of the static magnetic field in the fully relaxed state. These vectors can be affected by the application of RF pulses and show a precession within the transverse plane with a frequency that depends on the local magnetic field. The length of the vectors changes owing to the influence of relaxation processes. The amplitude and the shape of the signal peaks of some metabolites in H-MRS, such as NAA, creatine, and choline, can be completely explained with this simple model. Other metabolites, however, show a complex signal pattern, which not only depends on the chemical shift of the examined metabolites, but also on the interaction of the different protons within these metabolites. If more than one proton in a larger molecule contributes to the signal and if these protons are not magnetically equivalent, e.g., the two protons in water, then a coupling between these protons occurs. The main effect of this is a splitting of the resonance peak. Instead of only one signal, signals at two or more resonance frequencies are obtained. The distance between these frequencies is described by the coupling constant and, unlike the chemical shift, the coupling constant does not depend on the strength of the static magnetic field but is characteristic for a given molecule. The amplitude of the signals for each of the resonance frequencies, however, is strongly dependent on the measurement sequence. The shape of the complex signal patterns, obtained for molecules such as lactate, glutamate, glutamine, GABA, or glutathione can be calculated using a product operator formalism based on quantum mechanics principles (11) if the measurement parameters are exactly known. Most important for the shape of the signal pattern of a given metabolite is the timing of the used sequence (TE1 and TE2 in PRESS, and TE and TM in STEAM sequences) and the used flip angles. For a given combination of measurement parameters, the expected signal pattern can be calculated (12). In *in vivo* measurements, the sequence timing is known, but the exact values of the applied flip angles are often uncertain, as the actual flip angles of the RF pulses at a selected position are usually not equal to, but only near the nominal values. Further, it is not possible in volume-selective spectroscopy to apply pulses with an ideal slice profile. Although the real flip angle might be nearly the nominal flip angle in the center of a selected voxel,

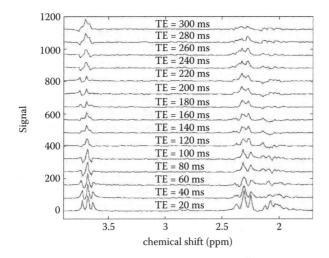

FIGURE 11.11 *In vitro* spectra of glutamate, acquired at different TE values with a STEAM sequence.

all possible flip angle values between that in the center of the voxel and the zero value occur between the center and the borders of the voxel. The signal pattern of a coupled system might, therefore, be different at different positions of the voxel, and only the mixture of all these signals can be obtained in SVS. The strong dependence of the signal pattern on the echo time of the used sequence was described in detail by Ernst and Hennig (13) and is shown in Figure 11.11 for a sample with a glutamate solution. The sample was measured with a STEAM sequence (TR 6 sec, TM 10 msec) using a whole body 3-T system (Siemens Trio) and with TE values between 20 msec and 300 msec. Three different groups of signals can be separated, and the frequencies of these groups are unchanged in all measurements, but especially for the multiplet at 2.1 ppm, the amplitude variations of the individual peaks lead to an almost complete cancellation of the signal at echo times larger than 70 msec. In *in vivo* measurements, additional superpositions with signals from other metabolites occur, which lead to an even stronger cancellation of signals. The observation of signals from molecules with coupled protons often requires, therefore, the use of short echo times in SVS but the superposition of signals will continue to make difficult the separation of signals from different metabolites. This problem can be partly overcome if editing sequences are used or series of sequences with varying measurement parameters are employed.

So-called homonuclear spectral editing techniques are characterized by additional frequency selective RF pulses within the sequence. These pulses can be used as a filter for signals from coupled spins if additional signal subtraction techniques are used (14). This technique can be combined with PRESS or STEAM sequences. Another possibility of improved separation of different metabolites is the two-dimensional (2-D)-spectroscopy. Here, the measurement is repeated several times with continuous change of a selected parameter, e.g., the echo time in PRESS or STEAM sequences. Whereas uncoupled spin shows the expected exponential signal decay with T2 as the time constant, the phase and amplitude of signals of coupled spins oscillate with a monotonously increasing echo time (15). With an additional Fourier transformation, the oscillating frequency, which is related to the coupling constant, can be obtained, and the peak of coupled spins occurs at different positions in a resulting 2-D spectrum (16). This correlation spectroscopy technique (COSY) can also be combined with conventional techniques of volume selection. A major disadvantage of this technique is the long measurement time, which is necessary to acquire data not only at one, but at many different echo times.

11.3 PRINCIPLES OF CSI

11.3.1 BASIC PRINCIPLES

As described in the preceding section, SVS provides information about metabolite composition in the selected volume of interest (voxel). However, in many cases knowledge of metabolite distribution over a large area in the examined sample is preferred. Typical examples are neurodegenerative diseases, in which lesions often have a diffuse character. Spectroscopic imaging (SI), also called CSI, is a method that encodes chemical shift and spatial distribution of metabolites simultaneously (17,18). Spectra from several voxels at different locations, instead of one, are measured during a single measurement. In this respect, the method combines features of both conventional MRI and SVS.

Because information about chemical shifts of individual metabolites present in the signal of each voxel has to be preserved, the classical frequency encoding known from conventional MR imaging cannot be applied. Instead, phase encoding is used exclusively in CSI measurement sequences to obtain information on the spatial distribution of signals.

Similar to MR imaging, depending on how many dimensions spectra are spatially resolved in, 1-D, 2-D, or 3-D CSI can be distinguished. A scheme of the simple 1-D spin-echo CSI sequence is shown in Figure 11.12. Following the excitation pulse, a phase-encoding gradient G along the $\mathbf{x}$ axis is switched on for the time τ. During this time, the precession frequency ω of all spins along the axis $\mathbf{x}$ is modified according to

$$\omega(x) = \gamma x G \tag{11.1}$$

giving rise to space-dependent phase shifts $\phi(\mathbf{x})$ at the end of the phase encoding

$$\phi(x) = -\gamma x G \tau \tag{11.2}$$

where $\mathbf{x}$ is the position of the spins along the $\mathbf{x}$ axis, assuming the gradient isocenter at $x = 0$.

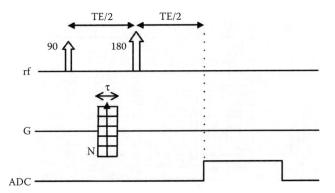

FIGURE 11.12 The scheme of the 1-D spin-echo CSI sequence.

After application of the refocusing pulse at time TE, data are sampled as a second part of the spin echo.

The whole sequence is repeated N times with repetition time TR, while the gradient strength G is changed in N equidistant steps ΔG from the value $G_{min} = -\Delta GN/2$ to $G_{max} = \Delta G(N/21)$ as depicted in Figure 11.12.

Denoting G_l as the gradient strength of the l-th phase-encoding step and introducing variable k_l

$$k_l = \frac{\gamma}{2\pi}\tau G_l = \frac{\gamma}{2\pi}\tau l\Delta G \quad l = -N/2....N/2-1 \tag{11.3}$$

the spatially dependent phase shift $\phi_l(x)$ corresponding to l-th phase-encoding step equals $\phi_l(x) = -2\pi k_l x$.

Because the overall measured signal S(t) is the sum of all elementary signals s(t,x) distributed along **x** axis, taking the additional phase into account, we can for a measured signal write $S(t, k_l)$ as a function of k_l

$$S(t, k_l) = \int_{sample} s(t,x)e^{-i2\pi k_l x}dx \tag{11.4}$$

The measured signal $S(t, k_l)$ is the continuous Fourier transform (FT) of the signals s(t, x) from elementary volumes. The positions of N voxels along the **x** axis can be reconstructed by the inverse discrete Fourier transform (DTF^{-1}).

The 1-D sequence can be easily extended to 2-D (Figure 11.13a) or 3-D (Figure 11.13b) variants. In 2-D and 3-D CSI, 2 and 3 orthogonal phase-encoding gradients are applied, respectively. In reality, as shown in Figure 11.13, the nonselective excitation pulse is often replaced by a frequency selective pulse

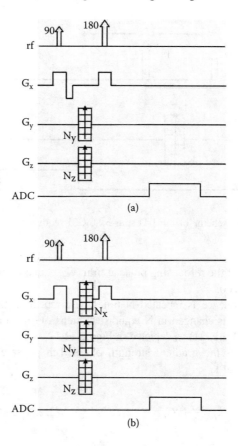

FIGURE 11.13 The scheme of (a) the standard 2-D CSI and (b) the standard 3-D CSI spin-echo sequence.

applied with the slice-selective gradient, resulting in exciting a slice (or slab in case of 3-D CSI). The phase-encoding gradients are then applied in the plane parallel to the slice in 2-D CSI, with the third gradient (called partition-encoding gradient) pointing perpendicular to the slice in 3-D CSI.

After the spatial reconstruction, spectra from each element of a matrix, which is called the spectroscopic grid, are available. In 2-D CSI, the size of the grid corresponds to the field of view (FOV) of the CSI experiment. In 3-D CSI, several grids with the same FOV, each corresponding to one partition of the excited slice thickness, are available. The number of voxels in the spectroscopic grid (and also the number of partitions) depends on the number of phase-encoding steps performed in the sequence along corresponding directions.

To encode positions of the voxels, the sequence with all combinations of phase-encoding increments in all directions has to be repeated. For 3-D CSI with N_x, N_y, N_z steps (voxels) along corresponding direction, and with N_{acq} representing the number of averages, the acquisition time T_{acq} becomes

$$T_{acq} = N_x N_y N_z N_{acq} TR \qquad (11.5)$$

Due to signal-to-noise and quantification requirements of the measured spectra, repetition time TR of the sequence has to be long enough (typically, TR = 1500 msec). For eight phase-encoding steps along each direction (which is the minimum used number, as discussed later), TR = 1500 msec, N_{acq} = 1, and the acquisition time T_{acq} = 12,8 min. Because in clinical measurements, more encoding steps are used to achieve better resolution, T_{acq} becomes too long. Therefore, CSI sequences are mostly used in 2-D mode.

Various fast CSI sequences, suitable for 3-D CSI, have been adopted (19), such as sequences using multiple echoes for phase encoding (20,21), sequences using time-varying gradients during the readout period (22–27), sequences derived from the steady-state MR imaging sequences (28,29), or the recently implemented parallel spectroscopic imaging techniques (30,31).

11.3.2 Avoiding Undesired Excitations

In many cases, examined tissue contains areas with spurious signals, such as areas with poor magnetic field homogeneity, bones, air-containing structures (sinuses), or fatty tissue. These signals are potential sources of the contamination of spectra in other regions. A typical example is ^{1}H CSI of the brain, in which strong lipid signals from extracranial subcutaneous fat can contaminate spectra within the brain. Therefore, methods eliminating signals from problematic regions are desirable. Two most-often-used methods are volume preselection and outer-volume suppression.

The idea of volume preselected CSI sequences is to incorporate a volume selection used in PRESS or STEAM SVS into a CSI sequence (32,33). In this case, only the desired part of the sample, the VOI is excited using the PRESS or STEAM sequence, while the position of the CSI voxels is coded by phase encoding (see Figure 11.14a, in which a 2-D PRESS CSI sequence is depicted). Apart from elimination of spurious signals, this approach has another advantage. As is known from MR imaging, to prevent aliasing artifacts, all regions of the sample contributing to the measured signal have to be inside the FOV (because only phase encoding is performed, frequency low-pass filters cannot be used). This fact dictates the minimum FOV size and also the minimum voxel size achievable per fixed time. Because by using volume preselected sequences only a restricted area of the sample is excited, FOV can be reduced correspondingly, resulting in smaller voxel size without the occurrence of aliasing artifact. However, due to imperfections of pulse profiles, areas outside the selected VOI are also partially excited and contribute to the measured signal. Therefore, the FOV should always extend beyond the VOI to encode positions of these signals (by how much depends on used pulse profiles and the chemical shift artifact), and

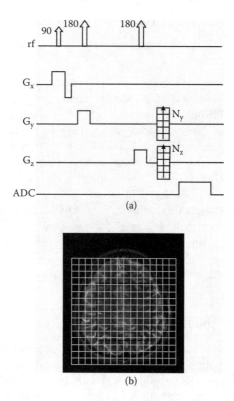

(a)

(b)

FIGURE 11.14 The scheme of (a) the volume preselected 2-D PRESS CSI sequence and (b) the size and the position of the preselected volume (depicted as a black rectangle) with respect to the spectroscopic grid (depicted as a white grid).

the VOI should be always positioned in the center of the spectroscopic grid. An example of VOI positioning with respect to the spectroscopic grid is shown in Figure 11.14b.

Spatial encoding of spectra is ensured by the additional phase shifts arising from application of the phase-encoding gradients. The phase-encoding gradients can be at different positions within the sequence. However, any signal present during data sampling and not experiencing this phase shift will not be spatially encoded and added to all points in k-space with the same phase. Because slice profiles are never ideal, every 180° pulse produces, apart from the desired refocusing effect, transverse magnetization also. This magnetization does not encounter encoding gradients whenever they are applied before the 180° pulse. In this case, the signal from this magnetization will not be encoded, and it will contribute to the central voxel. Therefore, performing of encoding after the last pulse in the sequence is preferred. Unfortunately, to achieve shorter echo times and avoid eddy currents in the measured spectra, phase encoding is often not realized after the last RF pulse of the sequence. Therefore, proper optimization

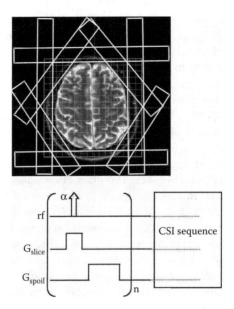

FIGURE 11.15 Outer-volume suppression technique. (a) The signal from the subcutane-ous fat is saturated using eight saturation slices depicted as white rectangles. The positions and the thicknesses of the slices are adjusted before the measurement. (b) The scheme of the OVS sequence. The OVS sequence precedes the CSI sequence.

of the sequence in terms of spoiling all unwanted transverse components of 180° pulses is necessary.

The disadvantage of volume preselection by PRESS or STEAM is the rectangular shape of the VOI. This restricts the size of the excited area, especially in the brain. Alternatively, by using 2-D pulses for excitation, a more general VOI shape can be selected (34). To avoid undesired signals, another method, called outer-volume suppression (OVS) (33,35,36), can also be used. In OVS, areas with spurious signals are saturated by slice selective pulses before the CSI sequence starts. Several saturation slices are available to cover regions to be suppressed, as shown in Figure 11.15a. After each saturation slice is excited, the generated transverse magnetization is spoiled by the crusher gradient as shown in Figure 11.15b. Because the profiles of the slices are never exactly rectangular (which is especially relevant for large slice thicknesses), saturation slices positioned too close to the area of interest can result in partial suppression of the signals inside the area. In the case when a very large area is to be suppressed, one thick slice can be replaced by two thinner adjoined slices, with the better profile. Because saturation pulses are applied in the pulse train, relaxation of saturated magnetizations during the time between the satu-ration pulse and the first excitation pulse should be taken into account, especially when many slices are used. This can be solved by varying flip angles of the saturation slices depending on their position in the saturation sequence.

Also, because magnetization is always partially recovered at the time of application of the following saturation pulses, multiple excitation in overlapping regions may result in unwanted interactions and the refocusing of spoiled magnetization. This leads to suboptimal signal suppression in the overlapping areas. Therefore, to assure complete spoiling of undesired signals, the OVS concept can be combined with volume preselection.

For removal of spurious lipid signals, additional methods can be used such as lipid nulling by means of preparation pulses (37) or using spectral-spatial selective pulses when both only the desired areas and the desired frequency bandwidth are excited at a time (34,38,39). Lipid signals can also be removed by postprocessing methods (40–42). The main disadvantage of lipid nulling is the T1 weighting of all metabolite signals in the spectra. Concerning spectral-spatial pulses, their length depends on the maximum achievable slew rate, resulting in rather long duration of the pulses, small excitation bandwidth, and also limited minimum echo time of the sequence. Generally, suppression of unwanted lipid resonances becomes less critical when long echo times are used, because T2-relaxation time of lipids is much shorter than that of metabolites. However, when quantitative analysis of signals from metabolites having short T2 is desired, volume preselected CSI sequence and outer-volume suppression are the methods of choice.

11.3.3 RECONSTRUCTION OF CSI DATA

In Subsection 11.3.1 the effect of 1-D phase encoding was demonstrated. In this subsection the generalized reconstruction of 3-D CSI will be reviewed and the concept of the point-spread function introduced.

Let us assume a 3-D CSI sequence with the phase-encoding gradients $\mathbf{G} = (G_x, G_y, G_z)$ applied along the orthogonal coordinate system described by unit vectors $\mathbf{i_x}$, $\mathbf{i_y}$, and $\mathbf{i_z}$. As shown in Figure 11.13b, gradient strengths along all three directions are incremented in N_x, N_y, and N_z steps of sizes ΔG_x, ΔG_y, and ΔG_z. Introducing corresponding increment indexes l, m, and n, then $\mathbf{G}_{l,m,n}$, representing the discrete value of the resulting applied gradient, equals

$$\mathbf{G}_{l,m,n} = l\,\Delta G_x \mathbf{i_x} + m\,\Delta G_y \mathbf{i_y} + n\,\Delta G_z \mathbf{i_z}$$
$$l = -N_x/2 \dots (N_x/2) - 1; \text{ m, n correspondingly} \qquad (11.6)$$

Each gradient combination can be associated with the discrete vector $\mathbf{k}_{l,m,n}$

$$\mathbf{k}_{l,m,n} = \gamma \int_0^\tau \mathbf{G}_{l,m,n}(t)dt = l\Delta k_x \mathbf{i_x} + m\Delta k_y \mathbf{i_y} + n\Delta k_z \mathbf{i_z} \qquad (11.7)$$

The integral in Equation 11.7 assumes the general case of the time-dependent gradient **G**. In this case, Equation 11.2 and Equation 11.3 have to be changed correspondingly.

From Equation 11.1 to Equation 11.3, Equation 11.4 can be generalized for 3-D phase encoding

$$S(t, \mathbf{k}_{l,m,n}) = \int_{coil} s(t, \mathbf{r}) e^{-i2\pi \mathbf{k}_{l,m,n}\mathbf{r}} \, d\mathbf{r} \tag{11.8}$$

where the integration is performed over the sensitive area of the coil (denoted as *coil*) and $\mathbf{r}$ represents the position vector relative to the gradient isocenter.

$$\mathbf{r} = x\mathbf{i}_x + y\mathbf{i}_y + z\mathbf{i}_z \tag{11.9}$$

$S(t, \mathbf{k}_{l,m,n})$ represents the analogy of k-space in MR imaging, with the difference that not only is one k-space acquired, as in the case of MR imaging, but one k-space for each time point t is sampled. Because gradient amplitudes are incremented discretely, $S(t, \mathbf{k}_{l,m,n})$ is a discrete function of $\mathbf{k}_{l,m,n}$. The reconstructed signal $s_{rec}(t, \mathbf{r})$ is usually calculated by the discrete Fourier transform (DFT)

$$s_{rec}(t, \mathbf{r}) = \frac{1}{N_x} \frac{1}{N_y} \frac{1}{N_z} \sum_{l,m,n} S(t, \mathbf{k}_{l,m,n}) e^{i2\pi \mathbf{k}_{l,m,n}\mathbf{r}} \tag{11.10}$$

The truncation of the k-space (due to the finite number N_x, N_y, N_z of phase-encoding steps) leads to $\mathbf{r}$-space blurring, resulting in finite spatial resolutions $\Delta x, \Delta y, \Delta z$ along $\mathbf{i}_x, \mathbf{i}_y$, and $\mathbf{i}_z$ (43).

$$\Delta u = \frac{FOV_u}{N_u} = \frac{1}{\Delta k_u N_u} \quad u = \{x, y, z\} \tag{11.11}$$

The number of voxels equals the number of corresponding phase-encoding steps. The second equality in Equation 11.11 gives the relation between the size of the FOV and the step Δk between k values as a consequence of discrete k-space sampling (Nyquist criterion).

Following Equation 11.10 and Equation 11.11, signals from voxels at positions $\mathbf{r}_{l',m',n'}$

$$\mathbf{r}_{l',m',n'} = l'\, \Delta x\, \mathbf{i}_x + m'\, \Delta y \mathbf{i}_y + n'\, \Delta z\, \mathbf{i}_z,$$
$$l' = -N_x/2 \, ... \, (N_x/2) - 1; \, m', n' \text{ correspondingly} \tag{11.12}$$

can be reconstructed by

$$s_{rec}(t, \mathbf{r}_{l',m',n'}) = \frac{1}{N_x} \frac{1}{N_y} \frac{1}{N_z} \sum_{l,m,n} S(t, \mathbf{k}_{l,m,n}) e^{i2\pi \mathbf{k}_{l,m,n}\mathbf{r}_{l',m',n'}} \tag{11.13}$$

Equation 11.13 represents the basic formula for the reconstruction of the voxel signals in the CSI experiment.

To see the relation between the true signal distribution $s(t, \mathbf{r})$ and the signal $s_{rec}(t, \mathbf{r})$, the inverse DFT has to be computed. Substituting $S(t, \mathbf{k}_{l,m,n})$ using Equation 11.8 and inverting the order of the integral and the summation, the reconstructed signal $s_{rec}(t, \mathbf{r})$ in Equation 11.13 can be expressed as a convolution of the true signal distribution $s(t, \mathbf{r})$ and the point-spread function (PSF)

$$s_{rec}(t, \mathbf{r}) = \int_{coil} s(t, \mathbf{u})PSF(\mathbf{r} - \mathbf{u})d\mathbf{u} \qquad (11.14)$$

where the PSF is defined as

$$PSF(\mathbf{r}) = \frac{1}{N_x}\frac{1}{N_y}\frac{1}{N_z} \sum_{l,m,n} e^{i2\pi \mathbf{k}_{l,m,n}\mathbf{r}} \qquad (11.15)$$

The PSF describes the signal of a hypothetical infinite small point object in the reconstructed image and characterizes the efficiency of the employed reconstruction method.

In the case of equally spaced Cartesian sampling, 3-D $PSF(\mathbf{r})$ can be separated into three 1-D PSF, each describing the corresponding dimension

$$PSF(x,y,z) \equiv PSF(\mathbf{r}) = psf_x(x)psf_y(y)psf_z(z) \qquad (11.16)$$

Depending on the number of phase-encoding steps (odd or even number) and the implementation of the gradient incrementing in the measuring sequence, the gradients (and, hence, the $\mathbf{k}_{l,m,n}$ vector) can be sampled symmetrically or asymmetrically with respect to the zero value. This symmetry influences the final shape of $psf_x(x)$, $psf_y(y)$, and $psf_z(z)$. Direct computation of Equation 11.15 for $psf_x(x)$, for example, yields the following results (44):

In the case of symmetric sampling about zero, when the index 1 ranges over $1 = -(N_x-1)/2,\ldots, 0,\ldots, (N_x-1)/2$ for an odd N_x or over $1 = -(N_x-1)/2, \ldots, -1/2, 1/2,\ldots, (N_x-1)/2$ for an even N_x

$$psf_x(x) = \frac{1}{N_x} \sum_{l=-(N_x-1)/2}^{(N_x-1)/2} e^{i2\pi lx/FOV_x} = \frac{1}{N_x}\frac{\sin(\pi N_x x/FOV_x)}{\sin(\pi x/FOV_x)} \qquad (11.17)$$

In the case of asymmetric sampling with an even N_x, when the index 1 ranges over $1 = -(N_x/2),\ldots, (N_x/2) -1$,

$$psf_x(x) = \frac{1}{N_x} \sum_{l=-(N_x/2)}^{(N_x/2)-1} e^{i2\pi lx/FOV_x} = \frac{1}{N_x}e^{-i\pi x/FOV_x}\frac{\sin(\pi N_x x/FOV_x)}{\sin(\pi x/FOV_x)} \qquad (11.18)$$

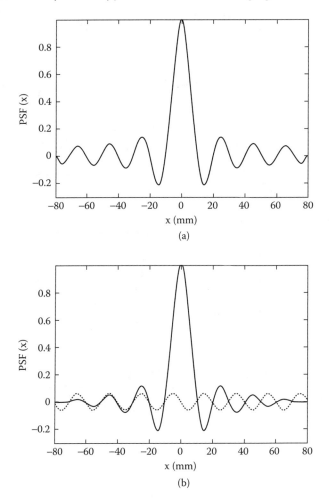

FIGURE 11.16 The shape of the 1-D point-spread function for (a) symmetrical and (b) asymmetrical k-space sampling. In the case of asymmetrical sampling, the PSF is a complex function. The real part is depicted by the solid and the imaginary part by the dotted line.

From the preceding expressions, it follows that symmetric sampling leads to a real PSF, and asymmetric sampling to a complex PSF. Both situations are depicted in Figure 11.16. In the case of a complex PSF, both real and imaginary parts of the FIDs are mixed, and CSI spectra show the phase difference with respect to each other. This may cause phasing problems. On the other hand, the PSF shape corresponding to asymmetric sampling has a slightly improved profile in terms of the diminished extent and amplitude of the side lobes.

In Figure 11.17 the influence of the number of phase-encoding steps on PSF shape is demonstrated. Clearly, PSF shape improves with the increasing number

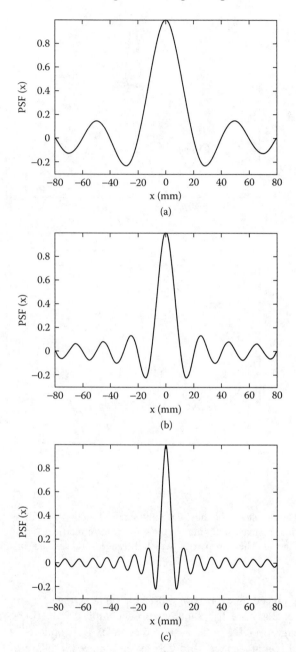

FIGURE 11.17 The dependence of 1-D point-spread functions on the number of symmetrically sampled k-space points. The field of view was assumed to have a fixed value of 160 mm. The number of k-space points was (a) 8, (b) 16, and (c) 32.

of phase-encoding steps. However, as shown in Equation 11.5, this improvement is at the expense of acquisition time. The minimum accepted number of encoding steps is usually considered to be eight (for the FFT algorithm, N must be a power of two, unless zero filling is used).

Because the PSF shape is not rectangular and the PSF extends over many voxels, Equation 11.14 implies that the signal measured at the given position (the given voxel, in reality) is always contaminated by the signals from other locations in the sample. This signal "bleeding" leads to the loss of localization precision, and it is the determining factor for the resulting resolution of the CSI experiment. In large samples with prevailing homogenous signal distribution, the effects of the side lobes of PSF may, owing to the oscillatory character of PSF, partially cancel each other out. However, because metabolites are generally distributed in the sample nonuniformly, the existence of PSF leads to the inaccurate determination of the signal distribution, and the reconstructed signal s_{rec} (t, $\mathbf{r}_{l',m',n'}$) does not equal the true one, s (t, $\mathbf{r}_{l',m',n'}$). Only a potential contamination can be predicted from the PSF shape, because the final contamination depends on the exact metabolite distribution.

The elimination of PSF by the deconvolution of the measured concentrations into the known PSF and the true metabolite signal is not possible, because all spectra in the CSI grid are never available with the sufficient quality and deconvolution methods are very prone to noise. Alternatively, *a priori* information from an MRI image can be used for more sophisticated reconstructions to diminish PSF effects to some extent (45,46).

The existence of PSF explains the need for the suppression of subcutaneous lipids in the scalp and other areas with spurious signals. Even if the relative contributions of distinct voxels are small, the contamination by lipid signal can be severe due to big differences in metabolite and fat concentrations ($1:10^3$). Moreover, because lipid resonances in distant areas are often shifted (owing to different magnetic field strengths), signal bleeding usually results in the spoiling of the whole spectral range.

11.3.4 K-SPACE WEIGHTING TECHNIQUES

For a fixed size of FOV, the profile of the PSF improves if the number of phase-encoding steps is increased as depicted in Figure 11.17. However, due to time constraints and sensitivity reasons, a compromise in the number of encoding steps is necessary.

Even if the number of phase-encoding steps is limited, other possibilities of improving the PSF shape are available. Because the FT of the product of two functions is the convolution of their FT, multiplying measured k-space data S (t, $\mathbf{k}_{l,m,n}$) with a filter function will influence the resulting PSF. This postacquisition k-space filtering, also called *apodization*, is realized by the multiplication of the measured k-space data with symmetrical filters having the maximal value in the center of the k-space and smoothly decreasing toward its edges. The operation is performed for all time points t. After the application of the filter w, Equation 11.10 can be rewritten as:

$$s_{rec}(t, \mathbf{r}) = \frac{1}{N_x}\frac{1}{N_y}\frac{1}{N_z}\sum_{l,m,n} w(l,m,n)S(t, \mathbf{k}_{l,m,n})e^{i2\pi k_{l,m,n}\mathbf{r}} \tag{11.19}$$

where w(l, m, n) describes the value of the filter function for the l-th, m-th, and n-th encoding steps.

From Equation 11.14 and Equation 11.15 the resulting PSF is given by

$$PSF(\mathbf{r}) = \frac{1}{N_x}\frac{1}{N_y}\frac{1}{N_z}\sum_{l,m,n} w(l,m,n)e^{i2\pi k_{l,m,n}\mathbf{r}} \tag{11.20}$$

Therefore, by using a proper filter function w(l, m, n), the PSF profile can be improved. Various filter functions can be used (47), such as the cosine filter

$$w(l) = \cos\left(\frac{\pi l}{2l_{max}}\right) \tag{11.21}$$

or the Hamming filter

$$w(l) = 0.54 + 0.46\cos\left(\frac{\pi l}{2l_{max}}\right) \tag{11.22}$$

where l ranges over N_x sampled values, and l_{max} stands for the maximal sampled value of l. 3-D extensions are, in the case of Cartesian sampling, given by the product of the corresponding 1-D expressions.

The effect of filter functions on PSF shape is shown in Figure 11.18. Generally, apodization is always a compromise between PSF side lobe reduction and the increase of the width of the main lobe and, hence, worsens resolution. In this respect, the Hamming function is the optimal filter (47). Because the real resolution of the CSI experiment is related to the width of the main lobe of the PSF, apodization influence the final resolution. Therefore, the details of the applied filter should be provided whenever k-space filtering is used.

Postacquisition filtering is not an efficient method of k-space apodization in terms of signal-to-noise ratio (SNR) or time, because the sampled signal from the edges of the k-space is eventually reduced. If more averages are needed, k-space apodization can be performed directly during the measurement by varying the number of averages A_{lmn} for each phase-encoding step (excitation) in proportion to the desired filter function w(l, m, n):

$$w(l,m,n) = A_{lmn}/N_{exc}; \; N_{exc} = \sum_{l,m,n} A_{lmn} \tag{11.23}$$

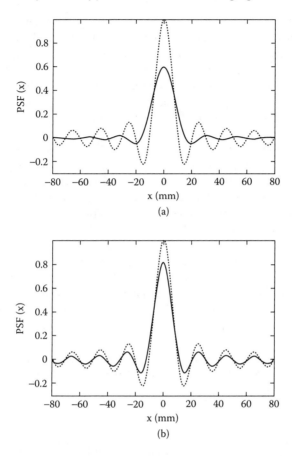

FIGURE 11.18 The influence of (a) a cosine filter and (b) a Hamming filter on the point-spread function shape (solid) in comparison to the original PSF (dashed) ($FOV_x = 160$ mm, $N_x = 16$, symmetrical sampling).

where N_{exc} is the total number of excitations. Analogous to postacquisition filtering, the value of A_{lmn} is maximal in the center of the k-space, where A_{lmn} equals the number of averages N_{acq} of the sequence.

It is obvious that apodization during the measurement can only be properly performed when the total number of excitations N_{exc} is sufficient to approximate filter function w(l,m,n) by fractional weights A_{lmn}. Because in the case of ^{1}H CSI usually only a few averages are needed, an exclusive acquisition weighting is used for nonhydrogen nuclei, which require more averages due to sensitivity reasons. When the number of total excitations N_{exc} is small, multiplication of k-space data by a correction smoothing filter, to prevent ringing from sharp digital transitions in weighting, should be performed (48).

Apodization can be generally accomplished by any combination of acquisition k-space weighting and the postacquisition filtering. This is the case in reality, because due to the integral number of averages A_{lmn}, the acquisition weighting will not exactly match the required value of $w(l, m, n)$ given by Equation 11.23. However, it can be shown (49) that pure acquisition weighting, according to Equation 11.23, represents an optimal method of producing the desired PSF shape in terms of the highest SNR in a given measurement time. Alternative methods of acquisition weighting have been proposed, such as weighting achieved by the variable repetition time (50) or, for spiral-based k-space sampling, by variable density of the sampled spiral in k-space (51).

The extreme case of weighted sampling is reduced k-space sampling, when some parts of the k-space are not sampled at all. A typical example is circular (spherical) sampling when only points of the k-space inside the circular (spherical) region are sampled and the remaining points are zero-filled (52).This leads to the reduction of the measurement time and also to the improvement of the PSF profile. The side lobes of PSF are reduced in circular sampling compared to rectangular sampling. This is, however, at the expense of a slight broadening of the central PSF lobe. Also, circular sampling leads to an isotropic PSF in comparison to rectangular sampling, in which PSF side lobes are propagated only along the principal axis. This can be important when potential signal contamination from problematic areas could be reduced by the proper orientation of the CSI grid. Variations of circular sampling to achieve further improvements have been suggested (53,54).

Reduced sampling can be combined with both acquisition k-space weighting and postacquisition filtering, resulting in various PSFs with different data collection efficiencies (48).

11.3.5 CSI PreProcessing

After spatial reconstruction of CSI data has been performed, spectra in all voxels of the spectroscopic grid are available. Performing N_x, N_y, and N_z phase-encoding steps along orthogonal directions results in $N_x * N_y * N_z$ spatially resolved spectra along corresponding axes. However, it is possible to increase the number of voxels artificially after the measurement. This operation, called *zero filling*, consists in appending zeros to $S(t, \mathbf{k}_{l,m,n})$ values prior to the FT (Figure 11.19). Zeros can be appended either symmetrically or asymmetrically, the latter leading to a phase shift in spectra among voxels. Zero filling represents an interpolation method and does not affect the PSF. Therefore, even if the voxel size is decreased, the real resolution of the CSI experiment is not improved.

Another unique feature of the FT reconstruction is the possibility of adjusting the exact position of the grid after the measurement. This operation is based on the shift theorem of FT

$$s(t, x_{l'} - x_{j'}) = DFT^{-1}\{S(t, k_l) \exp^{-i2\pi k_1 x_{j'}/N}\} \tag{11.24}$$

where x_j represents generally the subvoxel size shift.

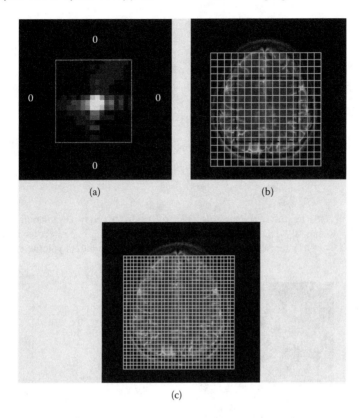

FIGURE 11.19 The effect of zero filling. Zero filling was performed from 16 to 32 data points in both dimensions in k-space (k-space of the first time point shown). (a) Area in the rectangle represents the original k-space, while zeros were added symmetrically, resulting in the increase of the number of voxels from the original value (b) 16 × 16 to (c) 32 × 32.

Equation 11.24 implies that a multiplication of all k-space values S (t, $k_{1,}$) by a proper phase factor before the DFT^{-1} results in shifting all positions of the voxels (the whole spectroscopic grid). This is very useful because partial volume effects play an important role due to the large voxel size, and by means of grid shifting, the area of interest can be centered in the voxel (Figure 11.20).

11.3.6 DISPLAY OF THE CSI DATA

Spectroscopic imaging data usually contain substantial amounts of information. To make use of this information efficiently, proper display and analysis of the data are required.

There are several ways of presenting spectroscopic imaging data sets. Similar to SVS, one or few selected spectra can be viewed as shown in Figure 11.21a. This provides spectra in high resolution and enables their reliable inspection. To view all spectra simultaneously, a grid of spectra overlaying the MR scout image

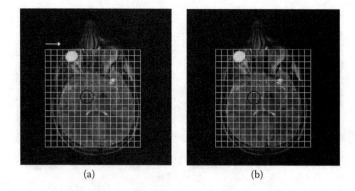

          (a)   (b)

FIGURE 11.20 The shift of the spectroscopic grid. (a) In the original evaluation, a selected region has no central voxel. (b) The shift was performed along one direction in order to center the object of interest in one of the voxels and to reduce partial volume effects.

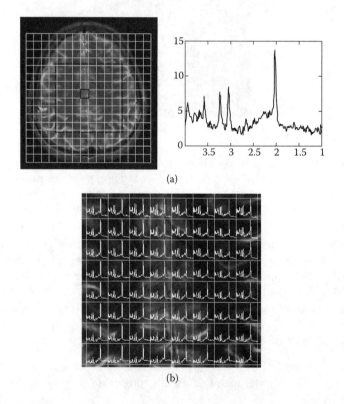

FIGURE 11.21 The display of acquired spectra. The spectrum from the desired voxel can be viewed (a) separately or (b) the grid of spectra overlaying the MR scout image can be drawn.

can be displayed (Figure 11.21b). Because the matrix contains many spectra, the size of the displayed spectra is often very small, preventing the operator from resolving all the spectral details. To overcome this, only subsets of all spectra can be displayed. The matrix of spectra provides an overview of spectral quality and trends in signal distribution; however, spectra contain complex and often redundant information. Some metabolites may be irrelevant or not well resolved for the study. To summarize the metabolite distribution at a glance, metabolite images (called also metabolite maps) can be computed. For each detectable metabolite in the spectrum, a metabolite image can be computed when the intensity in the metabolite image corresponds to the signal intensity of the selected metabolite in the given voxel. The resulting image, usually in a color palette, is overlaid on the MR scout image (Figure 11.22).

As pointed out earlier, due to sensitivity reasons, the voxel size is typically of the order of 1 cm^3. The image matrix of spectroscopic images is, therefore, coarse. As previously mentioned, zero filling resulting in a finer matrix can be performed. However, in this case, spectra from more voxels have to be processed. To improve the appearance of the images, interpolation in the image space (after the FT) can also be performed. For this purpose bilinear, cubic, or other interpolation methods are used. Even if the apparent resolution of images is improved and the images are better readable, similar to zero filling, the image interpolation does not change the PSF and the true resolution of images remains low.

Metabolite images are based on the results of spectra fitting and the assumption that spectral quality is sufficient for the analysis. If the fitting routine is not accurate or if spectra contain artifacts, the metabolic image may be misleading. A typical example can be the incorrect computation of resonance areas due to magnetic-field–inhomogeneity-induced frequency shifts of the spectra. If integration

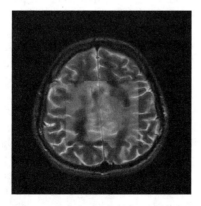

FIGURE 11.22 The metabolite image of the sum of glutamine and glutamate signals. The low signal values correspond to the blue color, whereas the high signal values correspond to the red color.

is performed over the fixed frequency range, wrong resonance areas will be computed. This can be overcome by using spectra-fitting algorithms that take spectra shifts into account.

In any case, for reliable analysis of CSI data, an inspection of spectra in considered areas is unavoidable.

11.3.7 COMPARISON OF SVS AND CSI TECHNIQUES

CSI offers the possibility of acquiring more spectra in the same time instead of just one spectrum as with SVS, but does not bring about any improvement in sensitivity. In terms of SNR per unit time and considering that the application of phase encoding is equivalent to spectra averaging, the CSI and SVS techniques are equally efficient (55,56). It should be pointed out that the equivalence of both methods, in terms of sensitivity, assumes that SVS and CSI voxels have the same size. As shown earlier, due to PSF effects, the area contributing to the signal of the selected voxel extends the nominal voxel size. This fact leads to the decreased sensitivity of the CSI experiment. This can be demonstrated by the integration of PSF over the nominal voxel size (57), which in the case of no filtering equals

$$\int_{-\Delta x/2}^{\Delta x/2} PSF(x')dx' = \int_{-\Delta x/2}^{\Delta x/2} \frac{\sin(\pi N_x x' / FOV_x)}{N_x \sin(\pi x' / FOV_x)}dx' \approx 0.873 \qquad (11.25)$$

The loss of the voxel signal compared to an ideal selective excitation method is apparent. The signal is not actually lost, but distributed among other voxels (signal bleeding). The overall signal measured in each voxel depends, therefore, on the detailed distribution of the measured signal in the tissue. This fact limits the equivalence of both methods.

The preference of one or the other method depends on the number of averages required to acquire spectra of the desired quality. In 2-D CSI using 16×16 matrix, signals from 256 voxels are measured in the same time and with the same SNR as opposed to one voxel in SVS using 256 averages. In this case, CSI becomes more efficient in terms of acquired information per unit time. On the other hand, if only few averages are needed, e.g., for a reference spectrum without water suppression, SVS is preferable.

In practical situations there are also other effects influencing the quality of acquired spectra. The need for shimming of the large volume in the case of CSI often means worse magnetic field homogeneity, broader signals, and, hence, lower SNR.

On the other hand, measurements with smaller voxel size are often performed with CSI. This leads to better spatial resolution and to improved magnetic field homogeneity within the voxel, resulting in the significant decrease of the widths when voxel sizes below 0.4 cm^3 are measured (58). When an absolute quantification of the signals should be performed, SVS sequences are preferred, because the PSF effects in CSI sequences make the quantification difficult.

11.4 DIFFERENCES IN SEQUENCES FOR MEASUREMENTS WITH NONPROTON NUCLEI

The measurement techniques for volume selection in single-volume spectroscopy sequences described in chapter 11.2 are widely used in *in vivo* H-MRS. Spectroscopy measurements can also be performed using other nuclei such as phosphorus-31, carbon-13, sodium-23, and fluorine-19. Because the resonance frequencies of these nuclei are different from those of protons, specific coils for the application of RF pulses and for signal receiving are necessary. A simple spatial selection of the acquired signal is possible by using small surface coils, which acquire only signals from the volume near the coil. In combination with a CSI measurement, good spatial resolution is obtainable (44), but difficulties arise from the spatial inhomogeneous coil sensitivity. The measurement with surface coils is, however, limited to volumes of interest in the proximity of the surfaces of volunteers or patients. If small volumes within the body have to be examined, a volume coil similar to the head coil has to be used and an image data set is necessary for the selection of the appropriate voxel. Therefore, so-called double-tuned coils are desirable, which allow measurements at the resonance frequencies of protons and the specific nucleus of interest. Although in principle the same measurement techniques used in H-MRS can be used for measurements with other nuclei (so-called heteronuclear spectroscopy), for some applications other methods have to be used. In measurements of phosphorus nuclei, e.g., signal contributions from resonances with very low T2 values should be examined in many applications. This is not possible if echo times longer than 20 msec are used as is usual for applications of PRESS and STEAM. For such applications, the ISIS technique (3) can be used as an alternative. In this measurement technique, up to three slice-selective inversion pulses are used as preparation pulses prior to a nonselective excitation, which is followed directly by the data acquisition (Figure 11.23). The delay time between the excitation and the beginning of the data acquisition can be shorter than 1 msec in this case. In such a measurement, signals from the whole volume excited by the 90° pulse are acquired. The selection of signals from chosen parts of this volume requires several measurements with different combinations of preparation pulses. For a selection of a cuboid, eight measurements are necessary, and the signal from the VOI can be calculated from a series of successive subtractions. This procedure has the disadvantage that small changes between the single measurement, e.g., patient movements, can lead to large errors in the spectrum calculation. Therefore, the applications of the ISIS technique are limited to those cases in which the use of single-shot techniques such as STEAM and PRESS is not possible.

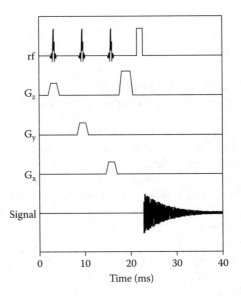

FIGURE 11.23 Sequence scheme of an ISIS sequence. In eight successive measurements, different combinations of the preceding inversion pulses are used, i.e., only single inversion pulse or a pair of pulses instead of all three pulses as shown in the scheme.

REFERENCES

1. Bottomley, P.A. (1987). Spatial localization in NMR spectroscopy in vivo. *Ann. NY Acad. Sci.* 508: 333–348.
2. Frahm, J., Merboldt, K.-D., and Hänicke, W. (1987). Localized proton spectroscopy using stimulated echoes. *J. Magn. Reson.* 72: 502–508.
3. Ordidge, R.J., Connelly, A., and Lohmann, J.A.B. (1986). Image-selected in vivo spectroscopy (ISIS). a new technique for spatially selective NMR spectroscopy. *J. Magn. Reson.* 66: 283–294.
4. Hahn, E.L. (1950). Spin echoes. *Phys. Rev.* 80: 580–594.
5. Payne, G.S. and Leach, M.O. (1995). On doubling the signal in localised stimulated echo measurements. *Magn. Reson. Imaging.* 13: 629–632.
6. Seeger, U., Klose, U., Lutz, O., and Grodd, W. (1999). Elimination of residual lipid contamination in single volume proton MR spectra of human brain. *Magn. Reson. Imaging.* 17: 1219–1226.
7. Ernst, T. and Chang, L. (1996). Elimination of artifacts in short echo time H MR spectroscopy of the frontal lobe. *Magn. Reson. Med.* 36: 462–468.
8. Haase, A., Frahm, J., Hanicke, W., and Matthaei, D. (1985). 1H NMR chemical shift selective (CHESS) imaging. *Phys. Med. Biol.* 30: 341–344.
9. Sijens, P.E., Knopp, M.V., Brunetti, A., Wicklow, K., Alfano, B., Bachert, P., Sanders, J.A., Stillman, A.E., Kett, H., Sauter, R., and Oudkerk, M. (1995). 1H MR spectroscopy in patients with metastatic brain tumors: a multicenter study. *Magn. Reson. Med.* 33: 818–826.

10. Ogg, R.J., Kingsley, P.B. and Taylor, J.S. (1994). WET, a T1 and B1 insensitive water-suppression method for in-vivo localized 1H NMR spectroscopy. *J. Magn. Reson.* B 104: 1–10.
11. Sorensen, O.W., Eich, G.W., Levitt, M.H., Bodenhausen, G., and Ernst, R.R. (1983). Product operator formalism for the description of NMR pulse experiments. *Prog. NMR Spectroscopy.* 16: 163–192.
12. Kimmich, R., Rommel, R., and Knüttel, A. (1989). *J. Magn. Reson.* 81: 333.
13. Ernst, T. and Hennig, J. (1991). Coupling effects in volume-selective 1H spectroscopy of major brain metabolites. *Magn. Reson. Med.* 21: 82–96.
14. Rothman, D.L., Behar, K.L., Hetherington, H.P., and Schulman, R.G. (1984). Homonuclear 1H double-resonance difference spectroscopy of the rat brain *in vivo. Proc. Natl. Acad. Sci. USA.* 81: 6330–6334.
15. Aue, W.P., Bartholdi, E., and Ernst, R.R. (1976). Two-dimensional spectroscopy. Application to nuclear magnetic resonance. *J. Chem. Phys.* 64: 2229–2246.
16. De Graaf, R. B. (1998). *In vivo NMR spectroscopy.* John Wiley Chichester, England.
17. Brown, T., Kincaid, B., and Ugurbil, K. (1982). NMR chemical shift imaging in three dimensions. *Proc. Natl. Acad. Sci. USA.* 79: 3523–3526.
18. Maudsley, A.A., Hilal, S.K., Perman, W.P., and Simon, H.E. (1983). Spatially resolved high-resolution spectroscopy by four dimensional NMR. *J. Magn. Reson.* 51: 147–152.
19. Pohmann, R., von Kienlin, M., and Haase, A. (1997). Theoretical evaluation and comparison of fast chemical shift imaging methods. *J. Magn. Reson.* 129: 145–160.
20. Duyn, J.H. and Moonen, C.T. (1993). Fast proton spectroscopic imaging of human brain using multiple spin-echoes. *Magn. Reson. Med.* 30: 409–414.
21. Duyn, J.H., Frank, J.A., and Moonen, C.T. (1995). Incorporation of lactate measurement in multi-spin-echo proton spectroscopic imaging. *Magn. Reson. Med.* 33: 101–107.
22. Adalsteinsson, E., Irarrazabal, P., Spielman, D.M., and Macovski, A. (1995). Three-dimensional spectroscopic imaging with time-varying gradients. *Magn. Reson. Med.* 33: 461–466.
23. Posse, S., DeCarli, C., and Le Bihan, D. (1994). Three-dimensional echo-planar MR spectroscopic imaging at short echo times in the human brain. *Radiology.* 192: 733–738.
24. Mansfield, P. (1984). Spatial mapping of the chemical shift in NMR. *Magn. Reson. Med.* 1: 370–386.
25. Norris, D.G. and Dreher, W. (1993). Fast proton spectroscopic imaging using the sliced k-space method. *Magn. Reson. Med.* 30: 641–645.
26. Matsui, S., Sekihara, K., and Kihno, H. (1986). Spatially resolved NMR spectroscopy using phase-modulated spin-echo trains. *J. Magn. Reson.* 67: 476–490.
27. Web, P., Spielman, D., and Macovski, A. (1989). A fast spectroscopic imaging method using a blipped phase encode gradient. *Magn. Reson. Med.* 12: 306–315.
28. Speck, O., Scheffler, K., and Hennig, J. (2002). Fast 31P chemical shift imaging using SSFP methods. *Magn. Reson. Med.* 48: 633–639.
29. Dreher, W., Geppert, C., Althaus, M., and Leibfritz, D. (2003). Fast proton spectroscopic imaging using steady-state free precession methods. *Magn. Reson. Med.* 50: 453–460.

30. Dydak, U., Weiger, M., Pruessmann, K.P., Meier, D., and Boesiger, P. (2001). Sensitivity-encoded spectroscopic imaging. *Magn. Reson. Med.* 46: 713–722.

31. Dydak, U., Pruessmann, K.P., Weiger, M., Tsao, J., Meier, D., and Boesiger, P. (2003). Parallel spectroscopic imaging with spin-echo trains. *Magn. Reson. Med.* 50: 196–200.

32. Moonen, C.T., Sobering, G., van Zijl, P.C., Gillen, J., von Kienlin, M., and Bizzi, A. (1992). Proton spectroscopic imaging of human brain. *J. Magn. Reson.* 98: 556–575.

33. Duyn, J.H., Matson, G.B., Maudsley, A.A., and Weiner, M.W. (1992). 3-D phase encoding 1H spectroscopic imaging of human brain. *Magn. Reson. Imaging.* 10: 315–319.

34. Spielman, D., Pauly, J., Macovski, A., and Enzmann, D. (1991). Spectroscopic imaging with multidimensional pulses for excitation: SIMPLE. *Magn. Reson. Med.* 19: 67–84.

35. Duyn, J.H., Gillen, J., Sobering, G., van Zijl, P.C., and Moonen, C.T. (1993). Multisection proton MR spectroscopic imaging of the brain. *Radiology.* 188: 277–282.

36. Posse, S., Schuknecht, B., Smith, M.E., van Zijl, P.C., Herschkowitz, N., and Moonen, C.T. (1993). Short echo time proton MR spectroscopic imaging. *J. Comput. Assist. Tomogr.* 17: 1–14.

37. Ebel, A., Govindaraju, V., and Maudsley, A.A. (2003). Comparison of inversion recovery preparation schemes for lipid suppression in 1H MRSI of human brain. *Magn. Reson. Med.* 49: 903–908.

38. Spielman, D., Meyer, C., Macovski, A., and Enzmann, D. (1991). 1H spectroscopic imaging using a spectral-spatial excitation pulse. *Magn. Reson. Med.* 18: 269–279.

39. Kiefer, C. and Klose, U. (2000). Presaturation of irregular spatial structures with two-dimensional waveforms corrected for B1-inhomogeneities. *Proceedings of the International Society for Magnetic Resonance in Medicine.* 8: 1429.

40. Hu, X., Patel, M., and Ugurbil, K. (1994). A new strategy for spectroscopic imaging. *J. Magn. Reson. B.* 103: 30–38.

41. Hu, X., Patel, M., Chen, W., and Ugurbil, K. (1995). Reduction of truncation artifacts in chemical shift imaging by extended sampling using variable repetition time. *J. Magn. Reson. B.* 106: 292–296.

42. Haupt, C.I., Schuff, N., Weiner, M.W., and Maudsley, A.A. (1996). Removal of lipid artifacts in 1H spectroscopic imaging by data extrapolation. *Magn. Reson. Med.* 35: 678–687.

43. Bracewill, R. (1978). *The Fourier Transform and its Applications.* McGraw-Hill, New York.

44. Murphy Boesch, J., Jiang, H., Stoyanova, R., and Brown, T.R. (1998). Quantification of phosphorus metabolites from chemical shift imaging spectra with corrections for point-spread effects and B1 inhomogeneity. *Magn. Reson. Med.* 39: 429–438.

45. Constantinides, C.D., Weiss, R.G., Lee, R., Bolar, D., and Bottomley, P.A. (2000). Restoration of low-resolution metabolic images with a priori anatomic information: 23Na MRI in myocardial infarction. *Magn. Reson. Imaging.* 18: 461–471.

46. Gao, Y. and Reeves, S.J. (2000). Optimal k-space sampling in MRSI for images with a limited region of support. *IEEE Trans. Med. Imaging.* 19: 1168–1178.

47. Ernst, R.R., Bodenhausen, G., and Wokaun, A. (1987). Principles of Nuclear Magnetic Resonance in One and Two Dimensions. Clarendon Press Oxford, England.

48. Hugg, J.W., Maudsley, A.A., Weiner, M.W., and Matson, G.B. (1996). Comparison of k-space sampling schemes for multidimensional MR spectroscopic imaging. *Magn. Reson. Med.* 36: 469–473.

49. Mareci, T.H. and Brooker, H.R. (1991). Essential considerations for spectral localization using indirect gradient encoding of spatial information. *J. Magn. Reson.* 92: 229–246.

50. Kuhn, B., Dreher, W., Norris, D.G., and Leibfritz, D. (1996). Fast proton spectroscopic imaging employing k-space weighting achieved by variable repetition times. *Magn. Reson. Med.* 35: 457–464.

51. Adalsteinsson, E., Star Lack, J., Meyer, C.H., and Spielman, D.M. (1999). Reduced spatial side lobes in chemical-shift imaging. *Magn. Reson. Med.* 42: 314–323.

52. Maudsley, A.A., Matson, G.B., Hugg, J.W., and Weiner, M.W. (1994). Reduced phase encoding in spectroscopic imaging. *Magn. Reson. Med.* 31: 645–651.

53. Ponder, S.L. and Twieg, D.B. (1994). A novel sampling method for 31P spectroscopic imaging with improved sensitivity, resolution, and sidelobe suppression. *J. Magn. Reson. B.* 104: 85–88.

54. Hetherington, H.P., Luney, D.J., Vaughan, J.T., Pan, J.W., Ponder, S.L., Tschendel, O., Twieg, D.B., and Pohost, G.M. (1995). 3-D 31P spectroscopic imaging of the human heart at 4.1 T. *Magn. Reson. Med.* 33: 427–431.

55. Granot, J. (1986). Selected Volume Spectroscopy (SVS) and chemical-shift imaging. A comparison. *J. Magn. Reson.* 66: 197–200.

56. Posse, S. and Aue, W. P. (1989). 1H spectroscopic imaging at high spatial resolution. *NMR Biomed.* 2: 234–239.

57. Maudsley, A.A. (1986). Sensitivity in Fourier imaging. *J. Magn. Reson.* 363–366.

58. Gruber, S., Mlynarik, V., and Moser, E. (2003). High-resolution 3-D proton spectroscopic imaging of the human brain at 3T: SNR issues and application for anatomy-matched voxel sizes. *Magn. Reson. Med.* 49: 299–306.

12 Data Processing Methods in MRS

Luca T. Mainardi and Sergio Cerutti

CONTENTS

12.1 INTRODUCTION

In the last 20 years, there has been an increasing interest in magnetic resonance spectroscopy (MRS) in different fields ranging from analytical chemistry through material sciences to biomedical applications. In particular, the well-known success and the widespread application of MRS techniques in biomedical research and medical practice have been supported by a number of inherent advantages of this technique: MRS performs repetitive, nondestructive measurements of metabolic processes *in situ* as they proceed in their own environment and it allows the extraction of valuable *in vivo* information on the physiological and pathological state of human tissues in different organs [1].

It is known that the metabolic information contained in the magnetic resonance (MR) signal (the free induction decay [FID] signal) is apparent in its spectrum. In fact, in MRS spectra, the different compounds appear as different resonance peaks: the position of resonance identifies the specific compound, peak line width indicates the transverse relaxation time of the nucleus (an index linked to the mobility of the molecule) and, finally, the total area of the resonance peak is proportional to the concentration of the detected substance. The quantification of these parameters leads to metabolic tissues' characterization [1,2].

Since MRS has been applied to humans, ^{1}H-MRS has attracted much attention. The reason is that a proton is the most sensitive and stable nucleus, and hydrogenous atoms are largely diffused in living tissues [3]. Nowadays, thanks to the recent technical advances in MR instrumentation, ^{1}H-MRS is routinely applied in clinical settings, especially in brain study, where it has been documented to be effective in the diagnosis, prognosis, and treatment selection of cerebral tumors [4,5], cerebral ischemia [6], epilepsy [7,8], and multiple sclerosis [9,10]. Changes in the ^{1}H-MRS resonance patterns were observed between normal brain and cerebral tumors, with potential applications for the grading and classification of different tumor types [11,12]. The metabolism of cerebral ischemia shows both acute and chronic changes, with relevant implications from a pathophysiological and a therapeutic point of view [16,17]. Finally, a few studies describe the possibility of investigating the metabolic characteristics of multiple sclerosis (MS) lesions classified in acute, subacute, and chronic cases, using MR spectroscopy [9,10,18]. In addition to brain studies, other body tissues have been investigated using MRS, including prostate, liver, and muscle. In particular, ^{31}P-MRS has been used in the diagnosis of muscular disease such as McArdle's syndrome [19] and Duchenne dystrophy [20].

Because of its clinical importance, the processing of *in vivo* MRS signals and the extraction of the relevant information is not a trivial task. Major problems that may limit the theoretical potentiality of the technique include the narrow chemical shift range of ^{1}H signals, which requires a precise shimming of the B_0 field, and the presence of unwanted water and lipids contributions, which overwhelm the small metabolites of medical interest [3,13]. Also, the presence of severe phase distortions [14] and consistent overlaps among spectral peaks [15] make it difficult to quantify the parameters of interest, especially in a clinical environment where short echo time and low-intensity magnetic fields are employed. Finally, good shimming and correct suppression of water contribution usually depend on the intervention of the experimenters, thus limiting the repeatability of the study [3].

For these reasons, there is a need for robust and reliable signal processing methods that make it possible to extract the relevant FID information. The methods should be fast, automatic, and operator independent.

In this chapter, we briefly introduce the basic principles of the methodology available for advanced quantitative FID analysis and for the extraction of metabolite parameters. It is not within the scope of this chapter to provide an exhaustive description of the wide range of signal processing methods proposed for MRS study. Rather, it presents some introductory concepts of signal processing that may be fruitfully applied for FID analysis, focusing the reader's attention on the potentiality and the flexibility of these techniques in metabolite quantification. More details can be found in recent reviews on these topics [21–23].

12.2 EXTRACTING INFORMATION FOR THE FID SIGNAL

The FID signal is usually approximated as a sum of K-complex damped sinusoids according to the following model [22]:

$$x(n) = \sum_{k=1}^{K} a_k e^{j\phi_k} e^{(-d_k + j2\pi f_k)n} + e(n) \tag{12.1}$$

where a_k is the amplitude of the k-th sinusoid, f_k its frequency, d_k the damping factor, $_k$ the phase, and $e(n)$ is assumed to be circular complex white Gaussian noise (i.e., the real and imaginary parts of the noise are not correlated and have equal variance). In Equation 12.1, $n = \{1, 2,...,N\}$ is the discrete-time index and N the number of observed data points. Each sinusoid is described by a set of parameters $v_k = [f_k, a_k, d_k, \phi_k]$, where the relative frequency f_k is used to identify the biochemical species and a_k and d_k are the relevant parameters for metabolite quantification and characterization. a_k is proportional to the number of nuclei contributing to the spectral component at the frequency f_k (number that depends on the metabolite concentration), and d_k may provide information about the mobility and macromolecular environment of the nucleus. When expressed in the frequency domain and ignoring the noise term, Equation 12.1 becomes

$$X(f) = \sum_{k=1}^{p} \frac{a_k e^{j\phi_k}}{d_k + j2\pi(f - f_k)} \tag{12.2}$$

It consists of a set of spectral peaks of Lorentzian shape. The real part of $X(\omega)$ is known as the *absorption-mode spectrum* (Figure 12.1). Taking the real part of Equation 12.2 and after correct phasing ($\phi_k = 0$), we obtain

$$\text{Re}\{X(f)\} = \sum_{k=1}^{p} \frac{a_k d_K}{d_k^2 + [2\pi(f - f_k)]^2} \tag{12.3}$$

It can be easily observed that

$$\int_{-\infty}^{+\infty} \frac{a_k d_k}{d_k^2 + [2\pi(f - f_k)]^2} df = a_k \tag{12.4}$$

Thus, the quantification of the metabolite concentrations can be easily obtained as the integral of spectral lines (Figure 12.1). Manually integrating the

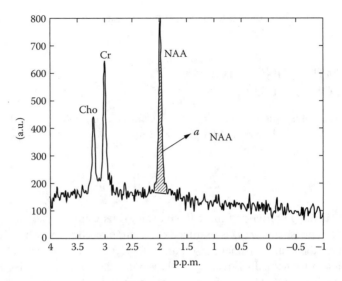

FIGURE 12.1 Quantification of *N*-acetyl-aspartate (NAA) concentration by manual integration of the absorption-mode ¹H MRS spectral profile. The presence of baseline wandering makes it difficult to define the area of integration. Also, coline (Cho) and creatine (Cr) peaks are partially overlapped; in this case it is difficult to correctly define the integration range.

area under the spectral peaks of interest is the oldest and simplest method used to extract quantitative metabolite concentrations from the FID. In practice, this approach is slightly inaccurate, and a bias is usually introduced by the operator and baseline correction technique. As shown by Equation 12.3 and Equation 12.4, the results rely on the correct phasing of the spectral components and depend on the width of the integration area (ideally, from $-\infty + \infty$; in practice, around the spectral peak). Defining the integration width cannot be a trivial task, especially when spectral peaks are partially superimposed or when acquisition artifacts or broadband resonances (i.e., water signal contribution in ¹H-MRS) may distort the spectrum baseline (Figure 12.2a). Acquisition noise and baseline waving may also mask the contributions of low-amplitude metabolite components, whose concentration is nonquantifiable (Figure 12.3).

Therefore, in the last decade, a wide range of signal processing methods have been proposed to obtain automatic, reliable quantification of metabolite concentrations from the recorded FID signals. Calculation of metabolite concentration can be carried out in both time and frequency domain, requiring accurate estimation of either the amplitude or the peak area.

This chapter will review the main signal processing methodologies used in this field, and it will survey the successive processing steps that lead to the calculation of molecular concentration of various metabolites contributing to the recorded FID.

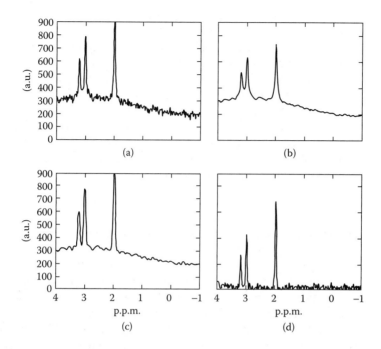

FIGURE 12.2 Effects of different time-domain preprocessing on the absorption-mode spectrum. (a) Original spectrum, (b) increased SNR by line broadening, (c) Lorentz–Gauss transformation, and (d) removal of broadband components.

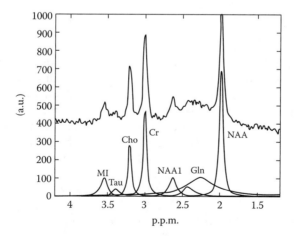

FIGURE 12.3 Simulated TE20 ¹H MRS spectrum (bold line). Decomposition obtained by LPSVD method. The contributions of small metabolite (myo-Inosytol [mI], taurine [Tau], or Glutamate+glutamine [Gln]), which were not clearly evident on the original spectrum, were identified and separated by the LPSVD methods.

12.3 TIME-DOMAIN PREPROCESSING

The acquired FID is noisy and not suitable for direct analysis. Due to several acquisition problems, the resulting FID may be distorted from the ideal curve of Equation 12.1. The causes of these distortions include, among others, field inhomogeneity, the truncation of the FID data before the signal has decayed below the noise level, unavoidable delay between the RF pulse and the first FID sample, as well as presence of unwanted broadband resonance. Therefore, acquired FID is usually preprocessed before the parameters of interest can be estimated. A brief description of common preprocessing methods applied to FID data is presented in the following subsections.

12.3.1 ZERO FILLING

This is a common procedure in signal processing known as *zero padding*. The number N of sampled points is increased by adding (padding) zeros at the end of data series. In general, the procedure is required to fit the fast Fourier transform (FFT) criterion (number of data points must be a power of two for computational efficiency); in MRS, zero filling is applied when the acquisition time has been kept short because of some practical constraints (in this case, few data points being available) or when signal terms have already decayed below the noise level and further sampling would add only noise. Adding zeros has the effect of interpolating extra points into the spectrum and improving its digital resolution (i.e., the frequency interval between data points). Because adding zeros does not add extra information to the data, the final result is merely an apparent improvement of spectral resolution used for visualization purposes, which does not increase the frequency resolution.

12.3.2 WINDOWING

Prior to FFT, the FID signal is usually windowed by multiplying it by a given function $g(t)$ with known characteristics. The aims of this procedure can be different — including either the improvement of signal-to-noise ratio (SNR), or a better spectral resolution, or the removal of truncation artifacts. In the following text, it will be shown how the selection of $g(n)$ may produce different effects on the absorption-mode spectra. Let us assume $g(n)$ to be an exponential function $g(n) = e^{-d_o n}$. Multiplying the acquired FID by $g(n)$ and excluding the noise term, we obtain

$$x(n)g(n) = \sum_{k=1}^{K} a_k e^{j\phi_k} e^{(-(d_k+d_0)+j2\pi f_k)n} \qquad (12.5)$$

Thus, the resulting curve maintains the original lineshape, but the spectral line width is increased by a factor d_0 (line broadening). The effect is to increase the SNR as shown in Figure 12.2b: data points at the beginning of FID, including higher-amplitude signal terms, are least affected by $g(n)$, whereas later points, in which signals are decayed below the noise level, are strongly attenuated. In this case, improvement of SNR is made by broadening spectral lines, thus reducing spectral resolution.

The windowing procedure can also be used for resolution improvement. In fact, if d_0 is chosen as negative, the decaying time of the FID is increased, producing a narrowing of spectral lines. However, use of this technique needs some care — if d_0 is too negative, the latest point in the FID is highly amplified, thus enhancing noise rather than signal and resulting in a prohibitively noisy spectrum.

A compromise between enhancement of SNR and resolution is obtained by combining the weighting function with negative d_0 with another function decaying to zero at the tail of the FID. The resulting weighting curve has a maximum located in the early points of the FID. Usually, a combination of the exponential and Gaussian function is used

$$g(n) = e^{-d_0 n} e^{-(d_g n)^2} \tag{12.6}$$

In practice, d_0 is chosen as close as possible to d_k's to generate delta spectral lines in the presence of true resonance, which are then convolved by the Gaussian curve of Equation 12.6. The resulting spectrum will be composed of a summation of Gaussian lines. For this reason, the latter operation is known as *Lorentz–Gauss Transform*. The advantage is that the Gaussian lineshape, with the same half-height line width of the Lorentzian one, has a narrow base, thus reducing overlap of peaks and increasing resolution. Results of the Lorentz–Gauss transformation are shown in Figure 12.2c. A weighting function decaying to zero at the end of the FID can also be useful to smooth the truncation at the end of FID, if it has not decayed to zero during acquisition (*apodization*).

Finally, windowing can be also used to reduce the influence of broad spectral components. Here, the selected weighting function is $g(n) = 1 - Ae^{-d_b n}$, where d_b is much greater than the d_k's of the narrow signals of clinical interest. Because $d_b \gg d_k$'s, the term $g(n) = e^{-d_b n}$ enhances broadband metabolites. A fraction A of these metabolite is then subtracted from the acquired FID data to remove them. Results are shown in Figure 12.2d.

12.3.3 REMOVAL OF UNDESIRED RESONANCE

The acquired MRS signal contains several resonance frequencies due to the complexity of living systems and metabolism. However, only few of them have clinical and diagnostic relevance. It is, therefore, desirable to remove unwanted peaks, thus improving the readability of the spectrum, accuracy in parameter estimation, and reducing the computational burden. The most straightforward example of undesired resonance is the water peak in proton MRS. In fact, water is largely diffused in living tissues and provides the most relevant contribution to the ^{1}H-MRS FID.

Depending on the characteristics of the undesired peaks, different signal processing methods can be applied to suppress them. These techniques range from time-domain filtering (in which unwanted peaks are separated in frequency from the peak of interest), discharging or weighting the FID data points in different ways (when nuisance peaks have larger line widths than the peak of interest) to a more or less complicated modeling of the spectrum baseline.

Time-domain filtering has been proposed to suppress the water peak, which is usually located at zero frequency. High-pass digital filters can be used in this regard [24,25]. Marion [25] proposed a linear-phase low-pass filter to extract purely water signals and subtract the filter output from the data. The proposed filter coefficients h_m were sine-bell-shaped

$$h_m = \cos\left(\pi \frac{(m - (M - 1)/2)}{M}\right) \tag{12.7}$$

or Gaussian-shaped

$$h_m = e^{-(4m - (M-1))^2/M^2} \tag{12.8}$$

with M being the filter order. Linear-phase filters are proposed to avoid phase distortion in the unfiltered components, whereas high-order M ($17 < M < 65$) values were suggested [25] to reduce the influence on the peak of interest. Because the first and last (M1)/2 values cannot be calculated, they must be extrapolated using linear predictions [25] or modeling [26]. Extrapolating the data may introduce distortions, and a careful compromise between ideal filter response and the number of extrapolated points has to be taken into account for filter design.

Another approach to water suppression is based on modeling of time-domain water signals [27,28]. Because of partial water suppression performed using special sequences, the water peak is far from the theoretical lineshape. Therefore, several damped exponentials are usually used to fit it. In modeling the water peak, user intervention is usually necessary to define the frequency region of the water peak and the number of fitting exponentials.

12.4 FREQUENCY-DOMAIN METHODS

Accurate quantification of *in vivo* spectra parameters can be obtained by fitting the observed MRS spectrum with known lineshapes. In the ideal case, the MRS spectrum consists of a superimposition of pure complex Lorentzian lines [29,30]. In the real case, due to acquisition imperfections, the ideal lineshape is distorted, and the model should usually include a mixture of Lorentzian (L) and Gaussian (G) curves [21,31]. The following is a general model of an MRS spectrum:

$$X(f) = \sum_{k=1}^{K} (\beta_L L_k(\mathbf{v}_k, f) + \beta_G G_k(\mathbf{v}_k, f)) + \sum_{p=1}^{P} c_p f^p + w(f) \tag{12.9}$$

where β_L and β_G are the weighted factors used to create the mixture of L and G curves and $w(f)$ is a white Gaussian noise. The values of β_L and β_G can be either fixed *a priori* or included in the parameters to be estimated. Also, the term $\sum_{p=1}^{P} c_p f^p$ is added to account for possible baseline distortions that may be derived from underlying signals or unwanted resonances. A polynomial can be used to model the baseline, p being the polynomial order and c_p its coefficients.

The estimation of model parameters v_k is obtained by solving a classical non-linear least-square (NLLS) problem, by minimizing the following figure of merit

$$
\sum_{n=0}^{N-1} |X(n\Delta f) - \hat{X}(n\Delta f)|^2
$$

$$
= \sum_{n=0}^{N-1} \left| X(n\Delta f) - \left(\sum_{k=1}^{K} (\beta_L L_k(\mathbf{v}_k, n\Delta f) + \beta_G G_k(\mathbf{v}_k, n\Delta f)) + \sum_{p=1}^{P} c_p f^p \right) \right|^2 \tag{12.10}
$$

where $X(n\Delta f) - \hat{X}(n\Delta f)$ is the difference between the actual and model values of the spectrum. Both real and imaginary parts of $X(f)$ are considered in the fitting. Using vector notation, J can be rewritten in a more compact form:

$$
J = \| \mathbf{X} - \hat{\mathbf{X}} \| \tag{12.11}
$$

where $\| \; \|$ represents the Euclidean norm, $X = [X(0), X(\Delta f), \ldots, X((N-1)\Delta f)]^T$, $\hat{X} = [\hat{X}(0), \hat{X}(\Delta f), \ldots, \hat{X}((N-1)\Delta f)]^T$ and ^{T} indicates the matrix transpose.

Because Equation 12.11 is nonlinear in the parameter, the Levenberg–Marquardt [32] algorithm is often used to solve the problem [15]. Several fitting algorithms operating in the frequency domain have been developed in recent years [15]. Some of them [33,34] allow for the inclusion of *a priori* knowledge between spectral components. In fact, known relationships sometimes do exist between spectral lines, such as known amplitude and damping ratios, and frequency and phase shifts. These relationships are translated into constraints among spectral parameters, which may be included in Equation 12.11, thus reducing the number of parameters to be fitted. This operation dramatically increases estimation accuracy and decreases computation time. Prior knowledge is particularly important to resolve overlapping peaks or to impose common line widths in noise spectra to improve accuracy of the other estimates.

12.5 TIME-DOMAIN METHODS

Quantitative evaluation of metabolite parameters may be obtained by processing the time-domain FID signals. Two main categories of methods have been proposed: black-box methods and interactive approaches [22]. The former are based on linear prediction (LP) methods and autoregressive (AR) modeling of the FID, whereas the latter are based on NLLS optimization to directly fit the damped sinusoid model on the data.

The first example of the black-box approach applied to the analysis of FID data is the so-called linear prediction singular value decomposition (LPSVD) method due to Barkhuijsen [35], who applied the well-known approach by Kumaresan and Tufts (KT) [36]. The aim is to estimate the amplitude and frequency of K-complex damped sinusoids embedded in noise when N data samples of the process are observed. $v_k = [f_k, a_k, d_k, \phi_k]$ are unknown parameters. The method is based on the observation that signal $x(n)$ can be modeled by an AR process according to the following backward equation system

$$x(n) = \sum_{k=1}^{K} b_k x(n+k) + w(n) \tag{12.12}$$

where the b_k's are the backward prediction coefficients. When Equation 12.12 is used to describe the process generated by Equation 12.1, an interesting relation does exist between b_k's and signal frequencies:

$$B(z) = 1 + \sum_{k=1}^{K} b_k z^{-k} = \prod_{k=1}^{K} (1 - e^{-d_k + j2\pi f_k} z^{-1}) \tag{12.13}$$

Thus, the unknown frequencies in Equation 12.1 are easily obtained by estimating the b_k's and by rooting the polynomial $B(z)$. The identification of b_k's is obtained by solving the following linear system derived from Equation 12.12

$$\mathbf{Ab} = -\mathbf{h} \tag{12.14}$$

where

$$\mathbf{A} = \begin{bmatrix} x^*(1) & x^*(2) & \cdots & x^*(L) \\ x^*(2) & x^*(3) & \cdots & x^*(L+1) \\ \vdots & \vdots & & \vdots \\ x^*(N-L) & x^*(N-L+1) & \cdots & x^*(N-1) \end{bmatrix}$$

$$\tag{12.15}$$

$$\mathbf{b} = \begin{bmatrix} b(1) \\ b(2) \\ \vdots \\ b(L) \end{bmatrix} \qquad \mathbf{h} = \begin{bmatrix} x^*(0) \\ x^*(1) \\ \vdots \\ x^*(N-L-1) \end{bmatrix}$$

and where $K < L < N$, L being the guessed order of prediction. From a theoretical point of view, in the absence of noise, the matrix A should have rank K. This fact is only "approximately" true when the signal is embedded in noise. Using this fact, Kumaresan and Tufts [36] proposed to replace the matrix A in Equation 12.14 by its K-rank approximant obtained by singular value decomposition (SVD). The approximant is built by retaining only the K dominant singular values of A and setting the others to zero. Therefore, instead of solving Equation 12.14 directly, a modified version (in which matrix A is substituted by its approximant) is used. In this case, the most significant contribution of the noise can be removed, and it can be demonstrated that frequencies and damping factors are obtained in a more robust way [35,36]. However, the amplitudes are still unknown. In the LPSVD approach, the problem is solved by substituting the estimated frequencies and damping factors in Equation 12.1 and writing this equation down for all N data points. In this way, we obtain a set of linear equations in the unknown parameters $\alpha = a_k e^{j\phi_k}$ [35]. The latter can therefore be estimated by solving another linear system analogous to Equation 12.14. An example of metabolite quantification using the LPSVD method is shown in Figure 12.3. It is worth noting that the method is able to detect and separate small metabolite contributions that were not detectable in the absorption-mode spectra.

Several variants of the method cited in the preceding text have been applied in the analysis of MRS data in an effort to reduce computation time and improve accuracy in parameter estimation. Details can be found in various articles [22]. These methods try to overcome the limitations imposed by the fixed truncation of SVD [37,38] or by polynomial rooting and root selection [39,40]. In Reference 41, frequency localization is improved by applying the KT approach in different subband signals obtained by wavelet packet decomposition of the original FID.

With interactive methods, the estimation of model parameters v_k is obtained by solving a classical NLLS problem. A maximum likelihood (ML) estimate of v_k is achieved under the hypothesis of white Gaussian noise by minimizing the following figure of merit (in analogy with the figure of merit introduced for frequency-domain fitting in Equation 12.10 and Equation 12.11):

$$J = \sum_{n=1}^{N} |x(n) - \hat{x}(n)|^2 = \sum_{n=1}^{N} \left| x(n) - \sum_{k=1}^{K} a_k e^{j\phi_k} e^{(-d_k + j2\pi f_k)n} \right|^2 = \| \mathbf{x} - \hat{\mathbf{x}} \| \quad (12.16)$$

where $x(n) - \hat{x}(n)$ is the so-called prediction error (i.e., the difference between actual sample and model-predicted sample). Minimization of Equation 12.16 can be obtained by using the classical Levenberg–Marquardt algorithm [32] or more sophisticated NLLS methods [42,43]. In addition, the problem can be simplified by using the variable projection method [44]. A visual inspection of Equation 12.1 reveals that the model can be split into linear and nonlinear parts, thus obtaining

$$\hat{x}(n) = \sum_{k=1}^{K} \alpha_k \kappa(f_k, d_k, n) \quad (12.17)$$

where $\alpha_k = a_k e^{j\phi_k}$ and $\kappa_k = e^{(-d_k+j2\pi f_k)n}$. Writing Equation 12.17 for each acquired sample and using matrix notation, we obtain

$$\hat{\mathbf{x}} = \mathbf{KA} \tag{12.18}$$

where

$$\mathbf{K} = \begin{bmatrix} \kappa(f_1,d_1,1) & \cdots & \kappa(f_K,d_K,1) \\ \vdots & \ddots & \vdots \\ \kappa(f_1,d_1,N) & \cdots & \kappa(f_K,d_K,N) \end{bmatrix} \qquad \mathbf{A} = \begin{bmatrix} \alpha_1 \\ \vdots \\ \alpha_K \end{bmatrix} \tag{12.19}$$

If we assume that the nonlinear parameters $(f_k, d_k, k = \{1, 2,...,K\})$ are known (or some initial guess about them is available), we can solve Equation 12.18 in the LS sense, thus obtaining

$$\mathbf{A} = \mathbf{K}^{\dagger}\hat{\mathbf{x}} \tag{12.20}$$

where $\mathbf{K}^{\dagger}$ is the pseudoinverse of $\mathbf{K}$. Having estimated the $\mathbf{A}$ parameters, we can get rid of it in Equation 12.16 and solve a simpler problem

$$\|\mathbf{x} - \mathbf{KK}^{\dagger}\mathbf{x}\| \tag{12.21}$$

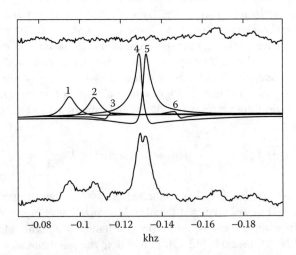

FIGURE 12.4 Example of application of AMARES for the analysis of *in vivo* citrate signal. From bottom to top: the FT spectrum of the original signal, Lorentzians (FT of fitted sinusoids), and the residual. (From Vanhamme, L., van den Boogaart, A. and van Huffel, S. [1997]. Improved method for accurate and efficient quantification of MRS data with use of prior knowledge. *J. Magn. Reson.* 129: 35–43.)

Solving Equation 12.21 or Equation 12.16 will lead to the same results. However, in Equation 12.21, a fewer number of variables are involved even if the functional is more complicated.

Following the same idea, *a priori* biochemical knowledge (e.g., known amplitudes and damping ratios, and frequency shifts and common phases) can be included in the functional as well. In fact, when prior knowledge can be expressed as a set of constraints among parameters, these constraints are substituted in the functional (either Equation 12.21 or Equation 12.16), thus reducing the dimension of the parameter space. In this way, regardless of the existence of constraints among parameters, an unconstrained NLLS optimization problem is always solved.

Several modified versions of the original approach have been presented, which differ for the selected minimization algorithm [42,45–48] or for the kind of prior knowledge included into the procedure [42,48,49]. An example of application of the method proposed in Reference 42 is shown in Figure 12.4.

12.6 CONCLUSIONS

A general overview of the common signal processing methods for metabolite quantification in MRS is presented. Fundamentals of both time-domain methods and frequency-domain approaches have been introduced in order to provide some theoretical basis that may help the reader to appreciate the power of these techniques and their correct applicability to the processing of MRS data. It is worth noting that an optimal approach does not exist. Conversely, due to the variety of MRS spectra characteristics, it is a common experience that different signal processing methods (and models) must be employed for analyzing data acquired from various tissues (brain, muscle, or liver) or different nuclei (^{1}H, ^{31}P, etc.). The accuracy of the results depends on the correct match between the selected model and characteristics of the experimental signals. A few multicenter studies and review articles [21,22,50] have deeply compared the various signal processing approaches and widely exploited the advantages and shortcomings of each method. It is not the aim of this chapter to describe them in detail, but we hope that the presented concepts could give a general picture of the peculiarities of each method, stressing those aspects that make them suitable and attractive for quantitative analysis of MRS data.

REFERENCES

1. Gandian, D. (1995). *NMR and Its Application to Living Systems*. 2nd ed. Oxford: Oxford University Press.
2. de Certaines, J.D., Bovée, W.M.M.J., and Podo, F. (1992). *Magnetic Resonance Spectroscopy in Biology and Medicine*. Oxford: Pergamon Press.
3. Howe, F.A., Maxwell, R.J., Saunders, D.E., Brown, M.M., and Griffiths, J.R. (1993). Proton spectroscopy *in vivo*, *Magn. Res. Q.* 9: 31–59.
4. Falini, A., Calabrese, G., Origgi, D., Lipari, S., Triulzi, F., Losa, M., and Scotti, G. (1996). Proton magnetic resonance spectroscopy in intracranial tumors: clinical perspectives *J. Neurol.* 243: 706–714.

5. Demaerel, P., Johannik, K., Van Hecke, P., van Ogeval, C., Verellen, S., Marchal, G., Plets, C., Goffin, J., Van Calenbergh, F., Lamens, M., and Baert, A.L. (1991). Localization of ¹H-MR spectroscopy in fifty cases newly diagnosed intracranial tumors, *J. Comput. Assist. Tomogr.* 15: 67–76.

6. Williams, S.R., Crockard, H.A., and Gaudian, D.G. (1989). Cerebral ischemia studied by nuclear magnetic resonance spectroscopy. *Cerebrovasc. Brain Metab. Rev.* 1: 91–114.

7. Matthews, P.M., Andermann, F., and Arnold, D.L. (1990). A proton magnetic resonance spectroscopy study of focal epilepsy in humans. *Neuroradiology,* 40: 985–989.

8. Hugg, J.W., Laxer, K.D., Matson, G.B., Maudseley, A.A., and Weiner, M.W. (1992). Neuron loss localizes human focal epilepsy by in-vivo proton MR spectroscopy imaging. *Ann. Neurol.* 42: 2011–2018.

9. Matthews, P.M., Francis, G., Antel, J., and Arnold, D.L. (1991). Proton magnetic resonance spectroscopy for metabolic characterization of plaques in multiple sclerosis. *Neurology,* 41: 1251–1256.

10. Wolinsky, J.S., Narayana, P.A., and Fenstermacher, M.J. (1990). Proton magnetic resonance spectroscopy in multiple sclerosis. *Neurology,* 40: 1764–1769.

11. Peeling, J. and Sutherland, G. (1992). High-resolution ¹H-NMR spectroscopy studies of extracts of human cerebral neoplasms. *Magn. Reson. Med.* 24: 123–136.

12. Preul, M.C., Olivier, A., Pokrupa, R., and Arnold, D.L. (1996). Accurate non-invasive diagnosis of human brain tumors by using ¹H magnetic resonance spectroscopy. *Nature Med.* 2: 323–325.

13. Garcia-Martin, M.L., Garcia-Espinosa, M.A., and Cerdan, S. (1998). Biochemistry detectable by MRS, in Syllabus. *Methodology, Spectroscopy and Clinical MRI.* 15th Annual Scientific Meeting. Geneva. September 17–20.

14. Emsley, J.W., Feeney, J., and Sutcliffe, L.H. (1991). Progress in nuclear magnetic resonance spectroscopy. *Progress in NMR spectroscopy.* 23: 211–258.

15. de Graaf, A.A. and Bovée, W.M.M.J. (1990). Improved quantification of *in vivo* ¹H NMR spectra by optimization of signal acquisition and processing and by incorporation of prior knowledge into spectral fitting. *Magnetic Resonance in Medicine.* 15: 305–319.

16. van Rijen, P.C., Verheul, H.B., and van Echteld, C.J.A. (1991). Effects of dextromethorphan on rat brain during ischemia and reperfusion assessed by magnetic resonance spectroscopy. *Stroke,* 22: 343–350.

17. Berkelbach van der Sprenkel, J.W., Luyten, P.R., van Rijen, P.C., Tulleken, C.A., and den Hollander, J.A. (1988). Cerebral lactate detected by regional proton magnetic resonance spectroscopy in a patient with cerebral infarction. *Stroke,* 19: 1556–1560.

18. Grossman, R.I., Lenkiski, R.E., Ramer, K.N., Gonzaales-Scarano, F., and Cohen, J.A. (1992). MR proton spectroscopy in multiple sclerosis. *AJNR.* 13: 1535–1543.

19. Ross, B.D., Radda, G.K., Gadian, D.G., Rocker, G., Esiri, M., and Falconer-Smith, J. (1981). Examination of a case of suspected McArdle's Syndrome by 31P nuclear magnetic resonance. *N. Engl. J. Med.* 304: 1338–1342.

20. Newman, R.J., Bore, P.J., Chan, L., Gadian, D.G., Styles, P., Taylor D., and Radda, G.K. (1982). Nuclear magnetic resonance studies of forearm muscle in Duchenne dystrophy. *Br. Med. J Clin. Res. Ed.* 284: 1072–1074.

21. Mierisova, S. and Ala-Korpela, M. (2001). MR spectroscopy quantitation: a review of frequency-domain methods, *NMR in Biomed.* 14: 247–259.

22. Vanhamme, L., Sundin, T., Ven Hecke, P., and Van Huffel, S. (2001). MR spectroscopy quantitation: a review of time-domain methods, *NMR in Biomed.* 14: 233–246.
23. Zandt, H., van der Graff, M., and Heerschap, A. (2001). Common processing of *in-vivo* MR spectra. *NMR in Biomed.* 14: 224–232.
24. Kuroda, Y., Wada, A., Yamazaki, T., and Nagayama, K. (1989). Postacquisition data processing method for suppression of the solvent signal. *J. Magn. Res.* 84: 604–610.
25. Marion, D., Ikura, M., and Bax, A. (1989). Improved solvent suppression in one- and two-dimensional NMR spectra by convolution of time-domain data. *J. Magn. Res.* 84: 425–430.
26. Sobering, G., Kienlin, M., Moonen, C., van Zijl, P., and Bizzi, A. (1991). Postacquisition reduction of water signals in proton spectroscopic imaging of the brain. in *Proceeding SMRM*, 10th annual Meeting, San Francisco, CA.
27. van den Boogaart, A., van Ormondt, D., Pijnappel, W.W.F., de Beer, R., and Ala-Korpela, M. (1994). Removal of water resonance from ^{1}H magnetic resonance spectra. in *Mathematics in Signal Processing III*, McWhirter, J.D. (Ed.). Clarendon Press: Oxford.
28. Vanhamme, L., Fierro, R.D., van Huffel, S., and de Beer, R. (1998). Fast removal of residual water in proton spectra. *J. Magn. Reson.* 132: 197–203.
29. van den Boogaart, A., Ala-Korpela, M., Jokisaari, J., and Griffiths, J.R. (1998). Time and frequency domain analysis of NMR data compared: an application to 1D ^{1}H spectra of lipoproteins. *Magn. Reson. Med.* 39: 899–911.
30. de Graaf, A.A., van Dijk, J.E., and Boveé, W.M.M.J. (1990). QUALITY: quantification improvement by converting lineshapes to the Lorentzian type. *Magn. Reson. Med.* 13: 343–357.
31. Marshall, I., Higinbothan, J., Bruce, S., and Freise, A. (1997). Use of Voigt lineshape for quantification of *in vivo* ^{1}H spectra. *Magn. Reson. Med.* 37: 343–357.
32. Marquardt, D.W. (1963). An algorithm for least-square estimation of non-linear parameters. *J. Soc. Ind. Appl. Math.* 11: 431–441.
33. Hiltunen, Y., Ala-Korpela, M., Jokisaari, J., Eskelinen, S., Kiviniitty, K., Savolainen, M., and Kesaniemi, Y.A. (1991). A lineshape fitting model for ^{1}H NMR spectra of human blood plasma. *Magn. Reson Med.* 21: 222–232.
34. Slotboom, J., Boesch, C., and Kreis, R. (1998). Versatile frequency domain fitting using time-domain models and prior knowledge. *Magn. Reson. Med.* 39: 899–911.
35. Barkhuijsen, H., de Beer, R., Boveé, W.M.M.J., and van Ormondt, D. (1985). Retrieval of frequencies, amplitudes, damping factors and phases from time-domain signals using a linear least-square procedure. *J Magn. Res.* 61: 465–481.
36. Kumaresan, R., and Tufts, D. (1982). Estimating the parameters of exponentially damped sinusoids and pole-zero modeling in noise, *IEEE Trans. Acoust. Speech Signal Proc.* 30: 833–840.
37. Kolbel, W. and Schafer, H. (1992). Improvement and automation of the LPSVD algorithm by continuous regularization of the singular values. *J. Magn. Reson.* 100: 598–603.
38. Chen, H., Van Huffel, S., Decanniere, C., and Van Hecke, P. (1994). A signal enhancement algorithm for time-domain data MR quantification. *J. Magn. Res. Series A.* 109: 46–55.
39. Barkhuysen, H., de Beer, R., and van Ormondt, D. (1987). Improved algorithm for noniterative time-domain model fitting to exponentially damped magnetic resonance signals. *J. Magn. Reson.* 73: 553–557.

40. Lin, Y.Y., Hodgkinson, P., Ernst, M., and Pines, A. (1997). A novel detection-estimation scheme for noisy NMR signals: applications to delayed acquisition data. *J. Magn. Reson.* 128: 30–41.

41. Mainardi, L.T., Origgi, D., Lucia, P., Scotti, G., and Cerutti, S. (2002). A wavelet packets decomposition algorithm for quantification of in vivo ^{1}H-MRS parameters, *Med. Eng. Phys.* 24: 201–208.

42. Vanhamme, L., van den Boogaart, A., and van Huffel, S. (1997). Improved method for accurate and efficient quantification of MRS data with use of prior knowledge. *J. Magn. Reson.* 129: 35–43.

43. Dennis, J.E. and Schnabel, R.B. (1983). *Numerical Methods for Unconstrained Optimization and Nonlinear Equations.* Prentice-Hall, Englewood Cliffs, NJ.

44. Golub, G.H. and Pereyra, V. (1973). The differentiation of pseudo inverses and nonlinear least-squares problems whose variables separate. *SIAM J. Numer. Anal.* 10: 413–432.

45. Miller, M.I. and Greene, A.S. (1989). Maximum-likelihood estimation for nuclear magnetic resonance spectroscopy, *J. Magn. Reson.* 83: 525–548.

46. Liu, Z.S., Li, J., and Stoica, P. (1997). RELAX-based estimation of damped sinusoidal signal parameters. *Signal Process.* 62: 311–321

47. van der Veen, J.W.C., de Beer, R., Luyten, P.R., and van Ormondt, D. (1988). Accurate quantification of in vivo ^{31}P NMR signals using the variable projection method and a priori knowledge. *Magn. Reson. Med.* 6: 92–98.

48. Chen, S.C., Schaewe, T.J., Teichman, R.S., Miller, M.I., Nadel, S.N., and Greene, A.S. (1993). Parallel algorithms for maximum-likelihood nuclear magnetic resonance spectroscopy. *J. Magn. Reson. Series A.* 102: 16–23.

49. Bi, Z., Bruner, A.P., Li, J., Scott, K.N., Liu, Z.S., Stopka, C.B., Kim, H.W., and Wilson, D.C. (1999). Spectral fitting of NMR spectra using an alternating optimization method with a priori knowledge. *J. Magn. Reson.* 140: 108–119.

50. de Beer, R., van den Boogaart, A., Cady, E., Graveron-Demilly, D., Knijn, A., Langenberger, K.W., Lindon, J.C., Ohlhoff, A., Serrai, H., and Wylezinska-Arridge, M. (1996). Multicentre quantitative data-analysis trial: the overlapping background problem. in Podo, F., Bovée, M.M.J., de Certaines, J.D., Heinrikesen, O.N., Leach, M.O. and Leibfritz, D. *Eurospin Ann. ISS.* pp. 341–365.

13 Image Processing of Diffusion Tensor MRI Data

Lauren O'Donnell, Steven Haker,
and Carl-Fredrik Westin

CONTENTS

13.1 INTRODUCTION

Diffusion magnetic resonance imaging (MRI) measures the random molecular motion (diffusion) of water in biological tissue. Tissues with an oriented structure, such as the fibers of muscle or the axons of nervous tissue, produce measurable diffusion anisotropy: a pattern is imposed on the diffusing molecules by the shape of the tissue. One such tissue is shown in Figure 13.1, an image of some major white matter fiber tracts in a sagittal view of the human brain. Fiber tracts are bundles of neurons whose membranes hinder water diffusion [5]. The orientation of cells in neural tissue is thus reflected in MRI measurements of water diffusion, which vary with direction. This is shown schematically in Figure 13.2.

427

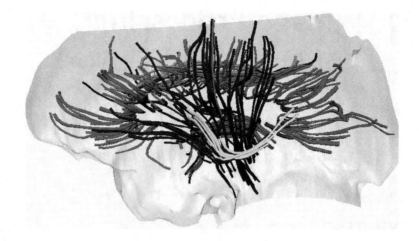

FIGURE 13.1 Selected fiber traces (tractography) from a DT-MRI data set, shown in a sagittal view. Note the presence of both long and short paths, and varying degrees of curvature. These factors complicate automated extraction of tracts from diffusion MRI data. This image was created with the 3-D Slicer DT-MRI visualization tool. (From Talos, I.-F., O'Donnell, L., Westin, C.-F., Warfield, S.K., Wells, W.M., Yoo, S.-S., Panych, L.P., Golby, A., Mamata, H., Maier, S.S., Ratiu, P., Guttmann, C.R., Black, P.M., Jolesz, F.A., and Kikinis, R. (2003). Diffusion tensor and functional MRI fusion with anatomical MRI for image guided neurosurgery. in *Conference on Medical Image Computing and Computer-Assisted Intervention (MICCAI)*. pp. 407–415. Toronto, Canada.)

One mathematical representation of this three-dimensional (3-D) diffusion pattern is the diffusion tensor, a 3×3 symmetric, positive definite matrix. In diffusion tensor MRI (DT-MRI) imagery of the brain, the eigensystem of the diffusion tensor gives a local coordinate system that approximates the local neural structure. The major eigenvector gives the direction of greatest diffusion (the most probable fiber direction). The eigenvalues of the diffusion tensor represent the diffusion coefficients in the principal directions of diffusion. Figure 13.3 shows the eigenvectors of a diffusion tensor in two dimensions, scaled by the

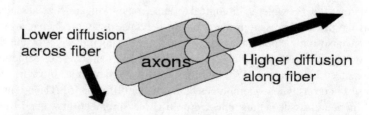

FIGURE 13.2 Idealized diagram showing the effect of axons on water diffusion, as measured by diffusion MRI.

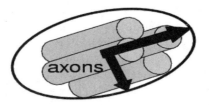

FIGURE 13.3 Idealized diagram showing the local coordinate system described by the eigensystem of the diffusion tensor, in two dimensions. Note that where tracts cross in a voxel, the major eigenvector is not likely to be parallel to either tract.

corresponding eigenvalues, and surrounded by an ellipse describing the shape of water diffusion.

The diffusion tensor model is reasonable but is limited when describing neuroanatomy. In the case where one voxel contains more complicated geometry than a single tract, it is misleading to only consider the information in the major eigenvector. In addition, the diffusion tensor model cannot describe fiber tract crossings or complicated patterns of tracts that may occur within a voxel. It is important to take these limitations into account when developing a data analysis method.

In this chapter, we first present background information on diffusion and diffusion tensor calculation. We continue with a description of diffusion tensor shape analysis and visualization methods. Finally, we introduce two techniques for extracting connectivity information from diffusion tensor data sets.

13.2 DIFFUSION AND DIFFUSION TENSOR CALCULATION

The process of diffusion is described by Fick's first law, which relates a concentration difference to a flux (a flow across a unit area). It states that the flux, j, is proportional to the gradient of the concentration, ∇u. The proportionality constant d is the diffusion coefficient.

$$j = -d\nabla u \qquad (13.1)$$

For an anisotropic material, the flow field does not follow the concentration gradient directly, because the material properties also affect diffusion. Consequently, the diffusion tensor, D, is introduced to model the material locally.

$$j = -D\nabla u \qquad (13.2)$$

The standard model of diffusion says that over time, the concentration of the solute will change as the divergence of the flux:

$$u_t = \nabla \cdot (D\nabla u). \qquad (13.3)$$

This is due to conservation of mass. Intuitively, it implies that, for example, fluid flow outward from a point (divergence) should decrease the concentration at that point while increasing the concentration at neighboring points. In the steady state, the concentration does not change; consequently the steady-state flux vector field is divergence free.

In diffusion MRI, magnetic field gradients are employed to sensitize the image to diffusion in a particular direction. In each resulting diffusion-weighted image, signal is lost wherever molecules diffuse in the direction of interest during imaging. By repeating the process of diffusion weighting in multiple directions, at each voxel a 3-D diffusion pattern can be estimated, which will reflect the shape of the underlying anatomy.

In DT-MRI, the diffusion tensor field is calculated from a set of diffusion-weighted images by solving the Stejskal–Tanner equation (Equation 13.4) [1]. This equation describes how the signal intensity at each voxel decreases in the presence of diffusion:

$$S_k = S_0 e^{-b \hat{g}_k^T D \hat{g}_k} \tag{13.4}$$

Here, S_0 is the image intensity at the voxel (measured with no diffusion gradient), and S_k is the intensity measured after the application of the kth diffusion-sensitizing gradient. $\hat{g}_k$ is a unit vector representing the direction of this diffusion-sensitizing magnetic field gradient. D is the diffusion tensor, and the product $\hat{g}_k^T D \hat{g}_k$ represents the diffusion coefficient in direction $\hat{g}_k$. In addition, b is LeBihan's factor describing the pulse sequence, gradient strength, and physical constants. For rectangular gradient pulses, the b-factor is defined by $b = \gamma^2 \delta^2 (\Delta - \frac{\delta}{3}) |g|^2$, where γ is the proton gyromagnetic ratio (42 MHz/Tesla), $|g|$ the strength of the diffusion-sensitizing gradient pulses, δ the duration of the diffusion gradient pulses, and Δ the time between diffusion gradient RF pulses [24]. For more information on the tensor calculation process, see for example Reference 24.

13.3 ANISOTROPY AND MACROSTRUCTURAL MEASURES

The geometric nature of the measured diffusion tensor within a voxel is a meaningful measure of fiber tract organization. Factors affecting the shape of the apparent diffusion tensor (shape of the diffusion ellipsoid) in the white matter include the density of fibers, the degree of myelination, the average fiber diameter, and the directional similarity of the fibers in the voxel. In addition, because MRI methods obtain a macroscopic measure of a microscopic quantity (which necessarily entails intravoxel averaging), the voxel dimensions influence the measured diffusion tensor at any given location in the brain.

The advent of robust diffusion tensor imaging techniques has prompted the development of quantitative measures for describing diffusion anisotropy. However, to relate the measure of diffusion anisotropy to the structural geometry of the tissue, a mathematical description of diffusion tensors and their quantification is necessary [1]. Several different measures of anisotropy have been proposed in the literature. Among the most popular are two that are based on the normalized variance of the eigenvalues: relative anisotropy (RA) and fractional anisotropy (FA) [3]. An advantage of these measures is that they can be calculated without first explicitly calculating any eigenvalues. Both anisotropy measures can be expressed in terms of the norm and trace of the diffusion tensor. The norm is calculated as the square root of the sum of the squared elements of the tensor, which equals the square root of the sum of the squared eigenvalues; and the trace is calculated as the sum of the diagonal elements, which equals the sum of the eigenvalues:

$$RA = \frac{1}{\sqrt{2}} \frac{\sqrt{(\lambda_1 - \lambda_2)^2 + (\lambda_2 - \lambda_3)^2 + (\lambda_1 - \lambda_3)^2}}{(\lambda_1 + \lambda_2 + \lambda_3)} = \frac{\sqrt{3}}{\sqrt{2}} \frac{|\mathbf{D} - \frac{1}{3}\mathrm{trace}(\mathbf{D})\mathbf{I}|}{\mathrm{trace}(\mathbf{D})} \quad (13.5)$$

$$FA = \frac{1}{\sqrt{2}} \frac{\sqrt{(\lambda_1 - \lambda_2)^2 + (\lambda_2 - \lambda_3)^2 + (\lambda_1 - \lambda_3)^2}}{\sqrt{\lambda_1^2 + \lambda_2^2 + \lambda_3^2}} = \frac{\sqrt{3}}{\sqrt{2}} \frac{|\mathbf{D} - \frac{1}{3}\mathrm{trace}(\mathbf{D})\mathbf{I}|}{|\mathbf{D}|} \quad (13.6)$$

where $\mathbf{I}$ is the identity tensor. The constants are inserted to ensure that the measures range from zero to one. In the next section, we will present alternatives to these measures based on the geometric properties of the diffusion ellipsoid.

13.3.1 GEOMETRICAL MEASURES OF DIFFUSION

The diffusion tensor can be visualized using an ellipsoid in which the principal axes correspond to the directions of the eigenvector system. Using the symmetry properties of this ellipsoid, the diffusion tensor can be decomposed into basic geometric measures [25], a concept that we will elaborate in this section.

Let $\lambda_1 \geq \lambda_2 \geq \lambda_3 \geq 0$ be the eigenvalues of the symmetric diffusion tensor $\mathbf{D}$, and let $\hat{\mathbf{e}}_i$ be the normalized eigenvector corresponding to λ_i. The tensor $\mathbf{D}$ can then be described by

$$\mathbf{D} = \lambda_1 \hat{\mathbf{e}}_1 \hat{\mathbf{e}}_1^T + \lambda_2 \hat{\mathbf{e}}_2 \hat{\mathbf{e}}_2^T + \lambda_3 \hat{\mathbf{e}}_3 \hat{\mathbf{e}}_3^T \quad (13.7)$$

Diffusion can be divided into three basic cases depending on the rank of the diffusion tensor:

Linear case ($\lambda_1 \gg \lambda_2 \simeq \lambda_3$): diffusion is mainly in the direction corresponding to the largest eigenvalue,

$$\mathbf{D} \simeq \lambda_1 \mathbf{D}_l = \lambda_1 \hat{\mathbf{e}}_1 \hat{\mathbf{e}}_1^T \quad (13.8)$$

Planar case ($\lambda_1 \simeq \lambda_2 \gg \lambda_3$): diffusion is restricted to a plane spanned by the two eigenvectors corresponding to the two largest eigenvalues,

$$\mathbf{D} \simeq \lambda_1 \mathbf{D}_p = \lambda_1 \left(\hat{\mathbf{e}}_1 \hat{\mathbf{e}}_1^T + \hat{\mathbf{e}}_2 \hat{\mathbf{e}}_2^T \right) \quad (13.9)$$

Spherical case ($\lambda_1 \simeq \lambda_2 \simeq \lambda_3$): isotropic diffusion,

$$\mathbf{D} \simeq \lambda_1 \mathbf{D}_s = \lambda_1 \left(\hat{\mathbf{e}}_1 \hat{\mathbf{e}}_1^T + \hat{\mathbf{e}}_2 \hat{\mathbf{e}}_2^T + \hat{\mathbf{e}}_3 \hat{\mathbf{e}}_3^T \right). \tag{13.10}$$

In general, the diffusion tensor $\mathbf{D}$ will be a combination of these cases. Expanding the diffusion tensor using these cases as a basis gives

$$\begin{aligned}
\mathbf{D} &= \lambda_1 \hat{\mathbf{e}}_1 \hat{\mathbf{e}}_1^T + \lambda_2 \hat{\mathbf{e}}_2 \hat{\mathbf{e}}_2^T + \lambda_3 \hat{\mathbf{e}}_3 \hat{\mathbf{e}}_3^T \\
&= (\lambda_1 - \lambda_2) \hat{\mathbf{e}}_1 \hat{\mathbf{e}}_1^T + (\lambda_2 - \lambda_3) \left(\hat{\mathbf{e}}_1 \hat{\mathbf{e}}_1^T + \hat{\mathbf{e}}_2 \hat{\mathbf{e}}_2^T \right) \\
&\quad + \lambda_3 \left(\hat{\mathbf{e}}_1 \hat{\mathbf{e}}_1^T + \hat{\mathbf{e}}_2 \hat{\mathbf{e}}_2^T + \hat{\mathbf{e}}_3 \hat{\mathbf{e}}_3^T \right) \\
&= (\lambda_1 - \lambda_2) \mathbf{D}_i + (\lambda_2 - \lambda_3) \mathbf{D}_p + \lambda_3 \mathbf{D}_s
\end{aligned}$$

where $(\lambda_1 - \lambda_2)$, $(\lambda_2 - \lambda_3)$ and λ_3 are the coordinates of $\mathbf{D}$ in the tensor basis $\{\mathbf{D}_i, \mathbf{D}_p, \mathbf{D}_s\}$. A similar tensor shape analysis has been shown to be useful in a number of computer vision applications [22,23].

The coordinates of the tensor in our new basis classify the diffusion tensor and describe how close the tensor is to the generic cases of line, plane, and sphere, and hence, can be used for classification of the diffusion tensor according to its geometry. Because the coordinates are based on the eigenvalues of the tensor, they are rotationally invariant, and the values do not depend on the chosen frame of reference. To obtain quantitative measures of the anisotropy, the derived coordinates have to be normalized, which, in turn, will lead to geometric shape measures. As in the case of fractional and relative anisotropy, there are several (rotationally invariant) choices of normalization. For example, the maximum diffusivity, λ_1, the trace of the tensor, $\lambda_1 + \lambda_2 + \lambda_3$, or the norm of the tensor, $\sqrt{\lambda_1^2 + \lambda_2^2 + \lambda_3^2}$, can be used as normalization factors.

By using the largest eigenvalues of the tensor, the following quantitative shape measures are obtained for the linear, planar, and spherical measures:

$$c_l = \frac{\lambda_1 - \lambda_2}{\lambda_1} \tag{13.11}$$

$$c_p = \frac{\lambda_2 - \lambda_3}{\lambda_1} \tag{13.12}$$

$$c_s = \frac{\lambda_3}{\lambda_1} \tag{13.13}$$

where all measures lie in the range from zero to one, and their sum is equal to one,

$$c_l + c_p + c_s = 1 \tag{13.14}$$

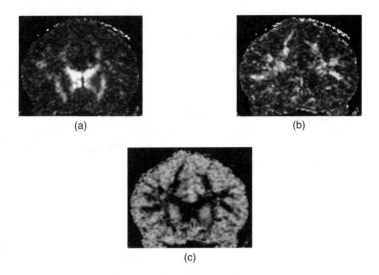

(a)

(b)

(c)

FIGURE 13.4 (a, b, and c) Coronal brain images showing the three geometrical measures. Note that most of the major fiber tracts are visible despite the low resolution of the data set (1.7 × 1.7 × 4 mm). (From Westin, C.-F., Maier, S., Mamata, H., Nabavi, A., Jolesz, F., and Kikinis, R. (2002). Processing and visualization of diffusion tensor MRI. *Med. Image Anal.* 6(2): 93–108.)

Figure 13.4 contains coronal brain images depicting these geometrical measures. Alternatively, the coordinates can be normalized with the norm of the tensor, giving:

$$c_l = \frac{\lambda_1 - \lambda_2}{\sqrt{\lambda_1^2 + \lambda_2^2 + \lambda_3^2}} \tag{13.15}$$

$$c_p = \frac{2(\lambda_2 - \lambda_3)}{\sqrt{\lambda_1^2 + \lambda_2^2 + \lambda_3^2}} \tag{13.16}$$

$$c_s = \frac{3\lambda_3}{\sqrt{\lambda_1^2 + \lambda_2^2 + \lambda_3^2}} \tag{13.17}$$

To ensure that the measures remain in the range from 0 to 1, and the sum is 1, the scaling factors 2 and 3 have been inserted for the planar and the spherical cases. A geometrical anisotropy measure that has a behavior similar to the FA measure (fractional anisotropy, Equation 13.6) describes the deviation from the spherical case

$$c_a = 1 - c_s = c_l + c_p \tag{13.18}$$

which is the sum of the linear and planar measures. By normalizing with the trace of the tensor instead of the norm, the measure will be more similar to the RA measure (relative anisotropy, Equation 13.5).

The presented measures quantitatively describe the geometrical shape of the diffusion tensor and, therefore, do not depend on the absolute level of the diffusion present. However, in low-signal regions, where the noise level dominates these shape measures, they make little sense. In practice, all shape measures should be regularized by adding a constant in the denominator of size similar to that of the noise level. For example, the λ_1 normalized linear measure (Equation 13.11) would be expressed as follows:

$$c_{l_\sigma} = \frac{\lambda_1 - \lambda_2}{\lambda_1 + \sigma} \tag{13.19}$$

where a suitable value for σ would be the expected value of λ_1 in a low-signal region. This expression has similarities to classical Wiener filtering, where the noise level σ has very little influence on signals larger than σ, but penalizes signals that are smaller. When the normalization is done using the trace or norm, σ should have the expected value of the trace or norm, respectively, in low-signal regions.

When applied to white matter, the linear measure, c_l, reflects the uniformity of tract direction within a voxel. In other words, it will be high only if the diffusion is restricted in two orthogonal directions. The anisotropy measure, c_a, indicates the relative restriction of the diffusion in the most restricted direction and emphasizes white matter tracts, which, within a voxel, exhibit at least one direction of relatively restricted diffusion.

13.3.2 MACROSTRUCTURAL TENSOR AND DIFFUSIVE MEASURES

In the previous section, we characterized the diffusion isotropy and anisotropy within a voxel. Here, we will discuss methods for examining the pattern or distribution of diffusion within a local image neighborhood. Basser and Pierpaoli proposed a scalar measure for macrostructural diffusive similarity based on tensor inner products between the center voxel tensor and its neighbors [3,16] (as in the vector case, the inner product between two tensors measures their degree of similarity). This intervoxel scalar measure is known as a *lattice index* and is defined by*

$$LI = \sum_k a_k \left(\frac{\sqrt{3}}{\sqrt{8}} \frac{\sqrt{\langle \mathbf{T}, \mathbf{T}_k \rangle}}{\sqrt{\langle \mathbf{D}, \mathbf{D}_k \rangle}} + \frac{3}{4} \frac{\langle \mathbf{T}, \mathbf{T}_k \rangle}{\sqrt{\langle \mathbf{D}, \mathbf{D} \rangle}\sqrt{\langle \mathbf{D}_k, \mathbf{D}_k \rangle}} \right) \tag{13.20}$$

---

* Corrected formula; personal communication from Pierpaoli, 1997.

where a_k is a spatial mask, for example, 3×3 voxels, with sum of coefficients equal to 1, and $\mathbf{T}$ is the anisotropic part of the diffusion tensor $\mathbf{D}$. The anisotropic part of the tensor has trace zero and can be written as

$$\mathbf{T} = \mathbf{D} - \frac{1}{3}\text{trace}(\mathbf{D})\mathbf{I} \tag{13.21}$$

$$= \mathbf{D} - \frac{1}{3}\langle \mathbf{D}, \mathbf{I} \rangle \tag{13.22}$$

$$= \mathbf{D} - \frac{\langle \mathbf{D}, \mathbf{I} \rangle}{\langle \mathbf{I}, \mathbf{I} \rangle} \tag{13.23}$$

where $\mathbf{I}$ is the identity tensor. By rewriting Equation 13.20, it becomes clear that all the tensor inner products are individually normalized

$$LI = \sum_k a_k \left(\sqrt{\frac{3}{8}} \sqrt{\frac{\langle \mathbf{T}, \mathbf{T}_k \rangle}{\langle \mathbf{D}, \mathbf{D}_k \rangle}} + \frac{3}{4} \sqrt{\frac{\langle \mathbf{T}, \mathbf{T}_k \rangle}{\langle \mathbf{D}, \mathbf{D} \rangle}} \sqrt{\frac{\langle \mathbf{T}, \mathbf{T}_k \rangle}{\langle \mathbf{D}_k, \mathbf{D}_k \rangle}} \right) \tag{13.24}$$

Because the components in the sum are normalized, small and large diffusion tensor components will have equal weight in determining the lattice index. Unfortunately, the smaller tensors are more influenced by noise and, hence, affect the index more than is desirable.

An alternative measure to the lattice index, which can be seen as an external measure as it is based on the tensors in a neighborhood, is to use an internal voxel-based measure ($RA, FA, c_{l,p,s}$) on a *filtered* version of the diffusion tensor field. For example, local averaging of the tensor field with a spatial mask a (normalized so that the coefficients sum to one):

$$\mathbf{D}_a = \sum_k a_k \mathbf{D}_k \tag{13.25}$$

describes the local average diffusion, where the rank of the average tensor $\mathbf{D}_a$ describes the complexity of the macroscopic diffusion structure. If the rank is close to one, the structure is highly linear, which will be the case in regions of bundles of fibers having the same direction. If the rank is 2, fibers are crossing in a plane, or the underlying diffusivity is planar. Applying the geometrical linear measure c_l in Equation 13.11 to this tensor gives a measure that is high in regions with coherent tensors.

Figure 13.5 compares the three original geometrical measures (top) and the same measures applied to identical tensor data averaged by a Gaussian mask (bottom). Major white matter tracts such as the corpus callosum show high linearity in the averaged data set, indicating high macrostructural organization.

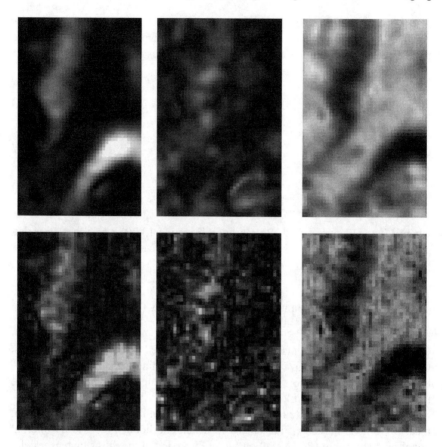

FIGURE 13.5 Axial brain images showing the three geometrical measures and corresponding macrostructural diffusion measures. Top: shows the geometrical measures derived from the original data; bottom: shows the corresponding macrostructural diffusion measures. The geometrical measures derived from the tensors were averaged with a $9 \times 9 \times 3$ Gaussian kernel. (From Westin, C.-F., Maier, S., Mamata, H., Nabavi, A., Jolesz, F., and Kikinis, R. (2002). Processing and visualization of diffusion tensor MRI. *Med. Image Anal.* 6(2): 93–108.)

It should be noted that averaging a diffusion tensor field and then deriving a scalar measure from the averaged field is not the same as averaging the scalar measure derived from the original field.

The relatively simple approach of averaging is useful because the rank of the tensors increases when lower-rank, noncollinear tensors are summed. This effect is illustrated in Figure 13.6 and compared to adding vectors that do not have the freedom to change rank. Adding the two vectors (a) and (b) results in a new vector (c), which is of the same order of complexity as the original vectors. However, adding 2 rank-1 tensors (d) and (e), e.g., diffusion tensors from 2 differently oriented white matter tracts, results in a rank-2 tensor (f), i.e., the

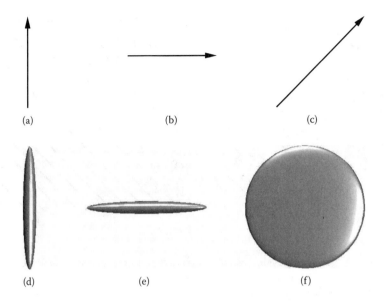

FIGURE 13.6 Vector and tensor summation. Two vectors, (a) and (b), and their sum (c). Two diffusion tensors, (d) and (e), of rank close to 1 visualized as ellipsoids with eigenvectors forming principal axes. The summation of the 2 tensors gives a rank-2 tensor (f). (From Westin, C.-F., Maier, S., Mamata, H., Nabavi, A., Jolesz, F., and Kikinis, R. (2002). Processing and visualization of diffusion tensor MRI. *Med. Image Anal.* 6(2): 93–108.)

output has more degrees of freedom than the input tensors and describes the plane in which diffusion is present. In this sense, averaging of tensors is different from averaging a vector field. The average of a set of vectors gives the "mean event," whereas the average of a set of tensors gives the "mean event" and the "range of the present events."

Figure 13.7 shows a 2-D example illustrating the effect of Gaussian filtering of a diffusion tensor field. The filtered areas that contain inconsistent data give a result of almost-round ellipses (upper right half of the image). Moreover, Gaussian filtering results in more stable estimates of the field directionality in the areas where there is a clear bias in one direction (lower left).

The macrostructural measure achieved (by averaging the tensor field using an isotropic mask) is essentially a feature extraction method rather than a restoration method; the latter aims at reducing the noise level in the data. Although our method does remove noise, the incorporation of more advanced regularization methods [14,18] should be explored, if noise reduction is the main target. Anisotropic filter masks are preferable because they reduce the risk of blurring edges. However, using an anisotropic mask for the macrostructural measure would limit its purpose, i.e., the description of the organization within an area. If the signal changes due to edges inside the local area of interest, this should be reflected in the measure.

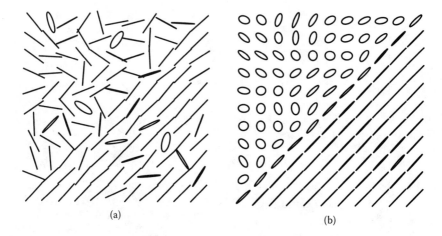

(a)   (b)

FIGURE 13.7 A 2-D diffusion tensor field (a) and the effect of relaxation using a Gaussian filter (b). Note how the tensors in an inconsistent region become rounder, whereas in consistent areas their orientation is stabilized. (From Westin, C.-F., Maier, S., Mamata, H., Nabavi, A., Jolesz, F., and Kikinis, R. (2002). Processing and visualization of diffusion tensor MRI. *Med. Image Anal.* 6(2): 93–108.)

13.4 VISUALIZATION OF DIFFUSION TENSORS

Several methods have been proposed for visualizing the information contained in DT-MRI data. Pierpaoli et al. [17] render ellipsoids to visualize diffusion data in a slice. Peled et al. [15] used headless arrows to represent the in-plane component of the principal eigenvector, along with a color-coded out-of-plane component. Recently, Kindlmann and Weinstein [10] applied our geometric shape indices [25] to opacity maps in volume rendering. They termed this method *barycentric opacity mapping*. They compare volume renderings using opacity maps based on the indices c_l, c_p, and c_a (Equation 13.15–Equation 13.18).

In Figure 13.8a, a diffusion tensor field from an axial slice of the brain is shown (using the visualization method presented in Reference 15) and the filtered tensor field (Figure 13.8b). Prior to visualization, the tensors were weighted with their linear diffusion measures.

The filtered images show the result of applying the macrostructural measure presented in Subsection 13.3.2. First a $5 \times 5 \times 3$ Gaussian window, a, with standard deviation 2 mm was applied to the data (Equation 13.25). Because the out-of-plane resolution is slightly less than half the in-plane resolution, there is almost no smoothing performed between the slices. The tensors resulting from the Gaussian filtering $\mathbf{D}_a$ (Equation 13.25) have been weighted with their linear diffusion measures, c_l (Equation 13.15), respectively. The result illustrates the fact that the filtering increases the rank of the tensors in nonstructured areas because the linear measures are decreased in those areas.

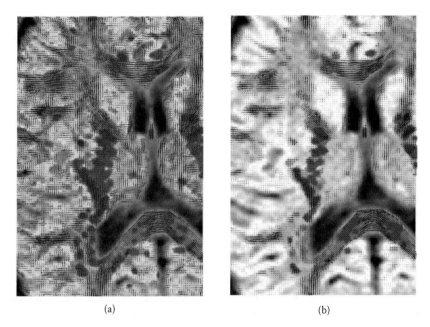

(a)   (b)

FIGURE 13.8 (a) Diffusion tensors, weighted with their linear measure c_l, from an axial slice of a human brain. (b) Averaged diffusion tensors using a $5 \times 5 \times 3$ Gaussian kernel weighted with their linear measure c_l, resulting in a macrostructural measure of fiber tract organization. (From Westin, C.-F., Maier, S., Mamata, H., Nabavi, A., Jolesz, F., and Kikinis, R. (2002). Processing and visualization of diffusion tensor MRI. *Med. Image Anal.* 6(2): 93–108.)

As mentioned in the preceding text, in three dimensions, a diffusion tensor can be visualized as using an ellipsoid in which the principal axes correspond to the tensor's eigenvector system. However, it is difficult to distinguish between an edge-on, flat ellipsoid and an oblong one using the surface-shading information. Similar ambiguity exists between a face-on, flat ellipsoid and a sphere. We propose a technique for the visualization of tensor fields that overcomes the problems with ellipsoids. Figure 13.9 compares the ellipsoidal representation of a tensor (left) with a composite shape of linear, planar, and spherical components (right). The components

FIGURE 13.9 Comparison of an ellipsoid and a composite shape depicting the same tensor with eigenvalues $\lambda_1 = 1$, $\lambda_2 = 0.7$, and $\lambda_3 = 0.4$ [24].

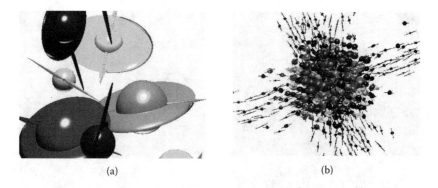

(a)   (b)

FIGURE 13.10 (a) Visualization of diffusion tensors. The tensors are color-coded according to the shape: the linear case is blue, planar case is yellow, and spherical case is red. The radius of the sphere is the smallest eigenvalue of the diffusion tensor; the radius of the disk is the second largest; and the length of the rod is twice the largest eigenvalue. (b) Simulated tensor field of three crossing white matter tracts. Due to partial voluming effects, the tensors in the area where the fibers are crossing have spherical shape. (From Westin, C.-F., Maier, S., Mamata, H., Nabavi, A., Jolesz, F., and Kikinis, R. (2002). Processing and visualization of diffusion tensor MRI. *Med. Image Anal.* 6(2): 93–108.)

are here scaled according to the eigenvalues, but can alternatively be scaled according to the shape measures c_l, c_p, and c_s.

Additionally, coloring based on the shape measures c_l, c_p, and c_s can be used for visualization of shape. Figure 13.10 shows a coloring scheme in which the color is interpolated between the blue linear case, the yellow planar case, and the red spherical case.

13.5 CONNECTIVITY ANALYSIS

Determination of neural fiber architecture from diffusion tensors measured in the brain is a complex problem with many potential applications in neurosurgery and neuroscience. The phrase "white matter connectivity" refers to a measure of the neural connection strength between points or regions in the brain. In animal research, connectivity measurements can be made using injected tracers in combination with histological analysis. These methods are not applicable to human neuroanatomical study, but a wealth of information can be acquired noninvasively through the analysis of diffusion MRI.

Initial work on DT-MRI connectivity focused on tractography [1,2,18], or the interpolation of paths through the principal eigenvector field. An extension of this method evolved a surface using a fast marching method, in which the speed function was dependent on the principal eigenvector field [13]. Another approach iteratively simulated diffusion in a 2-D tensor volume, and quantified connection strengths based on a probabilistic interpretation of the arrival time of the diffusion front [4]. A new level-set-based method evolved a surface in a field of vectors

FIGURE 13.11 A simple diagram to demonstrate the difficulty of making only local decisions when performing tractography through an ambiguous region. The ellipses represent tensors along two crossing tracts in two dimensions.

created perpendicular to the major eigenvector field, so that the surfaces will tend to enclose the tracts [8]. The tractography approach has also been extended to diffusion data with much higher angular resolution, and connectivity has been estimated using the most probable path between points [21].

The standard tractographic interpolation approach has one main drawback, which is that all decisions are made locally. Thus, errors can accumulate, and the tracing can be confounded by regions of crossing fibers (with high planar or spherical indices). This is demonstrated schematically in Figure 13.11. To avoid this problem, several approaches have been proposed that do probabilistic tracking, basically adding noise about the major eigenvector in an attempt to produce many probable paths. These methods model the tract direction at each voxel using a probability distribution whose mean is generally the direction of the major eigenvector. Instead of producing only one path, many paths are produced by sampling, resulting in a collection of likely paths starting at a point of interest [6,7].

Here, we present two novel global approaches to connectivity estimation [12]. Both approaches use the information from the whole tensor, not just the major eigenvector field, and can provide numerical measures of connectivity. Our first approach finds a steady-state concentration or heat distribution using the 3-D tensor field as diffusion or conductivity tensors. In this method, the steady-state flow along any path reflects connectivity. Our second approach casts the problem in a Riemannian framework, deriving from each tensor a local warping of space, and finding geodesic paths in the space. In this method, path lengths are related to connectivity.

13.6 METHOD ONE: DIFFUSION-BASED CONNECTIVITY

In this method, we use the anisotropic diffusion equation and simulate sources and sinks in the tensor field. We find a steady-state concentration or heat distribution using the 3-D tensor field as diffusion or conductivity tensors. The steady-state flow along any path reflects connectivity along that path. This method allows us to investigate many paths from one region (the source) to another (the sink).

Previous related work has employed an iterative technique to create time-of-arrival maps of a heat diffusion front [4]. Instead, we solve directly for the steady-state concentration, u, which can also be thought of as a heat distribution in the tensor field:

$$\nabla \cdot (D\nabla u) = 0 \qquad (13.26)$$

We use this information to create the flux vector field, $j = -\nabla u$, which describes the steady-state heat flow in the tensor volume (Equation 13.2). Paths in this divergence-free vector field can be compared using a connection strength metric that approximates the total flow along the path:

$$\int_P |j^T t|\, ds \qquad (13.27)$$

where j is the flux along the path, and t the unit tangent to the path. Normalization for the length of the path may also be included in the metric. To obtain an overall connection strength measure between two points, the value of the maximum flow path can be taken.

Of great interest in this method are the boundary conditions, or the locations of sources and sinks in the tensor field. One possibility is to set a region or regions of interest as the source, and simulate a sink at infinity. Another useful possibility is to choose one region of interest as the source, and another as the sink. In the experiments discussed in this chapter, we have simulated a sink at one point of interest, and a source at another, in order to estimate the flow between the regions.

13.6.1 EXPERIMENTS

Three experiments performed are described in the following text:

DT-MRI data acquisition: DT-MRI scans of normal subjects were acquired using line scan diffusion imaging [9] on a 1.5-T GE Echospeed system. The following scan parameters were used: rectangular 22-cm field of view (FOV; 256×128 image matrix, 0.86 mm by 1.72 mm in-plane pixel size); slice thickness = 4 mm; interslice distance = 1 mm; receiver bandwidth = ±6 kHz; echo time (TE) = 70 msec; repetition time (TR) = 80 msec (effective TR = 2500 msec); scan time (13.60 sec/section. Twenty axial slices were acquired, covering the entire brain. This protocol provides diffusion data in six gradient directions as well as on the corresponding T2-weighted image. All gradients and T2-weighted images are acquired simultaneously, and thus do not need any rigid registration prior to the tensor reconstruction process. Tensors are calculated as described in Reference 24.

Tensor preprocessing: We are interested in measuring connectivity in the white matter and, consequently, to de-emphasize other regions, we multiply the tensors by a soft mask. This is necessary to decrease the effect of the ventricles, where neural fiber tracts are nonexistent but water diffusion is relatively unrestricted and

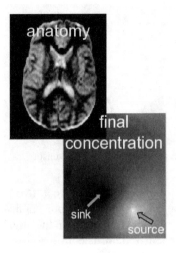
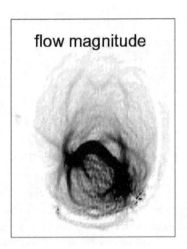

FIGURE 13.12 Results of solving Equation 13.26 for the steady-state heat distribution. The temperature (labeled as concentration) and the steady-state flow magnitude demonstrate the flow from the source to the sink. In the temperature image, the source is bright and the sink is dark; in the flow image, dark means high flow magnitude. The grayscale image on the left, a nondiffusion-weighted image, shows the corresponding anatomy.

has large magnitude. We calculate the weights in the mask as the linear shape measure at each voxel, which lies in the range of zero to one [24,25]

$$c_l = \frac{\lambda_1 - \lambda_2}{\lambda_1}. \tag{13.28}$$

In addition, we remove negative eigenvalues to ensure that each tensor is a positive definite matrix. We set a small positive lower bound for the eigenvalues to guarantee that the tensors are invertible, which is necessary when utilizing them as local metric descriptors as described later. Setting the negative eigenvalues to zero would give the closest positive semidefinite tensor in the least-squares sense, but would not ensure invertibility.

Concentration or heat flow between regions: In this experiment, we solve for the steady-state concentration or heat distribution in the tensor field, with boundary conditions of one source and one sink. The maximal flow is found as expected along the strong anatomical path between the source and sink, the corpus callosum. Figure 13.12 displays the steady-state concentration and flow.

13.7 METHOD TWO: DISTANCE-BASED CONNECTIVITY

This method allows us to relate geodesic paths to connectivity, and investigate paths from one point or region outward to the entire brain.

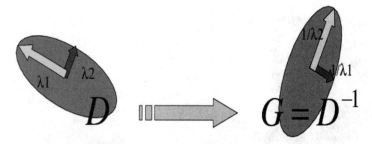

FIGURE 13.13 Relation between, the diffusion tensor, and the metric tensor.

A natural interpretation of the degree of connectivity between two points is the distance between the points in some metric space. For our application, the distance between two anatomical locations should depend on the alignment of their connecting paths with the diffusion tensor field. The diffusion operator (Equation 13.3) can naturally be associated with a Riemannian metric tensor G via the relation $G = D^{-1}$, allowing us to compute geometric quantities such as geodesic paths and distances between points in the brain. Unlike tractographic methods based on following the flow of principal eigenvectors of D, these geodesic paths are well defined even in regions where the tensor diffusion is isotropic.

The inverse relation between the diffusion and metric tensors is intuitive: large eigenvalues in the original tensor create short metric distances along the direction of the corresponding eigenvector (see Figure 13.13). Or in general, the path length is locally inversely proportional to the diffusion coefficient in the direction tangent to the path. So, short paths are those that are aligned with the tensor field (not just the major eigenvector field).

We will limit ourselves here to a brief discussion of the theory; see Reference 11 for a more rigorous and thorough treatment of the connection between diffusion and Riemannian geometry. The Laplace–Beltrami operator is the generalization of the Laplacian to manifolds. In matrix notation, the Laplace–Beltrami operator can be written as

$$\nabla_G^2 u = |G|^{-\frac{1}{2}} \nabla \cdot \left(|G|^{\frac{1}{2}} G^{-1} \nabla u \right) \tag{13.29}$$

Here, we seek the relation between isotropic diffusion on a manifold (Equation 13.29) and anisotropic diffusion in Euclidean space (Equation 13.3). The following relation exists between the diffusion operator in Equation 13.3 and a diffusion operator in the Riemannian space characterized by G:

$$\nabla \cdot (D\nabla u) = \nabla_G^2 u - \frac{1}{2} \langle \nabla \log |G|, \nabla_G u \rangle \tag{13.30}$$

where the second-order term on the right-hand side represents simple Laplacian smoothing in the tensor-warped space, i.e., isotropic diffusion associated with the heat equation.

13.7.1 MEASURING DISTANCES IN THE TENSOR-WARPED SPACE

Once we have the metric tensor G, we are able to apply results from Riemannian geometry to describe geometric objects such as geodesic paths and distances between points in the brain. Unlike tractographic methods based on following the flow of principal eigenvectors of D, these geodesic paths are well defined even in regions where the tensor diffusion is isotropic.

We have approached the measurement of distances in this space in two ways. First, we have implemented an Eikonal-type equation using level-set methods to produce a distance transform that respects the metric G. This required the derivation of a formula for the speed of an evolving front in the direction of its Euclidean normal. Second, we have implemented Dijkstra's algorithm using G to determine distances between neighboring voxels, employing the formula $(w^T G w)^{\frac{1}{2}}$, in which w is the vector from a voxel to its neighbor. Though it can suffer from discretization problems, Dijkstra's algorithm is fast and allows interactive display of return paths.

For our level-set [19] implementation, we seek a speed function F for use in the evolution equation

$$\phi_t = F \,|\, \nabla\phi \,|. \qquad (13.31)$$

This can be done using the following algorithm, which amounts to finding the length of the projection of the unit normal in the tensor-warped space onto the Euclidean normal:

1. Set $n = \frac{\nabla\phi}{|\nabla\phi|}$, the Euclidean normal to the level set.
2. Find any two linearly independent vectors t_1 and t_2 perpendicular to n. These are tangents that span the tangent space to the level set.
3. Set $w = (Gt_1) \times (Gt_2)$.
4. Set $\tilde{n} = \frac{w}{(w^T G w)^{\frac{1}{2}}}$. This is the unit normal with respect to G.
5. Set $F = |\tilde{n}^T n|$. This is the length of the projection of $\tilde{n}$ onto n.

13.7.2 EXPERIMENTS

The data acquisition and preprocessing experiments were the same as described in Subsection 13.6.1. Other experiments are described in the following text:

Tensor-warped distances: Figure 13.14 shows a slice through a 3-D distance transform with respect to the metric derived from the DT-MRI tensor field. The contours are isodistance contours.

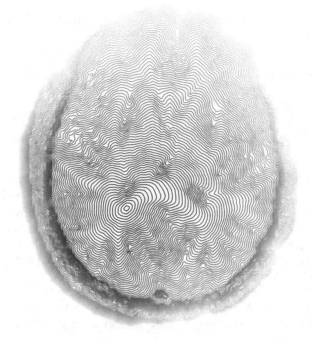

FIGURE 13.14 Tensor-warped distance map: this contour map shows metric distance from an initial point located in the posterolateral part of the corpus callosum. The image is a slice through a 3-D distance map, at the level of the initial point. The regions where neighboring contours are widely separated indicate low metric distance, or high connectivity. (From O'Donnell, L., Haker, S., and Westin, C.-F. (2002). New approaches to estimation of white matter connectivity in diffusion tensor MRI: Elliptic PDEs and geodesics in a tensor-warped space. in Dohi, T. and Kikinis, R. (Eds.). *Medical Image Computing and Computer-Assisted Intervention (MICCAI)*. pp. 459–466, Tokyo, Japan.).

> *Connectivity measure:* We would like to quantify path quality, but both long and short paths occur in the brain (see Figure 13.1). So we cannot rank paths by Euclidean or metric length alone. We choose to use one length to normalize the other, which is one way to address this issue.

By comparing the geodesic path length to the Euclidean length of the same path, we produce a measure of the degree of connectivity between any two points. We compute the ratio of Euclidean path length to geodesic path length for all paths outward from the initial point. Figure 13.15 displays the connectivity measure as calculated for the distance map shown in Figure 13.14.

13.8 CONCLUSION

In this chapter, we first gave background information on diffusion and diffusion tensor MRI, then presented tensor shape measures that quantify anisotropy or the lack thereof. Next, we described visualization techniques for diffusion tensor fields.

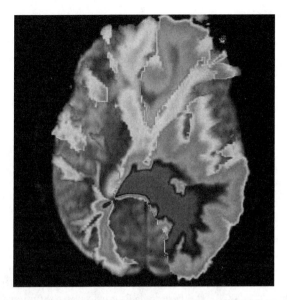

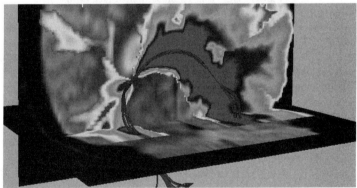

FIGURE 13.15 Degree of connectivity, measured as Euclidean path length over geodesic path length. Very low connectivity is not shown. Purple denotes the highest connectivity. Traditional tractography based on following the principal eigenvector direction, with seed locations around the initial point, is displayed in red (right). Visual inspection confirms that the trace lines agree well with the region of highest connectivity. (From O'Donnell, L., Haker, S., and Westin, C.-F. (2002). New approaches to estimation of white matter connectivity in diffusion tensor MRI: Elliptic PDEs and geodesics in a tensor-warped space. in Dohi, T. and Kikinis, R. (Eds.). *Medical Image Computing and Computer-Assisted Intervention (MICCAI)*. pp. 459–466, Tokyo, Japan.)

Finally, we introduced two novel image processing methods for connectivity estimation. In the first method, we solved for a steady-state heat distribution and flow field that reflect connectivity. In the second method, the introduction of a Riemannian metric allowed us to reformulate the connectivity/diffusion simulation problem as a search for geodesic paths.

ACKNOWLEDGMENTS

This work was funded in part by a National Science Foundation Graduate Fellowship, by the NIH/MEMP Neuroimaging Training Program, and by NIH grants P41-RR13218 and 1 R33-CA99015.

REFERENCES

1. Basser, P. (1995). Inferring microstructural features and the physiological state of tissues from diffusion-weighted images. *NMR Biomed.* 8: 333–344.
2. Basser, P., Pajevic, S., Pierpaoli, C., Duda, J., and Aldroubi, A. (2000). *In vivo* fiber tractography using DT–MRI data. *Magn. Reson. Med.* 44: 625–632.
3. Basser, P. and Pierpaoli, C. (1996). Microstructural and physiological features of tissues elucidated by quantitative-diffusion-tensor MRI. *J. Magn. Reson. Ser. B.* 111: 209–219.
4. Batchelor, P., Hill, D., Calamante, F., and Atkinson, D. (2001). Study of connectivity in the brain using the full diffusion tensor from MRI. in *Information Processing in Medical Imaging, 17th International Conference (IPMI)*. pp. 121–133.
5. Beaulieu, C. (2001). The basis of anisotropic water diffusion in the nervous system — a technical review. *NMR Biomed.* (15): 435–455.
6. Behrens, T., Johansen-Berg, H., Woolrich, M., Smith, S., Wheeler-Kingshott, C., Boulby, P., Barker, G., Sillery, E., Sheehan, K., Ciccarelli, O., Thompson, A., Brady, J., and Matthews, P. (2003). Non-invasive mapping of connections between human thalamus and cortex using diffusion imaging. *Nat. Neurosci.* 6: 750–757.
7. Björnemo, M., Brun, A., Kikinis, R., and Westin, C.-F. (2002). Regularized stochastic white matter tractography using diffusion tensor MRI. in *Fifth International Conference on Medical Image Computing and Computer-Assisted Intervention (MICCAI'02)*, pp. 435–442, Tokyo, Japan.
8. Campbell, J., Siddiqi, K., Vemuri, B., and Pike, G. (2002). A geometric flow for white matter fibre tract reconstruction. in *IEEE International Symposium on Biomedical Imaging*, pp. 505–508.
9. Gudbjartsson, H., Maier, S., Mulkern, R., Morocz, I., Patz, S., and Jolesz, F. (1996). Line scan diffusion imaging. *Magn. Reson. Med.* 36: 509–519.
10. Kindlmann, G. and Weinstein, D. (1999). Hue balls and lit-tensors for direct volume rendering of diffusion tensor fields. in *IEEE Visualization 1999, VIS1999.* Salt Lake City, Utah. October 8–13.
11. Lara, M.D. (1995). Geometric and symmetry properties of a nondegenerate diffusion process. *Ann. Probab.* 23(4): 1557–1604.
12. O'Donnell, L., Haker, S., and Westin, C.-F. (2002). New approaches to estimation of white matter connectivity in diffusion tensor MRI: Elliptic PDEs and geodesics in a tensor-warped space. in Dohi, T., Kikinis, R., (Eds.). *Medical Image Computing and Computer-Assisted Intervention (MICCAI)*. pp. 459–466, Tokyo, Japan.
13. Parker, G., Wheeler-Kingshott, C., and Barker, G. (2001). Distributed anatomical brain connectivity derived from diffusion tensor imaging. in *Information Processing in Medical Imaging 17th International Conference (IPMI)*, pp. 106–120.
14. Parker, G.J.M., Schnabel, J.A., Symms, M.R., Werring, D.J., and Barker, G.J. (2000). Nonlinear smoothing for reduction of systematic and random errors in diffusion tensor imaging. *J. Magn. Reson. Imaging.* 11:702–710.

15. Peled, S., Gudbjartsson, H., Westin, C.-F., Kikinis, R., and Jolesz, F. (January 1998). Magnetic resonance imaging shows orientation and asymmetry of white matter tracts. *Brain Res.* 780(1): 27–33.
16. Pierpaoli, C. and Basser, P.J. (1996). Toward a quantitative assessment of diffusion anisotropy. *Magn. Reson. Med.* 36: 893–906.
17. Pierpaoli, C., Jezzard, P., Basser, P.J., Barnett, A., and Chiro, G.D. (1996). Diffusion tensor MR imaging of the human brain. *Radiology* 201: 637.
18. Poupon, C., Clark, C.A., Frouin, F., R´egis, J., Bloch, I., Bihan, D.L., Bloch, I., and Mangin, J.-F. (2000). Regularization of diffusion-based direction maps for the tracking brain white matter fascicles. *NeuroImage* 12: 184–195.
19. Sethian, J. (1999). *Level Set Methods and Fast Marching Methods.* Cambridge University Press.
20. Talos, I.-F., O'Donnell, L., Westin, C.-F., Warfield, S.K., Wells, W.M., Yoo, S.-S., Panych, L.P., Golby, A., Mamata, H., Maier, S.S., Ratiu, P., Guttmann, C.R., Black, P.M., Jolesz, F.A., and Kikinis, R. (2003). Diffusion tensor and functional MRI fusion with anatomical MRI for image guided neurosurgery. in *Conference on Medical Image Computing and Computer-Assisted Intervention (MICCAI).* pp. 407–415. Toronto, Canada.
21. Tuch, D.S. (2002). Diffusion MRI of Complex Tissue Structure. Ph.D. thesis. Division of Health Sciences and Technology. Massachusetts Institute of Technology.
22. Westin, C.-F. and Knutsson, H. (September 1992). Extraction of local symmetries using tensor field filtering. in *Proceedings of 2nd Singapore International Conference on Image Processing.* IEEE Singapore Section.
23. Westin, C.-F. and Knutsson, H. (November 1994). Estimation of motion vector fields using tensor field filtering. in *Proceedings of IEEE International Conference on Image Processing.* pp. 237–242. Austin, Texas. IEEE.
24. Westin, C.-F., Maier, S., Mamata, H., Nabavi, A., Jolesz, F., and Kikinis, R. (April 2002). Processing and visualization of diffusion tensor MRI. *Med. Image Anal.* 6(2): 93–108.
25. Westin, C.-F., Peled, S., Gudbjartsson, H., Kikinis, R., and Jolesz. F. (1997). Geometrical diffusion measures for MRI from tensor basis analysis. in *ISMRM '97*, p. 1742. Vancouver, Canada.

14 Analysis of Dynamic Magnetic Resonance Elastography Data

Armando Manduca

CONTENTS

14.1 INTRODUCTION

For centuries, palpation has been used as an efficient detection tool by physicians, based on the fact that many diseases cause changes in the mechanical properties of tissues. Student physicians learn that the presence of a hard mass in the thyroid, breast, or prostate is suspicious for malignancy. Indeed, even today many tumors of these structures are first detected by touch. It is not uncommon for surgeons at the time of laparotomy to palpate tumors that were undetected in preoperative imaging by CT, MRI, or ultrasound. None of these modalities provides information about the elastic properties of tissues elicited by palpation. The elastic moduli of various human soft tissues are known to vary over a wide range. The Young's modulus of soft tissues can vary as much as four orders of magnitude [1,2] in healthy and diseased tissues. The literature on mechanical properties of abnormal tissues is limited, but it is known that the elastic modulus of the breast may differ from surrounding tissues by a factor of 90 [3,4]. It is also known that the shear modulus of many tissues can vary in response to changes in the physiologic state [1,5]. The elasticity of muscle in the relaxed and contracted states can differ by more than 100-fold [1]. In contrast, most of the other physical properties depicted by conventional medical imaging modalities are distributed over a much smaller numerical range. Over the last decade, the recognition of the potential diagnostic value of characterizing mechanical properties has led a number of investigators to seek methods for imaging tissue elasticity. For reviews of such work, see Reference 6 and Reference 7 [6,7].

Magnetic resonance elastography (MRE) is a technique that can directly image and quantitatively measure displacements due to propagating acoustic strain waves in tissue-like materials subjected to harmonic mechanical excitation [8,9]. A phase-contrast MRI technique is used to spatially map and measure the shear-wave displacement patterns. From this data, local quantitative values of the shear modulus can be calculated and images (elastograms) that depict tissue elasticity or stiffness can be generated. In this chapter, we describe the principles of MRE, consider the equations of harmonic motion in soft tissue, and describe approaches for reconstructing elastograms from MRE data and the assumptions inherent in each, and present a summary of some *ex vivo* and *in vivo* results.

14.2 ELASTIC PROPERTIES OF SOFT TISSUE

In isotropic Hookean materials, the proportionality constant that describes the amount of longitudinal deformation (expressed in terms of strain) that occurs in a given material in response to an applied longitudinal force (expressed in terms of stress) is known as the *Young's modulus* (E) of elasticity. The *shear modulus* (μ) relates transverse strain to transverse stress. The *bulk modulus* (K) of elasticity describes the change in volume of a material due to external stress. Poisson's ratio (v) is the ratio of transverse contraction per unit breadth divided by longitudinal extension per unit length. These parameters are interrelated so that knowledge of any two allows calculation of the other two.

Most soft tissues have mechanical properties that are intermediate between those of fluids and solids. The value of Poisson's ratio for soft tissues, which can be directly calculated from the ratio of longitudinal- to shear-wave speeds, is typically 0.499999, very close to the value for liquids ($v = 0.500$). In this case the Young's modulus and shear modulus differ only by a scaling factor ($E = 3\mu$). Another characteristic that soft tissues share with liquids is that they are nearly incompressible. In contrast to the many orders of magnitude over which the Young's and shear moduli are distributed, the bulk moduli of most soft tissues differ by less than 15% from that of water [10]. Also the density of soft tissues differs little from that of water [11]. These concepts represent a simplification of the mechanical behavior of soft tissues, which in general can be anisotropic, non-Hookean, and viscoelastic.

14.3 MR ELASTICITY IMAGING TECHNIQUES

Much of the pioneering work in elasticity imaging has been accomplished using ultrasound and either a quasi-static stress model [12–15] or a dynamic stress model [16–19]. Ultrasound elastography continues to be a very active area of research [20–22] but will not be further discussed here.

Other investigators have proposed several approaches for delineating tissue elasticity using MRI. Saturation-tagging methods have been used in applications as diverse as evaluating the local motion of cardiac muscle [23,24] and observing connective motion in vibrated granular material [25]. A grid of saturation tags can be applied to the tissue before it is deformed by an applied static stress [26]. The pattern of deformation of the grid of saturation tags can then be analyzed to provide a map of local strain. A mathematical model of stress distribution within the object could also be used, in principle, to convert this map of local strain into a quantitative depiction of the regional elastic modulus [27].

Another well-known MRI method for measuring local motion is motion-encoded phase-contrast imaging. This has been used clinically in applications such as assessing regional myocardial motion, CSF pulsation, and intravascular blood flow [28–30]. Plewes et al. proposed a method for elastography involving use of a phase-contrast imaging sequence to estimate the spatial strain distribution resulting from a small quasi-static longitudinal stress applied once each time the pulse sequence is repeated [31]. Often in these methods, the strain alone is used as a surrogate for stiffness, that is, low strain means high stiffness and high strain results in softer or low-stiffness regions. The distribution of strain can also be related to the predicted distribution of stress and the material parameters deduced through elasticity equations [32]. Such calculations are usually global in nature (and thus computationally intensive), and they require knowledge of boundary conditions. Dynamic methods, as described in the following text, rely on the wave equation, which in its differential form is local in character. Therefore, the distribution of dynamic displacement (a 3-D vector) and its second-order partial derivatives in time and space, due to a propagating shear wave in a small region of the tissue, are enough to completely characterize the shear moduli of the tissue in that region. Strictly speaking, quasi-static methods, mentioned in the preceding text, as well as the dynamic or harmonic methods, described in the following text, can be termed MRE; nevertheless this chapter will henceforth concentrate only on the latter.

14.4 MRE

Dynamic MRE uses propagating mechanical waves as a probe for the elastic properties of tissues. Shear waves of frequencies in the 50 to 1000 Hz range are suitable as probes because they are much less attenuated than those of higher frequencies, their wavelength in tissue-like materials is in the useful range of millimeters to tens of millimeters, and the shear modulus varies so widely in body tissues. High-frequency longitudinal acoustic waves (ultrasound) are not directly suitable for use as probes because their propagation is governed by the bulk modulus, which varies little in soft tissue. Longitudinal acoustic waves at lower frequencies are also not suitable because they have long wavelengths (on the order of meters for waves of frequency below 1 kHz) [10].

In MRE, a phase-contrast MRI technique is used to spatially map and measure displacement patterns corresponding to harmonic shear waves with amplitudes in the range of microns or less. A conventional MRI system is used with an additional motion-sensitizing gradient imposed along a specific direction and switched in polarity at an adjustable frequency (Figure 14.1). Trigger pulses synchronize an oscillator and amplifier unit that drives an electromechanical actuator coupled to the surface of the object to be imaged. The actuator induces shear waves in the object at the same frequency as the motion-sensitizing gradient. The harmonic motion of the spins at the frequency of the motion-sensitizing gradients causes a measurable phase shift in the received MR signal. From the measured phase shift, it is possible to calculate the displacement at each voxel and directly image the acoustic waves within the object.

The phase shift caused by a propagating mechanical wave with a wave vector $\vec{k}$ within a medium at a given frequency $(1/T)$ in the presence of a cyclic motion-encoding gradient at the same frequency is given by [8,9]

$$\phi(\vec{r},\theta) = \frac{\gamma NT\left(\vec{G}_0 \bullet \vec{\xi}_0\right)}{2}\cos(\vec{k} \bullet \vec{r} + \theta) \tag{14.1}$$

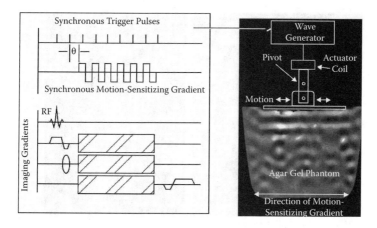

FIGURE 14.1 Schematic diagram of the MRE system. Conventional MRI gradients and RF pulses that encode spatial positions are shown at the bottom left. The electromechanical driver applies transverse acoustic waves to the object to be imaged via a surface plate (right). The cyclic motion-sensitizing gradients and the acoustic drive are synchronized using trigger pulses provided by the imager. The phase offset (θ) between the two can be varied. As shown by the shaded regions, the motion-sensitizing gradients can be superimposed along any desired axis to detect cyclic motion.

This accumulated phase shift is proportional to the relative phase θ of the mechanical and magnetic oscillations and the dot product of the displacement amplitude vector $\vec{\xi}_0$ and the motion-sensitizing magnetic gradient vector $\vec{G}$. Particles whose components of motion along the gradient vector are exactly in phase or out of phase with the magnetic oscillation have maximum phase shifts of opposing polarities. Particles whose components of motion along the gradient vector is 90° out of phase with the magnetic oscillation have no net phase shift. Because the response is also proportional to the number of gradient cycles (N) and the period of the gradient waveform (T), high sensitivity to small-amplitude synchronous motion can be achieved by accumulating phase shifts over multiple cycles of mechanical excitation and the motion-sensitizing gradient waveform. The quantity γ is the gyromagnetic ratio, and $\vec{r}$ is the spin position vector.

Generally, two acquisitions are made for each repetition in an interleaved fashion reversing the polarity of the motion-sensitizing gradients. This reduces systematic phase errors and doubles sensitivity to small displacements. Typical data-acquisition parameter ranges for 2-D sequences are: TR, 10 to 300 msec; TE, 10 to 60 msec; acquisition time, 20 to 120 sec; and flip angle, 10 to 60°. The number of gradient pulses (N) varies from 2 to 30 cycles, and the frequency of mechanical excitation ranges from 50 to 1000 Hz. Acquiring and processing 2-D slices captures only two of the three components of the wave propagation vector and may yield misleading results unless the shear wave is propagating in a plane, but in some cases, time considerations or other factors mandate their use. 3-D MRE pulse

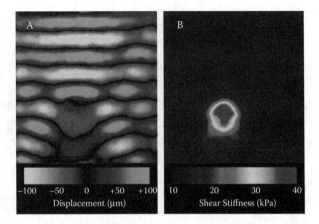

FIGURE 14.2 (a) Shear waves propagating in a phantom with an embedded 1.5-cm-diameter cylinder of stiffer gel. Shear waves at 300 Hz were applied at the top margin of the gel block, with transverse motion oriented orthogonal to the plane of the image. (b) The elastogram based on LFE processing clearly depicts the object even though it is significantly smaller than the wavelength in the stiff material.

sequences are also in common use but require longer acquisition times (several minutes to tens of minutes).

The phase images reflect the displacement of spins due to acoustic strain wave propagation in the medium and are termed *wave images*. Such an image of propagating acoustic waves in a tissue-simulating agarose gel phantom is shown in Figure 14.2a. The wave propagation depends on the elasticity of the material at each location in the object, so inversion of the data can yield elastograms as in Figure 14.2b. Experiments to assess the sensitivity of the shear-wave-imaging method at low amplitudes of mechanical excitation demonstrated that shear waves with displacements of less than 100 nm can be readily observed [8].

Mechanical excitation can be provided by a moving-coil driver with the imager providing the static magnetic field. Trigger pulses are provided by the sequencing computer of the MR imager and are fed to a function generator that produces a waveform that is amplified and applied to the coil of the actuator. Transverse stresses applied to a flat contact plate provide planar shear waves. Piezoelectric drivers can also be used for generating shear waves. Mechanical excitation can also be applied longitudinally, with mode conversion leading to shear waves being generated inside the object [33,34]. Focused ultrasound also can be used to generate shear waves that can be imaged with MRE [35]. The ultrasound beam is temporally modulated to create cyclic variation in acoustic radiation pressure at the focus of the ultrasound source, which can be located deep within an object.

By adjusting the phase offset between the mechanical excitation and the oscillating magnetic gradient (θ in Figure 14.1), acoustic-wave images can be obtained at various phase offsets (typically 4 to 8) regularly spaced in a cycle, which allows visualization of wave propagation as a cine loop. More importantly, it allows extraction of the harmonic component at the frequency of interest, giving the amplitude and the phase (relative to an arbitrary zero point) of the harmonic displacement at each point in space [36]. This extraction also provides some degree of noise reduction. The resulting complex displacement field is the input to all the processing techniques described in the following text.

A single MRE acquisition is sensitized to motion in a single direction. However, the experiment can be repeated and the sensitization direction varied in order to capture all three orthogonal components of displacement. Thus, MRE can acquire full 3-D cyclic displacement information at MR pixel resolution throughout a 3-D volume. In principle, this makes it feasible to estimate all components of the strain tensor, making it possible to probe the anisotropic mechanical properties of tissues [8,33,37]. This ability to capture full displacement information and the freely oriented field of view unencumbered by any acoustic window requirements represent the main advantages of MRE over ultrasound-based techniques. Conversely, the much longer acquisition times required by MRE are its single largest disadvantage.

MRE is highly sensitive only to motion that is precisely synchronized with the sensitization gradients and is no more sensitive to physiologic motion than a conventional gradient-echo sequence [38]. Sensitivity to nonsynchronous motion can be further reduced by explicitly nulling the individual moments of the gradient waveform. It is also possible to amplitude modulate the envelope of the motion-encoding waveform to further increase its spectral selectivity [39].

In summary, MRE offers direct visualization and quantitative measurement of tissue displacements, high sensitivity to very small motions, a field of view unencumbered by acoustic window requirements, and the ability to obtain full 3-D displacement information throughout a 3-D volume. As shown in the following text, under some assumptions this allows direct local inversion of the data to recover the elastic properties, with no need for boundary conditions or the estimation of a stress field.

14.5 DATA PROCESSING

A variety of approaches can be used to invert the displacement data to recover mechanical properties. These are characterized in the following text by the assumptions or simplifications made in their derivations. Unlike many biomedical inversions for which data are available only along a boundary, data in MRE are available everywhere in a 3-D volume. In favorable situations, it is possible to deduce quantitatively accurate values of properties such as the shear modulus. In general, however, despite the richness of the data set and the variety of processing techniques, it remains a challenge to extract accurate results from the intrinsically noisy data in complex, heterogeneous objects.

14.5.1 Equations of Motion

The mechanical quantities we wish to characterize are those that relate strain to stress, and because the displacements in MRE are very small (microns to tens of microns), a linear relationship can be assumed between these. In the general case, stress and strain are related by a rank 4 tensor, with up to 21 independent quantities [40]. If one assumes that the material is isotropic, this reduces to two independent quantities, the Lame constants λ and μ, related to longitudinal and shear deformation, respectively. The isotropic relation between stress and strain is given by

$$\sigma_{ij} = 2\mu e_{ij} + \lambda \delta_{ij} e_{nn} \qquad (14.2)$$

where e_{ij} is one component of the stress tensor, δ_{ij} is the Kronecker delta, and summation over repeated indices is assumed. The strain tensor e_{ij} is defined in terms of the displacement tensor u_{ij} as

$$e_{ij} = (u_{i,j} + u_{j,i})/2 \qquad (14.3)$$

where indices after a comma indicate differentiation. Substituting these into the equation of motion, we obtain the general equation for harmonic motion in an isotropic, linearly elastic medium [41]:

$$[\lambda u_{j,j}]_{,i} + [\mu(u_{i,j} + u_{j,i})]_{,j} = -\rho\omega^2 u_i \qquad (14.4)$$

with ρ being the density of the material and ω the angular frequency of the mechanical oscillation. The Lame constants can be considered to be complex quantities, with the imaginary parts representing attenuation for a viscoelastic medium. Solving this equation requires knowledge of the full 3-D displacement because the equations for the individual components are coupled. MRE phase difference measurements in all three spatial orientations are thus required.

Additional assumptions can be made to further simplify the equation. If one assumes local homogeneity, λ and μ become single unknowns instead of functions of position, and Equation 14.4 becomes an algebraic matrix equation that can be solved locally by direct inversion, as described by the following equation (terms in boldface are column vectors):

$$\mu\nabla^2\mathbf{u} + (\lambda+\mu)\nabla(\nabla\cdot\mathbf{u}) = -\rho\omega^2\mathbf{u} \qquad (14.5)$$

In soft tissues, $\lambda \gg \mu$ (typically by 10^4 or more). This makes it difficult to estimate both parameters simultaneously, and the longitudinal wavelength is so long in tissues (tens of meters) that accurate estimation of λ is very challenging in any case. It is possible to partially filter out the effects of the longitudinal wave because its contributions are at very low (near zero) spatial frequency. To remove

λ from consideration, the assumption can be made that displacements due to the longitudinal wave vary slowly and are thus negligible (this corresponds to assuming that $\lambda(\nabla \cdot \mathbf{u}) = 0$). The large difference between longitudinal and shear waves in tissue make this a reasonable assumption. The equation then simplifies to a single vector equation in μ, but all three components of motion are still required:

$$[\nabla(\nabla \cdot \mathbf{u}) \quad \nabla^2\mathbf{u}]\,\mu = -\rho\omega^2[\mathbf{u}] \tag{14.6}$$

Alternatively, one can assume incompressibility ($\nabla \cdot \mathbf{u} = 0$), and the equation then simplifies to the Helmholtz equation:

$$\mu\nabla^2\mathbf{u} = -\rho\omega^2\mathbf{u} \tag{14.7}$$

The terms involving components in the different orthogonal directions are now decoupled, and each component satisfies the equation separately. Thus, measurements in only one sensitization direction (and an estimate of the Laplacian of that component) suffice to determine μ. Experiments with tissue-simulating phantom data sets have shown little difference among inversion results that use Equation 14.5, Equation 14.6, and Equation 14.7, suggesting that the incompressibility assumption is valid in practice [42,43].

Filtering approaches can also be designed based on the fact that the displacement field corresponding to the longitudinal wave is curl free, whereas that corresponding to the shear wave is divergence free [43]. Taking the curl of Equation 14.5 leads directly to the Helmholtz equation (with the curl of the displacement replacing the displacement itself) but with no need for the incompressibility assumption. This technique can remove artifacts present in the standard inversion in certain situations, but it is also more susceptible to noise because it involves additional derivative operations [43].

14.5.2 Shear Modulus and Mechanical Frequency

In the earlier treatment, the Lame constants were complex quantities, with the imaginary parts representing attenuation for a viscoelastic medium. Because the damping term involves the time derivative of the strain, for harmonic motion this can be denoted as $\mu = \mu_r + i\mu_i = c + i\omega\eta$ [40]. The simplest case is an isotropic, homogeneous, and incompressible medium (Equation 14.7). With no attenuation, a simple shear wave propagates with a specific wavelength or spatial frequency f_{sp}. The shear modulus is $\mu = \rho f_{mech}^2 / f_{sp}^2 = \rho v_s^2$, where f_{mech} is the mechanical driving frequency and v_s is the wave speed (or phase velocity). We will henceforth assume that $\rho \sim 1.0$ for all soft tissues [11]. If there is attenuation, the wave speed and attenuation are functions of frequency and are given by

$$v_s^2 = \frac{2(c^2 + \omega^2\eta^2)}{c + (c^2 + \omega^2\eta^2)^{1/2}} \quad \text{and} \quad \alpha^2 = \frac{\omega^2}{2}\frac{(c^2 + \omega^2\eta^2)^{1/2} - c}{c^2 + \omega^2\eta^2} \tag{14.8, 14.9}$$

where α denotes attenuation by the factor e^k in the direction of propagation k. The attenuation can also be expressed as the attenuation per wavelength, which is the acoustic quality factor $Q = c/\omega\eta$ [36].

The "true" shear modulus is the real part μ_r or c, which describes the behavior of a static object in equilibrium. However, some processing techniques described in the following text calculate only the local wavelength and do not consider attenuation. These techniques essentially estimate the wave speed, and we can speak of an "effective" shear modulus or "shear stiffness" that is defined as the square of the wave speed by analogy to the lossless case. The results are usually presented in terms of this shear stiffness at a given frequency. Other techniques calculate both μ_r and μ_I (or c and η) directly, which can be converted to wave speed and attenuation using Equation 14.8 and Equation 14.9. This determination of μ_i or α tends to be very sensitive to noise. A more stable way to determine the c and η parameters is to calculate the shear stiffness at several different frequencies and fit the result to expressions derived from Equation 14.8 and Equation 14.9.

Spatial wavelength and attenuation decrease and increase, respectively, as the mechanical frequency increases. This has two competing effects on stiffness determination: (1) higher resolution because the wavelength is smaller, and (2) lower displacement and hence lower signal. The best frequency for a particular application depends on trade-off between these two effects.

14.5.3 PHASE GRADIENT

After extracting the harmonic component at the driving frequency, the amplitude and phase (relative to an arbitrary zero point) that characterize the harmonic oscillation at each pixel in the image are obtained. If the motion is a simple propagating shear wave, the gradient of this phase directly yields the change in phase per pixel, easily convertible to a local frequency and thus to shear stiffness. This analysis can have very high resolution, but is very sensitive to noise, and data smoothing is usually necessary. This technique yields inaccurate results when two or more waves are superimposed (e.g., reflected waves) or when the motion is complex because the phase values then do not represent a single propagating wave [44]. However, it is useful in specialized situations in which simple plane-wave propagation is a good approximation. The other approaches, described in the following text, do not suffer from this drawback; they correctly handle reflections and other complex interactions because they are based on the underlying equations of motion.

14.5.4 LOCAL FREQUENCY ESTIMATION (LFE)

The local spatial frequency of the shear-wave propagation pattern can be calculated using an algorithm that combines local estimates of instantaneous frequency over several scales [45]. These estimates are derived from filters that are a product of radial and directional components and can be considered to be oriented log-normal quadrature wavelets. The shear stiffness is then given by $\mu = f_{\text{mech}}^2 / f_{\text{spatial}}^2$, with the assumption that $\rho \sim 1.0$ for all soft tissues. It can be shown that this approach involves solving the Helmholtz equation obtained under the assumptions of local homogeneity, incompressibility, and no attenuation [43]. LFE allows estimation of μ from a single image, i.e., using displacement values for a single sensitization direction and a single phase offset. It is equally applicable to the complex harmonic displacement extracted from multiple phase offsets.

The LFE algorithm has proven to be a robust technique because of the multiscale data averaging in the estimation. It yields accurate and isotropic local frequency estimates and is relatively insensitive to noise [46]. One disadvantage is the limited resolution; at sharp boundaries the LFE estimate is blurred, and the correct estimate is obtained at approximately half a wavelength into a given region. If one considers a stiff object of size equal to an eighth of a spatial wavelength embedded in a less-stiff background material, the LFE estimate of μ for the object will never be the correct value. However, the object will be detectable; that is, the existence of a stiff object is evident even if the quantitative determination of its stiffness is inaccurate.

14.5.5 DIRECT INVERSION

Assuming local homogeneity, Equation 14.5 to Equation 14.7 can be solved separately at each pixel, using only data from a local neighborhood to estimate local derivatives [38,39]. Inversion of Equation 14.5 estimates both Lame constants for an isotropic material and requires all the components of motion. If we rewrite Equation 14.5 as

$$\mathbf{A} \begin{bmatrix} \lambda + \mu \\ \mu \end{bmatrix} = -\rho \omega^2 \begin{bmatrix} u_1 \\ u_2 \\ u_3 \end{bmatrix} \quad \text{where} \quad \mathbf{A} = \begin{bmatrix} A_{11} & A_{12} \\ A_{21} & A_{22} \\ A_{31} & A_{32} \end{bmatrix} = \begin{bmatrix} u_{i,i1} & u_{1,ii} \\ u_{i,i2} & u_{2,ii} \\ u_{i,i3} & u_{3,ii} \end{bmatrix} \quad (14.10)$$

then the solution is given by

$$\begin{bmatrix} \lambda + \mu \\ \mu \end{bmatrix} = -\rho \omega^2 (\mathbf{A}^* \mathbf{A})^{-1} \mathbf{A}^* \begin{bmatrix} u_1 \\ u_2 \\ u_3 \end{bmatrix} \quad (14.11)$$

where $\mathbf{A}^*$ is the conjugate transpose of the matrix $\mathbf{A}$. Assuming an incompressible material, Equation 14.7 gives

$$\mu = -\rho \omega^2 \frac{u_i}{\nabla^2 u_i} \quad (14.12)$$

which allows estimation of the shear modulus from a single polarization of motion. A separate assumption path can be used for 2-D imaging by assuming that all derivatives in the out-of-plane direction are negligible. The shear mode then decouples and can also be solved by 2-D Helmholtz inversion [42].

In practice, such direct inversion techniques require data smoothing and the calculation of accurate second derivatives from the noisy data. The resolution is essentially limited only by the noise level in the data. In a stiffer material, the shear wave has a longer wavelength, making the derivatives smaller and the effects of noise more serious. The relative performance of different filtering approaches for smoothing was studied in detail by Oliphant [43].

These techniques do not depend on planar shear-wave propagation but simply on the presence of motion (that satisfies the assumed physical model) in the region of interest. In particular, complex interference patterns from reflection, diffraction, etc., do not pose difficulties except that these patterns may contain areas of low amplitude and, hence, low signal-to-noise ratio (SNR). This is also true for the LFE algorithm, which, despite its origin as an image processing method, actually involves inverting the Helmholtz equation (with the additional assumption of no attenuation) and correctly handles superimposed waves.

14.5.6 VARIATIONAL METHOD

Romano et al. [47,48] have suggested using the weak (variational) integral form of Equation 14.4 and test functions to estimate the Lame constants. The test functions are chosen such that they and their first derivatives vanish at the local window boundaries, removing all effects of surface forces. Integration by parts is used to shift the derivative operations from the noisy data to the analytic derivatives of the smooth test functions and integrating these over local windows in product with the data. In practice, this is similar to calculating derivatives by filtering with the derivative of a smooth function. Their assumption of constant μ/ρ also is essentially equivalent to the local homogeneity assumption. In the incompressible case, this is equivalent to direct inversion with the specific conditions described earlier imposed on the smoothing filter.

14.5.7 MATCHED FILTER

The matched-filter algorithm uses an adaptive smoothed matched filter (i.e., a smoothed version of the data itself) and its Laplacian to perform the same division as direct inversion. It is motivated by theoretical considerations to minimize the uncertainty in the estimate of μ in the face of random noise [43,49]. The processing is computationally more intensive than direct inversion because a different filter is calculated and applied at each voxel.

14.5.8 REMOVING THE LOCAL HOMOGENEITY ASSUMPTION

The assumption of local homogeneity is used in all these techniques to simplify the equation of motion to an algebraic equation that can be solved locally (Equation 14.5).

This necessarily implies inaccurate results at or near the boundaries between regions and a limit of resolution on the order of the local window size. However, the local window can be as small as desired, subject only to increased noise in derivative estimates, and in practice, inversions "fail gracefully" (with a gradual transition) across regional boundaries. The resolution of MRE inversion is limited by the accuracy of the spatial derivative estimation and thus ultimately by the SNR (with noisy data, averaging over larger spatial windows may be required). Stiffer objects are more difficult to estimate accurately because their spatial derivatives of displacement change more slowly.

A different possibility is to not make this assumption and solve Equation 14.4 allowing the mechanical properties to vary in the physical model. This method has two confounding effects: (1) the equation remains a differential equation and not an algebraic one, and (2) the assumption of incompressibility does not decouple the equations of motion for shear modulus, and so all components of motion are once again necessary. This approach is computationally more challenging but in principle models more accurately the physics of motion for arbitrary materials.

14.5.9 FINITE ELEMENT ANALYSIS

Van Houten et al. [50,51] have described a finite-element-based subzone technique for solving Equation 14.4. In their approach, a solution is iteratively refined on small overlapping subzones of the overall domain by updating the solution based on differences between forward calculations of the displacement from the current solution and measured values. After an update is performed on one subzone, the subzone with the greatest residual error is determined and updated. Local homogeneity is not assumed. The approach is elegant, and good results have been demonstrated on synthetic and actual data sets. It is computationally very intensive, and it seems to be very sensitive to the data being acquired in a true steady state, i.e., with enough motion cycles before acquisition so that the wave field reaches equilibrium. Although this is technically a requirement on all harmonic inversion techniques, analysis and practice have shown that violation of these conditions — i.e., excitation with only a few cycles before acquisition such that the wave field has not reached equilibrium — has only minor effects on the other inversions [43]. No direct comparisons have been made between finite element inversions and the other methods mentioned earlier, but the results appear to be comparable in quality.

Finite element methods are also widely used for forward simulations of MRE experiments. Another approach termed *coupled harmonic oscillator simulation* has also been used to simulate and, to a limited extent, interpret and analyze MRE data [52,53].

14.5.10 ANISOTROPIC INVERSIONS

Certain tissues are far from being isotropic; for example, muscle tissue is highly anisotropic, and it is known that shear waves propagate preferentially along muscle fibers [22,37]. The analysis algorithms presented earlier all assumed tissues to be isotropic in order to reduce the large number of unknown parameters to be estimated. Similar algorithms can be developed without this assumption or by replacing it with a less-restrictive symmetry (e.g., transverse isotropy is probably the appropriate model for muscle fibers). Solving for the larger number of unknowns may require larger data sets, probably from a wider variety of experiments that excite different motions, and may be quite difficult. Sinkus et al. [33] have proposed a technique that attempts to solve, in a limited way, for anisotropic characteristics of tissue. They have applied this technique to the breast and suggest that this can help differentiate between benign tissue, which appears isotropic, and carcinoma, which appears to exhibit an increased degree of anisotropy.

14.5.11 HYPERELASTIC PARAMETER DETERMINATION

Tissues in general show nonlinear behavior, and the stress–strain curve for large displacements can deviate considerably from a straight line. In MRE, the displacements are usually small enough so behavior is linear, and the stiffness is represented by the slope of the stress–strain curve at particular experimental conditions. However, the amount of compression applied to the tissue (by the mechanical driver to couple it to the tissue or by other aspects of the experiment) can determine where the stress–strain curve is being probed, and different experiments on the same tissue can report different stiffness values. Samani et al. [54] modeled this behavior with hyperelastic parameters and have proposed an inversion scheme that attempts to recover these and use the entire stress–strain curve for material characterization.

14.5.12 SIGNAL-TO-NOISE CONSIDERATIONS

It is important to understand whether there is sufficient signal in a given region to yield an accurate stiffness estimate and what the uncertainty is in that estimate. In MRE, "signal" means not only MR signal but, more importantly, that the region is undergoing sufficient motion so that the induced phase shifts can be detected and well quantified. A simple model of how noise in the MR acquisition translates to noise in the phase difference or "wave" images can be derived. The noise level in the standard MR magnitude image reconstructed from the MRE acquisition can be determined from the background (after correction for the effect of rectification). In areas of significant magnitude, this noise can be considered to be Gaussian in both the real and imaginary components. Thus, it forms a Gaussian cloud about the true magnitude and phase, and the uncertainty in phase for a given noise level and magnitude can be calculated. In the wave images, this uncertainty in phase is the noise, and the signal is the accumulated phase shift due to motion. Local SNR in the wave images can thus be determined. Higher SNR is obtained by (1) a larger underlying MR magnitude signal and (2) a larger displacement amplitude, leading to a larger accumulated phase shift. The effect of a given SNR in the wave images on the uncertainty of shear modulus estimations depends on the actual shear modulus and on the processing algorithm used.

14.5.13 PHASE UNWRAPPING

Because shear waves attenuate quickly in certain tissues, large amplitudes near the surface may be required to achieve sufficient amplitude in a deep region of interest. Too much amplitude can cause phase wrap, i.e., large enough displacements can cause accumulated phase shifts outside the range $\pm\pi$, which are ambiguous. This sets an upper limit for the amplitude at which the tissue should be driven. However, standard phase unwrapping algorithms [55] can be applied to MRE data with good success, because the wrapped regions are usually fairly well localized. The upper limit on amplitude can then be increased as long as the phases can be reliably unwrapped. Alternatively, the increased amplitude can be traded off for higher resolution by operation at a higher mechanical frequency.

14.5.14 DIRECTIONAL FILTERING

Most of the inversion algorithms described in the preceding text are derived from the fundamental equations of motion and, in principle, correctly handle complex wave fields with interference patterns due to reflection and refraction. However, they tend to be sensitive to areas of low displacement amplitude (and hence low SNR) that result from such situations. A spatiotemporal directional filter has been described that can be applied as a preprocessing step to separate complex wave fields into components propagating in different directions, each of which can be analyzed separately [56] (see Subsection 6.1). Areas of low motion due to standing waves or cancellation between interfering waves are minimized, and weighted combinations of inversions from such directionally separated data sets significantly improve reconstructions of shear moduli and attenuation.

14.6 RESULTS

To illustrate the different noise sensitivities, resolutions, and accuracies of the various processing techniques, we present results on physical phantoms having known parameters and *ex vivo* and *in vivo* results on both human and animal tissues. Quantitative measurements of shear moduli with MRE have shown high correlation with measurements using biomechanical testing devices [8,57,58].

14.6.1 PHANTOM OBJECT

Figure 14.3 illustrates a sample wave image from an agar gel phantom with four stiff cylindrical inclusions (ranging from 5 to 25 mm in diameter) perpendicular to the slice, acquired with mechanical vibrations of 100 Hz. The figure depicts the out-of-plane displacement component, with the largest displacement being approximately 10 μm. The acoustic shear waves are introduced from the top of the image and propagate downward, but they reflect off the cylinders and the

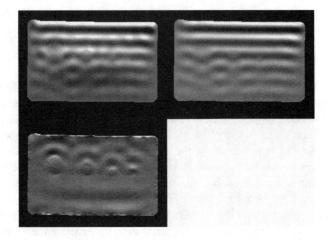

FIGURE 14.3 Left: Original band-pass-filtered wave image (for one of eight phase offsets). Center: Data after top-down directional filtering. Right: Data after bottom-up directional filtering (contrast increased by 5X).

boundaries of the phantom, giving rise to interference patterns. The effect of selecting waves propagating in the top-down and bottom-up directions using the direction-filtering approach described earlier is shown in the center and right panels. The simplification of the wave field is further evidenced by the amplitude maps in Figure 14.4, which show the amplitude of the harmonic motion at the driving frequency at each voxel (from the first positive temporal frequency plane, extracted by the Fourier transform of the phase offsets, as described earlier). The amplitude map for the nondirectionally filtered data (left) contains many nodes, i.e., areas of low motion due to cancellation between the main top-down wave and reflections from the inclusions and the bottom wall. Again, the top-down filtered data is far smoother, and the nodes evident in the original wave field are greatly reduced (right). The result of the four inversion algorithms described in the preceding text are shown in Figure 14.5. Profiles through the reconstructions are shown in Figure 14.6. All four algorithms clearly show the four inclusions, with differing levels of artifacts in the lower regions of the phantom. The reference

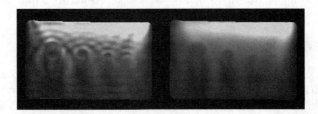

FIGURE 14.4 The amplitude of the harmonic motion at the driving frequency for the band-pass-filtered (left) and the top-down direction-filtered (right) data sets.

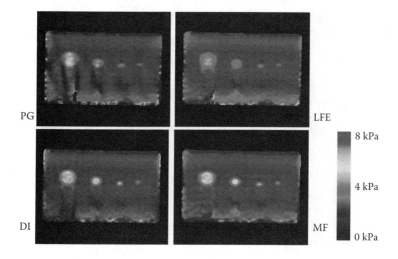

FIGURE 14.5 The PG, LFE, DI, and MF reconstructions of shear modulus from the top-down directionally filtered data.

values are 2.9 kPa for the background gel and 6.4 kPa for the inclusions. All the algorithms reconstruct the background value correctly, and all but LFE reach approximately correct values for the two larger inclusions. The stiffness of the smaller inclusions, which are less than a wavelength in size, is underestimated by all four algorithms.

14.6.2 ANIMAL TISSUES

A variety of *ex vivo* experiments have demonstrated that MRE can quantitatively assess the viscoelastic properties of real tissues and detect changes in stiffness

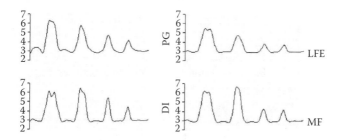

FIGURE 14.6 Profiles through the inclusions for the results in Figure 14.5 for the PG, LFE, DI, and MF reconstructions, respectively. The reference values are 2.9 kPa for the background gel and 6.4 kPa for the inclusions. All the algorithms reconstruct the background value correctly, and all but LFE reach approximately correct values for the two larger inclusions. The stiffness of the smaller inclusions is underestimated by all the algorithms.

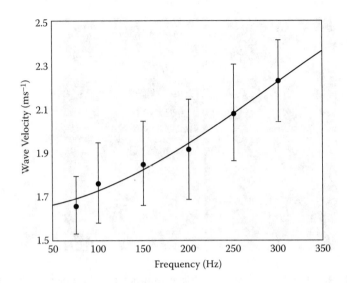

FIGURE 14.7 Shear-wave velocity in specimens of porcine liver tissue obtained from 5 different animals, measured at frequencies of 75 to 300 Hz. The velocity increases systematically with frequency. The curve represents a best-fit line that accounts for viscoelastic behavior. (From Kruse, S.A., Smith, J.A., Lawrence, A.J., Dresner, M.A., Manduca, A., Greenleaf, J.F., Ehman, R.L. (2000). Tissue characterization using magnetic resonance elastography: preliminary results. *Phys. Med. Biol.* 45: 1579–1590.)

with frequency and temperature [37,59]. For example, Figure 14.7 shows an example of the calculated shear stiffness of porcine liver tissue at different mechanical frequencies. The data are well fit by a viscoelastic model (Equation 14.8 and Equation 14.9). Because of the lack of metabolic activity, homeostasis, and *in situ* preloading in specimens, the observed mechanical properties are likely to be different from those that would be measured *in vivo*.

14.6.3 BREAST

Various groups have reported MRE results on *in vivo* human breasts [31,33,34,54,60,61]. Clear distinction has been found between fat and glandular tissues [34,61], in rough agreement with earlier results on excised tissue, and between normal breast tissue and carcinoma [60,61]. Figure 14.8 illustrates the differentiation between adipose and glandular tissue in the elastogram of a normal volunteer. Recent work [60] has included *in vivo* imaging of six normal volunteers and six patients with cancer. In patients with tumors, the stiffest regions in the elastogram corresponded to the known tumor locations and were 5 to 20 times stiffer than normal tissue (Figure 14.9). No regions of such high stiffness were found in the normal volunteers. The stiffness values found for the tumors are expected to be an underestimate in. all cases because of the limitations of the processing technique and the scans being 2-D instead of 3-D. Sinkus et al. [33]

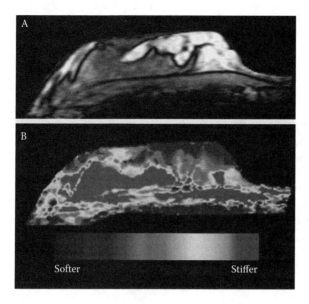

FIGURE 14.8 (a) Axial T1-weighted spin-echo image of the breast of a normal volunteer. (b) MR elastogram obtained for this volunteer with shear waves at 100 Hz applied to the anterior part of the breast. The elastogram depicts clear differentiation between soft adipose tissue and stiffer fibroglandular tissue.

have reported an anisotropic analysis of 3-D MRE data of the breast and have suggested that this can help differentiate between benign tissue, which appears isotropic, and carcinoma, which appears to exhibit an increased degree of anisotropy.

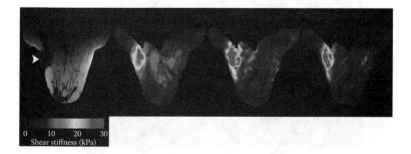

FIGURE 14.9 T1-weighted image (left) and LFE, DI, and MF elastograms respectively of the breast of a patient with 4-cm diameter biopsy-proven breast cancer. The image was obtained with shear waves of 100 Hz applied to the skin of the medial and lateral aspects of the breast. The field of view is approximately 16 cm and the section thickness is 5 mm. The elastograms all indicate that the shear stiffness of the tumor in the posterolateral aspect of the breast (arrowhead) is substantially higher than that of normal fibroglandular and adipose tissues in the breast.

14.6.4 BRAIN

Although there is no clinical precedent for "brain palpation," it is possible that measurements of elastic properties might be useful for characterizing brain disease. In addition, such measurements are necessary prerequisites for finite element analysis studies of brain trauma and surgical simulation. However, the estimates of the shear modulus of brain tissue available in the literature are inconsistent and do not even agree on the relative stiffness of gray and white matter [62]. These estimates were obtained *ex vivo* from specimens without blood pressure and metabolic activity. This may explain why they span several orders of magnitude and disagree on whether gray matter is softer or harder than white matter. Cerebral elastography studies have been performed to date in 19 normal volunteers [62]. Waves are clearly observed to propagate throughout the brain, and the elastograms (Figure 14.10) demonstrate that the *in vivo* shear stiffness at this frequency of white matter (average value 14.2 kPa) is higher than that of gray matter (average value 5.3 kPa). The difference is statistically significant. No discernible relationship between age and shear modulus has been found.

14.6.5 MUSCLE

MRE has been applied to skeletal muscle to quantify the change in stiffness with muscle loading [63]. Five volunteers supported varying loads during MRE examinations to assess the biceps brachii muscle during active force generation. The

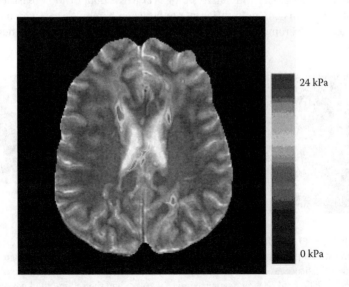

FIGURE 14.10 An elastogram of the brain of a normal volunteer overlaid on the MR magnitude image.

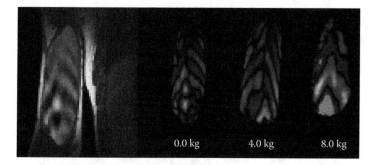

FIGURE 14.11 Wave images obtained in a coronal plane passing through the biceps brachii muscle of a volunteer under the loading conditions indicated. The superior end of the muscle is at the top of the image. Shear waves of 150 Hz were generated in the muscle by an electromechanical driver applied to the skin over the distal biceps tendon (outside the field of view). The section thickness was 7 mm, and the field of view of the image was approximately 16 cm. The shear-wave image, sensitized for wave motion orthogonal to the plane of section, demonstrates propagating waves, which have a characteristic chevron-like pattern. The displacement amplitudes are on the order of 30 μm. The shear wavelength clearly increases with load.

wavelength of the shear wave clearly increased with load for each volunteer, as shown in Figure 14.11, and in each case the shear stiffness of the muscle increased approximately linearly with force. The slope of the stiffness–force relation varied among the volunteers and was proportional to the inverse of the muscle size. The muscle is strongly anisotropic, and these results were based on a simple 1-D analysis along the muscle fiber, using a damped sinusoid fit to line profiles in the direction of wave propagation. Simulation studies of the characteristic chevron-shaped wave patterns observed in muscles have been performed by Sack et al. [53].

14.6.6 Ultrasound Wave Field Visualization

Walker et al. [64] have used specially constructed apparatus to image with MRE, the ultrasound wave fields in a tissue-equivalent agar gel medium. Nanometer motions at ultrasonic frequencies were clearly detected and visualized, and direct measurements of absolute pressure, intensity, and speed of sound were obtained. Although the magnetic field gradients required are an order of magnitude greater than the recommended limits for human imaging, this technique allows a detailed study of ultrasound propagation and scattering in heterogeneous *ex vivo* tissue samples. No other technique can directly and noninvasively visualize the nanometer-scale displacements due to ultrasound in tissue or tissue-equivalent materials.

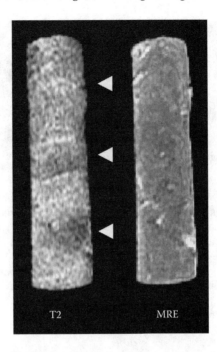

FIGURE 14.12 Bovine muscle specimen with three areas of thermal coagulation, perceptible in the T2 image. MRE (superimposed on a T1-w image) indicates zones of coagulation as areas of high shear stiffness (red).

14.6.7 CHARACTERIZATION OF THERMALLY ABLATED TISSUE

MR-guided focused ultrasound (FUS) tissue ablation is a procedure in which FUS is used to treat tumors by heating the tumor tissue and coagulating it, while sparing the surrounding normal tissue as much as possible. Existing methods for assessing the spatial extent of tissue coagulation obtained with FUS have limitations because it is difficult to assess the exact location actually being heated and the extent to which the heating has coagulated the tissue during the procedure. Studies have demonstrated that MRE delineates thermally coagulated tissues as areas of increased shear stiffness (see Figure 14.12) [35,65]. Further work in which elastography was performed at multiple times during heating and cooling of bovine tissue revealed a gradual softening of tissue as the temperature was raised from 20°C to 60°C and then a large, irreversible increase in tissue shear stiffness upon further heating that appears to correspond to tissue coagulation due to irreversible protein denaturation (Figure 14.13). These results suggest that MRE may be well suited to assess the results of tissue ablation procedures. In addition, because the shear waves required by MRE can be generated by FUS radiation pressure at the focal spot (as described earlier), it may be possible to use MRE for guiding tissue ablation procedures in real time, using FUS for both

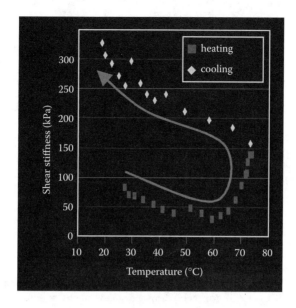

FIGURE 14.13 MRE of thermally treated bovine muscle tissue demonstrates a major, irreversible change in shear stiffness as the tissue is heated above 60°C.

the ablation and for shear-wave generation, and assessing tissue stiffness and coagulation as the tissue is being heated during the procedure itself.

14.7 CONCLUSION

MRE shows great potential for noninvasive *in vivo* determination of mechanical properties of a variety of tissues. The detection of propagating acoustic waves has been demonstrated *in vivo* in the breast, brain, and muscle and *ex vivo* in numerous animal and human tissues. Reconstruction algorithms have been tested and characterized, and although far from perfect, they yield quantitative measures of elasticity that clearly demarcate differences between tissue types and identify tumors as areas of higher stiffness. Challenges remain in pulse sequence design, delivering sufficient signal to all areas of the body, and improving processing algorithms to generate more accurate, higher-resolution elasticity and attenuation maps. We speculate that MRE may prove to be useful in tumor detection, diseased tissue characterization, and the evaluation of rehabilitation.

ACKNOWLEDGMENTS

This research has been supported by NIH grants CA75552, CA91959, EB00812, and EB01981.

REFERENCES

1. Duck, F.A. (1990). *Physical Properties of Tissues — A Comprehensive Reference Book*. 6th ed. Sheffield, England: Academic Press.
2. Sarvazyan, A. (1993). Shear acoustic properties of soft biological tissues in medical diagnostics [abstr]. *J. Acoust. Soc. Am. Proc. 125th Mtg.* 93(2): 2329.
3. Krouskop, T.A., Wheller, T.M., Kallel, F., Garra, B.S., and Hall, T. (1998). Elastic moduli of breast and prostate tissues under compression. *Ultrasound Imaging.* 20(4): 260–274.
4. Sarvazyan, A., Goukassian, D., and Maevsky, G. (1994). Elasticity imaging as a new modality of medical imaging for cancer detection. *Proc. Intl. Workshop Interaction Ultrasound with Biol. Media* pp. 69–81.
5. Sarvazyan, A., Rudenko, O.V., Swanson, S.D., Fowlkes, J.B., and Emelianov, Y. (1998). Shear-wave elasticity imaging: A new ultrasonic technology of medical diagnostics. *Ultrasound Med. Biol.* 24(9): 1419–1235.
6. Gao, L., Parker, K.J., Lerner, R.M., and Levinson, S.F. (1996). Imaging of the elastic properties of tissue — a review. *Ultrasound Med. Biol.* 22: 959–977.
7. *Physics in Medicine and Biology* (June 2000).
8. Muthupillai, R., Lomas, D.J., Rossman, P.J., Greenleaf, J.F., Manduca, A., and Ehman, R.L. (1995). Magnetic resonance elastography by direct visualization of propagating acoustic strain waves. *Science* 269: 1854–1857.
9. Muthupillai, R., Rossman, P.J., Lomas, D.J., Greenleaf, J.F., Riederer, S.J., and Ehman, R.L. (1996). Magnetic resonance imaging of transverse acoustic strain waves. *Magn. Reson. Med.* 36: 266–274.
10. Goss, S.A., Johnston, R.L., and Shnol, S.E. (1978). Comprehensive compilation of empirical ultrasonic properties of mammalian tissues. *J. Acoust. Soc. Am.* 64(2): 423–457.
11. Burlew, M.M., Madsen, E.L., Zagzebski, J.A., Banjavic, R.A., and Sum, S.W. (1980). A new ultrasound tissue-equivalent material. *Radiology* 134(2): 517–520.
12. Ophir, J., Cespedes, I., Ponnekanti, H., Yazdi, Y., and Li, X. (1991). Elastography: A quantitative method for imaging the elasticity of biological tissues. *Ultrasound Imaging* 13: 111–134.
13. O'Donnell, M., Skovoroda, A.R., Shapo, B.M., and Emellanov, S.Y. (1994). Internal displacement and strain imaging using ultrasonic speckle tracking. *IEEE Trans. Ultrasonics Ferroelect. Freq. Control* 41: 314–325.
14. Cespedes, I., Ophir, J., Ponnekanti, H., and Maklad, N. (1993). Elastography: elasticity imaging using ultrasound with application to muscle and breast *in vivo*. *Ultrasound Imaging* 15: 73–88.
15. Garra, B.S., Cespedes, E.I., Ophir, J., Spratt, S.R., Zuurbier, R.A., Magnant, C.M., and Pennanen, M.F. (1997). Elastography of breast lesions: Initial clinical results. *Radiology* 202(1): 79–86.
16. Gao, L., Parker, K.J., and Slam, S.K. (1995). Sonoelasticity imaging: Theory and experimental verification. *J. Acoust. Soc. Am.* 97: 3875–3885.
17. Lerner, R.M., Huang, S.R., and Parker, K.J. (1990). Sonoelasticity images derived from ultrasound signals in mechanically vibrated tissues. *Ultrasound Med. Biol.* 16: 237–239.
18. Parker, K. and Lerner, R. (1992). Sonoelasticity of organs: Shear waves ring a bell. *J. Ultrasound. Med.* 11: 387–392.

19. Rubens, D.J., Hadley, M.A., Alam, S.K., Gao, L., Mayer, R.D., and Parker, K.J. (1995). Sonoelasticity imaging of prostate cancer: *in vitro* results. *Radiology* 195: 379–383.

20. Nightingale, K.R., Soo, M.S., Nightingale, R.W., and Trahey, G.E. (2002). Acoustic radiation force impulse imaging: *in vivo* demonstration of clinical feasibility. *Ultrasound Med. Biol.* 28(2): 227–235.

21. Sandrin, L., Tanter, M., Catheline, S., and Fink, M. (2002). Shear modulus imaging with 2-D transient elastography. *IEEE Trans. Ultrasonics Ferroelect. Freq. Control* 49: 426–435.

22. Gennisson, J.L., Catheline, S., Chaffai, S., and Fink, M. (2003). Transient elastography in anisotropic medium: Application to the measurement of slow and fast shear-wave speeds in muscles. *J. Acoust. Soc. Am.* 114: 536–541.

23. Axel, L. and Dougherty, L. (1989). MR imaging of motion with spatial modulation of magnetization. *Radiology* 171: 841–845.

24. Zerhouni, E.A., Parish, D.M., Rogers, W.J., Yang, A., and Shapiro, E.P. (1988). Human heart: Tagging with MR imaging — a method for noninvasive assessment of myocardial motion. *Radiology* 169: 59–63.

25. Ehrichs, E.E., Jaeger, H.M., Karczmar, G.S., Knight, J.B., Kuperman, V.Y., and Nagel, S.R. (1995). Granular convection observed by magnetic resonance imaging. *Science* 267: 1632–1634.

26. Fowlkes, J.B., Emelianov, S.Y., Pipe, J.G., Skovoroda, A.R., Carson, P.L., Adler, R.S., and Sarvazyan, A.P. (1995). Magnetic resonance imaging techniques for detection of elasticity variation. *Med. Phys.* 22(11): 1771–1778.

27. Sumi, C. and Nakayama, K. (1998). A robust numerical solution to reconstruct the globally relative shear modulus distribution from strain measurements. *IEEE Trans. Med. Imaging* 17: 419–428.

28. O'Donnell, M. (1985). NMR blood flow imaging using multiecho, phase-contrast sequences. *Med Phys.* 12: 59–64.

29. Bernstein, M.A. and Ikezaki, Y. (1991). Comparison of phase-difference and complex-difference processing in phase-contrast MR angiography. *J. Magn. Reson. Imaging.* 1: 725–729.

30. Pelc, N.J., Shimakawa, A., and Glover, G.H. (1989). Phase-contrast cine MRI [abstr]. *Ann. Mtg. Soc. Magn. Reson. Med.* 1: 101.

31. Plewes, D.B., Bishop, J., Samani, A., and Sciaretta, J. (2000). Visualization and quantification of breast cancer biomechanical properties with magnetic resonance elastography. *Phys. Med. Biol.* 45: 1591–1610.

32. Samani, A., Bishop, J., and Plewes, D.B. (2001). A constrained modulus reconstruction technique for breast cancer assessment. *IEEE Trans. Med. Imaging.* 20: 877–885.

33. Sinkus, R., Lorenzen, J., Schrader, D., Lorenzen, M., Dargatz, M., and Holz, D. (2000). High-resolution tensor MRE for breast tumor detection. *Phys. Med. Biol.* 45: 1649–1664.

34. Van Houten, E.E.W., Doyley, M.M., Kennedy, F.E., Weaver, J.B., and Paulsen, K.D. (2003). Initial *in-vivo* experience with steady-state subzone-based MR elastography of the human breast. *J. Magn. Reson. Imaging.* 17: 72–85.

35. Wu, T., Felmlee, J.P., Greenleaf, J.F., Riederer, S.J., and Ehman, R.L. (2000). MR imaging of shear waves generated by focused ultrasound. *Magn. Reson. Med.* 43: 111–115.

36. Manduca, A., Smith, J.A., Muthupillai, R., Rossman, P.J., Greenleaf, J.F., and Ehman, R.L. (1997). Image analysis techniques for magnetic resonance elastography [abstr]. *Proc. ISMRM* 5: 1905.

37. Kruse, S.A., Smith, J.A., Lawrence, A.J., Dresner, M.A., Manduca, A., Greenleaf, J.F., and Ehman, R.L. (2000). Tissue characterization using magnetic resonance elastography: Preliminary results. *Phys. Med. Biol.* 45: 1579–1590.

38. Muthupillai, R., Rossman, P.J., Greenleaf, J.F., Riederer, S.J., and Ehman, R.L. (1996). MRI visualization of acoustic strain waves: Effect of linear motion [abstr]. *Proc. ISMRM* 4: 1515.

39. Muthupillai, R. and Ehman, R.L. (1997). Amplitude modulated cyclic gradient waveforms: Applications in MRE [abstr]. *Proc. ISMRM* 6: 1904.

40. Auld, B.A. (1990). *Acoustic Fields and Waves in Solids*, Malabar, FL: Krieger Publishing Company.

41. Kallel, F. and Bertrand, M. (1996). Tissue elasticity reconstruction using linear perturbation method. *IEEE Trans. Med. Imaging* 15(3): 299–313.

42. Oliphant, T.E., Manduca, A., Ehman, R.L., and Greenleaf, J.F. (2001). Complex-valued stiffness reconstruction for magnetic resonance elastography by algebraic inversion of the differential equation. *Magn. Reson. Med.* 45: 299–310.

43. Oliphant, T.E. (2001). Direct Methods for Dynamic Elastography Reconstructions: Optimal Inversion of the Interior Helmholtz Problem. Ph.D. thesis, Mayo Graduate School, Rochester, MN.

44. Catheline, S., Wu, F., and Fink, M. (1999). A solution to diffraction biases in sonoelasticity: The acoustic impulse technique. *J. Acoust. Soc. Am.* 105: 2941–2950.

45. Knutsson, H., Westin, C.J., and Granlund, G. (1994). Local multiscale frequency and bandwidth estimation. *Proc. IEEE Intl. Conf. Image Proc.* 1: 36–40.

46. Manduca, A., Muthupillai, R., Rossman, P.J., Greenleaf, J.F., and Ehman, R.L. (1996). Image processing for magnetic resonance elastography. *SPIE Med. Imaging.* 2710: 616–623.

47. Romano, A.J., Shirron, J.J., and Bucaro, J.A. (1998). On the noninvasive determination of material parameters from a knowledge of elastic displacements: Theory and numerical simulation. *IEEE Trans. Ultrasonics Ferroelect. Freq. Control* 45: 751–759.

48. Romano, A.J., Bucaro, J.A., Ehman, R.L., and Shirron, J.J. (2000). Evaluation of a material parameter extraction algorithm using MRI-based displacement measurements. *IEEE Trans. Ultrasonics Ferroelect. Freq. Control* 47: 1575–1581.

49. Oliphant, T.E., Manduca, A., Dresner, M.A., Ehman, R.L., and Greenleaf, J.F. (2001). Adaptive estimation of shear modulus for MR elastography [abstr]. *Proc. ISMRM* 9: 1642.

50. Van Houten, E.E.W., Paulsen, K.D., Miga, M.I., Kennedy, F.E., and Weaver, J.B. (1999). An overlapping subzone technique for MR-based elastic property reconstruction. *Magn. Reson. Med.* 42: 779–786.

51. Van Houten, E.E.W., Miga, M.I., Weaver, J.B., Kennedy, F.E., and Paulsen, K.D. (2001). Three-dimensional subzone-based reconstruction algorithm for MR elastography. *Magn. Reson. Med.* 45: 827–837.

52. Braun, J., Buntkowsky, G., Bernarding, J., Tolxdorff, T., and Sack, I. (2001). Simulation and analysis of magnetic resonance elastography wave images using coupled harmonic oscillators and Gaussian local frequency estimation. *Magn. Reson. Imaging.* 19: 703–713.

53. Sack, I., Bernarding, J., and Braun, J. (2002). Analysis of wave patterns in MR elastography at skeletal muscle using coupled harmonic oscillator simulations. *Magn. Reson. Imaging* 20: 95–104.

54. Samani, A. and Plewes, D.B. (2004). A method to measure the hyperelastic parameters of *ex vivo* breast tissue samples. *Phys. Med. Biol.* 49: 4395–4405.

55. Ghiglia, D.C. and Pritt, M.D. (1998). *Two-Dimensional Phase Unwrapping: Theory, Algorithms and Software,* New York: John Wiley & Sons.

56. Manduca, A., Lake, D.S., Kruse, S.A., and Ehman, R.L. (2003). Spatio-temporal directional filtering for improved inversion of MR elastography images, *Med. Image Anal.* 7: 465–473.

57. Hamhaber, U., Grieshaber, F.A., Nagel, J.H., and Klose, U. (2003). Comparison of quantitative shear-wave MR-elastography with mechanical compression tests. *Magn. Reson. Med.* 49: 71–77.

58. Ringleb, S.I., Chen, Q., Lake, D.S., Manduca, A., Ehman, R.L., and An, K.-N. (2004). Quantitative shear-wave magnetic resonance elastography: Comparison to a dynamic shear material test. *Magn. Reson. Med.* (in press).

59. Bishop, J., Poole, G., Leitch, M., and Plewes, D.B. (1998). Magnetic resonance imaging of shear-wave propagation in excised tissue. *J. Magn. Reson. Imaging.* 8: 1257–1265.

60. McKnight, A.L., Kugel, J.L., Rossman, P.J., Manduca, A., Hartmann, L.C., and Ehman, R.L. (2002). MR elastography of breast cancer: Preliminary results. *Am. J. Roentgen.* 178: 1411–1417.

61. Lawrence, A.J., Rossman, P.J., Mahowald, J.L., Manduca, A., Hartmann, L.C., and Ehman, R.L. (1999). Assessment of breast cancer by MR elastography [abstr]. *Proc. ISMRM* 7: 525

62. Kruse, S.A., Dresner, M.A., Rossman, P.J., Felmlee, J.P., Jack, C.R., and Ehman, R.L. (1999). "Palpation of the brain" using MR elastography [abstr]. *Proc. ISMRM* 7: 258.

63. Dresner, M.A., Rose, G.H., Rossman, P.J., Muthupillai, R., Manduca, A., and Ehman, R.L. (2001). Magnetic resonance elastography of skeletal muscle. *J. Magn. Reson. Imaging.* 13(2): 269–276.

64. Walker, C.L., Foster, F.S., and Plewes, D.B. (1998). Magnetic resonance imaging of ultrasonic fields. *Ultrasound Med. Biol.* 24: 137–142.

65. Wu, T., Felmlee, J.P., Greenleaf, J.F., Riederer, S.J., and Ehman, R.L. (2001). Assessment of thermal tissue ablations with MRE. *Magn. Reson. Med.* 45: 80–87.

Part V

BOLD Contrast MR Imaging and fMRI Signal Analysis

15 Fundamentals of Data Analysis Methods in Functional MRI

Elia Formisano, Francesco Di Salle, and Rainer Goebel

CONTENTS

15.1 INTRODUCTION

Since its invention in the early 1990s [1–3], functional magnetic resonance imaging (fMRI) has rapidly assumed a leading role among the techniques used to localize brain activity. The spatial and temporal resolution provided by state-of-the-art MR technology and its noninvasive character, which allows multiple studies of the same subject, are some of the main advantages of fMRI over the other functional neuroimaging modalities that are based on changes in blood flow and cortical metabolism [4].

In a typical fMRI study, a measurement session includes (1) the acquisition of one or multiple time series of "functional" volumes while a subject performs in a predefined sensory, motor, or cognitive stimulation paradigm and (2) the acquisition of an "anatomical" volume covering a certain region of interest (possibly the whole brain). Functional time series are acquired using fast or ultrafast MR sequences sensitive to blood oxygenation level dependent (BOLD) contrast (conventionally, T2/T2*-weighted echo planar imaging [EPI] sequences). Anatomical volumes, conversely, are acquired using slow MR sequences in which the contrast between gray and white matter is enhanced (conventionally, high-spatial-resolution three-dimensional [3-D] T1-weighted sequences) and serve as a structural reference for the visualization of the functional information obtained through the analysis of the functional time series.

The aim of this chapter is to provide a basic overview of the data analysis pipeline that is typically employed in fMRI.

Given the small amplitude (1–5%) of the stimulus-related MR signal changes and the presence of many confounding effects, the localization and characterization of the brain regions that respond to the various conditions of the stimulation protocol is a nontrivial process that involves several processing steps. Figure 15.1 shows a flowchart of these steps. Some of the steps aim to reduce the influence of the artifactual signal fluctuations and enhance the functional contrast-to-noise ratio (realignment, spatial, and temporal filtering; see Reference 2). Others aim to detect localized task-dependent signal changes and to visualize them by means of activation maps (see Reference 3). The step of spatial normalization refers to the transformation of anatomical and functional data in conventional reference

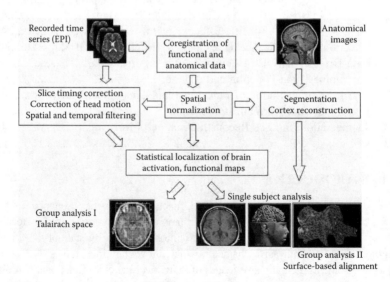

FIGURE 15.1 Schematic flowcharts of the data analysis steps involved in an fMRI study. Input data are the four-dimensional (4-D, space [3-D] × time [1-D]) functional time series (left) and the 3-D anatomical reference volumes (right). See text for a detailed description of each block.

spaces, and is essential to compare the results between different subjects or different groups of subjects and to facilitate communication among laboratories (see Reference 4). fMRI, however, as opposed to PET, also allows for the analysis of single-subject data. Additional steps can thus be included to display the relation between functional information and individual brain anatomy with more efficacy compared to the conventional serial 2-D slice representation (see Reference 7).

15.2 PREPROCESSING OF FUNCTIONAL TIME SERIES

15.2.1 SLICE TIMING CORRECTION

With 2-D EPI sequences conventionally used to collect the functional time series, volumes are formed by collecting one slice at a time. This implies that different parts of the brain are measured at slightly different moments in time, and that a functional volume cannot be considered an "instantaneous" temporal sample during which the brain activity is observed simultaneously (Figure 15.2). In Figure 15.2, for example, slice 5 is collected 400 msec after slice 1 (assuming an interval between the acquisitions of 2 subsequent slices of 100 msec). This systematic temporal "bias" in sampling the brain activity may induce fitting errors and misinterpretation of the results, especially in fMRI studies in which event-related designs are employed, in fMRI mental chronometry studies [5], and in the investigation of directed interactions [6]. In order to correct for this, data can be adjusted by appropriately shifting backward or forward each voxel's time series. This operation is referred to as *slice timing correction* and is achieved

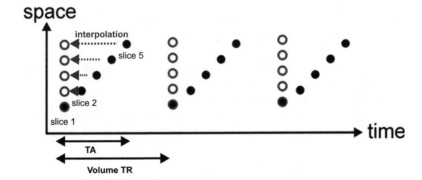

FIGURE 15.2 Slice timing correction. In fMRI different parts of the brain are sampled at different moments in time. The temporal offset between the acquisitions of two generic slices within the same volume depends on the time required to collect one slice (acquisition time, TA), and on the number of slices and the order in which the slices are collected. The error introduced by the presence of this small offset can be corrected by means of temporal interpolation of the time courses. Note that in order to apply this type of correction it is necessary to know the exact order in which the slices are collected.

with 1-D (sinc) interpolation in the temporal domain or by Fourier transforming the time series into a 1-D frequency representation, applying a phase shift to this data, and then recovering the corrected data by applying a reverse 1-D Fourier transform [7].

15.2.2 Motion Correction

The subject's motion poses a severe problem for the analysis of functional data. Despite the use of physical constraints, head movements cannot be completely eliminated during functional scanning. Head movements can be identified by viewing successive volumes of the functional time series as a "movie." Functional time series with gross motion of the head (greater than the voxel size) can be severely corrupted and, because they cannot be easily corrected with postprocessing algorithms, they should be discarded from further analysis. Small head movements (less than the voxel size) also produce effects that can mask the relatively small BOLD signal changes and should be corrected using realignment algorithms. In the following text, we describe the basic steps of these algorithms [8–15].

Let us consider $I_i(\mathbf{x})$ and $I_k(\mathbf{x})$ as two images (2-D or 3-D) collected at times i and k within a series of T repeated functional measurements. Let us suppose that $I_i(\mathbf{x})$ and $I_k(\mathbf{x})$ are related by a geometric transformation $\mathbf{T}[\mathbf{x}]$, so that

$$I_k(\mathbf{T}[\mathbf{x}]) \approx I_i(\mathbf{x}). \tag{15.1}$$

Realignment algorithms deal with the problem of finding the transformation $\mathbf{T}$ that minimizes the differences between the two images due to the subject's motion.

The most commonly adopted algorithms are based on iterative computation of the rotation-translation parameters that reduce the mismatch between a reference image (e.g., the T/2 scan of the time series) and the other images of the time series [8–11]. These realignment procedures are based on the following steps:

- Measurement of the spatial discrepancy between the transformed image $I_k(\mathbf{T}[\mathbf{x}])$ and the reference image $I_{T/2}(\mathbf{x})$
- Evaluation of the parameters that define $\mathbf{T}$
- Evaluation of the new values of I_k after $\mathbf{T}$ has been determined (interpolation method)

A robust method, commonly adopted in fMRI data analysis, considers $\mathbf{T}[\mathbf{x}]$ to be a rotation-translation transformation based on the rigid-motion hypothesis [8]. With this hypothesis, the transformation $\mathbf{T}[\mathbf{x}]$ is defined by three parameters in the case of realignment of 2-D images (two translation offsets and one rotation angle) and, by six parameters in the case of 3-D images (three translation offsets and three rotation angles).

Defining

$$\mathbf{r}(\mathbf{x}) = I_K(\mathbf{T}[\mathbf{x}])/I_{T/2}(\mathbf{x}) \tag{15.2}$$

as the voxel-by-voxel ratio between the two images, the algorithm then estimates the degree of misregistration between them by considering the mean m_r and the standard deviation σ_r of $\mathbf{r}(x)$ over all the voxels for which $\{I_{T/2}(\mathbf{x}) > = 0.215 \max(I_{T/2})\}$ (i.e., over all the intrabrain voxels)

The ratio

$$E = \sigma_r/m_r \tag{3}$$

is then used to measure the degree of misregistration between the template and the target image. When $I_K(\mathbf{T}[\mathbf{x}])$ is realigned to $I_{T/2}(\mathbf{x})$, then $\mathbf{r}(x)$ is constant and, consequently, the ratio E is small; conversely, when $I_k(\mathbf{T}[\mathbf{x}])$ is not realigned to $I_{T/2}(\mathbf{x})$, then the ratio E is large and new iterations are computed.

In the original version of the algorithm, Newton's method [15] was used separately for each parameter of $\mathbf{T}$ to minimize the ratio E, and trilinear interpolation was used to calculate the new values of I_k on the grid defined at each iteration. Other implementations of Wood's algorithm use more complex multidimensional minimization schemes and different interpolation methods (e.g., sinc interpolation).

The algorithm in [9], for example, differs from the Wood's algorithm in that it utilizes the Euclidean norm in L^2 as the mismatch function:

$$E = \sum_{x \in brain} (I_k(T[x]) - I_{T/2}(x))^2 \tag{15.4}$$

and the Levenberg–Marquardt algorithm [15] for the optimization of the rotation-translation parameters.

Other realignment algorithms emphasize the importance of removing the additional effects of the subject's movements on the magnetic spin excitation history (e.g., by correcting with an autoregression moving average (ARMA) model [11]), and other residual effects remaining after image realignment [14].

15.2.3 SPATIAL AND TEMPORAL FILTERING

Spatial and temporal filtering of fMRI time series aims to reduce the effects of the confounding factors that arise from instrumentation and spontaneous physiological activity on the detection of brain activation.

The high-spatial-frequency noise, mainly from the scanner devices, can be attenuated by spatially "smoothing" the fMRI time series with low-pass filters (Gaussian, Hamming, and Fermi filters) [16].

Let us express the acquired data as

$$I_i(k) = S_i(k) + E_i(k) \tag{15.5}$$

where $\mathbf{k}$ is the 3-D spatial-frequency span, $S_i(\mathbf{k})$ the spatial-frequency domain representation of a functional volume at scan i, and $E_i(\mathbf{k})$ the noise contribution (physiological and electronic). The underlying assumption of spatial smoothing is that $S_i(\mathbf{k})$ is a monotonically decreasing function of $\mathbf{k}$ and, thus, there exists some frequency $\mathbf{k}_c$ such that

$$S_i(\mathbf{k}) \ll E_i(\mathbf{k}) \quad \text{for } \mathbf{k} > \mathbf{k}_c \qquad (15.6)$$

Thus, a function $\mathbf{H}(\mathbf{k})$ such that

$$\begin{aligned} &1 \text{ for } \mathbf{k} < \mathbf{k}_c \\ \mathbf{H}(\mathbf{k}) \approx {}&0 \quad \text{for } \mathbf{k} > \mathbf{k}_c \end{aligned} \qquad (15.7)$$

would be an ideal filter. Indeed, multiplying $\mathbf{H}(\mathbf{k})$ by $\mathbf{I}_i((\mathbf{k})$ would result in noise suppression with minimal effect on the signal $S_i(\mathbf{k})$. However, the presence of regionally specific activation implies that high-spatial-frequency components of the signal are present in $S_i(\mathbf{k})$ as well and, thus, besides reducing the noise contribution, spatial smoothing will also decrease the effective spatial resolution of the functional analysis.

These two contrasting effects influence the detection of activation regions and have to be balanced. Intuitively, when the activated brain regions extend over clusters of several voxels, spatial smoothing will strengthen the signal relative to the noise (see Figure 15.3). Conversely, when focal regions of the brain are activated, they might no longer be discernible after spatial smoothing. Furthermore, according to the matched filter theorem, the signal is best detected by smoothing with a filter whose width matches that of the signal. In practical cases, however, because both focal and broad activation regions may be present in the same data set and their real extent cannot be known, the width and type of spatial filter are chosen on the basis of a trade-off between the desired spatial resolution and the expected enhancement of the functional contrast-to-noise ratio [14]. High effective spatial resolution is especially important in individual studies and, thus, little or no spatial smoothing is suggested. In multisubject studies, on the other hand, a high degree of spatial smoothing is necessary even after normalization to a standard stereotaxic space [17] in order to reduce the anatomical differences between subjects and to allow the correct use of statistical tools [10].

Thermal noise and high-temporal-frequency fluctuations arising from spontaneous activity can be attenuated by temporal smoothing of each voxel's time series [7]. In the case of temporal smoothing, the choice of the bandwidth of the filter is driven by a trade-off between the expected enhancement of functional contrast-to-noise ratio and the loss of temporal resolution (see Figure 15.4) unavoidably caused by the filtering. Indeed, despite the fMRI response being governed by slow hemodynamics, neural information on the order of a few hundreds of milliseconds may be present in BOLD signals collected using event-related protocols, and may be washed out by the temporal smoothing. Further, it should be noted that temporal smoothing introduces a high degree of dependency between subsequent samples of the time courses (temporal autocorrelation). This temporal autocorrelation has to be appropriately taken into account when performing the statistical analysis (see the following section), in order not to artificially inflate the significance levels of the tests.

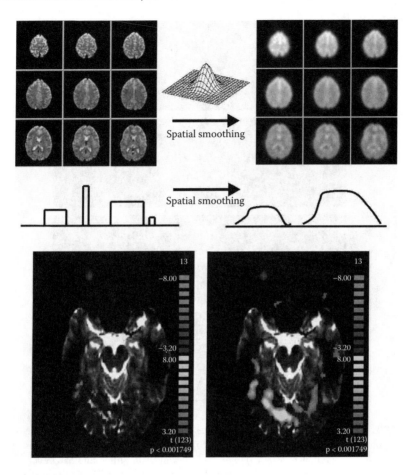

FIGURE 15.3 Spatial smoothing. Top row: Functional time series before (left) and after (right) spatial filtering with a 3-D Gaussian smoothing kernel (FWHM = 3 voxels). Middle row: Schematic representation of the effects of spatial smoothing on brain activation. For simplicity, space is represented on a 1-D axis. Note that "activated" regions with a small spatial extension may disappear after spatial smoothing. Lower row: Activation map before (left) and after (right) spatial smoothing with a 3-D Gaussian smoothing kernel (FWHM = 3 voxels).

MR signal drifts and physiological fluctuations at low temporal frequencies are always present in fMRI time series and have to be corrected by linear detrending or, more generally, by high-pass temporal filtering (see Figure 15.4) [18–21]. If the temporal sampling rate used for functional scanning (volume repetition time) is sufficiently short, systematic physiological noise linked to the cardiac and respiratory cycles can be eliminated using band-reject [20] or least-mean-square (LMS) adaptive filters [21].

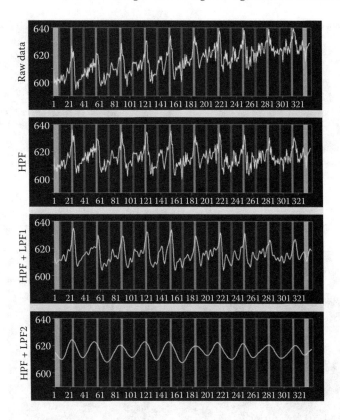

FIGURE 15.4 Temporal filtering. (a) Unfiltered time course from a region in the visual cortex during an event-related fMRI study. (b), (c), and (d) show filtered versions of the same time course obtained after (b) high-pass filtering (HPF, filter cutoff = 5 cycles/run), (c) HPF and moderate low-pass filtering (LPF1, filter cutoff = 75 cycles/run), and (d) HPF and heavy low-pass filtering (LPF2, filter cutoff = 25 cycles/run).

15.3 STATISTICAL LOCALIZATION OF BRAIN ACTIVATION

The most important step in the analysis of the functional time series is the detection of task-related BOLD activation and the creation of activation maps.

In the first fMRI studies, a simple method based on image subtraction was used to create descriptive images of the task-dependent brain areas [1–3]. According to the "pure insertion" hypothesis, voxels with a high gray level in the "difference" image, formed by subtracting the "control" from the "task" condition images, reflect the areas with a task-induced differential activation. With this method, the value of the intensity threshold between activated and nonactivated voxels is arbitrarily chosen. Furthermore, image subtraction is very sensitive to movement-related effects and to other unexpected signal changes. More reliable activation maps are produced by using univariate parametric or nonparametric statistical methods. The commonly used parametrical z- or t-test maps are formed by computing on a voxel-by-voxel basis the value of statistical significance for the difference of the means between two conditions. Voxels with a significance value below a given threshold (e.g., $P < 0.001$) are considered activated by the task (see e.g., Reference 22). In so-called correlation maps, each voxel is associated with the value of the linear cross-correlation coefficient (r) between the time course of the voxel intensity and a reference function [19]. Separation of activated and nonactivated voxels is achieved by imposing a threshold value for r. Correlation maps measure the similarity between the shape of the gray-level time course of a voxel and the expected hemodynamic response; thus, the sensitivity and specificity of this method strongly depends on the resemblance between the reference function and the "real" shape of the BOLD response. A simple on–off response [19], and the convolution of this ideal function with the impulse response of a linear model of the hemodynamics [18,23], are commonly used.

It can be shown that the t-test, correlation, and most other parametric tests can be regarded as special cases of the general linear model (GLM). The GLM is a standard statistical tool, which was introduced to imaging data analysis by Friston and coworkers [18,24]. The method allows the analysis of factorial designs that are expressed in a "design matrix" containing the description of all factors of interest as well as confounders (e.g., a linear trend) of an experiment [18,24–26]. The GLM is the most commonly used method for the voxel-by-voxel statistical analysis of the functional time series and is thus reviewed in more detail in the following subsection.

15.3.1 THE GLM

The GLM "explains" or "predicts" the variation of the observed time courses in terms of a linear combination of several regressor variables (or predictors) plus an error term

$$y_t = X_{t1} \beta_1 + \cdots + X_{tl} \beta_l + \cdots + X_{tL} \beta_L + e_{t.} \qquad (15.8)$$

In Equation 15.8, y_t ($t = 1,\ldots,T$; T = number of measurements) is the observed signal time course at a given voxel, $X_{tl}(l = 1,\ldots,L; \ L < T)$ are a set of L "explanatory" variables or "predictors" (functions of measurements), β_l's are the unknown weights (or regressor values)—one for each predictor, and the e_t's denote error terms that are assumed to be independent and identically normally distributed with zero mean and variance σ^2.

Writing Equation 15.8 for each observation t gives the equation system:

$$y_1 = X_{11} \beta_1 + \cdots + X_{1l} \beta_l + \cdots + X_{1L}\beta_L + e_1$$
$$\ldots\ldots\ldots\ldots\ldots\ldots\ldots\ldots\ldots\ldots\ldots\ldots\ldots\ldots\ldots\ldots$$
$$y_t = X_{t1} \beta_1 + \cdots + X_{tl} \beta_l + \cdots + X_{tL}\beta_L + e_t$$
$$\ldots\ldots\ldots\ldots\ldots\ldots\ldots\ldots\ldots\ldots\ldots\ldots\ldots\ldots\ldots\ldots$$
$$y_T = X_{T1} \beta_1 + \cdots + X_{Tl} \beta_l + \cdots + X_{TL}\beta_L + e_T$$

$$\beta_1 x \ + \ \beta_2 x \ +..+ \ \beta_n x$$

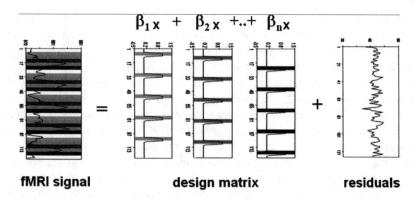

fMRI signal   **design matrix**   **residuals**

FIGURE 15.5 Schematic illustration of a general linear model of an fMRI experiment. Each voxel's time course is modeled as the linear combination of condition-specific predictors obtained by convolution of the ideal on–off response with a realistic model of the hemodynamic response (see text).

or in matrix notation:

$$\mathbf{y} = \mathbf{X}\boldsymbol{\beta} + \mathbf{e}. \tag{15.9}$$

Here, $\mathbf{y}$ is the $T \times 1$ column vector of the observations, $\mathbf{X}$ is the $T \times L$ matrix of the predictors (one row per observation, one column per model parameter), $\boldsymbol{\beta} = [\beta_1, \ldots, \beta_1, \ldots, \beta_L]^T$ is the $L \times 1$ column vector of parameters, and $\mathbf{e}$ is the $T \times 1$ column vector of error terms. The matrix $\mathbf{X}$ is conventionally referred to as the design matrix of the experiment. Most of the fMRI studies (with a single subject or multiple subjects) that contain a baseline condition as well as several repetitions of one or more experimental conditions may be easily expressed as a multiple regression problem by defining an appropriate form for $\mathbf{X}$ (Figure 15.5). For instance, for an experimental design with a baseline condition and five different stimulation conditions, the design matrix $\mathbf{X}$ has five columns and one row for each measurement time point. Each predictor is obtained by convolution of the ideal box-car (on–off) response with a realistic model of the hemodynamic response (hemodynamic response function; see, e.g., Reference 18 and Reference 21). Effects other than the expected task-related BOLD changes may also be modeled in the design matrix. Normally, $\mathbf{X}$ includes a column consisting of 1's for all measurements to account for the mean value of the voxel time course. Similarly, $\mathbf{X}$ may include a column with linearly increasing values to account for linear trends in the voxels' time courses.

Once the design matrix has been defined, the next step in the GLM analysis consists of the estimation of the regression weights $\boldsymbol{\beta}$ such that the predicted values $\mathbf{y}'$ are as close as possible to the measured values $\mathbf{y}$ at each time point. Let us denote with $\mathbf{y}'$ the estimate of the time course $\mathbf{Y}$ for the regression values $\boldsymbol{\beta}'$

$$\mathbf{y}' = \mathbf{X}\boldsymbol{\beta}' \tag{15.10}$$

and with

$$\mathbf{e} = [e_1, \ldots, e_j, \ldots, e_N]^T = \mathbf{y} - \mathbf{y}' = \mathbf{y} - \mathbf{X}\boldsymbol{\beta}' \qquad (15.11)$$

the residuals errors. The least-squares estimates $\boldsymbol{\beta}'$ of the regression values are the ones that minimize the residual sum of squares:

$$\mathbf{e}^T\mathbf{e} = (\mathbf{y} - \mathbf{X}\boldsymbol{\beta})^T (\mathbf{y} - \mathbf{X}\boldsymbol{\beta}) - >\mathbf{min}$$

This criterion leads to the normal equations:

$$(\mathbf{X}^T \mathbf{X})\boldsymbol{\beta}' = \mathbf{X}^T\mathbf{y} \qquad (15.12)$$

If $(\mathbf{X}^T \mathbf{X})$ is invertible, i.e., if the design matrix is of full rank, then the least-squares estimates are given by

$$\boldsymbol{\beta}' = (\mathbf{X}^T \mathbf{X})^{-1} \mathbf{X}^T\mathbf{y} \qquad (15.13)$$

If the model is correct and the errors are normally distributed, the least-squares estimates are also the maximum likelihood estimates and the best linear unbiased estimates. The mean value and the variance of $\boldsymbol{\beta}'$ are respectively:

$$\mathbf{E}\{\boldsymbol{\beta}'\} = \boldsymbol{\beta} \quad \text{and} \quad \text{Var}\{\boldsymbol{\beta}'\} = \sigma^2(\mathbf{S}^T\mathbf{S})^{-1} \qquad (15.14)$$

The estimation of the regressor values allows testing multiple linear hypotheses and creating different types of statistical maps. These maps are used to assess the effects of the various conditions included in the stimulation protocol and to draw inferences regarding the differential responses of different locations of the brain.

15.3.1.1 Overall Effects (R^2 Maps, *F* Maps)

A first type of map that can be obtained within the GLM framework is a map of the overall fit of the model to the data. This map is obtained by computing at each voxel the squared multiple regression coefficient (R^2):

$$R^2 = \frac{\text{var}\{\mathbf{y}'\}}{\text{var}\{\mathbf{y}\}} = \frac{\text{var}\{\mathbf{X}\beta\}}{\text{var}\{\mathbf{y}\}} \qquad (15.15)$$

R^2 represents the portion of variance in the measured signal $\mathbf{y}$ (as measured about its mean) that is accounted for by variations in the estimated signal $\mathbf{y}'$. In voxels with $R^2 \approx 1$, the variance of the observed signal is well explained by the estimated model. Conversely, in voxels with $R^2 \approx 0$, most of the observed variance remains unexplained after fitting the model.

More generally, within the GLM framework it is possible to create maps reflecting the locations of significant effects of interest after the other modeled effects have been taken into account. Let us consider an experiment whose design matrix can be partitioned into two subsets

$$\mathbf{X} = [\mathbf{X}_a \mid \mathbf{X}_b] \tag{15.16}$$

with corresponding partition of the regression values

$$\boldsymbol{\beta} = [\boldsymbol{\beta}_a^T \mid \boldsymbol{\beta}_b^T]. \tag{15.17}$$

where $\mathbf{X}_a(\boldsymbol{\beta}_a)$ indicates the predictors (regression values) corresponding to the confounds (e.g., mean level, low-frequency fluctuations) and $\mathbf{X}_b(\boldsymbol{\beta}_b)$ indicates the effects of interest. Detecting the locations of the brain in which there is a significant effect of interest corresponds to testing voxel-by-voxel the hypothesis $\{H_b : \boldsymbol{\beta}_b = \mathbf{0}\}$ and selecting those voxels in which this hypothesis can be safely rejected. The extra sum of squares principle provides a means to perform these tests [27]. Under H_b, the model in Equation 15.7 reduces to:

$$\mathbf{y} = \mathbf{X}_a\, \boldsymbol{\beta}_a + \mathbf{e}. \tag{15.18}$$

The extra sum of squares due to $\boldsymbol{\beta}_b$ after $\boldsymbol{\beta}_a$ is defined as:

$$X_r(\boldsymbol{\beta}_a \mid \boldsymbol{\beta}_b) = X_r(\boldsymbol{\beta}_a) - X_r(\boldsymbol{\beta}). \tag{15.19}$$

where $X_r(\boldsymbol{\beta})$ and $X_r(\boldsymbol{\beta}_a)$ denote respectively the residual sum of squares for the full model and for the reduced model. Under H_b, $X_r(\boldsymbol{\beta}_a \mid \boldsymbol{\beta}_b) \sim \sigma^2 \chi^2$ independently of $X_r(\boldsymbol{\beta})$, with $L_b = \mathrm{rank}(\mathbf{X}) - \mathrm{rank}(\mathbf{X}_a)$ degrees of freedom. Therefore, under H_b, the ratio

$$F = \frac{(X_r(\beta_a) - X_r(\beta))/L_b}{X_r(\beta)/(T - L_b)} \tag{15.20}$$

has a central F distribution with $n_1 = L_b$ and $n_2 = T - L_b$ degrees of freedom [27]. The desired map can thus be computed using the following steps:

1. Calculate the statistic F of Equation 15.20 for each voxel.
2. For a fixed value of false alarm p determined by

$$p = \int_{F_0}^{\infty} f_f(u)\, du. \tag{15.21}$$

compare F with F_0 ($f_F(u)$ is an F distribution with n_1 and n_2 degrees of freedom.

3. Color-code the voxels where $F > F_0$.

It is important to remark that if there is significant temporal autocorrelation in the data, the degrees of freedom n_2 are to be "corrected" to take appropriately into account the serial dependency between the adjacent samples of each voxel's time course. Without this correction, the resulting value may be overestimated (see Reference 18 and Reference 25). An alternative solution is to perform the statistical tests on the time courses after they have been "prewhitened" (i.e., after the autocorrelation has been removed; see Reference 28 and Reference 29] for possible prewhitening methods).

Note that if $\mathbf{X}_b = \mathbf{X}$, there is a one-to-one correspondence between F in Equation 15.20 and R^2 in Equation 15.15 and it is given by

$$F = \frac{R^2}{1-R^2} \cdot \frac{n_2}{n_1} \tag{15.22}$$

15.3.1.2 Relative Contribution (RC) Maps

Statistical maps generated as described in the preceding text only provide information about the effects of all the predictors in $\mathbf{X}_b$. The contribution of specific predictors (or subsets of predictors) within $\mathbf{X}_b$ to explaining the variance of the observed time course can be highlighted in RC maps (see e.g., Reference 30 and Reference 31). These maps are obtained by calculating at each voxel the index given by

$$RC = (\mathbf{b}_{si} - \mathbf{b}_{s2})/(\mathbf{b}_{si} + \mathbf{b}_{s2}) \tag{15.23}$$

In Equation 15.23 $\mathbf{b}_{si}$ is the sum of the estimates of the standardized regression coefficients of all conditions in subset s_i (we assume, for simplicity, that $\mathbf{X}_b$ only includes two subsets of conditions). For voxels with F greater than a given threshold, RC is color-coded using a double-color scale (e.g., red–green color scale). An RC value of 1 (green) indicates that a voxel time course is solely explained with predictor set b_1, whereas an RC value of -1 (red) indicates that a voxel time course is explained solely with predictor set b_2. An RC value of 0 indicates that a voxel time course is explained with equal contribution of both predictor sets.

15.3.1.3 Specific Effects, Contrasts (t Maps)

In fMRI studies, researchers are often interested in testing the effects of a specific condition or in comparing statistically the effects of two or more experimental conditions. Within the GLM framework, this comparison is done by using contrast vectors $c = [c_1\ c_2 \ldots c_L]$ and computing contrast (t) maps. It can be shown that for any contrast c the ratio

$$t = \frac{c^T \beta'}{\sqrt{\hat{\sigma}^2 c^T (X^T X)^{-1} c}} \qquad (15.24)$$

follows a Student's t distribution with $T - L$ degrees of freedom [24].

Thus, a map that highlights brain locations in which there is a significantly greater activity in the first condition as compared to the second can be easily obtained in the following way:

1. Define a contrast vector $c = [1\ {-1}\ 0\ 0 \ldots 0]$.
2. Calculate the statistic t of Equation 15.24 for each voxel.
3. Color-code the voxels with a value $t > t_0$, with t_0 being the value that corresponds to a prefixed significance value p_0.

15.4 SELECTION OF SIGNIFICANCE THRESHOLDS IN FMRI STATISTICAL MAPS

A common problem of the methods based on the voxel-by-voxel statistical analysis is in the selection of a correct threshold value for segment activated and nonactivated voxels. This value may be chosen with reference to the "uncorrected" single pixel significance derived by comparing the statistical values obtained at each voxel with the values of an assumed [18] or empirically derived [32] null-hypothesis distribution. The optimum selection of this threshold, however, is not straightforward. High (conservative) threshold values lower the probability of incorrect detection of activation (false positives), but they also increase the probability of failing to detect "true-but-noisy" activated regions (false negatives). Low threshold values, on the other hand, result in maps in which regions unrelated to the task may appear as activated. Further, the uncorrected value of significance is the result of statistical tests performed separately for a large number of time courses. Thus, this value has to be "corrected" for multiple comparisons. Given the very large amount of sampled voxels, the simple statistical approach of preventing false positives by adjusting the significance level p using Bonferroni correction ($p' = p/N$) is too conservative and leads to a substantial loss of statistical power [33].

One of the causes for such loss of power is that in the application of the Bonferroni correction, it is assumed that each voxel is an independent comparison, whereas voxels may be spatially correlated and, thus, the number of effectively independent comparisons is less than the total number of voxels in the data set. Another cause is the failure of this approach to exploit the spatial extent of activations. "Real" cortical activation is expected to cover clusters of adjacent pixels; conversely, there is a low probability that a given number of pixels exceeding a threshold will be contiguous simply by chance. This distinction between signal (which tends to cluster) and noise (which does not tend to cluster) may be exploited to reduce the false positive probability without decreasing the statistical power [33–37]. A natural way to do this is to use the detection criteria that rely on the use of a cluster-size threshold in conjunction with the intensity threshold, i.e., voxels are considered as activated by the stimulus if the uncorrected false positive rate is below the fixed threshold and the same condition is verified for a minimum number of adjacent voxels.

When this is the case, to quantify the statistical significance of an activated region, it becomes necessary to determine the probability with which clusters of various sizes occur by chance and to determine the likelihood of detecting such clusters when activation is really present [33]. Several authors approached this problem assuming that the fMR images can be approximated by a continuous random field, where the voxel values are considered to be the realizations of a random field sampled on an equally spaced grid. With this approach, the significance of activated clusters is determined on the basis of explicit expressions for the probability of excursion sets of random fields derived from the theory of Gaussian fields [34–36]. Although elegant and quantitative, this approach has the drawback of requiring fMR images to be smoothed with spatial Gaussian filters with broad full widths at half maximum (FWHM, FWHM/pixel size > 2). As discussed earlier, this inevitably reduces the effective spatial resolution of images and is especially undesirable for single-subject studies. Alternative approaches are based on the generation of null-hypothesis probability distributions through Monte Carlo simulations [33] or randomization tests [37]. Forman et al. [33] provided probability distributions of cluster sizes as a function of the uncorrected false positive rate and for different values of the Gaussian spatial smoothing filter. These distributions, obtained with Monte Carlo simulations and verified with fMRI studies, provide approximated values for false positive rates that are associated with combinations of the uncorrected false positive rates and a minimum cluster-size threshold. These methods require fewer assumptions but are more time consuming.

A solution that mitigates the problem of multiple comparisons in fMRI is to limit the number of statistical tests to those voxel time courses that are indexed by a reconstructed cortical sheet (gray matter voxels) and to use surface-based 2-D cluster-size thresholds [38].

Recently, a new approach has been proposed by Genovese et al. [39] to deal with the problem of the multiple comparisons in fMRI. The approach is based on the control of the false discovery rate (FDR), i.e., of the proportion of false positives (incorrect rejections of the null hypothesis) among those tests for which the null hypothesis is rejected. One advantage of this approach is that it offers an objective way to automatically select "adaptive" thresholds across subjects (for details see Reference 39).

15.5 DATA-DRIVEN ANALYSIS OF FUNCTIONAL TIME SERIES

Data analysis with the aforementioned voxel-based methods is limited to the detection of cortical activity with a strictly task-related temporal behavior. With other processing strategies, such as principal component analysis (PCA) [40] and independent component analysis (ICA) [41, 42], information contained in time courses is extracted without strong *a priori* hypotheses about the time profile or the spatial extension of the cortical areas (see Chapter 18). These methods are based on the decomposition of the intrinsic spatiotemporal structure of the fMRI time series in orthogonal spatial patterns or eigenimages (PCA), and in independent component maps (ICA) with different time courses. The drawback of PCA and ICA consists in the difficulty of giving a physiological interpretation to the great number of different components [41, 42].

15.6 COMBINING BRAIN FUNCTION AND ANATOMY

15.6.1 COREGISTRATION OF FUNCTIONAL AND ANATOMICAL DATA SETS

As mentioned in the introduction, a typical fMRI measurement session includes, before or after the collection of the functional time series, the collection of anatomical images covering a region of interest or the whole brain. These images are used for the spatial normalization of the data in a standard space and also for a better visualization of the statistical maps. Anatomical images may be collected using 2-D T1-weighted sequences (2-D anatomical reference) with the same spatial parameters (position, field of view, thickness) as the functional volumes. In this case, assuming that there is no significant subject motion in the interval between the functional and anatomical acquisition, coregistration between functional and anatomical volumes can be obtained simply by superimposing, for each volume of the time series, the stack of functional slices on the stack of coplanar anatomical slices. Functional maps can thus be overlaid either onto the anatomical stack of slices or functional stack of slices.

However, if more sophisticated visualizations of functional maps are to be obtained (e.g., using folded or morphed reconstructions of the subject's cortex; see Subsection 15.6.3), anatomical reference images of the subject's whole brain are typically collected using 3-D (e.g., with three encoding gradients) T1-weighted sequences. These sequences provide very good spatial resolution (1 mm × 1 mm × 1 mm) and high contrast between gray and white matter in a relatively short acquisition time (8–20 min for whole brain imaging). When these 3-D anatomical images are collected, coregistration of functional and anatomical data sets is obtained using information on the MR-scanner slice position parameters of the T2*-weighted measurements (number of slices, slice thickness, interslice gap, in-plane resolution, field of view, angles and offsets in the readout, phase and z directions) and on analogous parameters of the T1-weighted 3-D measurement. Again, the underlying assumption is that subjects do not move between the functional and anatomical scans. However, because small head movements cannot be excluded, the accuracy of the coregistration obtained in this way by means of various landmark-based, edge-matching or other automatic registration algorithms. These algorithms produce a spatial transformation that ensures an accurate coregistration between anatomical and functional data. By applying this spatial transformation to each volume of the functional time series, it is possible to generate a new 4-D representation of the functional data in which the functional time courses are directly linked to the anatomical reference volume. This 4-D representation also makes possible (after spatial normalization [see Subsection 15.6.2]) the concatenation of the time series of multiple subjects for a group analysis, e.g., within the GLM framework.

15.6.2 Spatial Normalization

The comparison of spatial locations of functional activation among subjects is commonly made in both fMRI and PET studies by normalizing the individual brains in a standard space. The most widely used standard anatomical reference is the stereotaxic space, which was defined in the Talairach and Tournoux atlas [17]. In our approach, the transformation in Talairach space is performed semi-automatically using the 3-D anatomical volume of each subject and following the procedure defined in the atlas:

1. In the first step, the 3-D anatomical data set of each subject is rotated in order to align it with the stereotaxic axes. For this step, the location of the anterior commissure (AC) and the posterior commissure (PC) and the two rotation parameters for midsagittal alignment have to be specified manually in the 3-D data set.
2. In the second step, the extreme points of the cerebrum (anterior, posterior, superior, inferior, left and right) are specified. Planes encapsulating these points together with the vertical frontal plane (VAC, the plane established along the AC and bisecting the AC–PC line orthogonally) and the vertical posterior plane (VPC, the plane established along the PC and bisecting the AC–PC line orthogonally) divide the brain into 12 subvolumes.
3. The 3-D data sets are scaled into the dimensions defined in the Talairach and Tournoux [17] atlas by applying separately to each of the 12 subvolumes a piecewise affine and continuous transformation.

Note that normalization in the volumetric Talairach space only ensures a coarse spatial correspondence between brains of different subjects. More advanced algorithms, based on the realignment of the subjects' cortices (see the following section), have been recently developed to address the problem of defining a spatial correspondence between different brains, which is the basis of all intersubject comparisons and group analyses in functional neuroimaging.

15.7 SEGMENTATION, SURFACE RECONSTRUCTION, AND MORPHING

Projection of the functional data onto a standardized or an individual anatomical 3-D volume not only has certain advantages (ease of use and widespread acceptance) but also several drawbacks. For example, on a 3-D volume, the distance between two activated regions on the cortical surface is in most of the cases substantially underestimated compared to the true distance along the cortical sheet. This is due to the intrinsic topology of the cerebral cortex, a bidimensional sheet with a highly folded and curved geometry. Furthermore, some of the features and organizational principles that distinguish cortical areas (e.g., retinotopy, tonotopy, somatotopy) are better analyzed with a 2-D surface-based representation. Finally, individual cortical surfaces can be used as anatomical constraints for hypothesis- [38, 43] and data-driven [44] statistical analysis of the functional time series. Therefore, representation of functional maps on the folded and morphed surface reconstruction of individual brains often reveals topographical information that may remain hidden in the conventional slices or 3-D volumetric visualization. In the following text, the steps required to obtain these types of representation are briefly described.

The first step in obtaining a reconstruction of a cortical surface is to derive, for each hemisphere, the border between white and gray matter from the set of slices of a 3-D anatomical volume (Figure 15.6). In general, this can be done by using one of the many existing segmentation algorithms that allow separating gray matter, white matter, and the other structures of the brain [45]. In our approach, the segmentation algorithm also ensures that the following tessellation will lead to a topologically correct representation of the cortex (i.e., without "bridges" or "holes"; see Reference 46 for details). After segmentation, the high-resolution, voxel-based partition of each hemisphere is transformed by triangularization of the outside voxel faces to a vertex-based surface $S_0 = (V_0, K)$, where V_0 is the $N_{V_0} \times 3$ matrix of vertex coordinates, K is an $N_K \times 3$ matrix of vertex indices, and N_{V_0} and N_K are the number of vertices and faces. Because the surface S_0 reflects the coarse voxel-based discretized approximation to the (real) underlying surface, which is assumed to be spatially smooth (i.e., local curvature values are bound by some maximum value), the coordinates described by V_0 are spatially smoothed with respect to the local vertex neighborhood (100–200 iterations). V_0 is thus transformed to a smooth representation of the white matter surface $S_W = (V_W, K)$. In the next step, a surface lying within the gray matter sheet is identified by translating the vertices in V_W using an interactive morphing algorithm. This gives a representation of the underlying gray matter surface $S_G = (V_G, K)$ that may be used as the reference mesh for the visualization of functional data (Figure 15.6).

The iterative morphing algorithm may be further used to compute an "inflated" surface $S_I = (V_I, K)$ of each hemisphere (Figure 15.7). The inflated representation of the cortical surface aims to provide a representation of the cortical hemisphere that retains much of the shape and metric properties of the original surface, but allows the visualization of functional activity occurring

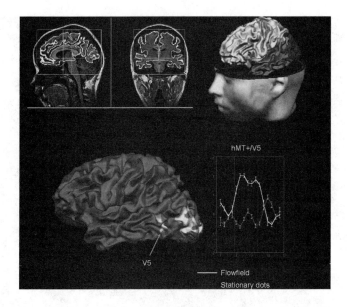

FIGURE 15.6 Advanced visualization of the functional data. Top row: (left) Sagittal and coronal cuts showing sections of the border between gray matter and white matter as defined by the segmentation; (right) reconstruction of the subject's head and cortex obtained from the 3-D anatomical images. Lower row: functional activation in hMT+/V5 during a visual motion experiment represented on the mesh reconstruction of the subject's cortex.

within sulci [47, 48]. Cortex inflation uses two forces, the corrective smoothing and the distortion reduction force, which are iteratively computed and applied to each vertex of the surface mesh. The "smoothing force" acts on the surface so that points that lie in concave regions move outward over time, whereas points in convex regions move inward. The "distortion reduction force" constrains the evolving surface to retain as many of the original metric properties as possible. It is possible to control the relative strength of the two forces by means of two coefficients, λ_s for the smoothing force and λ_d for the distortion reduction force, respectively. During the first $200 - 500$ iterations, λ_s takes on values much larger than λ_d (0.4–0.8 vs 0.01–0.05), and gradually decreases over time as the surface successfully inflates. Minimum geometric distortions are obtained by "linking" the morphing surface to a folded reference representation (S_W or S_G), and using a distortion reduction force that keeps the area of each triangle of the inflated hemisphere as close as possible to the value of the reference mesh. In this way, the inflated hemisphere also possesses a link to the functional data, as these are coregistered with the 3-D data set and the folded surface. A functional map may be, therefore, shown at the correct position of the inflated representation. The display of functional maps on an inflated hemisphere allows the topographic

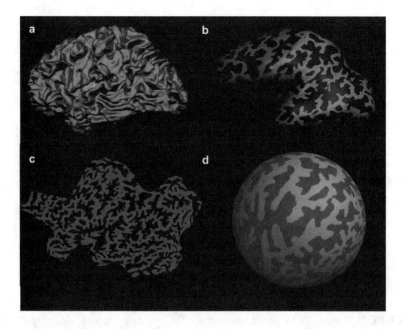

FIGURE 15.7 Morphing of the cortical surface. (a) Folded, (b) inflated, (c) flattened, and (d) sphered representation of the cortex.

representation of the 3-D pattern of cortical activation without loss of the lobular structure of the telencephalon (see, e.g., Reference 30 and Reference 32). Inflated representations can be further processed to obtain flattened representations of the cortex, which are used in the study of retinotopy [49–52] and tonotopy [53], or to obtain "spherical" representations of the cortex, which are used in advanced approaches of cortex-based normalization and realignment of the brains of different subjects [54–56] (Figure 15.7).

REFERENCES

1. Ogawa, S., Tank, D., Menon, R., Ellermann, J.M., Kim, S.G., Merkle, H., and Ugurbil, K. (1992). Intrinsic signal changes accompanying sensory stimulation: functional brain mapping using MRI. *Proc. Natl. Acad. Sci. USA* 89: 5951–5955.
2. Kwong, K.K., Belliveau, J.W., Chesler, D.A., Goldberg, I.E., Weisskoff, R.M., Poncelet, B.P., Kennedy, D.N., Hoppel, B.E., Cohen, M.S., Turner, R., Cheng, H.M., Brady T.J., and Rosen, B.R. (1992). Dynamic magnetic resonance imaging of human brain activity during primary sensory stimulation. *Proc. Natl. Acad. Sci. USA* 89: 5675–5679.
3. Bandettini, P.A., Wong, E.C., Hinks, R.S., Tikofsky, R.S., and Hyde, J.S. (1992). Time course EPI of human brain function during task activation. *Magn. Reson. Med.* 25: 390–398.

4. Di Salle, F., Formisano, E., Linden, D.E., Goebel, R., Bonavita, S., Pepino, A., Smaltino, F., and Tedeschi, G. (1999). Exploring brain function with magnetic resonance imaging. *Eur. J. Radiol.* 30: 84–94.

5. Smith, S.M. (2001). Preparing fMRI data for statistical analysis. in Jezzard, P., Matthews, P.M., and Smith, S.M. (Eds.). *Functional MRI — An Introduction to Methods.* Oxford, Great Britain: Oxford University Press, pp. 230–241.

6. Formisano, E. and Goebel, R. (2003). Tracking cognitive processes with functional MRI mental chronometry. *Curr. Opin. Neurobiol.* 13: 174–81.

7. Goebel, R., Roebroeck, A., Kim, D.-S., and Formisano, E. (2003). Investigating directed cortical interactions in time-resolved fMRI data using vector autoregressive modeling and Granger causality mapping. *Magn. Reson. Imaging.* 21: 1251–1261.

8. Woods, R., Mazziotta, J., and Cherry, S. (1992). Automated algorithm for aligning and reslicing PET images. *J. Comput. Assist. Tomogr.* 16: 620–633.

9. Hajnal, J.V., Saeed, N., Soar, E.J., Oatridge, A., Young, I.R., and Bydder, G. (1995). A registration and interpolation procedure for subvoxel matching of serially acquired MR images. *J. Comput. Assist. Tomogr.* 19: 289–296.

10. Friston, K.J., Ashburner, J., Frith, J.D., Poline, J.B., Heather, J.D., and Frackowiak, R.S.J. (1995). Spatial registration and normalization of images. *Hum. Brain Mapp.* 3: 165–189.

11. Friston, K.J., Williams, S., Howard, R., Frackowiak, R.S.J., and Turner, R. (1996). Movement-related effects in fMRI time-series. *Magn. Reson. Med.* 35: 346–355.

12. Alexander, M.E. and Somorjai, R.L. (1996). The registration of MR images using multiscale robust methods. *Magn. Reson. Imaging.* 14: 453–468.

13. Biswal, B.B. and Hyde, J.S. (1997). Contour-based registration technique to differentiate between task-activated and head motion-induced signal variations in fMRI. *Magn. Reson. Med.* 38: 470–476.

14. Brammer, M.J. (2001). Head motion and its correction. in Jezzard, P., Matthews, P.M., Smith, S.M., Eds., *Functional MRI — An Introduction to Methods.* Oxford, Great Britain: Oxford University Press, pp. 244–250.

15. Press, W., Flannery, B., Teukolsky, S., and Vetterling, W. (1992). *Numerical Recipes in C.* Cambridge University Press.

16. Lowe, M.J. and Sorenson, J.A. (1997). Spatially filtering functional magnetic resonance imaging data. *Magn. Reson. Med.* 37: 723–729.

17. Talairach, J. and Tournoux, P. (1988). *Co-Planar Stereotaxic Atlas of the Human Brain.* Stuttgart, New York: Thieme Medical Publishers.

18. Friston, K.J., Jezzard, P., and Turner, R. (1994a). The analysis of functional MRI time-series. *Hum. Brain Mapp.* 1: 153–171.

19. Bandettini, P.A., Jesmanowicz, A., Wong, E.C., and Hyde, J.S. (1993). Processing strategies for time-course data sets in functional MRI of the human brain. *Magn. Reson. Med.* 30: 161–173.

20. Biswal, B.B., DeYoe, E.A., and Hyde, J.S. (1996). Reduction of physiological fluctuations in fMRI using digital filters. *Magn. Reson. Med.* 35: 107–113.

21. Buonocore, M.H. and Maddock, R.J. (1997). Noise suppression digital filter for fMRI based on image reference data. *Magn. Reson. Med.* 38: 456–469.

22. Baudendistel, K., Shad, L., Friedlinger, M., Wenz, F., Schroder, J., and Lorenz, W. (1995). Postprocessing of fMRI data of motor cortex stimulation measured with a standard 1.5T imager. *Magn. Reson. Imaging.* 5: 701–707.

23. Boynton, G.M., Engel, S.A., Glover, G.H., and Heeger, D.J. (1996). Linear systems analysis of functional magnetic resonance imaging in human V1. *J. Neurosci.* 16: 4207–4221.

24. Friston, K.J., Holmes, A.P., Worsley, K.J., Poline, J.P., Frith, C.D., and Frackowiak, R.S.J. (1995). Statistical parametric maps in functional imaging: a general linear approach. *Hum. Brain Mapp.* 2: 189–210.

25. Friston, K.J., Holmes, A.P., Poline, J.B., Grasby, P.J., Williams, S.C.R., Frackowiak, R.S.J., and Turner, R. (1995). Analysis of fMRI time-series revisited. *Neuroimage* 2: 45–53.

26. Zarahn, E., Aguirre, G.K., and D'Esposito, M. (1997). Empirical analyses of BOLD fMRI statistics. *Neuroimage* 5: 179–197.

27. Holmes, A.P., Poline, J.B., and Friston, K. (1997). Characterising brain images with the general linear model. in *Human Brain Function*, Frackowiack, R.S.J., Friston, K.J., Frith, C.D., Dolan, R.J., Mazziotta, J.C., Eds., San Diego: Academic Press, pp. 59–84.

28. Bullmore, E., Long, C., Suckling, J., Fadili, J., Calvert, G., Zelaya, F., Carpenter, T.A., and Brammer, M. (2001). Colored noise and computational inference in neurophysiological (fMRI) time-series analysis: resampling methods in time and wavelet domains. *Human Brain Mapping.* 12: 61–78.

29. Woolrich, M.W., Ripley, B.D., Brady, M., and Smith, S.M. (2001) Temporal autocorrelation in univariate linear modeling of fMRI data. *Neuroimage* 14: 1370–1386.

30. Trojano, L., Grossi, D., Linden, D.E., Formisano, E., Hacker, H., Zanella, F.E., Goebel, R., and Di Salle, F. (2000). Matching two imagined clocks: the functional anatomy of spatial analysis in the absence of visual stimulation. *Cereb. Cortex.* 5: 473–81.

31. Formisano, E., Linden, D.E., Di Salle, F., Trojano, L., Esposito, F., Sack, A.T., Grossi, D., Zanella, F.E., and Goebel R. (2002). Tracking the mind's image in the brain I: time-resolved fMRI during visuospatial mental imagery. *Neuron.* 35: 185–94.

32. Nichols, T.E. and Holmes, A.P. (2002). Nonparametric permutation tests for functional neuroimaging: a primer with examples. *Hum. Brain Mapp.* 15: 1–25.

33. Forman, S.D., Cohen, J.D., Fitzgerald, M., Eddy, W.F., Mintun, M.A., and Noll, D.C. (1995). Improved assessment of significant activation in fMRI: Use of a cluster-size threshold. *Magn. Reson. Med.* 33: 636–647.

34. Friston, K.J., Worsley, K.J., Frackowiak, R.S.J., Mazziotta, J.C., and Evans, A.C. (1994). Assessing the significance of focal activations using their spatial extent. *Hum. Brain Mapp.* 1: 210–220.

35. Xiong, J., Gao, J.H., Lancaster, L., and Fox, P.T. (1995). Clustered pixels analysis for MRI — Activation studies of the human brain. *Hum. Brain Mapp.* 3: 207–301.

36. Worsley, K.J., Marrett, S., Neelin, P., Vandal, A.C., Friston, K.J., and Evans, A.C. (1996). A unified statistical approach for determining significant signals in images of cerebral activation. *Hum. Brain Mapp.* 4: 74–90.

37. Bullmore, E., Brammer, M., Williams, S.C.R., Rabe-Hesketh, S., Janot, N., David, A., Mellers, J., Howard, R., and Sham, P. (1996). Statistical methods of estimation and inference for functional MR image analysis. *Magn. Reson. Med.* 35: 261–277.

38. Goebel, R. and Singer, W. (1999). Cortical surface-based statistical analysis of functional magnetic resonance imaging data. Abstract at Human Brain Mapping meeting.

39. Genovese, C.R., Lazar, N.A., and Nichols, T. (2002). Thresholding of statistical maps in functional neuroimaging using the false discovery rate. *Neuroimage* 15: 870–8.

40. Mitra, P.P., Thomson, D.J., Ogawa, S., Hu, X., and Ugurbil, K. (1997). The nature of spatiotemporal changes in cerebral hemodynamics as manifested in fMRI. *Magn. Reson. Med.* 37: 511–518.

41. McKeown, M.J., Jung, T.P., Makeig, S., Brown, G., Kindermann, S.S., Lee, T.W., and Seinowski, T.J. (1998). Spatially independent activity patterns in functional MRI data during the Stroop color-naming task. *Proc. Natl. Acad. Sci. USA* 95: 803–810.

42. McKeown, M.J., Makeig, S., Brown, G., Jung, T.P., Kindermann, S.S., Bell, A.J., and Seinowski, T.J. (1998). Analysis of fMRI data by blind separation into independent spatial components. *Hum. Brain Mapp.* 6: 160–188.

43. Kiebel, S.J., Goebel, R., and Friston, K.J. (2000). Anatomically informed basis functions. *Neuroimage* 11: 656–67.

44. Formisano, E., Esposito, F., Kriegeskorte, N., Tedeschi, G., Di Salle, F., and Goebel, R. (2002). Spatial independent component analysis of functional magnetic resonance imaging time-series: characterization of the cortical components. *Neurocomputing* 49: 241–254.

45. Pham, D.L., Xu, C., and Prince, J.L. (2000). Current methods in medical image segmentation. *Annu. Rev. Biomed. Eng.* 2: 315–337.

46. Kriegeskorte, N. and Goebel, R. (2001). An efficient algorithm for topologically correct segmentation of the cortical sheet in anatomical MR volumes. *Neuroimage* 14: 329–46.

47. Dale, A.M., Fischl, B., and Sereno, M.I. (1999). Cortical surface-based analysis, I: segmentation and surface reconstruction. *Neuroimage* 9: 179–194.

48. Fischl, B., Sereno, M., and Dale, A.M. (1999). Cortical surface-based analysis, II: inflation, flattening and a surface-based coordinate system. *Neuroimage* 9: 195–207.

49. Sereno, M.I., Dale, A.M., Reppas, J.B., Kwong, K.K., Belliveau, J.W., Brady, T.J., Rosen, B.R., and Tootell, R.B. (1995). Borders of multiple visual areas in humans revealed by functional magnetic resonance imaging. *Science* 268: 889–893.

50. Goebel, R., Khorram-Sefat, D., Muckli, L., Hacker, H., and Singer, W. (1998). The constructive nature of vision: direct evidence from functional magnetic resonance imaging studies of apparent motion and motion imagery. *Eur. J. Neurosci.* 10: 1563–1573.

51. Goebel, R., Muckli, L., Zanella, F.E., Singer, W., and Stoerig, P. (2001). Sustained extrastriate cortical activation without visual awareness revealed by fMRI studies of hemianopic patients. *Vision Res.* 41: 1459–74.

52. Linden, D.E., Kallenbach, U., Heinecke, A., Singer, W., and Goebel, R. (1999). The myth of upright vision. A psychophysical and functional imaging study of adaptation to inverting spectacles. *Perception* 28(4): 469–81.

53. Formisano, E., Kim, D.-S., Di Salle, F., van de Moortele, P.-F., Ugurbil, K., and Goebel, R. (2003). Mirror-symmetric tonotopic maps in human primary auditory cortex. *Neuron* 40: 859–869.

54. Fischl, B., Sereno, M.I., Tootell, R.B., and Dale, A.M. (1999). High-resolution intersubject averaging and a coordinate system for the cortical surface. *Hum. Brain Mapp.* 8: 272–84.

55. Goebel, R. (2003). Cortex-Based Intersubject Realignment of fMRI Data. Abstract at Human Brain Mapping Meeting.

56. Van Atteveldt, N., Formisano, E., Goebel, R., and Blomert, L. (in press). Integration of letters and speech sounds in the human brain. *Neuron.*

16 Exploratory Data Analysis Methods in Functional MRI

Nicola Vanello and Luigi Landini

CONTENTS

16.1 INTRODUCTION

The study of brain function with magnetic resonance imaging (MRI), which is sensitive to changes in blood flow and oxygenation [1,2], is a widely used technique, and its applications are growing rapidly—from the early attempts with simple block-designed paradigms to the study of more complex cognitive functions until the study of emotions and behavior [3,4]. Moreover, functional MRI (fMRI) is becoming increasingly important in clinical applications, for example, in neurology and in planning surgical intervention of the brain. The utility of an exploratory data analysis approach is important in order to improve knowledge about the brain function as more complex processes are studied and because it allows the detection and characterization of unexpected phenomena that are not modeled or cannot be modeled *a priori*. Several components may affect signal generation and the experimenter's model, such as subject movement, physiological changes such as heartbeat and respiration, and noise due to the instrumentation. All these components will bias the results of a model-driven approach that relies on as good a model as possible of the signal as good as possible [5–7]. The knowledge obtained by an explorative approach can be used in confirmatory data analysis (CDA) methods that rely on a precise model of the expected activations. In this framework, exploratory data analysis methods can be seen as hypotheses-generating tools. Moreover, in clinical applications these methods are thought to play an increasingly important role because in these kinds of applications the brain responses typically cannot be modeled in advance. Even if the BOLD signal has been demonstrated to be correlated with the underlying neural activity, several aspects remain to be understood, and exploratory analysis may play a vital role in this. The strength of these exploratory data analysis methods is that information is extracted from the data [8] using only general assumptions, and there is no need of specifying in advance the shape and the extent of a phenomenon. These can be achieved by taking advantage of the multivariate nature of the fMRI data set [9] and the fact that both physiological phenomena of interest, due to the principles of localization and integration of the neural processes [10], and artifacts, may concern measurements in different brain regions. In this chapter we will introduce some methods applied in exploratory data analysis of fMRI data, such as clustering techniques [11–26], principal-component analysis (PCA) [27–32], and independent component analysis (ICA) [33–46]. We will show that even if these methods are powerful tools, in order to improve the knowledge about the brain function the experimenter is required to make some fundamental choices during their applications that can heavily influence the final results.

16.2 MULTIVARIATE APPROACHES

fMRI data are composed by a time sequence of p images or volumes, made of n volume elements (voxels) each. The data set can be arranged in matrix form, where, for example, in a $n \times p$ matrix X whose rows are the voxels time series, the jth column, the jth image in the time sequence, is written as a vector

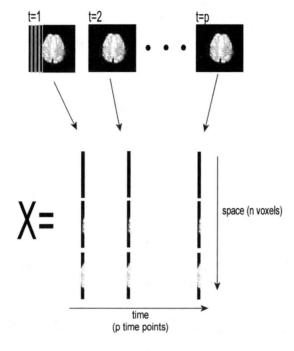

FIGURE 16.1 Schematic representation of the data matrix. Each image is vectorized and assigned to a column of the data matrix X.

(see Figure 16.1). In data clustering, each voxel time series is seen as an individual element to be classified, so these methods can be seen as *individual-directed methods*. The ith voxel's time course can be written as $x_i(t)$ with $t = 1, 2, \ldots, p$ and can be seen as a p-dimensional vector belonging to R^p, $\mathbf{x}_i = x_i(1)$, $x_i(2), \ldots, x_i(p)\}$ [5]. In Figure 16.2 the procedure is shown for a time course with three observations. The entire data set, consisting of n time series corresponding to n brain voxels, can be seen as a collection of n vectors in p-dimensional space. In other multivariate analysis methods, such as PCA or ICA, each voxel time series can be seen as a set of time-domain observations of a variable. These methods are called *variable-directed methods* because they try to find the relationships among variables. These approaches can be applied to fMRI data both in the temporal and spatial domains: in the temporal domain the variables are the voxels time series as described earlier, and in the spatial domain the variables are the time points and the observations are the voxels values at each time point. Within this framework the time series extracted from the ith brain voxel can be seen as a variable x_i with p time observations: the entire data set can be seen as n observed variables that can be written as a random n-dimensional vector $\boldsymbol{x} = \{x_1, x_2, \ldots, x_n\}$. The dual approach considers the time points as variables and the rows of X, the voxels values, as observations. Common preprocessing strategies involve the operation of slice timing correction and registration of images in order to

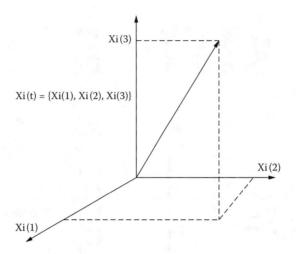

FIGURE 16.2 A 3-point-long time series, depicted as a vector in R^3. The same procedure can be extended to n-dimensional time series.

correct for movement (see Chapter 17). Spatial filtering in order to increase contrast-to-noise ratio (CNR) or temporal filtering in order to remove low-frequency fluctuations can be used: low-frequency drifts due to slow subject movements or instrumentation changes can be removed by means of linear regression detrending. In general linear model (GLM) approaches [6,7], spatial filtering is usually performed by means of an isotropic spatial filter that ignores the anatomical structure of the brain. This operation can be required in order to satisfy the hypothesis of spatial smoothness of the activated regions as in Gaussian random field theory [47] and are used in the successive inferential steps. In the case of exploratory methods this is not required and different approaches can be used in order to preserve anatomical information, such as nonlinear filtering, for example, the SUSAN filter [48], or spatiotemporal approaches, for example, those in Reference 49.

16.3   DATA CLUSTERING APPROACHES

Clustering methods try to group data set elements following a similarity criterion: the elements belonging to a cluster should be similar. The clustering procedure results in a classification of the data set in order to distinguish between different signal sources, both physiological, such as task-related activations, and artifactual, such as movement-related effects. Clustering techniques are suitable for fMRI data analysis because interesting phenomena, such as task-related activation, will involve several voxels, which can be grouped together without any *a priori* knowledge about the shape or the extension of the activations. Experimenters who want to use clustering techniques have to be aware that there are many possible choices for the algorithm and the preprocessing steps. Because the clustering techniques try to group similar objects together, the results will depend on the definition of this similarity. Moreover, some clustering approaches such as partitioning approaches require the experimenter to decide in advance the number of expected clusters. In the context of fMRI data analysis, the elements to be classified are the time series extracted from volume elements of the brain: an n-step-long time series, describing the signal changes in a voxel, can be seen as an n-dimensional vector, i.e., a point in n-dimensional space.

16.3.1 SIMILARITY

The definition of similarity is a crucial concept, and different possible choices can be made that can lead to different results. A simple definition of similarity is the Euclidean distance in an n-dimensional space between two points (x, y) given by

$$d(x,y)=\| x - y \|=\sqrt{\sum_{i=1}^{n}(x(i) - y(i))^2} \qquad (16.1)$$

where $x(i)$ is the ith element of vector x. The Mahalanobis distance is a generalization of the Euclidean distance and can be written as

$$d(x,y) = \sqrt{(x - y)^T B^{-1}(x - y)} \qquad (16.2)$$

where $(\cdot)^T$ is the transpose operator. If B is chosen to be the identity matrix, then the Euclidean distance is obtained. It is possible to choose B as a diagonal matrix with the elements in the diagonal as the variances of each coordinate; multiplying by the inverse of B is equivalent to weighting each coordinate by the inverse of its variance, resulting in a normalization process. It is then possible to choose

$$B = T^T T \qquad (16.3)$$

In such a case the Mahalanobis distance is equivalent to the Euclidean after the data have been transformed by T. This method was used in Reference 11, in which the transformation T was used to find the correlation coefficient with the stimulus reference function. This operation can be seen as focusing on the similarity with the expected activation. In this work, interesting connections with principal-component analysis preprocessing are outlined. Other metrics can be defined such as the ones proposed for fMRI data in Reference 12, in which decreasing functions of the Pearson's correlation coefficient are used. This metric can be written as

$$d(x,y) = f(cc(x,y)) \qquad (16.4)$$

where $cc(\mathbf{x}, \mathbf{y})$ is the *Pearson's correlation coefficient* given by

$$cc(\mathbf{x}, \mathbf{y}) = \frac{\sum_{i=1}^{n}(x_i - \mu_x)(y_i - \mu_y)}{S_x S_y} \qquad (16.5)$$

with μ_x and μ_y the mean values of x and y and S_x and S_y their standard deviations.

16.3.2 CLUSTERING TECHNIQUES

The clustering procedure consists in finding k clusters and assigning each element of the data set to a cluster. Each cluster may be individuated by a cluster centroid that is a time course representative of the cluster. The goal is to find homogeneous clusters, i.e., minimizing within-group variability, and at the same time separable clusters, i.e., maximizing between-group dissimilarities. If we define a within-class inertia as

$$I_W = \frac{1}{N}\sum_{k=1}^{K}\sum_{j \in C_k} d^2(x_j, c_k) \qquad (16.6)$$

and the between-class inertia as

$$I_B = \frac{1}{N}\sum_{k=1}^{K}|C_k| d^2(c_k, \bar{c}) \qquad (16.7)$$

where $\bar{c}$ is the center of gravity of the cluster centers c_k, this goal can be seen as minimizing within-class inertia while maximizing between-class inertia. Several clustering techniques have been applied in the analysis of functional data sets such as hierarchical clustering [11,13–15], k-means [11], fuzzy clustering [16–21], and self-organizing maps [22,23]. A comparison can be found in Reference 24.

These clustering techniques can be first divided into hierarchical methods and partitioning methods, because the former do not need to specify in advance the number of clusters whereas the latter need this preliminary information.

16.3.2.1 Hierarchical Methods

These can be classified into agglomerative methods, which start with N clusters of N objects and end with one cluster of N objects, and divisive methods, which use the inverse process. Both these iterative procedures result in a treelike structure called the *dendrogram*. In an agglomerative approach, all the N different elements (N voxels time series) are first classified into N different groups or clusters. The distance or dissimilarity matrix between the N elements is computed,

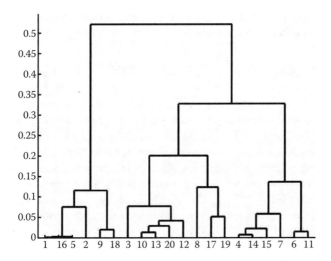

FIGURE 16.3 A dendrogram obtained starting from 20 elements or clusters.

resulting in an $N \times N$ array. After this operation, the two nearest clusters are merged together, and a new distance matrix is computed. This operation ends when there is only one cluster remaining (see Figure 16.3) Different clusters can be identified by this method depending upon the level of the dendrogram chosen: in order to decide at which level to stop and how many clusters to consider, it is possible to find a good compromise between between-class inertia minimization and the number of classes. Hierarchical clustering methods differ in the way the cluster distance is computed: in single-linkage methods, the distance between two clusters is computed as the minimum of all the distances of any element of the first cluster to any element of the second. in complete-linkage methods, the distance is measured as the maximum distance of all the distances of any element of one cluster to any element of the other. The complete-linkage method finds compact tightly bound clusters, whereas the single-linkage method finds clusters that are elongated and suffers a chaining effect. Dimitriadou et al. [24] found that the complete-linkage method outperformed the single-linkage method. The reason is the large number of clusters that do not show any activation and cause a bias in the clusters. In Reference 15 the problem of spatial separation of overlapping clusters in single-linkage methods was addressed, and a sharpening procedure of the dendrogram was proposed. Another criterion is the Ward method [50], which consists in merging every possible cluster pair and choosing the one that minimizes information loss. In order to estimate this quantity, the error sum of squares, defined as

$$ESS(C_i) = \sum_{x \in C_i} (x - m_i)$$ (16.8)

where m_i is the mean of cluster C_i and x are the data points, is used. The distance between two clusters is then given by

$$d_{Ward}(C_1, C_2) = ESS(C_1 \cup C_2) - ESS(C_1) - ESS(C_2) \qquad (16.9)$$

The Ward method merges two groups such that their heterogeneity does not change too much. Several applications of hierarchical clustering to fMRI data analysis can be found [11,13–15]. As an interesting application in Reference 13, a hierarchical clustering approach was proposed on resting-state data, by means of a single-link method on time series. As a distance measure between two voxels, the correlation coefficient between the corresponding time series was used, using only frequencies below 0.1 Hz. This frequency range was found to contribute to resting-state connectivity [51]; this method requires a high sampling rate in order to isolate the signal contributions related to heart and respiratory activities.

16.3.2.2 Hard Partitioning Methods

Partitioning methods result in a single partition instead of a treelike structure: the main difference is that the number of clusters must be specified in advance, and each element of the data set will be evaluated individually and assigned to a cluster in order to minimize a cost criterion that can be evaluated locally or globally. This is accomplished through an iterative process; starting from an initial guess for cluster centers, the clusters can be modified until a stable solution is found. The iterative process can be started from a random cluster initialization and may depend on this first step. Several solutions with different random initializations can be performed, or a first hierarchical clustering step can be applied. Partitioning methods are more efficient than hierarchical ones, which are computationally demanding for large data sets. One of the most used partitioning algorithms is the k-means method [52], which iteratively minimizes the within-class inertia. The algorithm starts with an initial guess for the k centers chosen randomly from the data set. In the second step, each element in the data set is assigned to the nearest center using the definition of a distance metric. The centers are then recomputed as the average of the elements in the cluster. The algorithm repeats these last two steps until the partition does not significantly change. In Figure 16.4 the results of the k-means algorithm for two-dimensional objects grouped in three clusters are shown. Although this method is fast and results in homogeneous clusters, its main drawback is that the number of clusters must be specified in advance. The algorithm can be modified to overcome this problem: this method was applied in the ISODATA algorithm [53] and consists of splitting a cluster when its variance is above a threshold and merging two clusters when the distance between their centers' is smaller than a threshold. In the context of fMRI data analysis, a comparison between a hierarchical algorithm and the k-means can be found in Reference 11. The two approaches are seen as complementary: the k-means is powerful and results in homogeneous clusters but requires that the number of clusters be chosen *a priori*, whereas the hierarchical approach allows

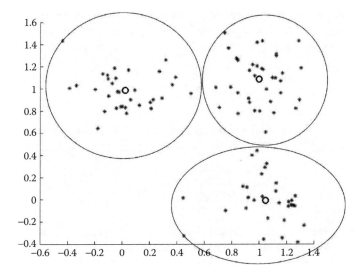

FIGURE 16.4 Two-dimensional objects grouped in three clusters by the k-means algorithm. The cluster centers are the bold circles.

the number of clusters to be chosen based on the evolution of within-class inertia. In Reference 14 the two approaches are combined in a hierarchical k-means approach: in this chapter, the combined approach is used to find the number of clusters and the initial guess of their centers is used in a final k-means algorithm. The procedure consists of iteratively splitting clusters that are found to have some structure left into two groups by means of a k-means algorithm. The crucial point here concerns the identification of the clusters to split, because it is not straightforward to define a threshold for a cluster to have some structure. In this paper, it was proposed to use visual inspection, eigenvalues decomposition of the data set belonging to each cluster, within-cluster sum of squares, and statistical tests such as nonparametric Kolmogorov–Smirnov. When the splitting part is finished, a merging step is performed by comparing the distances between all the centers: if two clusters show a smaller distance with respect to the other pairs, then they are merged together. The final centers, as stated before, are the initial guesses of the final k-means algorithm. The reason is that because of low CNR or SNR, some objects could have been wrongly assigned to a cluster.

16.3.2.3 Fuzzy Clustering

Besides hard partitioning algorithms, one of the most used approaches for clustering fMRI data is the fuzzy clustering approach [54], which introduces the concept of *fuzziness* [55]: each member of the data set may belong to several clusters, and the degree of belonging is described by a membership index. This approach differs from hard clustering, which allows each element to belong to one and only one group. If we define a membership value u_{ji} that describes the membership degree of the element i to the cluster j, hard clustering corresponds to setting u_{ji} equal to 0 or 1, whereas in a fuzzy context it can vary from 0 to 1 depending on the growing membership degree. For each data element, we have

$$\sum_{i=1}^{K} u_{ji} = 1 \qquad (16.10)$$

where K is the cluster number. The partition can be found using the fuzzy c-means algorithm [56,57], which consists of minimizing the objective function

$$J_m = \sum_{i=1}^{N} \sum_{j=1}^{K} (u_{ji})^m d_{ji}^2 \qquad (16.11)$$

where d_{ji} is the distance of the ith element to the jth center and m is the membership degree, which is related to the fuzziness of the algorithm. This function shows a minimum if

$$u_{jk} = \frac{1}{\displaystyle\sum_{j=1}^{K} \left(\frac{d_{ik}}{d_{jk}} \right)^{2/(m-1)}} \qquad (16.12)$$

and

$$v_j = \frac{\displaystyle\sum_{i=1}^{N} (u_{ji})^m x_j}{\displaystyle\sum_{i=1}^{N} (u_{ji})^m} \qquad (16.13)$$

where v_j is the center of the jth cluster. The algorithm starts from an initial set of membership values for the data set elements, named *fuzzy partition*, collected in a matrix form and given by

$$U^{(0)} = \left(1 - \frac{\sqrt{2}}{2} \right) U + \frac{\sqrt{2}}{2} V \qquad (16.14)$$

with $U = [1/K]$ and V a matrix of randomly chosen centers. In the second step, the new fuzzy centers are computed using Equation 16.13, and then the new membership values are computed using Equation 16.12. These two steps are repeated until

$$\left\| U^{(l+1)} - U^{(l)} \right\| \le \varepsilon \qquad (16.15)$$

When the parameter m equals 1, we get hard partitioning, and the algorithm is similar to k-means. When m tends to plus infinity, the membership values tend to $1/K$. The values usually used in fMRI data analysis range from 1.3, as in Reference 25, to 2, as in Reference 21. However, there are no precise indications in the literature for this choice. Another issue is the fact that in fMRI the activated voxels are a small fraction of the entire population [21], so the clusters found can be biased by voxels containing artifacts such as movement-related artifacts or large-vein contributions. This can be resolved by focusing on a region of interest (ROI) determined with *a priori* anatomical or functional knowledge or by selecting the voxels of interest using a first processing stage with statistical tools. In the study in Reference 19, the fuzzy cluster analysis (FCA) was repeated in a region of interest, found in a first step, whose centroid showed a correlation with the task. This second clustering step identified two clusters in the one detected previously: one contained few voxels with large signal changes and was thought to have originated from medium-to-large veins running perpendicular to the acquisition plane and the other contained more voxels with less signal change and was thought to be related to cortical activation. The problem of orientation of large vessels with respect to the acquisition plane was pointed out. Another issue is the noise present in the time series that can affect, as stated before, clustering techniques in general. Both Toft [26] and Goutte [11], by means of hierarchical and k-means clustering, tried to overcome this problem by the introduction of a distance based on a function of the correlation coefficient with a stimulus function, rather than the raw time series. This approach limited the use of the explorative method to a more selective exploration and was used also in the study in Reference 12, in which the significance of the membership values and the problem of the threshold were addressed. These values, in fact, do not represent absolute statistical information about the probability of a time series being correlated to a cluster, but they represent a relative measure taking into account also the relationship with the other clusters. In the same work, results were shown with the membership values thresholded at 0.8. The same value was indicated in Reference 19 as a result of a comparison with a correlation analysis. Fuzzy clustering for fMRI data has been extensively applied [12,16–21]. Preprocessing steps can heavily affect the results. Some of the necessary steps have been already mentioned in Section 16.2 and are common preprocessing techniques. The removal of the baseline level from each time series is usually applied in order not to allow the clustering algorithm to classify the voxels based on the underlying anatomical structure. Other preprocessing strategies involve subtracting the mean value and then dividing by it in order to achieve a percentage signal change, or the normalization obtained by subtracting the mean and dividing by the standard deviation. A comparison of different preprocessing strategies can be found in Reference 12.

16.3.2.4 Artificial Neural Networks

Artificial neural networks have been extensively used for clustering and classification purposes. In particular, competitive neural networks are useful because similar patterns in the input space are grouped together and associated with a single neuron. We will briefly review the use of self-organizing maps (SOM) [58]. These are single-layered networks constituted by a two-dimensional lattice of nodes: each node is a pattern in the input space. At the beginning of the training process of the network, the nodes are initialized randomly, and then the centers are updated through an iterative process consisting of randomly selecting an element of the input space and finding the winner node, i.e., the node whose distance from the input data element is smaller, and updating all the centers of the lattice. The rule is that the centers that are modified heavily are those of the winner node and its neighbors. The update function for all the cluster centers c_i is

$$c_i(t+1) = c_i(t) + h(t, c_i)(x - c_i) \qquad (16.16)$$

where x is the input data element, and $h(t, c_i)$ is the neighborhood function that takes into account the centers' distance in the lattice and allows the selection and the modification of the neighboring centers. The neighborhood function has to decrease as much as the learning process evolves in order to achieve convergence. The SOM algorithm results in a topology-preserving algorithm, facilitating the merging operation of neighboring nodes to create super clusters from smaller clusters whose distances are small [22]. This operation allows the separation of both small and large clusters from the same data set, overcoming some limitations of other algorithms such as for example the k-means, which result in homogeneous clusters. Moreover, this reduces the importance of the initial choice regarding the number of expected clusters. In Reference 23 the algorithm was modified in order to take into account spatial proximity among the voxels in the original image.

16.4 PCA

This is a multivariate technique that decomposes the data into a set of linearly independent components ordered according to explained data variance. This method is related to the Karhunen–Loève transform or the Hotelling transform [59], and was first proposed by Pearson [60]. PCA has found applications in data compression, image and statistical data analysis, and has been used in fMRI data analysis in order to explore and decompose the correlations in spatial or temporal domains present in the data set [27]. This analysis results in eigenimages and associated time vectors. The images can be seen as maps of functional connectivity [28,61,62] because they share the same temporal pattern. PCA in functional neuroimaging studies was first applied to PET data in Reference 62: the temporal resolution of PET allowed the acquisition of 12 images, alternating between a letter repetition task and a word generation task. The analysis of regional cerebral blood flow (rCBF) by means of PCA resulted in an eigenimage, or spatial distribution of voxel values, with positive loadings in regions involved in verbal fluency. The associated time pattern can be seen as a modulating function of the loadings and showed high levels during the verbal fluency task and low levels during the letter repetition task, confirming the hypothesis of correlations among the regions individuated by the eigenimage and the verbal fluency task. Another eigenimage, with positive loadings in anterior cingulate, showed a monotonic decreasing temporal pattern and was thought to be related to some attentional or perceptual change. It is important to stress that this component could not be easily detected by a model-driven approach, which could have been used for detecting task-related changes, because it may not be expected in advance.

16.4.1 SPATIAL AND TEMPORAL PCA

As stated earlier, the duality inherent in fMRI data sets can be found in the multivariate approaches: fMRI data are arranged in a matrix where each image or volume of the sequence is transformed in a row or in a column, depending on whether they are considered as a time sequence of images or volumes or as a spatial distribution of time courses. In multivariate methods, the data matrix X can be seen as a collection of measurements of some observed variables x_i. These can be the time points or the voxels values. PCA aims to find a smaller set of variables, as a linear combination of the original ones, that accounts for the variance in the data set. The first principal component (PC) is the one that has the greatest variance of all the possible linear combinations of the original variables, with the constraint that the combination weights form a unit vector. The second PC has the largest variance of all the possible linear combinations under the constraint of being orthogonal to the previous one. This procedure can be repeated up to the maximum number of variables r $\min(n, p)$, where r is the data matrix rank. In data reduction methods, a fewer number of variables that explain the largest portion of variance in the data set can be chosen. This reduction operation can be performed also in the field of fMRI data analysis even if the local variability or low-intensity signal changes may be not represented [29]. In fact, low-percentage signal variations may be discarded in the reduction process because their influence on the overall signal variance is small. In the same way, the influence of small activation regions may result in percentages that are small with respect to other sources of variance, both physiological or artifactual, distributed across brain voxels. These issues will be addressed further later. If we consider the voxels as variables, the first principal component can be written as

$$y_1 = \sum_{k=1}^{r} u_{i1} x_i \qquad (16.17)$$

where $\mathbf{u}_1 = \{u_{11}, u_{21}, \ldots, u_{n1}\}$ is a unity vector of the weights, and x_i's are the original variables. The weights, or loadings, of vector $\boldsymbol{u}_1$ can be found as the eigenvector of the covariance matrix of the data

$$\mathbf{C_x} = E\{(\mathbf{x} - \mu_x)(\mathbf{x} - \mu_x)^T\} \qquad (16.18)$$

associated with the largest eigenvalue s_1, where $E\{\cdot\}$ is the expectation operator, $\boldsymbol{x} = \{x_1, x_2, \ldots, x_n\}$ is the variables' random vector, and $\boldsymbol{\mu}_x$ is the mean vector of $\boldsymbol{x}$. Instead of using the expectation, we can estimate the covariance matrix as a sample covariance matrix

$$\hat{\mathbf{C}}_x = \frac{1}{p}\mathbf{XX}^T \qquad (16.19)$$

where the sample mean has been removed from each variable, and p is the number of observations. The remaining PCs are found by means of the eigenvectors of the covariance matrix ordered such that their associated eigenvalues are in decreasing order. The eigenvalues equal the variance explained by the corresponding PC and the sum of the eigenvalues equals the variance in the original observations. So, we can write $\Sigma_{i=1}^n \sigma_i^2 = \Sigma_{l=1}^r s_l$ where σ_i^2 is the variance of the ith variable. In Figure 16.5, a bidimensional distribution of data points is shown, along with the directions of the eigenvectors of the data matrix. In the same figure, the eigenvalues associated with each PC are shown. If we consider the voxels as random variables, then the matrix $\hat{C}_x$ accounts for the spatial covariance structure in the data set, i.e., the covariances among the voxels. The n-dimensional eigenvectors $\boldsymbol{u}_i$ are eigen-images and span the columns' space of the data matrix. On the other hand, if we consider the time points as variables, the covariance matrix

$$\frac{1}{n}X^T X \qquad (16.20)$$

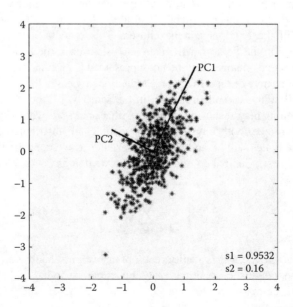

FIGURE 16.5 Bidimensional distribution of data points is shown along with the principal components and the associated eigenvalues.

accounts for the temporal covariance in the data set, i.e., covariances among the time points. The associated eigenvectors v_i span the row space of the data. It is important to note that if we want to study the covariances among the time points, the mean value has to be subtracted from each image; otherwise, we study the correlation matrix. The two approaches, the study of the spatial covariance matrix or temporal covariance matrix, are called, respectively, *temporal* and *spatial principal-component analysis* [30]. Depending on which features we are more interested in, we can choose between the two approaches. In the case of fMRI data analysis, these were not proved to exhibit great difference. In Reference 31 these approaches are referred to as R and Q analysis, respectively, whereas subtracting the mean both from time series and images is referred to as M analysis. It can often be found that instead of normalization by the number of observations, the matrices XX^T or X^TX are used. If the mean from each row (time series) is subtracted, then we have the matrix XX^T proportional to the sample covariance matrix for the voxels

$$\hat{\mathbf{C}}_{\mathbf{V}} \propto \mathbf{XX^T} \qquad (16.21)$$

If the mean from each column (images) is subtracted, the matrix X^TX is proportional to the sample covariance matrix for the time points

$$\hat{C}_t \propto X^TX \qquad (16.22)$$

The results in terms of eigenvectors are the same, whereas the eigenvalues differ by a scaling term. PCA can be achieved by means of singular value decomposition (SVD)[63] of the data matrix

$$\mathbf{X} = \mathbf{USV}^T \qquad (16.23)$$

where U is an orthogonal matrix of $n \times r$ dimension whose columns are the vectors u_i, S is an $r \times r$ diagonal matrix with elements $s_i^{1/2}$, and V is an orthogonal matrix whose columns are the vectors v_i. Reorganizing Equation 16.23 it is possible to see that

$$\mathbf{XX}^T = \mathbf{US}^2\mathbf{U}^T \qquad (16.24)$$

confirming that the columns u_i are eigenvectors of XX^T and

$$\mathbf{X}^T\mathbf{X} = \mathbf{VS}^2\mathbf{V}^T \qquad (16.25)$$

so that the columns of V are eigenvectors of the matrix X^TX. Equation 16.23 can be written as

$$\mathbf{X} = \sum_{k=i}^{r} \sqrt{s_i}\mathbf{u}_i\mathbf{v}_i^T \qquad (16.26)$$

with s_i being the eigenvalues. PC scores for the time points, which represent the contribution of eigenimages to each time point, are given by $X^T U$. These scores can be interpreted as the original data in the space of u_i. From Equation 16.26 it is possible to see that these are the loadings, apart from a scaling factor given by the eigenvalues, of the eigenvectors v_i. The same can be done for the PC scores for the voxels that are given by XV. These can be interpreted as the contribution of each eigenvector in each voxel. It is easy to show that the loadings of the eigenvectors u_i equal the scores, apart from a scaling factor, of the eigenimages. Although these approaches, temporal and spatial covariance based, seem equivalent, they differ largely in fMRI data analysis applications in the maximum number of computations needed to get the eigenimages starting from the matrix XX^T. In fact, in fMRI data, the number of voxels is far larger than the number of scans. However, when applied in the temporal domain, i.e., looking at the time series in each voxel as observations, the analysis is performed on a subset of all brain voxels. In Reference 62 this method was applied to a subset of voxels that showed significant differences across the alternating conditions. This allowed a reduction in the computational costs, justified by the fact that voxels that do not contribute to the measured variance cannot contribute to the covariance. In Reference 28, PCA by means of SVD was applied to an fMRI data set. In this work, the analysis was applied to activated voxels found by the regression method. PCA was applied considering the time points as variables and the voxels as observations. This resulted in eigenvectors whose loadings were the time evolution of the component, and a set of scores for each PC in each voxel.

16.4.1.1 Preprocessing of fMRI Data before PCA

Besides the usual preprocessing step mentioned in Section 16.2, several other preprocessing steps can be performed: we have already mentioned that it is possible to center the time series or the images, or both. Another preprocessing step that can be performed is the normalization of the time series, i.e., subtracting the mean and dividing by the standard deviation, in order to reduce the effect in the overall variance of spatially varying random signal fluctuations. Linear regression can be used to remove low-frequency fluctuations or signal drifts as in Reference 31, whereas in Reference 28, in order to reduce the effect of the noise present in the data, PCA was applied to a set of fitted time series using six sinusoidal regression parameters, a drift term, and a constant term.

16.4.2 INTERPRETATION OF THE PCA DECOMPOSITION

Once a decomposition is found, the experimenter has to decide which components are interesting: this can be done using information about the experimental design, for example, the correlations between the paradigm and the time course of a PC or using a hypothesis about the activation regions. Results from PCA can be reported as loadings of the eigenimages on anatomical images or as scores on the PC of interest, depending upon the approach chosen. Another method makes use of PC plots: these plots are used to identify each individual, which in this context means each observation, by means of the appropriate score on the PC of interest. Usually, these are bidimensional plots with the first two PCs as orthogonal axes, but if other components are relevant, more plots may be needed. The relevance of PCs is usually quantified by the percentage of variance explained by the components, determined using the eigenvalues spectrum. A plot of

$$\frac{\sum_{i=1}^{k} s_i^2}{\sum_{i=1}^{r} s_i^2} \qquad (16.27)$$

shows the percentage of total variance explained by the first k principal components. Usually, the first PCs account for most of the variance, and the noise level can be evaluated by examining the flattening of the plot. However, it is important to keep in mind that a classification based on variance may be of no interest. In fact, although some components may be related to interesting physiological phenomena, others may have originated from movement effects or other physiological sources [64]. Moreover, movement-related components can cause large signal changes in the data set, contributing heavily to the overall variance, whereas activation-related signal changes may explain a smaller percentage of the overall variance. Usually, out-of-brain voxels are masked in order not to influence the PC transform and hide some interesting activations. In order to enhance the extraction of an activation that may be located in a specific region, it is possible to select an ROI. In Reference 29, a finger-tapping study is reported: a PCA of all brain voxels resulted in a second PC highly correlated with the task. The percentage of the variance explained by the component was expressed as $\sqrt{(s_2/s_1)}$, where s_2 is the eigenvalue associated with the second PC, and s_1 is the eigenvalue of the PC explaining the larger variance fraction and was found to be 0.01. The selection of a smaller ROI centered around the activation yielded a fractional value of 0.038. In order to find the relevant number of PCs, information theoretic criteria such as Akaike's information criteria (AIC) [65], minimum description length (MDL) [66], and Bayesian model selection [67] can be used. These methods can also be used in combination with PCA before applying ICA to fMRI data in order to perform dimensionality reduction. However, the PC found could not be representative of real physical quantities [32]: because the decomposition is based on variance partitioning, this general assumption cannot be used to interpret the final results. Some attempts can be made to identify the sources of interest, such as applying rotations to the components found by the PCA approach [68]. The eigenvectors, or equivalently, the eigenimages, can be regarded as a new coordinate system for the data set. Along the directions of the first eigenvector (or the first eigenimage) the measurements show larger variability; the second direction is the orthogonal direction, along which the remaining variability is maximally explained, and so on. The rotation operation can be mathematically described by a transformation matrix T, so that the rotated components are

$$\mathbf{U}^T = \mathbf{UT} \qquad (16.28)$$

or, dually, $V^T = VT$. The transformation matrix can be an orthogonal transformation, such that the rotated components are still linearly independent or oblique. This operation is a search among the linear combinations of the PCs and can be seen as a projection pursuit task. The interesting projections of the data can be found manually or by using some graphical description of the data, by means of information theoretic criteria or by using some *a priori* information about the expected results. In Reference 32 the rotation matrix was applied to eigenimages and was estimated from the data set after a first processing step using PCA. This step was useful to obtain information about the images of interest, called *factor images*. In Reference 31 the rotation was applied to PC found by the analysis of fMRI data from a primate during the pharmacological stimulation of the dopaminergic nigrostriatal system. The first two PCs were associated with an effect due to pharmacological stimulation but also contained abrupt changes, probably due to movement. An oblique rotation was performed manually to separate the motion-related component, obtaining a smooth drug-response profile and a component with only the abrupt changes. The optimal rotation was also found by inspecting the associated maps: expected brain regions that respond to drug stimulation for the first component, and regions with rapid spatial changes in mean signal values for movement.

16.5 ICA

This [69] is one of the exploratory methods that has given rise to greater interest. The reason lies in the assumption of statistical independence of the components, and this guiding principle has proved to be useful in several applications ranging from telecommunications to image feature extraction, from financial time series analysis to artifacts separation in brain-imaging applications [70]. Similar to PCA, the ICA model is generative and models the data as a linear combination of components. The main difference with PCA is that the components are thought to be statistically independent of each other instead of linearly independent. The extraction of the independent components is based on an information theoretic criterion and not on the maximization of variance explained by the orthogonal components. As we have seen in Subsection 16.3.2, a problem with the application of PCA to fMRI data is that the components related to brain activity of interest may contribute very little to the overall variance. Moreover, a method based only on second-order statistics may not reveal more complex patterns of activation; these can be detected by methods such as ICA, which takes into account higher-order statistics also, to assess the hypothesis of statistical independence [71]. ICA was first used to solve the blind source separation (BSS) problem [72,73], in which a set of sources is mixed linearly to form the observations. Both the sources and the mixing process are unknown. If we denote the unknown sources by a vector $s = \{s_1(t), \ldots, s_n(t)\}$, the mixing matrix by A, and the data by $x = \{x_1(t), \ldots, x_n(t)\}$, the model can be written as

$$x = As \tag{16.29}$$

The problem is to find an unmixing matrix W, which can be thought of as the inverse of the unknown matrix A (here we assume we can invert the system) such that

$$\tilde{s} = Wx \qquad (16.30)$$

are an approximation of the sources s. In this model, each time series $s_i(t)$ is assumed to be the set of observations of a random variable s_i and the vectors s and x are random vectors. The assumption of statistical independence means that the knowledge about the value of a signal at a certain time cannot give us information about the other sources' values. Hence, the joint probability density of s factorizes, and we have

$$p(s) = \prod_{i=1}^{n} p_i(s_i) \qquad (16.31)$$

Note that this method does not require any *a priori* assumptions about the shape or extent of the activations. Instead, it does require the underlying sources be statistically independent. In the following sections we will see when this may happen with fMRI data set. Another restriction is that the unmixing matrix is square: This means that the number of unknown sources equals the number of observed variables. Working with fMRI data, this restriction would lead to wrong modeling, and the dimensionality reduction of the data has to be performed.

16.5.1 SPATIAL AND TEMPORAL ICA

As with PCA, ICA of fMRI data can be carried out in temporal or spatial domains [33]. The data set can be decomposed into a set of spatial patterns of activations associated with their own time courses, assuming statistical independence among the spatial patterns or among the time courses. In spatial ICA the data matrix X is $p \times n$, where p is the number of time points and n is the voxels number. The generative model of the data in (Equation 16.29) shows that the independent components are a set of statistically independent images or spatial patterns of activation, mixed linearly by the matrix A whose columns are the time courses associated with each independent component. In temporal ICA, the data matrix is $n \times p$ and its rows are the signal time courses in each voxel of the acquired volume. The independent components are temporally independent time courses, and the columns of the matrix A represent the spatial distribution of the temporal components. In the first application of ICA to fMRI data, McKeown [34] suggests that the spatial distribution of voxels whose activation is related to a task of interest should be unrelated to the spatial distribution of artifacts that affect the signal, such as physiological pulsations, subject-movement-related effects, and scanner noise. Several other assumptions are made: the model (Equation 16.29) assumes that the spatial distribution of the independent components does not change with time and that they are linearly mixed [35]. Temporal ICA assumes that there are statistically independent temporal processes [36]. These two approaches seem to give similar results in the case of one expected task-related component [37]. In Reference 38 it was shown that they give similar results in decomposing data with two components that are both spatially and temporally independent. In general, the two methods give divergent results depending on the agreement with the hypothesis of spatial or temporal independence. In this study [38], different spatiotemporal patterns of activation were simulated, and it was shown that the two methods produce similar results if the two components are both spatially and temporally uncorrelated. If the components are correlated in time, time-domain ICA will not give good correct results; similarly, spatial ICA will not give correct results in the case of spatially dependent patterns of activation. From these considerations, it emerges that the ICA model must be applied carefully, and the experimenters have to know that it can lead to incorrect results. Even if a paradigm is supposed to give temporally independent activations, there may be some interesting components temporally correlated to each other that are not supposed in advance. One way to proceed is to perform both spatial and temporal ICA and then use consensus methods [74] to find regions that are activated independently of the chosen model. Stone [39] introduced a spatiotemporal approach that maximizes simultaneously the independence in spatial and temporal domains. Spatial ICA has been the most used approach mainly because of fewer computational demands owing to the fact that in fMRI data sets the spatial dimension is much larger than the temporal one. In fact, in spatial ICA the variables are represented by the time points, whereas the observations are the voxels values; hence, the computational benefit of the spatial ICA approach is clear.

16.5.2 METHODS FOR ICA

16.5.2.1 Historical Background

One of the first solutions to the BSS problem was given by Cardoso [71], who used higher-order moments. The work on higher-order cumulant tensors led to the development of the JADE algorithm [75]. The work by Pham, using a maximum likelihood criterion [76], was further developed by Cardoso and led to the EASI method [77]. A great improvement to ICA was due to the algorithm developed by Bell and Sejnowski based on the InfoMax approach [78]. This algorithm was then modified by Amari, using the natural gradient [79]. It has been shown that this method has connections with the maximum likelihood approach and with the earlier work of Cichocki et al. [80]. A very popular algorithm, widely used for its computational efficiency, is the fastICA algorithm based on a fixed-point iteration scheme [81]. A good review of these algorithms and their relations can be found in Reference 82. Now, we will briefly review some of the principles used to extract the independent component from a data set.

16.5.2.2 Nonlinear Decorrelation

The data matrix X of dimension $N \times M$ is a set of M observations of N variables such that it is possible to write $x = \{x_1(v),...,x_N(v)\}$, where v represents a time index if we have N time series, or a spatial index if we have N images. The independent components that we have to find can be written as $s = \{s_1(v),...,s_N(v)\}$. Each observed variable x_i can be considered as a linear combination of the unknown statistically independent sources s_i, so that we can write

$$x_i = a_{i1}s_1 + a_{i2}s_2 + \cdots + a_{iN}s_N \qquad (16.32)$$

or, in matrix notation

$$\mathbf{x} = \mathbf{As} \qquad (16.33)$$

The goal is to find an unmixing matrix, such that

$$\tilde{s} = \mathbf{Wx} \qquad (16.34)$$

are an estimate of the original sources. A first general assumption is that both the original sources and the observed variables have zero mean. Because we are interested in signal changes, the mean value does not carry any information, and hence we can remove it by means of a centering stage so that x becomes $x - E\{x\}$. In compact notation

$$\mathbf{x} \leftarrow \mathbf{x} - \mathbf{E}\{\mathbf{x}\} \qquad (16.35)$$

where $E\{\cdot\}$ is the expectation operator. One principle that can be used to find the independent components is nonlinear decorrelation. We can consider two statistical variables x and y to be nonlinearly uncorrelated if

$$E\{g(x)h(y)\} = 0 \qquad (16.36)$$

where at least one function between $g(\cdot)$ and $h(\cdot)$ is a nonlinear one. The problem is how to choose these functions such that this condition implies statistical independence between x and y. It can be demonstrated that if two random variables x and y are statistically independent, then we have, for any absolutely integrable function of x and y, $g(\cdot)$ and $h(\cdot)$,

$$E\{g(x)h(y)\} = E\{g(x)\}E\{h(y)\} \qquad (16.37)$$

The problem was faced by Herault and Jutten [72] and then by Cichocki and Unbenhaven [83]. Herault and Jutten proposed using smooth functions that can be expanded in a Taylor series around zero, i.e.,

$$g(x) = g(0) + g'(0)x + \frac{1}{2}g''(0)x^2 + \cdots \tag{16.38}$$

The expectation operator applied to the Taylor expansion of these functions introduces higher-order moments of the random variables $E\{x_i\}$, with $i = 1, 2, ..., \infty$. A sufficient condition for Equation 16.36 to hold and for the variables to be nonlinearly uncorrelated is that x and y be independent so that Equation 16.37 is valid and either $E\{x^i\}$ or $E\{y^i\}$ is zero for each i. For this property to be satisfied, either $g(x)$ or $h(y)$ must be an odd function with zero mean. With these assumptions, looking for a matrix W such that the estimated sources s_i and s_j as well as $g(s_i)$ and $h(s_j)$ are uncorrelated for any $i \neq j$ may lead to finding independent components. Herault and Jutten proposed the use of $g(x) = x^3$ and $h(y) = \arctan(y)$. However, this algorithm showed some convergence problems and cannot be used to separate many sources. Cichocki and Unbenhaven proposed an extension of this algorithm that makes use of a feedforward network. The criterion used is that of nonlinear decorrelation, and it is the same as that used in the Amari natural gradient algorithm [79].

16.5.2.2.1 Whitening as a Preprocessing Step

It is noteworthy that nonlinear decorrelation is stronger than a linear decorrelation. In fact, uncorrelation is derived from Equation 16.37, where $g(x)$ and $h(y)$ are linear functions. For example, even if $x = sin(t)$ and $y = cos(t)$ are uncorrelated, it can be easily shown that x^2 and y^2 are correlated; in fact $x^2 + y^2 = 1$. Although independence implies uncorrelation, the converse does not hold, and a (linear) decorrelation method will not give independent components. However, a decorrelation step is often used as a preprocessing stage in ICA. In order to simplify successive algorithmic steps, a whitening operation is often performed: whitening means that the zero mean observed variables x_i are transformed into a new set of variables that are uncorrelated and have unit variance. After this operation, $E\{x_i x_j\} = \delta_{ij}$ holds and the variables are said to be whitened or sphered. After the data have been sphered, the search for the independent components is simplified because the new mixing

$$\hat{A} = W_S A \tag{16.39}$$

where W_S is the whitening or sphering matrix, becomes orthogonal and so an estimate of $N(N - 1)/2$ elements is needed to solve the ICA problem, as against an estimate of N^2 elements of the matrix A. The whitening matrix can be computed as the inverse square root of the covariance matrix $C_x = E\{xx^T\}$, which is written as $C_x^{-1/2}$. Given the eigenvalue decomposition of the covariance matrix

$$C_x = EDE^T \qquad (16.40)$$

with E the eigenvectors matrix and D the diagonal matrix of the eigenvectors, the whitening matrix can be written as

$$W_S = ED^{-1/2}E^T \qquad (16.41)$$

This whitening step is often performed in conjunction with the dimensionality reduction operation by means of PCA, as we shall see later.

16.5.2.3 Information Maximization and Maximum Likelihood Approaches

Another approach to find the independent components from observed mixtures is the information maximization approach, named *InfoMax*, which consists in maximizing the joint output entropy of a neural network whose inputs are the observed variables. The entropy of a variable can be seen as a measure of the information that its observation gives: the more random the variable, the more information we have from its observation, and so the higher its entropy. The outputs of the neural network can be written as $y_i = g_i(w_i^T x)$. Maximizing the joint entropy of the outputs of this neural network is found to be equivalent to minimizing the mutual information among the estimated components $w_i^T x$. The mutual information between a set of variables is an information theoretic criterion, which shows the amount of information that the knowledge about a variable carries about the other. The mutual information of a set of random variables y_i can be written as

$$I(y_1, y_2, \ldots, y_n) = \sum_{i=1}^{n} H(y_i) - H(y) \qquad (16.42)$$

where $H(\cdot)$ is the entropy. The first term is related to the amount of information we get from the observation of the variables separately, whereas the second term is related to the amount of information we get from the observation of all the variables together. If the variables are statistically independent, we do not have any additional information about any variable from the observation of any other, and the entropy of the complete variable vector is the sum of the entropies of the individual variables. In this case the mutual information equals zero. If there is some redundancy in the variable set, it means that we can get some information about some variable from the observation of the others, and the entropy of the complete vector is lower than the sum of the individual entropies. This results in a mutual information greater than zero: minimizing the mutual information

among a set of variables is equivalent to maximizing their statistical independence. It is possible to show that the mutual information of the estimated
components is

$$I(\mathbf{s}) = -H(\mathbf{g}(\mathbf{s})) + E\left\{\sum_i \log \frac{|g_i'(s_i)|}{p_i(s_i)}\right\} \qquad (16.43)$$

where $p_i(s_i)$ is the probability density function of the independent component s_i. It is
important to choose the proper shape of the nonlinearities g_i such that $g_i'(s_i) = p_i(s_i)$.
This means that it is necessary to have some prior knowledge about the statistical
distribution of the independent components, for example, if they have a super-
Gaussian distribution, such as images with small activation foci in a large number
of voxels, or sub-Gaussian, such as the distribution of time-domain components
that may be strongly related to a block-designed task. The reason why it is not
possible to estimate Gaussian-distributed components will be shown using intuitive reasoning afterward. This method was extended to allow the separation of
both super- and sub-Gaussian-distributed components [84]. The weights are estimated using a stochastic gradient descent algorithm [78] such that at each step
the weights are updated following the relations

$$\Delta W = [W^T]^{-1} + f(Wx)x^T \qquad (16.44)$$

and $f_i = (\log(p_i))'$. It can be shown that this algorithm is equivalent to the maximum
likelihood approach for the estimation of components with known distribution
densities. A simplification and optimization of this method was given by Amari
[79], using the natural gradient method, which can be obtained by multiplication
of Equation 16.44 by WW^T resulting in

$$\Delta W = [I + f(Wx)x^T W^T]W \qquad (16.45)$$

The algorithm stabilizes when

$$f(Wx)x^T W^T = -I \qquad (16.46)$$

The minus sign is given by the functions f_i. Typical functions are $f(s) = -2$
$\tanh(s)$ for super-Gaussian components and $f(s) = \tanh(s) - s$ for sub-Gaussian
components. The relation in Equation 16.46 shows that this method can be viewed
as nonlinear decorrelation, and it is interesting to note that if we perform a Taylor
expansion of the nonlinear functions, we get higher-order correlations of the
variables. These are the measures we have to take into account if we want to
estimate statistical independence.

16.5.2.4   Non-Gaussianity and Negentropy

Another approach is related to the non-Gaussianity of the components and can be understood if we introduce the central limit theorem. This theorem states that the linear combination of random variables approaches a Gaussian distribution as more variables are added. From this theorem, we can state that because the data set is supposed to be a linear mixing of statistically independent random variables, their linear mixing is supposed to be more Gaussian than the original ones. When we look for original independent components, we look for a linear combination of the mixtures $s_i = \Sigma W_i X_j$. The new variable is maximally non-Gaussian when it equals one of the independent components. It is now clear that it is not possible to estimate statistically independent components that are Gaussian distributed; moreover, higher-order moments of Gaussian-distributed variables equal zero, so they cannot be estimated by nonlinear decorrelation methods. In order to estimate the degree of non-Gaussianity of a random variable, it is possible to use different measures such as kurtosis. However, negentropy has proven to be more robust, even if computationally expensive to estimate. Negentropy is defined as $J(y) = H(y_{gauss}) - H(y)$ where $H(y)$ is the entropy of the y variable and y_{gauss} is a Gaussian variable with the same covariance matrix as y. Because Gaussian variables have the largest entropy among all variables with equal variance, negentropy is always greater than or equal to zero. It can be shown that if the estimated independent components are constrained to have unit variance, estimating the weights W such that the new variables have maximum non-Gaussianity is the same as minimizing their mutual information. The fastICA algorithm can be used to estimate the independent components by using of this principle. It is appealing because instead of using a gradient descent approach to find the solution, it employs a fast fixed-point iteration scheme. Moreover, the independent components can be found using a deflationary scheme, which means estimating the independent components one by one. This operation is simplified if we are working on whitened data because in the whitened space the directions w_i that maximize the non-Gaussianity of $w_i^T x$ are orthogonal. In fact, because the independent components are uncorrelated it follows that

$$E\left\{s_i s_j^T\right\} = \delta_{ij} = E\left\{\left(w_i^T x\right)\left(w_j^T x\right)^T\right\} = E\left\{w_i^T x x^T w_j\right\}$$

$$= w_i^T E\left\{x x^T\right\} w_j = w_i^T w_j \tag{16.47}$$

as for whitened data we have $E\{x x^T\} = I$. The first independent component can be estimated by calculating the direction w_1 that maximizes the non-Gaussianity of $w_i^T x$. The successive independent components are found as the directions that maximize the non-Gaussianity of $w_i^T x$ with the constraint that w_i lie in the subspace orthogonal to the one individuated by the directions found in the previous steps. Another approach consists in estimating the independent component in a single step: this symmetric approach has the advantage of reducing the propagation of estimation error from the first components to the last and requires a symmetric orthogonalization of the matrix $W = (w_1, w_2, \ldots, w_n)^T$.

16.5.2.5 Ambiguities in the ICA Model

The model (Equation 16.33) implies the existence of some ambiguities. In fact, because both the mixing matrix and the sources are unknown, it is not possible to determine the energies and the sign of the independent components. It is then possible to overcome this ambiguity by constraining the independent components to have unit variance, while the sign remains ambiguous. It can be shown that introducing the constraint of unit variance of the estimated independent components, along with the whitening operation, results in constraining w_i to lie in the unit sphere, so that the algorithms are modified accordingly. After a new estimation step of the directions of w_i, a normalization step has to be performed. Another ambiguity concerns the order of the independent components, which cannot be determined *a priori*: for this reason, we are not able to inter the significance of a component just looking at its extraction order; that is ambiguous.

16.5.3 PREPROCESSING

Because spatial-filtering as well as temporal-filtering operations do not affect the validity of the ICA model [85], some filtering stages such as spatial smoothing or temporal filtering can be applied. High-pass temporal filtering can be used to remove low-frequency signal changes or drift. Spatial smoothing is usually performed in order to enhance signal-to-noise ratio and reduce movement-related effects. The filtering procedure may have a strong influence on the results because some information in the data set may be lost: low-pass filtering in the time domain may lead to loss of independence of the components, whereas high-pass filtering may enhance independence because it allows the removal of low-frequency fluctuations or drifts in the signal that may bias the independence of the components. Low-pass filtering in the spatial domain is usually performed in order to enhance signal-to-noise ratio and reduce motion-related effects. Centering of the variables, i.e., the time points if we are applying spatial ICA or voxels in a temporal IC model, is also carried out. The centering operation does not alter the mixing matrix A, so that the mean can be added back to the independent components s by means of this operation: $s \leftarrow s + A^{-1}E\{x\}$, where x here refers to the observed data before mean removal. A data reduction stage is usually included; in fact, the basic ICA model assumes that the number of the sources that generate the data equals the number of the observed variables, which means that applying spatial ICA decomposition to a data set consisting of 100 time acquisitions of a volume will result in the extraction of 100 components. This may not be true in general, and the number of underlying sources is often supposed to be less than the number of observed mixtures. In this case, the mixing matrix would be rectangular, and so the basic ICA model would not hold. Moreover, trying to estimate more sources than the actual number may cause overlearning. This phenomenon is related to the use of too many parameters in the model with respect to the amount of data available. Data reduction is usually achieved by applying PCA to the observed data and retaining the PCs that account for most of the variance in the data, i.e., more than 90% of the overall variance. The discarded components may be related to noise only. As this may be not true in general, the PCA reduction process is very critical. This operation allows simultaneous whitening of the observed data. Given the data covariance matrix $C_x = E\{xx\}^T$, the matrix of its eigenvectors $U = (u_1, u_2, u_n)$ and the diagonal matrix of the eigenvalues $\lambda_1, \lambda_2, ..., \lambda_n$, the whitening transform can be written as

$$W_S = D^{-1/2}U^T \qquad (16.48)$$

The data reduction operation can be performed by retaining m of the n eigenvectors, usually the first ones that take into account most of the data variance. The first m whitened PCs are given by

$$Y = D_m^{-1/2}U_m^T X \qquad (16.49)$$

where U_m^T is formed by the first m columns of U, and the matrix D_m is the diagonal matrix with the first m eigenvalues of C_x. This initial preprocessing step is a very important and critical stage because using fewer dimensions than the actual number or underestimating the model order may cause information loss, whereas overestimating the model order may cause overlearning and generate spurious components. If we try to extract more independent components than the real number, we may find components with a single spike or sparsle distributed. These solutions can be seen as extreme cases of non-Gaussianity. The dimension reduction process requires caution, because though reduction may be preferable from the point of view of computational demands, information loss may result. As stated before, it is possible to use the number of components depending on the percentage of the overall variance explained. If the noise is supposed to contaminate all the observed mixtures in the same way, it may be supposed that the eigenvalues will be affected by it. It will be possible to detect a threshold above which all the eigenvalues will be statistically equal: this threshold can be determined by visual inspection of a scree plot for a change in the steepness of the plot. Other methods to determine the dimension are information theoretic criteria such as AIC [65,86] and MDL [66] and Bayesian approaches [87,88]. In Reference 43 a clustering approach based on the k-means algorithm was proposed for spatial ICA. The dimensionality reduction consists of applying the clustering operation to the observed variables and using the mean of each cluster as reduced data. This latter approach does not take advantage of second-order statistics and has proved to be superior to PCA for higher values of CNR, whereas for lower CNR, PCA was found to outperform clustering. In Reference 40 ICA was applied to voxels belonging to the cortex.

This method, defined as *cortex-based ICA*, was proposed to remove from the analysis the signal dynamics due to the voxels belonging to white matter, cerebrospinal fluid, ventricles, and other uninteresting structures. The method may enhance the localization power of ICA decomposition because it can work on reduced voxel numbers with the same number of independent components. Moreover, it speeds up the computation operation.

16.5.4 MODEL VALIDATION

In Reference 35 the validity of the ICA model in the spatial domain was investigated. The validity of the hypotheses of linear mixing of the components, and the effect of dimension reduction by means of PCA were studied. In each voxel v_i, the minus log-likelihood of the data given in the model was calculated, resulting in a value $u(v_i)$. A spatial-smoothing operation was performed to achieve a map of the goodness of fit of the model in the different brain regions. Once the unmixing matrix W is found as well as the estimate of the original sources, and if we consider the unmixing matrix invertible, it is possible to reconstruct the data from the components by means of

$$X = W^{-1}S \tag{16.50}$$

where S are the estimated independent components. The likelihood of observing the data under the model specified by W^{-1} and S can be written as $P(X_i|W) = \det(W)P(S_i)$ where X_i is the ith column or the ith voxel's time series and S_i is the ith column of the S matrix. Using the statistical independence among the components contained in the rows of S, it is possible to write $P(X_i|W) = \det(W)P(S_i)$ $= \det(W)\prod_{k=1}^{n}P_k(S_{ki})$. The quantity $P_k(S_{ki})$ is the probability of the ith point in the kth component maps and is derived from an estimation of the probability density functions of the kth component. In Reference 35 a smoothed version of the histogram of each component was used. The minus log-likelihood function can be written as $-\log(P(X_i|W)) \approx -\log(\det(W)) - \sum_{k=1}^{n}\log(P_k(S_{ki})) = u(v_i)$. The logarithm is usually used because of the exponential form of several probability density functions. In the work of McKeown it is verified that the ICA model fits better for white matter than for gray matter in real data sets. This is supposed to be related to the difference in the number of spatially independent components for the different regions or to a nonlinear mixing of the sources in the gray matter. Moreover, the validity of the model in simulated and real data sets against different degrees of dimensionality reduction by means of PCA was investigated. The simulated data sets were created using 50 eigenimages and mixed randomly. Gaussian random noise was then added. The method showed different behavior in real and simulated data sets: in the simulated data set the $u(v_i)$ function decreases steeply if more components than the actual number are chosen, and in the real data set this function decreases slowly showing that the smallest variance components are unlikely to have a Gaussian structure.

16.5.5 Interpretation of the Results

16.5.5.1 Thresholding the Maps

The thresholding operation of each map is usually performed by scaling the intensity values to the z score. Within each component map, the voxels that contribute significantly to the map are those having a z score whose absolute value is greater than a threshold. Voxels whose time series are modulated opposite to the time course of the component show a negative z score. It is important to stress that the z score has no statistical significance, but it is used only for descriptive purposes. In order to make statistical inferences about these maps, some hypothesis about the distribution of the noise or the signal is needed. In Reference 41 a probabilistic ICA model that takes noise into account is introduced. The use of a z score for inferential purposes is not recommended here because of the non-Gaussian distribution of the intensity values in each map. A different z score normalization is proposed, using the estimate of the voxel-wise noise standard deviation, evaluated as residuals of the IC model. Each resulting map is then is fitted with a Gaussian mixture model (GMM) by means of an expectation maximization algorithm. The Gaussian that identifies the background noise is typically thought to coincide with the dominant mode of the histogram, and the probability density function of the background noise can be evaluated, as well as the probability that any voxel belongs to the background noise. The Gaussian mixtures that do not belong to background noise can be used to estimate the probability of the hypothesis of activation related to the relevant time course.

16.5.5.2 Task-Related Activations

McKeown [34] was the first to apply ICA to fMRI data. In this early application, ICA was applied in the spatial domain. In the application of spatial ICA to fMRI data, McKeown grouped the components found in different classes based on the shapes of the associated time courses. Block-designed experiments were analyzed in which two conditions, task and control, alternate in time. Even if no information about the shape of the activation or its location is used in the decomposition process, it can be used after the independent components are estimated for classification purposes. Usually in order to detect task-related components, the correlation coefficient between the time courses of each component (in spatial ICA the columns of the mixing matrix A) and a reference function depicting the task is evaluated [42]. The components whose associated time courses highly correlate with the paradigm are considered task-related components or consistently task related (CTR), whereas components whose activation is related only partly to the paradigm are called transiently task related (TTR). These components were thought to be related to a complex spatiotemporal structure of the activation. In fact, they may be the decomposition results of different neural processes with overlapping areas or related to transient neural processes such as arousal, habituation, or learning.

These components can be considered super-Gaussian distributed in the spatial domain because they are focused activations in small clusters compared with the whole volume. In Reference 34 and Reference 35 the ICA decomposition was applied without PCA reduction and usually one component for each trial was found to highly correlate with the task ($r > 0.6$). Some components that showed abrupt changes in the associated time courses or ring-like spatial distributions were thought to be related to movement effects. Other components may have been diffuse and noisy. Quasi-periodic components, probably due to the physiological pulsations, heartbeat, and respiratory effects, were found. Because sampling times are often less than a second, aliasing often cannot be avoided, and these quasi-periodic effects can also be derived from spin excitation history effects. Several TTR components were also found, showing a correlation with the task only for one or two repetitions of the task blocks. Although the advantages of exploratory analysis performed with ICA were stressed because these TTR maps could not be detected by correlation analysis that computes the average over all the cycles, the question was if these components could be modified by the requirement of spatial independence of the CTR maps. In order to perform this test, ICA was performed again on the data with the CTR removed, showing that even if some TTR components were unaffected by the CTR removal, others changed sensibly, suggesting a spatial dependence among these components. The component removal can be performed by multiplying the components matrix, found by ICA, by a copy of the mixing matrix whose columns corresponding to the components to be removed are zeroed. If W is the unmixing matrix and W_a^{-1} is its inverse with the columns zeroed, then the reconstructed data matrix can be written as $X_a = W_a^{-1}WX$. In Reference 41 it is suggested that the presence of TTR activations may originate from interesting physiological processes but may originate also from an overfitting problem and the lack of a significance test for the components. In this work a probabilistic PCA model order reduction was used, which, starting from the covariance matrix of the data, estimated the posterior distribution of the model order. In Reference 40 the problem of identification and characterization of the maps was outlined. In this work the maps obtained after spatial ICA were classified according to three descriptive measures: the kurtosis of the map values, the degree of spatial clustering of each map, and the one-lag autocorrelation of each map time series. The kurtosis takes into account the distribution properties of each map intensity. The degree of spatial clustering of each map, after a thresholding operation by means of a z score, was chosen because activation maps usually have a defined spatial structure. The one-lag autocorrelation of each map time series was chosen in order to detect a temporal structure in the maps. This procedure was applied to voxels belonging to the cortex individuated by means of the segmentation of a high-resolution T1 anatomical image coregistered with the functional images. The method showed that the simultaneous inspection of these values could reveal potentially meaningful phenomena because in different tasks the interesting components show similar combinations of these parameters.

16.5.6 Simulation and Algorithm Comparison

The analysis of fMRI data by means of independent components is a complex process, and the final result may depend not only on the processing strategies but also on the algorithm that has been chosen. Because the real sources are unknown, it is important to perform some tests on simulated data sets: these may be simple data sets such as time series with varying degrees of Gaussian noise and with superimposed simulated activations. The noise can be extracted from real data sets, as in Reference 44, and may be white or correlated. It is then possible to use images such as phantoms to simulate the spatial properties of the activation maps such as the spatial correlation of the sources due to the vascular point-spread function. A model with a simplified distribution of baseline voxel values was used, with artificially added movement in Reference 45 to test the behavior of spatial ICA. In order to have realistic noise models, it is possible to use resting-state data (null data) and superimpose the activations using self-generated masks or masks obtained from real activation thresholded maps. These maps can be used as masks in order to modulate the null data set with simulated activation. In Reference 44 the InfoMax and the fastICA algorithms were compared along with some preprocessing strategies. The Kullback—Leibler divergence between the estimated and real source distributions was found. The results reported in this article show some preference for the PCA reduction stage at higher noise levels in a completely simulated data set, and always with hybrid data sets obtained by adding simulated sources to a real fMRI data set. In this work, it was found that InfoMax seems to outperform the fastICA algorithm because it does not constrain the sources to be orthogonal to each other. In Reference 46 receiver operating characteristics (ROC) are used to test for the detection of spatial accuracy of the InfoMax and fastICA algorithms. A likelihood analysis was used to test the reliability of the model. The filtering effect of each algorithm was tested using a Gaussian mixture model to fit each map because from this fit it is possible to estimate the variance or the residual noise. The spatial structure of the maps was estimated by means of a cluster-sizing function. The test was performed on a simulated data set, obtained by adding the activated regions to resting-state data, and on real data. In the latter case, the results of a linear correlation test were used as a benchmark for the ICA analysis. The results from the simulated data set were aligned with those from real data sets and it was confirmed that although fastICA exhibited better overlap with linear correlation results, the likelihood analysis and noise results showed the better performance of InfoMax, confirming its superiority in global model estimation and filtering capabilities. This was thought to be due to the adaptive nature of the InfoMax algorithm.

REFERENCES

1. Ogawa, S., Menon, D.W., Ellermann, J.M., Kim, S.G., Merkle, H., and Ugurbil, K. (1992). Intrinsic signal changes accompanying sensory stimulation: functional brain mapping with magnetic resonance imaging. *Proc. Natl. Acad. Sci.* 89: 5951–5955.

2. Kwong, K.K., Belliveau, J.W., Chesler, D.A., Goldberg, I.R.M., Poncelet, B.P., Kennedy, D.N., Hoppel, B.E., Cohen, M.S., and Turner, R. (1992). Imaging of human brain activity during primary sensory stimulation. *Proc. Natl. Acad. Sci.* 89: 5675–5679.

3. Davidson, R. J. and Irwin, W.(2000). Functional MRI in the study of emotion. in Moonen, C.T. W., Bandettini, P.A., Eds., *Functional MRI.* Berlin: Springer-Verlag, pp. 487–499.

4. Ricciardi, E., Gentili, C., Rizzo, M., Vanello, N., Sani, L., Landini, L., Guazzelli, M., and Pietrini, P. (2004). Brain Activity Associated with Forgiving and Unforgiving Behaviour in Humans as Assessed by fMRI. in 10th Intl. Conf. on Functional Mapping of Human Brain. Budapest, Hungary, June 13–17.

5. Bandettini, P.A., Jesmanowicz, A., Wong, E.C., and Hyde, J.S. (1993). Processing strategies for time-course data sets in functional MRI of the human brain. *Magn. Reson. Med.* 30: 161–173.

6. Friston, K.J., Frith, C.D., Turner, R., and Frackowiak, R.S.J. (1995).Characterising evoked hemodynamics with fMRI. *NeuroImage* 2: 157–165.

7. Worsley, K. J. and Friston, K. J. (1995). Analysis of fMRI time-series revisited— again. *NeuroImage* 2: 173–181.

8. Tukey, J.W. (1962). Exploratory data analysis. *Ann. Stat.* 33: 1–67.

9. Fletcher, P.C., Dolan, R.J., Shallice, T., Frith, C.D., Frackowiak, R.S.J., and Friston, K.J. (1996). Is multivariate analysis of PET data more revealing than the univariate approach? Evidence from a study of episodic memory retrieval. *Neuroimage* 3: 209–215.

10. Philips, C.G., Zeki, S., and Barlow, H.B. (1984). Localization of function in the cerebral cortex. Past, present and future. *Brain* 107: 327–361.

11. Goutte, C., Toft, P., Rostrup, E., Nielsen, F.Å., and Hansen, L.K. (1999). On clustering fMRI time series. *NeuroImage* 9: 298–310.

12. Golay, X., Kollias, S., Stoll, G., Meier, D., Valavanis, A., and Boesiger, P. (1998). A new correlation-based fuzzy logic clustering algorithm for fMRI. *Magn. Reson. Med.* 40: 249–260.

13. Cordes, D., Haughton, V., Darew, J.D., Arfanakis, K., and Maravilla, K. (2002). Hierarchical clustering to measure connectivity in fMRI resting-state data. *Magn. Reson. Imaging.* 20: 305–317.

14. Filzmoser, P., Baumgartner, R., and Moser, E. (1999). A hierarchical clustering method for analyzing functional MR images. *Magn. Reson. Imaging.* 17(6): 817–826.

15. Stanberry, L., Nandy, R., and Cordes, D. (2003). Cluster analysis of fMRI data using dendrogram sharpening. *Human Brain Mapping* 20: 201–219.

16. Scarth, G., McIntyre, M., Wowk, B., and Somorjai, R. (1995). Detection of novelty in functional images using fuzzy clustering. in *Proc. of the Annual Meeting of the Soc. of Magn. Reson. and Europ. Soc. for Magn. Reson. Med. and Biol.* Nice, France,1: 238.

17. Barth, M., Diemling, M., and Moser, E. (1997). Modulation of signal changes in gradient-recalled echo fMRI with increasing echo time correlate with model calculations. *Magn. Reson. Imaging.* 15(7): 745–752.

18. Baumgartner, R., Scarth, G., Teichtmeister, C., Somorjai, R., and Moser, E. (1997). Fuzzy clustering of gradient-echo functional MRI in the human visual cortex. Part I: reproducibility. *J. Magn. Reson. Imaging.* 7: 1094–1101.

19. Moser, E., Diemling, M., and Baumgartner, R. (1997). Fuzzy clustering of gradient-echo functional MRI in the human visual cortex. Part II: quantification. *J. Magn. Reson. Imaging.* 7: 1102–1108.

20. Baumgartner, R., Windischberger, C., and Moser, E. (1998). Quantification in functional magnetic resonance imaging: fuzzy clustering vs. correlation analysis. *Magn. Reson. Imaging.* 16: 115–125.

21. Fadili, M. J., Ruan, S., Bloyet, D., and Mazoyer, B. (2000). A multistep unsupervised fuzzy clustering analysis of fMRI time series. *Human Brain Mapping* 10: 160–178.

22. Fischer, H. and Hennig, J. (1999). Neural network-based analysis of MR time series. *Magn. Reson. Med.* 41: 124–131.

23. Ngan, S.-C. and Hu, X. (1999). Analysis of functional magnetic resonance imaging data using self-organizing mapping with spatial connectivity. *Magn. Reson. Med.* 41: 939–946.

24. Dimitriadou, E., Barth, M., Windischberger, C., Hornik, K., and Moser, E. (2004). A quantitative comparison of functional cluster analysis. *Artif. Intell. Med.* 31: 57–71.

25. Scarth, G., McIntyre, M., Wowk, B., and Somorjai, R. (1995). Detection of novelty in functional images using fuzzy clustering. in *Proc. of the Annual Meeting of the Soc. of Magn. Reson. and Europ. Soc. for Magn. Reson. in Med. and Biol.* Nice, France, 1: 238.

26. Toft, P., Hansen, L.K., Nielsen, F.Å., Goutte, C., Strother, S., Lange, N., Mørch, N., Svarer, C., Paulson, O.B., Savoy, R., Rosen, B., Rostrup, E., and Born, P. (1997). On clustering of fMRI time series. in Third International Conference on Functional Mapping of the Human Brain. *Neuroimage.* 3(3): S456.

27. Mitra, P.P., Ogawa, S., Hu, X., and Ugurbil, K. (1997). The nature of spatiotemporal changes in cerebral hemodynamics as manifested in functional magnetic resonance imaging. *Magn. Reson. Med.* 37: 511–518.

28. Bullmore, E.T., Rabe-Hesketh, S., Morris, R.G., Williams, S.C.R., Gregory, L., Gray, J. A., and Brammer, M.J. (1996). Functional magnetic resonance image analysis of a large-scale neurocognitive network. *Neuroimage.* 4: 16–33.

29. Sychra, J.J., Bandettini, P.A., Bhattacharya, N., and Lin, Q. (1994). Synthetic images by subspace transforms I. Principal component images and related filters. *Med. Phys.* 21(2): 193–201.

30. Dodel, S., Herrmann, J.M., and Geisel, T. (2000). Localization of brain activity-blind separation for fMRI data. *Neurocomputing.* 32–33: 701–708.

31. Andersen, A. H., Gash, D. M., and Avison, M. J. (1999). Principal component analysis of the dynamic response measured by fMRI: a generalized linear systems framework. *Magn. Reson. Imaging.* 17(6): 795–815.

32. Backfrieder, W., Baumgartner, R., Sámal, M., Moser, E., and Bergmann, H. (1996). Quantification of intensity variations in functional MR images using rotated principal components. *Phys. Med. Biol.* 41: 1425–1438.

33. Calhoun, V., Adali, T., Hansen, L.K., Larsen, J., and Pekar, J. (2003). ICA of functional MRI data: an overview. In: Fourth Int. Symp. on ICA and BSS. Nara, Japan, 281–288. http://www.kecl.ntt.co.jp/icl/signal/ica2003/cdrom/data/0219.pdf. (accessed 2004 November 29).

34. McKeown, M.J., Makeig, S., Brown, G., Jung, T.-P., Kindermann, S.S., Bell, A.J., and Sejmowski, A.J. (1998). Analysis of fMRI data by blind separation into independent spatial components. *Human Brain Mapping* 6: 160–188.

35. McKeown, M.J. and Sejnowski, T.J. (1998). Independent component analysis of fMRI data: examining the assumptions, *Human Brain Mapping* 6: 368–372.

36. Biswal, B.B. and Ulmer, J.L. (1999). Blind source separation of multiple signal sources of fMRI data sets using independent component analysis. *J. Comput. Assist. Tomogr.* 23(2): 265–271.

37. Peterson, K.S., Hansen, L.K., Kolenda, T., Rostrup, E., and Strother, S.C. (2000). On the independent components of functional neuroimages. in *Proc. Inter. Conf. on ICA and BSS*. Hesinkin, Finland, 615–620.

38. Calhoun, V. D., Adali, T., Pearlson, G. D., and Pekar, J. J. (2001). Spatial and temporal independent component analysis of functional MRI data containing a pair of task-related waveforms. *Human Brain Mapping* 13: 43–53.

39 Stone, J.V., Porrill, J., Buchel, C., and Friston, K. (1999). Spatial, temporal, and spatiotemporal independent component analysis of fMRI data. in *18th Leeds Statistical Research Workshop on Spatial-Temporal Modelling and its Applications*. University of Leeds: July.

40. Formisano, E., Esposito, F., Kriegeskorte, N., Tedeschi, G., Di Salle, F., and Goebel, R. (2002). Spatial independent component analysis of functional magnetic resonance imaging time-series: characterization of the cortical components. *Neurocomputing*. 49: 241–254.

41. Beckmann, A. and Smith, S.M. (2004). Probabilistic independent component analysis for functional magnetic resonance imaging. *IEEE Trans. Med. Imaging*. 23(2): 137–152.

42. Nakada, T., Suzuki, K., Fujii, Y., Matsuzawa, H., and Kwee, I.L. (2000). Independent component-cross correlation-sequential epoch (ICS) analysis of high field fMRI time series: direct visualization of dual representation of the primary motor cortex in human. *Neurosci. Res*. 37: 237–244.

43. Calhoun, V., Adali, T., and Pearlson, G. (2001). Independent components analysis applied to fMRI data: A natural model and order selection. in *Proc. NSIP*, Baltimore.

44. Calhoun, V., Adali, T., and Pearlson, G. (2001). Independent components analysis applied to fMRI data: a generative model for validating results. in *Proc. NNSP*, Falmouth, MA.

45. Vanello, N., Positano, V., Ricciardi, E., Santarelli, M.F., Guazzelli, M., Pietrini, P., and Landini, L. (2003). Separation of movement and task-related fMRI signal changes in a simulated data set by independent component analysis. in *9th Int. Conf. on Func. Mapp. of the Hum. Brain*. New York, NY, June 19–22. Available on CD-Rom in *NeuroImage*, 19(2).

46. Esposito, F., Formisano, E., Seifritz, E., Goebel, R., Morrone, R., Tedeschi, G., and Di Salle, F. (2002). Spatial independent component analysis of functional MRI time-series: to what extent do results depend on the algorithm used? *Human Brain Mapping* 16: 146–157.

47. Worsley, K. J., Marrett, S., Neelin, P., Vandal, A. C., Friston, K. J., and Evans, A. C. (1996). A unified statistical approach for determining significant signals in images of cerebral activation. *Hum. Brain Mapp*. 4: 58–73.

48. Smith, S. and Brady, J. (1997). SUSAN-A new approach to low-level image processing. *Int. J. Comput. Vis*. 23(1): 45–78.

49. Somorjai, R.L., Vivanco, R., and Pizzi, N. (2002). A novel, direct spatio-temporal approach for analyzing fMRI experiments. *Art. Intell. Med*. 25: 5–17.

50. Ward, J.H., Jr. (1963). Hierarchical grouping to optimize an objective function. *J. Am. Stat. Assoc*. 58: 236–244.

51. Biswal, B., Yetkin, F.Z., Haughton, V.M., and Hyde, J.S. (1995). Functional connectivity in the motor cortex of resting human brain using echo-planar MR imaging. *Magn. Reson. Med*. 34: 537–541.

52. Hartigan, J.A. and Wong, M.A. (1979). A *k*-means clustering algorithm. *Appl. Stat*. 28: 100–108

53. Tou, J.T. and Gonzalez, R.C. (1974). *Pattern Recognition Principles.* Number 7 in Applied Mathematics and Computation, Addison-Wesley, Reading, MA.
54. Ruspini, E.H. (1969). A new approach to clustering. *Inf. Control.* 15: 22–32. Ruspini, E.H. (1970). Numerical methods for fuzzy clustering. *Inf. Sci.* 2: 319–350.
55. Zadeh, L.A. (1965). Fuzzy sets. *Inf. Control.* 8: 338–353.
56. Bezdek, J.C. (1981). *Pattern Recognition with Fuzzy Objective Function Algorithms.* New York. Plenum Press.
57. Bezdek, J.C., Ehrlich, R., and Full, W. (1984). FCM: the fuzzy c-means clustering algorithm. *Comput. Geosci.* 10: 191–203.
58. Kohonen, T. (1995). *Self-Organizing Maps.* New York: Springer-Verlag.
59. Hotelling, H. (1933). Analysis of a complex of statistical variables into principal components. *J. Educ. Psychol.,* 24: 417–441.
60. Pearson, K. (1901). On lines and planes of closest fit to systems of points in space. *The London, Edinburgh and Dublin Philosophical Magazine and Journal of Science,* 2: 559–572.
61. Friston, K.J., Frith, C.D., and Frackowiak, R.S.J. (1993). Time-dependent changes in effective connectivity measured with PET. *Human Brain Mapping* 1: 69–80.
62. Friston, K.J., Frith, C.D., Liddle, P.F., and Frackowiak, S.J. (1993). Functional connectivity: the principal-component analysis of large (PET) data sets. *J. Cereb. Blood Flow Metab.* 13: 5–14.
63. Golub, G.H. and Van Loan, C.F. (1991). *Matrix Computations.* (2nd ed.). The Johns Hopkins University Press: Baltimore and London, pp. 241–248.
64. Friston, K.J., Williams, S., Howard, R., Frackowiak, S.J., Turner, R. (1996). Movement-related effects in fMRI time-series. *Magn. Reson. Imaging.* 35: 346–355.
65. Akaike, H. (1969). Fitting autoregressive models for regression. *Annula Institute Statist. Math.* 21: 243–247.
66. Rissanen, J. (1978). Modeling by shortest data description. *Automatica.* 14: 465–471.
67. Minka, T. P. (2000). Automatic Choice of Dimensionality for PCA. Technical Report 54. MIT Media Laboratory. Perceptual Computing Section, Cambridge.
68. Jackson, J. E. (1991). *A User's Guide to Principal Components.* New York: Wiley.
69. Comon, P. (1994). Independent component analysis, a new concept? *Signal Process.* 36: 287–314.
70. Hyvärinen, A., Karhunen, J., and Oja, E. (2001). *Independent Component Analysis.* New York: Wiley.
71. Stone, J.V. (2002). Independent component analysis: an introduction. *Trend. Cogn. Scis.* 6(2): 59–64.
72. Cardoso, J.-F. (1989). Blind identification of independent signals. in *Proc. Workshop on Higher Order Spectral Analysis.* Vail, CO.
73. Jutten, C. and Hérault, J. (1991). Blind separation of sources, part I: an adaptive algorithm based on neuromimetic architecture. *Signal Process.* 24: 1–10.
74. Herault, J. and Jutten, C. (1986). Space or time adaptive signal processing by neural network models, in Denker, J.S. (Ed.), *Neural Networks for Computing: AIP Conference Proceedings* 11, American Institute for Physics, New York.
75. Hansen, L.K., Nielsen, F.Å., Strother, S.C., and Lange, N.L. (2001). Consensus inference in neuroimaging. *Neuroimage.* 13: 1212–1218.

76. Cardoso, J.F. and Souloumiac, A. (1993). Blind beamforming for non-Gaussian Signals, *IEE Proceedings-F.* 140(6): 362–370.
77. Pham, D.-T., Garrat, P., and Jutten, C. (1992). Separation of mixtures of independent sources through a maximum likelihood approach. In: Proc. EUSIPCO, 771–774.
78. Cardoso, J.F. and Laheld, B.H. (1996). Equivariant adaptive source separation, *IEEE Transactions on Signal Processing.* 44: 3017–3030.
79. Bell, A.J. and Sejnowski, T.J. (1995): An information-maximization approach to blind separation and blind deconvolution. *Neural Comput.* 7: 1129–1159.
80. Amari, S.-I., Cichocki, A., and Yang, A.A. (1996). A new learning algorithm for blind source separation. in *Advances in Neural Information Processing Systems 8,* MIT Press, 757–763.
81. Cichocki, A., Unbehauen, R., and Rummert, E. (1994). Robust learning algorithm for blind separation of signals. *Electron. Lett.* 30(17): 1386–1387.
82. Hyvarinen, A. and Oja, E. (1997). A fast fixed-point algorithm for independent component analysis. *Neural Comput.* 9: 1483–1492.
83. Hyvärinen, A., Karhunen, J., and Oja, E. (2001). *Independent Component Analysis.* New York: Wiley.
84. Cichocki, A. and Unbehauen, R. (1996). Robust neural networks with on-line learning for blind identification and blind separation of sources. *IEEE Trans. On Circuits and Systems.* 43(11): 894–906.
85. Lee, W.T., Girolami, M., and Sejnowski, T.J. (1999). Independent component analysis using an extended InfoMax algorithm for mixed sub-Gaussian and super-Gaussian sources. *Neural Comput.* 11: 417–441.
86. Hyavrynen, A. and Oja, E. (2000). Independent component analysis: algorithms and applications. *Neural Netw.* 13: 441–430.
87. Karhunen, J., Cichocki, A., Kasprzak, W., and Pajunen, P. (1997). On neural blind separation with noise suppression and redundancy reduction. *Intl. J. Neural Syst.* 8(2): 219–237.
88. Roberts, S. (1998). Independent component analysis: source assessment and separation, a Bayesian approach. *IEEE Proceedings — Vision, Image and Signal Processing.* 145: 149–154.
89. Kaas, R. and Raftery, A. (1993). Bayes Factors and Model Uncertainty. Technical Report 254, University of Washington, Seattle.

17 Classical and Bayesian Inference in fMRI

William D. Penny and Karl J. Friston

CONTENTS

17.1 INTRODUCTION

A general issue in the analysis of functional magnetic resonance imaging (fMRI) data is the relationship between the neurobiological hypothesis one posits and the statistical models adopted to test that hypothesis. One key distinction is between functional specialization and integration. Briefly, fMRI was originally used to provide functional maps showing which regions are specialized for specific functions, a classic example being the study by Zeki et al. (1) who identified

V4 and V5 as being specialized for the processing of color and motion, respectively. More recently, these analyses have been augmented by functional integration studies, which describe how functionally specialized areas interact and how these interactions depend on changes of context. A recent example is the study by Buchel et al. (2) who found that the success with which a subject learned an object-location association task was correlated with the coupling between regions in the dorsal and ventral visual streams (3). In this chapter, we will address the design and analysis of neuroimaging studies from these two distinct perspectives but note that they have to be combined for a full understanding of brain mapping results.

In practice, the general linear model (GLM) is used to identify functionally specialized brain responses and is the most prevalent approach to characterizing functional anatomy and disease-related changes. GLMs are fitted to fMRI time series at each voxel resulting in a set of voxel-specific parameters. These parameters are then used to form statistical parametric maps (SPMs) or posterior probability maps (PPMs) that characterize regionally specific responses to experimental manipulation. (Figure 17.4 and Figure 17.5, for example, show PPMs and SPMs highlighting regions that are sensitive to visual motion stimuli).

Analyses of functional integration are implemented using multivariate approaches that examine the changes in multiple brain areas induced by experimental manipulation. Although there are a number of methods for doing this, we focus on a recent approach called dynamic causal modeling (DCM).

In order to assign an observed response to a particular brain structure or cortical area, the data must conform to a known anatomical space. Before considering statistical modeling, this chapter, therefore, deals briefly with how a time series of images (from single or multiple subjects) are realigned and mapped into some standard anatomical space (e.g., a stereotactic space).

A central issue in this chapter is the distinction between classical and Bayesian estimation and inference. Historically, the most popular and successful method for the analysis of fMRI is SPM. This is based on voxelwise GLM and Gaussian random field (GRF) theory. More recently, a number of Bayesian estimation and inference procedures have appeared in the literature. A key reason behind this is that as our models become more realistic (and, therefore, complex) they need to be constrained in some way. A simple and principled way of doing this is to use priors in a Bayesian context. In this chapter we will see Bayesian methods being used in spatial normalization (Subsection 17.2.3), posterior probability mapping (Section 17.5) and DCM (Section 17.6). One should not lose sight, however, of the simplicity of the original SPM procedures (Section 17.4) as they remain attractive, both from an interpretive and computational perspective.

The analysis of functional neuroimaging data involves many steps that can be broadly divided into: (1) spatial processing, (2) estimating the parameters of a statistical model, and (3) making inferences about those parameter estimates with appropriate statistics. This data processing stream is shown in Figure 17.1.

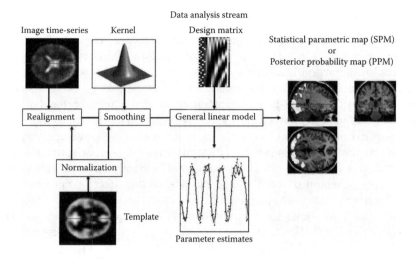

FIGURE 17.1 This schematic depicts the transformations that start with an imaging data sequence and end with a statistical parametric map (SPM) or posterior probability map (PPM). SPMs can be thought of as "x-rays" of the significance of an effect, whereas PPMs reflect our confidence that the effect is larger than a certain specified size. Voxel-based analyses require the data to be in the same anatomical space: This is effected by realigning the data (and removing movement-related signal components that persist after realignment). After realignment, the images are subject to nonlinear warping so that they match a template that already conforms to a standard anatomical space. After smoothing, the general linear model is employed to (1) estimate the parameters of the model and (2) derive the appropriate univariate test statistic at every voxel.

17.2 SPATIAL TRANSFORMATIONS

The analysis of neuroimaging data generally starts with a series of spatial transformations. These transformations aim to reduce unwanted variance components in the voxel time series that are induced by movement or shape differences among a series of scans. Subsequent analyses assume that the data from a particular voxel all derive from the same part of the brain. Violations of this assumption will introduce artifactual changes in the voxel values that may obscure changes or differences of interest. Even single-subject analyses proceed in a standard anatomical space, simply to enable reporting of regionally specific effects in a frame of reference that can be related to other studies.

The first step is to realign the data to "undo" the effects of subject movement during the scanning session. After realignment, the data are then transformed using linear or nonlinear warps into a standard anatomical space. Finally, the data are usually smoothed spatially prior to analysis with a GLM.

17.2.1 Realignment

Changes in signal intensity over time, from any one voxel, can arise from head motion and this represents a serious confound for fMRI studies. Despite physical restraints on head movement, subjects can still show displacements of upto several millimeters. Realignment involves estimating the six parameters of an affine "rigid-body" transformation that minimize the (sum of squared) differences between each successive scan and a reference scan (usually the first or the average of all scans in the time series), and applying the transformation by resampling the data using trilinear, sinc, or spline interpolation. Estimation of the affine transformation is usually effected with a first-order approximation of the Taylor expansion of the effect of movement on signal intensity using the spatial derivatives of the images (see the following subsection). This allows for a simple iterative least square solution that corresponds to a Gauss–Newton search (4). Even if this realignment were perfect, other movement-related signals (see the following text) could still persist. This calls for a further step in which the data are adjusted for residual movement-related effects.

17.2.2 Adjusting for Movement-Related Effects in fMRI

In extreme cases, as much as 90% of the variance in fMRI time series can be accounted for by the effects of movement even after realignment (5). Causes of these movement-related components are due to movement effects that cannot be modeled using a linear affine model. These nonlinear effects include: (1) subject movement between slice acquisition, (2) interpolation artifacts (6), (3) nonlinear distortion due to magnetic field inhomogeneities (7), and (4) spin-excitation history effects (5). The latter can be pronounced if the repetition time (TR) approaches T_1, making the current signal a function of movement history. These multiple effects render the movement-related signal (y) a nonlinear function of displacement (x) in the nth and previous scans $y_n = f(x_n, x_{n-1},...)$. By assuming a sensible form for this function, its parameters can be estimated using the observed time series and the estimated movement parameters x from the realignment procedure. The estimated movement-related signal is then simply subtracted from the original data. This adjustment can be carried out as a preprocessing step or embodied in model estimation during the GLM analysis. The form for $f(x)$, proposed in (5), was a nonlinear autoregression model that used polynomial expansions to second order. This model was motivated by the spin-excitation history effects and allowed displacement in previous scans to explain the current movement-related signal. However, it is also a reasonable model for many other sources of movement-related confounds. Generally, for TRs of several seconds, interpolation artifacts predominate (6) and first-order terms, comprising an expansion of the current displacement in terms of periodic basis functions, are sufficient.

17.2.3 NORMALIZATION

After realigning the data, a mean image of the series or other coregistered (e.g., a T_1-weighted) image, is used to estimate some warping parameters that map it onto a template that already conforms to some standard anatomical space (8). This estimation can use a variety of models for the mapping, including: (1) a 12-parameter affine transformation, where the parameters constitute a spatial transformation matrix, (2) low-frequency spatial basis functions (usually a discrete cosine set or polynomials), where the parameters are the coefficients of the basis functions employed and, (3) a vector field specifying the mapping for each control point (e.g., voxel). In the latter case, the parameters are vast in number and constitute a vector field that is bigger than the image itself. Estimation of the parameters of all these models can be accommodated in a simple Bayesian framework, in which one is trying to find the deformation parameters θ that have the maximum posterior probability $p(\theta|y)$ given the data y, where $p(\theta|y)p(y) = p(y|\theta)p(\theta)$. Put simply, one wants to find the deformation that is most likely, given the data. This deformation can be found by maximizing the probability of getting the data (assuming the current estimate of the deformation is true) times the probability of that estimate being true. In practice, the deformation is updated iteratively using a Gauss–Newton scheme to maximize $p(\theta|y)$. This involves jointly minimizing the likelihood and prior potentials $H(y|\theta) = -\ln p(y|\theta)$ and $H(\theta) = -\ln p(\theta)$. The likelihood potential is generally taken to be the sum of squared differences between the template and deformed image and reflects the probability of actually getting that image if the transformation was correct. The prior potential can be used to incorporate prior information about the likelihood of a given warp. Priors can be determined empirically or motivated by constraints on the mappings. Priors play a more essential role as the number of parameters specifying the mapping increases and are central to high-dimensional warping schemes (9).

In practice, most people use an affine or spatial basis function warps and iterative least squares to minimize the posterior potential. A nice extension of this approach is that the likelihood potential can be refined and taken as the difference between the index image and the best (linear) combination of templates (e.g., depicting gray, white, CSF, and skull tissue partitions). This models intensity differences that are unrelated to registration differences and allows different modalities to be coregistered (see Figure 17.2).

17.2.4 COREGISTRATION OF FUNCTIONAL AND ANATOMICAL DATA

It is sometimes useful to coregister functional and anatomical images. However, with echo-planar imaging, geometric distortions of T_2^* images, relative to anatomical T_1-weighted data, are a particularly serious problem because of the very low frequency per point in the phase-encoding direction. Typically, for echo-planar fMRI magnetic field inhomogeneity, sufficient to cause dephasing of 2 through the slice, corresponds to an in-plane distortion of a voxel. "Unwarping"

Spatial normalization

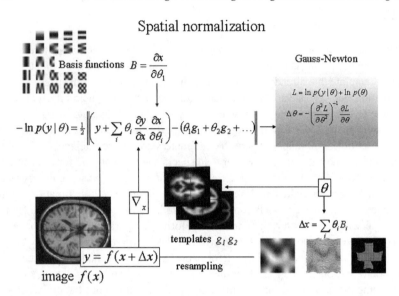

image $f(x)$

FIGURE 17.2 Schematic illustrating a Gauss–Newton scheme for maximizing the posterior probability $p(\theta|y)$ of the parameters required to spatially normalize an image. This scheme is iterative. At each step the conditional estimate of the parameters is obtained by jointly minimizing the likelihood and prior potentials. The former is the difference between a resampled (i.e., warped) version y of the image f and the best linear combination of some templates g. These parameters are used to mix the templates and resample the image to progressively reduce both the spatial and intensity differences. After convergence the resampled image can be considered normalized.

schemes have been proposed to correct for the distortion effects (10). However, this distortion is not an issue if one spatially normalizes the functional data.

17.2.5 SPATIAL SMOOTHING

The motivations for smoothing the data are fourfold:

1. By the matched filter theorem, the optimum smoothing kernel corresponds to the size of the effect that one anticipates. The spatial scale of hemodynamic responses is, according to high-resolution optical imaging experiments, about 2 to 5 mm. Despite the potentially high resolution afforded by fMRI an equivalent smoothing is suggested for most applications.
2. By the central limit theorem, smoothing the data will render the errors more normal in their distribution and ensure the validity of inferences based on parametric tests.
3. When making inferences about regional effects using GRF theory (see the following subsection), the assumption is that the error terms are a

reasonable lattice representation of an underlying and smooth Gaussian field. This necessitates smoothness to be substantially greater than voxel size. If the voxels are large, then they can be reduced by sub-sampling the data and smoothing (with the original point spread function) with little loss of intrinsic resolution.

4. In the context of intersubject averaging it is often necessary to smooth more (e.g., 8 to 12 mm) to project the data onto a spatial scale where homologies in functional anatomy are expressed among subjects.

17.3 GENERAL LINEAR MODEL

Statistical analysis of fMRI data entails (1) modeling the data to partition observed neurophysiological responses into components of interest, confounds, and error, and (2) making inferences about the interesting effects in relation to the error variance. This can be regarded as a direct comparison of the variance attributable to an interesting experimental manipulation to the variance attributable to the error. These comparisons can be made with T or F statistics.

A brief review of the literature may give the impression that there are numerous ways to analyze fMRI time series with a diversity of statistical and conceptual approaches. This is, however, not the case. With few exceptions, every analysis is a variant of the general linear model (GLM). This includes: (1) simple T tests on scans assigned to one condition or another, (2) correlation coefficients between observed responses and boxcar stimulus functions in fMRI, (3) inferences made using multiple linear regression, (4) evoked responses estimated using linear time invariant models, and (5) selective averaging to estimate event-related responses. Mathematically, they are formally identical and can be implemented with the same equations and algorithms. The only thing that distinguishes among them is the design matrix encoding the experimental design. The use of the correlation coefficient deserves special mention because of its popularity in fMRI (11). The significance of a correlation is identical to the significance of the equivalent T statistic testing for a regression of the data on the stimulus function. The correlation coefficient approach is useful but the inference is effectively based on a limiting case of multiple linear regression that obtains when there is only one regressor. In fMRI, many regressors usually enter into a statistical model. Therefore, the T statistic provides a more versatile and generic way of assessing the significance of regional effects.

17.3.1 DESIGN MATRIX

The GLM is an equation $Y = X\beta + \varepsilon$ that expresses the observed response variable Y in terms of a linear combination of explanatory variables X plus a well-behaved error term. The GLM is variously known as "analysis of covariance" or "multiple regression analysis" and subsumes simpler variants like the "T test" for a difference in means to more elaborate linear convolution models such as finite impulse response (FIR) models. The matrix X that contains the explanatory variables (e.g., designed effects or confounds) is called the *design matrix*. Each column of the design matrix corresponds to some effect built into the experiment or that may confound the results. These are referred to as explanatory variables, covariates, or regressors.

The design matrix can contain both covariates and indicator variables. Each column of X has an associated unknown parameter. Some of these parameters will be of interest (e.g., the effect of a particular sensorimotor or cognitive condition, or the regression coefficient of hemodynamic responses on reaction time). The remaining parameters will be of no interest and pertain to confounding effects (e.g., the effect of being a particular subject or the regression slope of voxel activity on global activity).

The example in Figure 17.1 relates to a fMRI study of visual stimulation under four conditions. The effects on the response variable are modeled in terms of functions of the presence of these conditions (i.e., boxcars smoothed with a hemodynamic response function) and constitute the first four columns of the design matrix. There then follows a series of terms that are designed to remove or model low-frequency variations in signal due to artifacts such as aliased biorhythms and other drift terms. The final column is whole brain activity. The relative contribution of each of these columns is assessed using standard least squares or Bayesian estimation. Classical inferences about these contributions are made using T or F statistics, depending upon whether one is looking at a particular linear combination (e.g., a subtraction), or all of them together. Bayesian inferences are based on the posterior or conditional probability that the contribution exceeded some threshold, usually zero.

Due primarily to the presence of aliased biorhythms and unmodeled neuronal activity, the errors in the GLM will be temporally autocorrelated. To accommodate this, the GLM has been extended (12) to incorporate intrinsic nonsphericity, or correlations among the error terms. This generalization brings with it the notion of *effective degrees of freedom*, which are less than the conventional degrees of freedom under i.i.d. assumptions (see footnote). They are smaller because the temporal correlations reduce the effective number of independent observations. More recently, a restricted maximum likelihood (ReML) algorithm for estimation of the autocorrelation, variance components, and regression parameters has been proposed (13).

17.3.2 CONTRASTS

To assess effects of interest that are spanned by one or more columns in the design matrix one uses a contrast (i.e., a linear combination of parameter estimates). An example of a contrast weight vector would be [−1 1 0 0 ...] to compare the difference in responses evoked by two conditions, modeled by the first two condition-specific regressors in the design matrix. Sometimes several contrasts of parameter estimates are jointly interesting. For example, when using polynomial (14) or basis function expansions (see Subsection 17.3.1) of some experimental factor. In these instances, a *matrix* of contrast weights is used that can be thought of as a collection of effects that one wants to test together. Such a contrast may look like,

$$\begin{bmatrix} -1 & 0 & 0 & 0 & \dots \\ 0 & 1 & 0 & 0 & \dots \end{bmatrix}$$

which would test for the significance of the first or second parameter estimates. The fact that the first weight is -1 as opposed to 1 has no effect on the test because F statistics are based on sums of squares.

17.3.3 TEMPORAL BASIS FUNCTIONS

Functional MRI using blood oxygen level dependent (BOLD) contrast provides an index of neuronal activity indirectly via changes in blood oxygenation levels. For a given impulse of neuronal activity, the fMRI signal peaks some 4–6 sec later, then after 10 sec or so drops below zero and returns to baseline after 20 to 30 sec. This response varies from subject to subject and from voxel to voxel and this variation can be captured using temporal basis functions.

In Reference 15, the form of the hemodynamic impulse response function (HRF) was estimated using a least squares deconvolution and a time invariant model, where evoked neuronal responses are convolved with the HRF to give the measured hemodynamic response (16). This simple linear framework is the cornerstone for making statistical inferences about activations in fMRI with the GLM. An impulse response function is the response to a single impulse, measured at a series of times after the input. It characterizes the input–output behavior of the system (i.e., voxel) and places important constraints on the sorts of inputs that will excite a response. The HRFs, estimated in (15) resembled a Poisson or gamma function, peaking at about 5 sec.

Knowing the forms that the HRF can take is important for several reasons, not the least of which is because it allows for better statistical models of the data. The HRF may vary from voxel to voxel, and this has to be accommodated in the GLM. To allow for different HRFs in different brain regions, the notion of temporal basis functions to model evoked responses in fMRI was introduced (17) and applied to event-related responses in (18,19). The basic idea behind temporal basis functions is that the hemodynamic response induced by any given trial type can be expressed as the linear combination of several (basis) functions of peristimulus time. The convolution model for fMRI responses takes a stimulus function encoding the supposed neuronal responses and convolves it with an HRF to give a regressor that enters into the design matrix. When using basis functions, the stimulus function is convolved with all the basis functions to give a series of regressors. The associated parameter estimates are the coefficients or weights that determine the mixture of basis functions that best models the HRF for the trial type and voxel in question. We find the most useful basis set to be a canonical HRF and its derivatives with respect to the key parameters that determine its form (e.g., latency and dispersion).

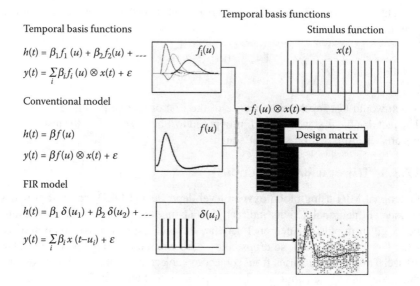

Temporal basis functions

$$h(t) = \beta_1 f_1(u) + \beta_2 f_2(u) + \text{---}$$

$$y(t) = \sum_i \beta_i f_i(u) \otimes x(t) + \varepsilon$$

Conventional model

$$h(t) = \beta f(u)$$

$$y(t) = \beta f(u) \otimes x(t) + \varepsilon$$

FIR model

$$h(t) = \beta_1 \delta(u_1) + \beta_2 \delta(u_2) + \text{---}$$

$$y(t) = \sum_i \beta_i x(t - u_i) + \varepsilon$$

FIGURE 17.3 Temporal basis functions offer useful constraints on the form of the estimated response that retain the flexibility of FIR models and the efficiency of single regressor models. The specification of these models involves setting up stimulus functions $x(t)$ that model expected neuronal changes (e.g., boxcars of epoch-related responses or spikes [delta functions] at the onset of specific events or trials). These stimulus functions are then convolved with a set of basis functions $f_i(u)$ of peristimulus time u that model the HRF in some linear combination. The ensuing regressors are assembled into the design matrix. The basis functions can be as simple as a single canonical HRF (middle), through to a series of delayed delta functions (bottom). The latter case corresponds to a FIR model and the coefficients constitute estimates of the impulse response function at a finite number of discrete sampling times. Selective averaging in event-related fMRI is mathematically equivalent to this limiting case.

The advantage of this approach is that it can partition differences among evoked responses into differences in magnitude, latency, or dispersion, which can be tested for using specific contrasts (20).

Temporal basis functions are important because they enable a graceful transition between conventional multilinear regression models with one stimulus function per condition and FIR models with a parameter for each time point following the onset of a condition or trial type. Figure 17.3 illustrates this graphically (see figure legend). In summary, temporal basis functions offer useful constraints on the form of the estimated response that retain the flexibility of FIR models and the efficiency of single regressor models. The advantage of using several temporal basis functions (as opposed to an assumed form for the HRF) is that one can model voxel-specific forms for hemodynamic responses and formal differences (e.g., onset latencies) among responses to different sorts of events. The advantages of using basis functions over FIR models are that the parameters are estimated more efficiently and stimuli can be presented at any point in the interstimulus interval. The latter is important because time locking stimulus presentation and data acquisition gives a biased sampling over peristimulus time and can lead to differential sensitivities in multislice acquisition over the brain.

17.4 STATISTICAL PARAMETRIC MAPPING

Statistical parametric mapping (SPM) entails the construction of spatially extended statistical processes to test hypotheses about regionally specific effects (21). SPMs are image processes with voxel values that are, under the null hypothesis, distributed according to a known probability density function, usually the Student's t- or f- distributions. These are known colloquially as t- or f-maps. The success of statistical parametric mapping is due largely to the simplicity of the idea. One analyzes each and every voxel using any standard (univariate) statistical test. The resulting statistical parameters are assembled into an image — the SPM. SPMs are interpreted as spatially extended statistical processes by referring to the probabilistic behavior of Gaussian fields (22–25). GRF model both the univariate probabilistic characteristics of a SPM and any nonstationary spatial covariance structure. "Unlikely" excursions of the SPM are interpreted as regionally specific effects, attributable to the sensorimotor or cognitive process that has been manipulated experimentally.

Over the years, statistical parametric mapping has come to refer to the conjoint use of GLM and GRF theory to analyze and make classical inferences about spatially extended data through SPMs. The GLM is used to estimate some parameters that could explain the spatially continuous data in exactly the same way as in conventional analysis of discrete data. GRF theory is used to resolve the multiple comparison problem that ensues when making inferences over a volume of the brain. GRF theory provides a method for correcting p values for the search volume of an SPM and plays the same role for continuous data (i.e., images) as the Bonferroni correction for the number of discontinuous or discrete statistical tests. The approach was called SPM for three reasons: (1) to acknowledge "significance probability mapping", the use of interpolated pseudomaps of p values used to summarize the analysis of multichannel ERP studies; (2) for consistency with the nomenclature of parametric maps of physiological or physical parameters (e.g., regional cerebral blood flow rCBF or volume rCBV parametric maps); and (3) In reference to the *parametric statistics* that comprise the maps. Despite its simplicity, there are some fairly subtle motivations for the approach that deserve mention. Usually, given a response or dependent variable comprising many thousands of voxels, one would use *multivariate analyses* as opposed to the *mass-univariate approach* that SPM represents. The problems with multivariate approaches are they do not support inferences about regionally specific effects, they require more observations than the dimension of the response variable (i.e., number of voxels) and even in the context of dimension reduction, they are less sensitive to focal effects than mass-univariate approaches. A heuristic argument for their relative lack of power is that multivariate approaches (in their most general form) estimate the model's error covariances using a number of parameters (e.g., the covariance between the errors at all pairs of voxels). In general, the more parameters an estimation procedure has to deal with, the more variable the estimate of any one parameter becomes. This renders any single estimate less efficient.

Multivariate approaches consider voxels as different levels of an experimental or treatment factor and use classical analysis of variance, not at each voxel (c.f. SPM), but by considering the data sequences from all voxels together, as replications over voxels. The problem here is that regional changes in error variance, and spatial correlations in the data, induce profound nonsphericity* in the error terms. This nonsphericity would require large numbers of parameters to be estimated for each voxel using conventional techniques. In SPM, the nonsphericity is parameterized in a very parsimonious way with just two parameters for each voxel. These are the error variance and smoothness estimators. This minimal parameterization lends SPM a sensitivity that surpasses multivariate approaches. SPM can do this because GRF theory implicitly imposes constraints on the nonsphericity implied by the continuous and (spatially) extended nature of the data. This is something that conventional multivariate and equivalent univariate approaches do not accommodate, to their cost.

Some analyses use statistical maps based on nonparametric tests that eschew distributional assumptions about the data e.g., nonparametric approaches (26). These approaches are generally less powerful (i.e., less sensitive) than parametric approaches (27). However, they have an important role in evaluating the assumptions behind parametric approaches and may supersede in terms of sensitivity when these assumptions are violated (e.g., when degrees of freedom are very small and voxel sizes are large in relation to smoothness).

17.4.1 RANDOM FIELD THEORY

Classical inferences using SPMs can be of two sorts depending on whether one knows where to look in advance. With an anatomically constrained hypothesis about effects in a particular brain region, the uncorrected p value associated with the height or extent of that region in the SPM can be used to test the hypothesis. With an anatomically open hypothesis (i.e., a null hypothesis that there is no effect anywhere in a specified volume of the brain), a correction for multiple dependent comparisons is necessary. The theory of random fields provides a way of adjusting the p value that takes into account the fact that neighboring voxels

---

* Sphericity refers to the assumption of identically and independently distributed error terms (i.i.d.). Under i.i.d., the probability density function of the errors, from all observations, has spherical isocontours, hence, sphericity. Deviations from either of the i.i.d. criteria constitute nonsphericity. If the error terms are not identically distributed then different observations have different error variances. Correlations among error terms reflect dependencies among the error terms (e.g., serial correlation in fMRI time series) and constitute the second component of nonsphericity. In fMRI both spatial and temporal nonsphericity can be quite profound issues.

are not independent by virtue of continuity in the original data. Provided the data are sufficiently smooth, the GRF correction is less severe (i.e., is more sensitive) than a Bonferroni correction for the number of voxels. As noted in the preceding text, the GRF theory deals with the multiple comparisons problem in the context of continuous, spatially extended statistical fields, in a way that is analogous to the Bonferroni procedure for families of discrete statistical tests. There are many ways to appreciate the difference between GRF and Bonferroni corrections. Perhaps the most intuitive is to consider the fundamental difference between an SPM and a collection of discrete T values. When declaring a connected volume or region of the SPM to be significant, we refer collectively to all the voxels that comprise that volume. The false positive rate is expressed in terms of connected (excursion) sets of voxels above some threshold, under the null hypothesis of no activation. This is not the expected number of false positive voxels. One false positive region may contain hundreds of voxels, if the SPM is very smooth. A Bonferroni correction would control the expected number of false positive voxels, whereas GRF theory controls the expected number of false positive regions. Because a false positive region can contain many voxels, the corrected threshold under a GRF correction is much lower, rendering it much more sensitive. In fact the number of voxels in a region is somewhat irrelevant because the correction is a function of smoothness. The GRF correction discounts voxel size by expressing the search volume in terms of smoothness or resolution elements (resels). This intuitive perspective is expressed formally in terms of differential topology using the Euler characteristic (23). At high thresholds the Euler characteristic corresponds to the number of regions exceeding the threshold.

There are only two assumptions underlying the use of the GRF correction: (1) The error fields (but not necessarily the data) are a reasonable lattice approximation to an underlying random field with a multivariate Gaussian distribution and (2) these fields are continuous, with a differentiable and invertible autocorrelation function. A common misconception is that the autocorrelation function has to be Gaussian. It does not. The only way in which these assumptions can be violated is if the data are not smoothed (with or without subsampling to preserve resolution), violating the reasonable lattice assumption or the statistical model is misspecified so that the errors are not normally distributed. Early formulations of the GRF correction were based on the assumption that the spatial correlation structure was wide-sense stationary. This assumption can now be relaxed due to a revision of the way in which the smoothness estimator enters the correction procedure (28). In other words, the corrections retain their validity, even if the smoothness varies from voxel to voxel.

17.5 POSTERIOR PROBABILITY MAPPING

Despite its success, SPM has a number of fundamental limitations. In SPM, the p value, ascribed to a particular effect, does not reflect the likelihood that the effect is present but simply the probability of getting the observed data in the effect's absence. If sufficiently small, this p value can be used to reject the null hypothesis that the effect is negligible. There are several shortcomings in this classical approach. Firstly, one can never reject the alternate hypothesis (i.e., say that an activation has not occurred) because the probability that an effect is exactly zero is itself zero. This is problematic, for example, in trying to establish double dissociations or indeed functional segregation; one can never say one area responds to color but not motion and another responds to motion but not color. Secondly, because the probability of an effect being zero is vanishingly small, given enough scans or subjects one can always demonstrate a significant effect at every voxel. This fallacy of classical inference is becoming relevant practically, with the thousands of scans entering into some fixed-effect analyzes of fMRI data. The issue here is that a trivially small activation can be declared significant if there are sufficient degrees of freedom to render the variability of the activation's estimate small enough. A third problem that is specific to SPM is the correction or adjustment applied to the p values to resolve the multiple comparison problem. This has the somewhat nonsensical effect of changing the inference about one part of the brain in a way that is contingent on whether another part is examined. Put simply, the threshold increases with search volume, rendering the inference very sensitive to what it encompasses. Clearly, the probability that any voxel has activated does not change with the search volume and yet the classical p value does.

All these problems would be eschewed by using the probability that a voxel had been activated or indeed, that its activation was greater than some threshold. This sort of inference is precluded by classical approaches, which simply give the likelihood of getting the data, given no activation. What one would really like is the probability distribution of the activation, given the data. This is the posterior probability used in Bayesian inference. The posterior distribution requires both the likelihood, afforded by assumptions about the distribution of errors, and the prior probability of activation. These priors can enter as known values or can be estimated from the data, provided we have observed multiple instances of the effect we are interested in. The latter is referred to as empirical Bayes. A key point here is that we do assess repeatedly the same effect over different voxels, and we are, therefore, in a position to adopt an empirical Bayesian approach (29).

17.5.1 EMPIRICAL EXAMPLE

In this subsection, we compare and contrast Bayesian and classical inference using PPMs and SPMs based on real data. The data set comprised data from a study of attention to visual motion (30). The data used here came from the first subject studied. This subject was scanned at 2-T to give a time series of 360 images comprising 10-block epochs of different visual motion conditions. These conditions included a fixation condition, visual presentation of static dots, visual presentation of radially moving dots under attention, and no-attention conditions. In the attention, condition subjects were asked to attend to changes in speed (which did not actually occur). This attentional manipulation was validated post-hoc using psychophysics and the motion after-effect. Further details of the data acquisition are given in the caption to Figure 17.8. These data were analyzed

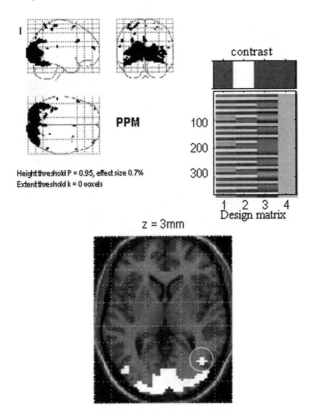

FIGURE 17.4 PPM for the fMRI study of attention to visual motion. The display format in the lower panel uses an axial slice through extrastriate regions, but the thresholds are the same as employed in maximum intensity projections (upper panels). The activation threshold for the PPM was 0.7. As can be imputed from the design matrix, the statistical model of evoked responses comprised of boxcar regressors convolved with a canonical hemodynamic response function.

using a conventional SPM procedure and the empirical Bayesian approach described in the previous section. The ensuing SPMs and PPMs are presented in Figure 17.4 and Figure 17.5. We used a contrast that tested for the effect of visual motion above and beyond that due to photic stimulation with stationary dots.

The difference between the PPM and SPM is immediately apparent on inspection of Figure 17.4 and Figure 17.5. Here, the threshold for the PPM was 0.7% (equivalent to percentage whole brain mean signal). Only voxels that exceed 95% confidence are shown. These are restricted to visual and extrastriate cortex involved in motion processing. The critical thing to note is that the corresponding SPM identifies a smaller number of voxels than the PPM. Indeed, the SPM appears to have missed a critical and bilaterally represented part of the V5 complex

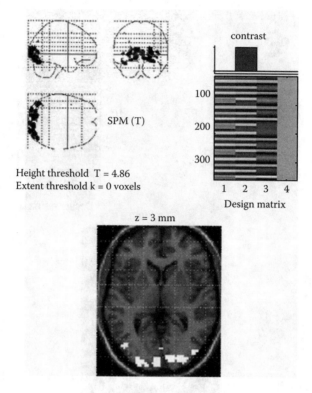

FIGURE 17.5 Same as for Figure 17.4, but this time showing the corresponding SPM using a corrected threshold at $p = 0.05$.

(circled cluster on the PPM in the lower panel of Figure 17.4). The SPM is more conservative because the correction for multiple comparisons in these data is very severe, rendering classical inference relatively insensitive. It is interesting to note that dynamic motion in the visual field has such widespread (though small) effects at a hemodynamic level.

17.6 DYNAMIC CAUSAL MODELING

Dynamic causal modeling (DCM) (31) is used to make inferences about functional integration from fMRI time series. The term "causal" in DCM arises because the brain is treated as a deterministic dynamical system in which external inputs cause changes in neuronal activity, which in turn cause changes in the resulting BOLD signal that is measured with fMRI. This is to be contrasted with a conventional GLM where there is no explicit representation of neuronal activity. The second main difference to the GLM is that DCM allows for interactions between regions. Of course, it is this interaction which is central to the study of functional integration.

Current DCMs for fMRI comprise a bilinear model for the neurodynamics and an extended balloon model for the hemodynamics. These are shown in Figure 17.6 and Figure 17.7. The neurodynamics are described by the multivariate differential equation shown in Figure 17.6. This is known as a bilinear model because the dependent variable, $\dot{z}$, is linearly dependent on the product of z and u. That u and z combine in a multiplicative fashion endows the model with "nonlinear" dynamics, which can be understood as a nonstationary linear system that changes according to experimental manipulation u. Importantly, because u is known, parameter estimation is relatively simple.

Connectivity in DCM is characterized by a set of "intrinsic connections," A, that specify which regions are connected and whether these connections are unidirectional or bidirectional. We also define a set of input connections, C, that specify which inputs are connected to which regions, and a set of modulatory or bilinear connections, B^j, that specify which intrinsic connections can be changed by which inputs. The overall specification of input, intrinsic, and modulatory connectivity comprise our assumptions about model structure. This in turn represents a scientific hypothesis about the structure of the large-scale neuronal network mediating the underlying sensorimotor or cognitive function.

In DCM, neuronal activity gives rise to hemodynamic activity by a dynamic process described by an extended balloon model. This involves a set of hemodynamic state variables, state equations and hemodynamic parameters shown in Figure 17.7. Together, these equations describe a nonlinear hemodynamic process that may be regarded as a biophysically informed generalization of the linear convolution models used in the GLM. It is possible to describe the second-order behavior of this process (i.e., how the response to one stimulus is changed by a preceding stimulus) using Volterra kernels.

17.6.1 Empirical Example

We now return to the visual motion study described in Subsection 17.5.1 so as to make inferences about functional integration. Figure 17.8b shows the location of the regions that entered the DCM (Figure 17.8b). These regions were based on maxima from conventional SPMs testing for the effects of photic stimulation, motion, and attention. Regional time courses were taken as the first eigenvariate of spherical volumes of interest centered on the maxima shown in Figure 17.8. The inputs, in this example, comprise one sensory perturbation and two contextual inputs. The sensory input was simply the presence of photic stimulation, and the first contextual one was presence of motion in the visual field. The second contextual input, encoding an attentional set, was unity during attention to speed changes and zero otherwise. The outputs corresponded to the four regional eigenvariates in (Figure 17.8b). The intrinsic connections were constrained to conform to a hierarchical pattern in which each area was reciprocally connected to its supraordinate area. Photic stimulation entered at, and only at, $V1$. The effect of motion in the visual field was modeled as a bilinear modulation of the $V1$ to $V5$

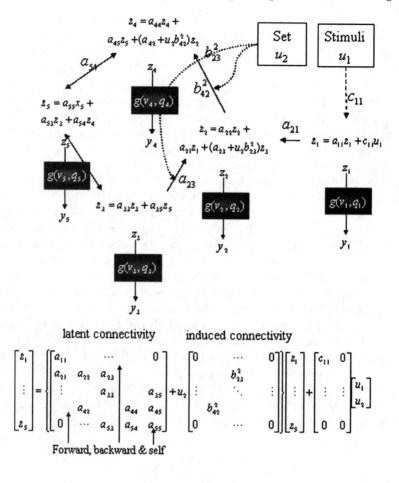

$$\dot{z} = \left(A + \sum_j u_j B^j\right)z + Cu$$

The bilinear model

FIGURE 17.6 This schematic (upper panel) outlines the differential equations implied by a bilinear model. The equations in each of the white areas describe the change in neuronal activity z_i in terms of linearly separable components that reflect the influence of other regional state variables. Note particularly how the second contextual inputs enter these equations. They effectively increase the intrinsic coupling parameters (a_{ij}) in proportion to the bilinear coupling parameters (b_{ij}^k). In this diagram, the hemodynamic component of the DCM illustrates how the neuronal states enter a region-specific hemodynamic model to produce the outputs y_i that are a function of the region's biophysical states reflecting deoxyhemoglobin content and venous volume (q_i and v_i). The lower panel reformulates the differential equations in the upper panel into a matrix format. These equations can be summarized more compactly in terms of coupling parameter matrices A, B^j and C.

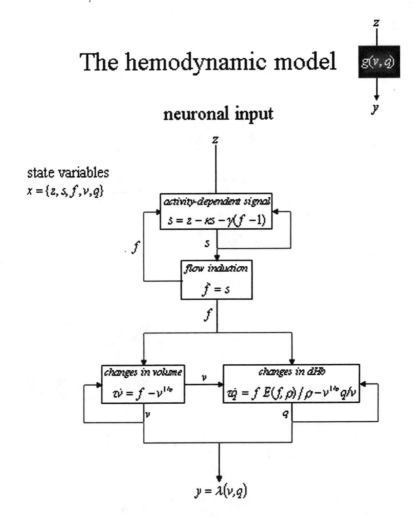

FIGURE 17.7 This schematic shows the architecture of the hemodynamic model for a single region (regional subscripts have been dropped for clarity). Neuronal activity induces a vasodilatory and activity-dependent signal s that increases the flow f. Flow causes changes in volume and deoxyhemoglobin (v and q). These two hemodynamic states enter an output nonlinearity to give the observed BOLD response y. This transformation from neuronal states z_i to hemodynamic response y_i is encoded graphically by the dark-gray boxes in the previous figure and that inserted in this figure.

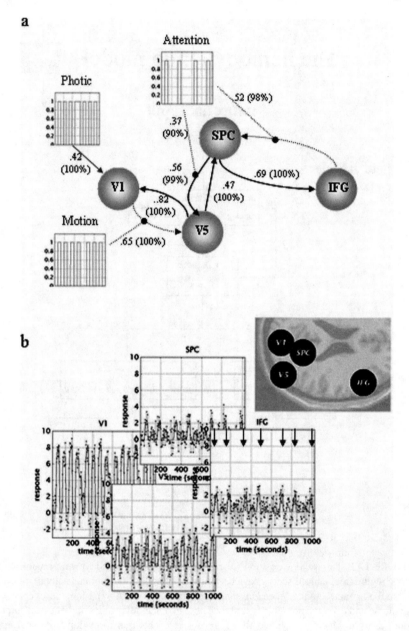

FIGURE 17.8 DCM for the fMRI study of attention to visual motion. The most interesting aspects of this connectivity involve (a) the role of motion and attention in exerting bilinear effects. Critically, the influence of motion is to enable connections from V1 to the motion-sensitive area V5. The influence of attention is to enable backward connections from the inferior frontal gyrus (IFG) to the superior parietal cortex (SPC). Furthermore, attention increases the latent influence of SPC on V5. Dotted arrows connecting regions represent

(continued)

FIGURE 17.8 (continued) significant bilinear effects in the absence of a significant intrinsic coupling. Number in brackets represent the posterior probability, expressed as a percentage, that the effect size is larger than 0.17. This cutoff corresponds to a time constant of 4 sec or less — in DCM stronger effects have faster time constants. (b) Fitted responses based upon the conditional estimates and the adjusted data. The insert shows the location of the regions centered on the primary visual cortex V1: 6, −84, −6 mm; motion-sensitive area V5: 45, −81, 5 mm. SPC: 18, −57, 66 mm; IFG: 54, 18, 30 mm. The volumes from which the first eigenvariates were calculated corresponded to 8 mm radius spheres centered on these locations. Subjects were studied with fMRI under identical stimulus conditions (visual motion subtended by radially moving dots) while manipulating the attentional component of the task (detection of velocity changes). The data were acquired from normal subjects at 2-T using a Magnetom VISION (Siemens, Erlangen) whole body MRI system, equipped with a head volume coil. Here, we analyze data from the first subject. Contiguous multislice T2*-weighted fMRI images were obtained with a gradient echo-planar sequence (TE = 40 msec, TR = 3.22 sec, matrix size = $64 \times 64 \times 32$, voxel size $3 \times 3 \times 3$ mm). Each subject had 4 consecutive 100-scan sessions comprising a series of 10-scan blocks (D F A F N F A F N S) under 5 different conditions. The first condition (D) was a dummy condition to allow for magnetic saturation effects. F (fixation) corresponds to a low-level baseline where the subjects viewed a fixation point at the center of a screen. In condition A (attention), subjects viewed 250 dots moving radially from the center at 4.7° per sec and were asked to detect changes in radial velocity. In condition N (no attention), the subjects were asked simply to view the moving dots. In condition S (stationary), subjects viewed stationary dots. The order of A and N was swapped for the last two sessions. In all conditions, subjects fixated the center of the screen. In a prescanning session, the subjects were given five trials with five speed changes (reducing to 1%). During scanning there were no speed changes. No overt response was required in any condition.

connectivity and attention was allowed to modulate the backward connections from **IFG** and **SPC**.

The results of the DCM are shown in Figure 17.8a. Of primary interest here is the modulatory effect of attention that is expressed in terms of the bilinear coupling parameters for this third input. As hoped, we can be highly confident that attention modulates the backward connections from **IFG** to **SPC** and from **SPC** to **V5**. Indeed, the influences of **IFG** on **SPC** are negligible in the absence of attention (dotted connection in Figure 17.8a). It is important to note that the only way that attentional manipulation could affect brain responses was through this bilinear effect. Attention-related responses are seen throughout the system (attention epochs are marked with arrows in the plot of **IFG** responses in Figure 17.8b). This attentional modulation is accounted for, sufficiently, by changing just two connections. This change is, presumably, instantiated by the instructional set at the beginning of each epoch. The second point this analysis illustrates is how the functional segregation is modeled in DCM. Here one can regard **V1** as a "segregating" motion from other visual information and distributing it to the motion-sensitive area **V5**. This segregation is modeled as a bilinear "enabling" of **V1** to **V5** connections when, and only when, motion is present. Note that in the absence of motion the intrinsic **V1** to **V5** connection was trivially small (in fact, the MAP estimate was −0.04). The key advantage of entering motion through a bilinear effect, as opposed to a direct effect on **V5**, is that we can finesse the inference that **V5** shows motion-selective responses with the assertion that these responses are mediated by afferents from **V1**.

The two bilinear effects given in the preceding text represent two important aspects of functional integration that DCM was designed to characterize.

17.7 CONCLUSION

Due to the concise nature of this review, we have been unable to cover a number of related topics. These include computational neuroanatomy, analysis of group data (whether structural or functional) using either fixed- or random-effect analysis. In the context of the GLM, we omitted discussion of event-related vs. block designs, parametric and factorial designs, and the factors underlying an efficient experimental design. We refer interested readers to the recent volume entitled "human brain function" (32) that builds upon the basic issues introduced here. These methods and all of the procedures covered in this review have been implemented in a public domain software package called SPM2 available from http://www.fil.ion.ucl.ac.uk/spm/.

REFERENCES

1. Zeki, S., Watson, J.D., Lueck, C.J., Friston, K.J., Kennard, C., and Frackowiak, R.S. (1991). A direct demonstration of functional specialization in human visual cortex. *J. Neurosci.* Vol. 11, pp. 641–649.
2. Buchel, C., Coull, J., and Friston, K.J. (1999). The predictive value of changes in effective connectivity for human learning. *Science* 283 (5407), 1538–41.
3. Ungerleider, L.G. and Mishkin, M. (1982). in *Analysis of Visual Behavior,* Eds. Ingle, D.J., Goodale, M.A., and Mansfield, R.J.W. MIT Press, Cambridge, MA., 549–586.
4. Friston, K.J., Ashburner, J., Frith, C.D., Poline, J.-B., Heather, J.D., and Frackowiak, R.S.J. (1995). Spatial registration and normalization of images. *Hum. Brain Mapp.* **2**: 165–189.
5. Friston, K.J., Williams, S., Howard, R., Frackowiak, R.S.J., and Turner, R. (1996). Movement-related effects in fMRI time series. *Magn. Reson. Med.* **35**: 346–355.
6. Grootoonk, S., Hutton, C., Ashburner, J., Howseman, A.M., Josephs, O., Rees, G., Friston, K.J., and Turner, R. (2000). Characterization and correction of interpolation effects in the realignment of fMRI time series. *NeuroImage.* **11**: 49–57.
7. Andersson, J.L., Hutton, C., Ashburner, J., Turner, R., and Friston, K. (2001). Modeling geometric deformations in EPI time series. *NeuroImage.* **13**: 903–19.
8. Talairach, P. and Tournoux, J. (1988). *A Stereotactic Coplanar Atlas of the Human Brain.* Stuttgart Thieme.
9. Ashburner, J., Neelin, P., Collins, D.L., Evans, A., and Friston, K. (1997). Incorporating prior knowledge into image registration. *NeuroImage* **6**: 344–352.
10. Jezzard, P. and Balaban, R.S. (1995). Correction for geometric distortion in echoplanar images from B0 field variations. *Mag. Reson. Med.* **34**: 65–73.
11. Bandettini, P.A., Jesmanowicz, A., Wong, E.C., and Hyde, J.S. (1993). Processing strategies for time course data sets in functional MRI of the human brain. *Magn. Reson. Med.* **30**: 161–173.
12. Worsley, K.J. and Friston, K.J. (1995). Analysis of fMRI time-series revisited — again. *NeuroImage.* **2**: 173–181.

13. Friston, K.J., Penny, W., Phillips, C., Kiebel, S., Hinton, G., and Ashburner, J. (2002). Classical and Bayesian inference in neuroimaging: theory. *NeuroImage.* 16: 465–483.
14. Büchel, C., Wise, R.J.S., Mummery, C.J., Poline, J.-B., and Friston, K.J. (1996). Nonlinear regression in parametric activation studies. *NeuroImage. 4: 60–66.*
15. Friston, K.J., Jezzard, P.J., and Turner, R. (1994). Analysis of functional MRI time series. *Hum. Brain Mapp.* **1**: 153–171.
16. Boynton, G.M., Engel, S.A., Glover, G.H., and Heeger, D.J. (1996). Linear systems analysis of functional magnetic resonance imaging in human V1. *J. Neurosci.* **16**: 4207–4221.
17. Friston, K.J., Frith, C.D., Turner, R., and Frackowiak, R.S.J. (1995c). Characterizing evoked hemodynamics with fMRI. *NeuroImage* **2**: 157–165.
18. Josephs, O., Turner, R., and Friston, K.J. (1997). Event-related fMRI. *Hum. Brain Mapp.* **5**: 243–248.
19. Lange, N. and Zeger, S.L. (1997). Nonlinear Fourier time-series analysis for human brain mapping by functional magnetic resonance imaging (with discussion). *J. Roy. Stat. Soc. Ser C.* **46**: 1–29.
20. Friston, K.J., Fletcher, P., Josephs, O., Holmes, A., Rugg, M.D., and Turner, R. (1998b) Event-related fMRI: Characterizing differential responses. *NeuroImage.* **7**: 30–40.
21. Friston, K.J., Frith, C.D., Liddle, P.F., and Frackowiak, R.S.J. (1991). Comparing functional (PET) images: the assessment of significant change. *J. Cereb. Blood Flow Metab.* **11**: 690–699.
22. Adler, R.J. (1981). in *The Geometry of Random Fields.* Wiley. New York.
23. Worsley, K.J., Evans, A.C., Marrett, S., and Neelin, P. (1992). A three-dimensional statistical analysis for rCBF activation studies in human brain. *J Cereb. Blood Flow Metab.* **12**: 900–918.
24. Friston, K.J., Worsley, K.J., Frackowiak, R.S.J., Mazziotta, J.C., and Evans, A.C. (1994). Assessing the significance of focal activations using their spatial extent. *Hum. Brain Mapp.* **1**: 214–220.
25. Worsley, K.J., Marrett, S., Neelin, P., Vandal, A.C., Friston, K.J., and Evans, A.C. (1996). A unified statistical approach for determining significant signals in images of cerebral activation. *Hum. Brain Mapp.* **4**: 58–73.
26. Nichols, T.E. and Holmes, A.P. (2001). Nonparametric permutation tests for functional neuroimaging: a primer with examples. *Hum. Brain Mapp.* **15**: 1–25.
27. Aguirre, G.K., Zarahn, E., and D'Esposito, M. (1998). A critique of the use of the Kolmogorov-Smirnov (KS) statistic for the analysis of BOLD fMRI data. *Magn. Reson. Med.* **39**: 500–505.
28. Kiebel, S.J., Poline, J.B., Friston, K.J., Holmes, A.P., and Worsley, K.J. (1999). Robust smoothness estimation in statistical parametric maps using standardized residuals from the general linear model. *NeuroImage.* **10**: 756–766.
29. Friston, K.J. and Penny, W. (2003). Posterior probability maps and SPMs. *NeuroImage.* 19(3): 1240–1249.
30. Büchel, C. and Friston, K.J. (1997). Modulation of connectivity in visual pathways by attention: Cortical interactions evaluated with structural equation modeling and fMRI. *Cereb. Cortex.* **7**: 768–778.
31. Friston, K.J., Harrison, L., and Penny, W. (2003). Dynamic causal modeling. *NeuroImage.* 19(4): 1273–1302.
32. Frackowiak, R., Friston, K.J., Frith, C., Dolan, R., Price, C., Zeki, S., Ashburner, J., and Penny, W. (2003). *Human Brain Function.* 2nd ed. Elsevier Academic Press.

18 Modeling and Nonlinear Analysis in fMRI via Statistical Learning

Yongmei Michelle Wang

CONTENTS

18.1 INTRODUCTION

This chapter focuses on functional magnetic resonance imaging (fMRI) data analysis and modeling using statistical learning techniques. fMRI is a powerful technique for mapping brain function by using the blood oxygenation level dependent (BOLD) effect (32); however, the small signal change due to the BOLD effect is very noisy and susceptible to artifacts such as those caused by scanner drift, head motion, and cardiorespiratory effects. Although a task or stimulus can be repeated over and over again, there are limits due to time constraints, habituation effects, etc. Therefore, refined techniques from statistics, biosignal analysis, and image processing and analysis is required for sensitive and robust detection and characterization of functional activity.

This chapter is organized as follows: Section 18.2 provides some background about fMRI and its data analysis. Section 18.3 gives the concepts and theory of the statistical learning methods, support vector machines (SVMs) and support vector regression (SVR). The proposed framework and its features are introduced and described in detail in Section 18.4, with results on both simulated and real fMRI data. Section 18.5 concludes the chapter with further discussions on this work.

18.2 BACKGROUND

18.2.1 NONLINEARITIES IN fMRI

The BOLD signal is a complex function of neural activity, oxygen metabolism, cerebral blood volume, cerebral blood flow (CBF), and other physiological parameters. The dynamics underlying neural activity and hemodynamic physiology are believed to be nonlinear (3,12,16). The observed fMRI response to a stimulus consists of two chain reactions: The stimulus first triggers a neural response, which sequentially triggers a hemodynamic response that is recorded by BOLD fMRI. The nonlinearity in fMRI could arise from either a nonlinearity in the neural response or a nonlinearity in the hemodynamics, or both (4,30,41). The spatial heterogeneity of the nonlinear characteristics of BOLD signals has also been reported in the literature (3,21).

The cascade of neuronal and hemodynamic nonlinearities in the system would make the determination of variations in neuronal activity difficult. For simplicity, most existing fMRI data analyses assume a linear convolution model and primarily rely on linear methods or general linear models (GLMs). However, as fMRI experiments have grown more sophisticated, the role of nonlinearities is becoming more important under certain situations. Some authors have investigated the physiological mechanisms that reveal the relationship between synaptic activation and vascular or metabolic controlling systems (22,27). Accordingly, initial attempts that model the BOLD signal at macroscopic levels have been made by using differential equations, linking the hemodynamical variations with physiological sense (7,16,34). Although these theoretical models have high impact on fMRI analysis, solid validation from real data is still needed in order to justify their practical use. Because of the complexity of the human brain, we propose to approach the whole brain through a more flexible and general model, and filter the noisy fMRI signals using a nonlinear statistical learning method, support vector regression (SVR). The restored signals are considered nonlinear functions or responses of the stimulus reference function, which agrees with recent findings about the presence of nonlinearities in fMRI.

18.2.2 FEATURES OF fMRI DATA

Two features peculiar to fMRI make its analysis more challenging. First, fMRI data have intrinsic spatial and temporal correlations (42). Second, fMRI data tend to have clustered activations. Spatial and temporal correlations affecting fMRI signal measurements are typically not considered simultaneously in statistical methods dedicated to detecting brain activation. In order to improve the detection of activated areas, common approaches usually smooth the data spatially with a Gaussian kernel in a preprocessing step. Spatial smoothing enables effective detection of a certain size of clustered activation. However, smoothing may produce a biased estimate by displacing activation peaks and underestimating their height. To address this issue, spatial modeling has been proposed (11,20) to take the spatial activation pattern into consideration. Recently, spatiotemporal linear regression methods have also been applied to fMRI data analysis (2,23). These methods use the time series of neighboring voxels together with their own, and thus take simultaneously the spatial and temporal correlations into account, which is also one of the benefits of the regression method to be introduced in this chapter.

18.2.3 FMRI DATA MODELING

In general, techniques for analyzing fMRI data can be divided into model-driven, e.g., standard general linear model (GLM) (14), and data-driven methods, e.g., principal component analysis (PCA) (1), independent component analysis (ICA) (29), or fuzzy cluster analysis (FCA) (13). In model-driven methods, a model of the expected response is generated and compared with the data. These methods require prior knowledge of event timing, from which an anticipated hemodynamic response can be modeled. However, for brain responses that are not directly locked to the paradigm, model-driven analysis may not be adequate (8). Data-driven methods, however, explore the fMRI data statistically without any assumption about the paradigm or the hemodynamic response function. This flexibility is desirable especially in cases in which it is difficult to generate a good model; however, there are drawbacks. For example, the assumption implicit in PCA is that different modes are Gaussian and uncorrelated, whereas ICA assumes that different modes are non-Gaussian and independent. In addition, a significance estimate for each component is usually not available. Given the advantages and disadvantages, a new approach is discussed in this chapter to merge data-driven methods with prior time course modeling by adjusting a model coefficient.

18.2.4 OVERVIEW

Despite the progress in fMRI analysis, there is still a need for robust and unified analysis methods because of the many limitations with existing techniques, as described in the preceding text. In this chapter, we present a novel, general, and reliable nonlinear approach for fMRI analysis based on statistical learning method, i.e., spatiotemporal SVR (ST-SVR), so that existing difficulties resulting from noise, low resolution, and inappropriate smoothing and modeling can be addressed. In summary, SVR provides a comprehensive but parsimonious mapping from a set of input vectors to a scalar output. Its ability to handle highly nonlinear mappings, in an unconstrained way, makes it a natural candidate for the analysis of biological time series and in particular fMRI. Here, we show how this nonlinear mapping can be used to "restore" data that can then be subject to conventional statistical analysis. The basic idea is to treat the input vectors as explanatory variables and the output as the observed BOLD response. This allows us to characterize the mapping between the explanatory and response variables in a way that is analogous to linear mapping between the design matrix and response used in conventional analyses with the GLM. Critically, the mapping obtained with SVR can be arbitrarily complicated and nonlinear. The predicted responses can represent any systematic relationship between the input variables and observed signal, and they can, therefore, be regarded as having been "restored" or "denoised." In this work, we start with inputs that encode where and when a brain response was measured. Using SVR, one can then estimate the response for any brain position at any time. In practice, this involves using a local clique of neighboring voxels over the entire time series. The inputs are then augmented to include regressors of the sort used in conventional analyses of fMRI time series. The predicted response is then used as a data surrogate that enters classical analyses.

18.3 STATISTICAL LEARNING THEORY

Statistical learning plays a key role in the fields of statistics, data mining, artificial intelligence, engineering, and other disciplines. SVM, introduced by Vapnik (37,38) and studied by others (9,36), is a new and powerful learning methodology that can deal with nonlinear classification (support vector classification [SVC]) and regression (SVR). It is systematic and principled, and it has begun to be widely applied in the machine learning community. The idea is to learn a function f between the input vector $\bar{x}$ and the output scalar y from M examples. When the output y takes binary values, the problem is SVC; when y takes continuous real values, it is SVR problem. The main feature of SVM is to map the input data to a high-dimensional feature space through a nonlinear mapping Φ. Then we learn the function between the mapped data $\Phi(\bar{x})$ and the output y. Typically, this function is nonlinear in the input data space but linear in the feature space. Classification or regression is performed in this feature space. An intuitive diagram of this mapping and its advantage in the SVC mode is shown in Figure 18.1.

In order to model the continuous fMRI signal, we use SVR. Here we sketch the ideas behind SVR; a more detailed description of SVR can be found in Smola (36). Given M input sample points $\bar{x}_1, \bar{x}_2, \ldots, \bar{x}_i, \ldots, \bar{x}_M$, where $\bar{x}_i \in \Re^z$, and M corresponding scalar output values $y_1, y_2, \ldots, y_i, \ldots, y_M$, the aim is to find an approximation or a regression function of the form

$$y = f(\bar{x}) = \sum_{i=1}^{M} \alpha_i K(\bar{x}_i, \bar{x}) + b \tag{18.1}$$

to learn this input–output mapping from the set of training examples with high generalizability. Here K is the kernel function, which is going to be explained later.

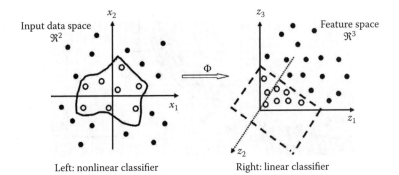

Left: nonlinear classifier   Right: linear classifier

FIGURE 18.1 Diagram of nonlinear mapping in SVC. In order to classify the two classes of points—black and white dots—in the original $\mathfrak{R}^2$ space, the red nonlinear classifier is needed, as shown on the left. However, if we use the nonlinear mapping Φ to map the original data space $\mathfrak{R}^2$ to a high-dimensional feature space $\mathfrak{R}^3$, that is mapping $\bar{x}$ to $\Phi(\bar{x}) = \bar{z}$ through Φ, (for example, $\Phi(\bar{x}) = (z_1, z_2, z_3) = (x_1^2, \sqrt{2}x_1x_2, x_2^2)$), the points would be able to be separated with a linear hyperplane, as shown on the right.

The SVR training process can be formulated as a problem of finding an optimal set of Lagrange multipliers α_i $\forall i \in [1, M]$, by maximizing the SVR objective function

$$O = -\varepsilon \sum_{i=1}^{M} |\alpha_i| + \sum_{i=1}^{M} y_i \alpha_i - \frac{1}{2} \sum_{i=1}^{M} \sum_{j=1}^{M} \alpha_i \alpha_j K(\bar{x}_i, \bar{x}_j) \qquad (18.2)$$

subject to

1. linear constraints,

$$\sum_{i=1}^{M} \alpha_i = 0, \qquad (18.3)$$

and

2. box constraint,

$$-C \le \alpha_i \le C, \quad \forall i \in [1, M] \qquad (18.4)$$

Here, α_i is the Lagrange multiplier associated with each training example $\bar{x}_i$, and ε in Equation 18.2 is the insensitivity value, meaning that training error below ε is ignored. Figure 18.2 depicts the situation graphically for an ε-insensitivity loss function L_ε given by:

$$L_\varepsilon = \begin{cases} 0 & \text{for } |f(\bar{x}) - y| < \varepsilon \\ |f(\bar{x}) - y| - \varepsilon & \text{otherwise} \end{cases}$$

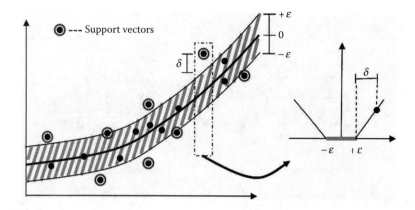

FIGURE 18.2 Diagram of SVR, showing the ε-insensitivity region (shaded area) and the support vectors.

Only the points outside the shaded region contribute to the cost, and the deviations are penalized in a linear fashion. C in Equation 18.4 is the trade-off constant between the smoothness of the SVR function and the total training error. When the approximation function cannot be linearly regressed, the kernel function maps training examples from the input space to a high-dimensional feature space $\Im$ by $\bar{x} \rightarrow \Phi(\bar{x}) \in \Im$, in such a way that the function f between the output and the mapped input data points can now be linearly regressed in the feature space. K describes the inner product in the feature space:

$$K(\bar{x}_i, \bar{x}_j) = \Phi(\bar{x}_i) \cdot \Phi(\bar{x}_j) \qquad (18.5)$$

There are different types of kernel functions. A commonly used kernel function is the Gaussian radial basis function (RBF):

$$K(\bar{x}_i, \bar{x}_j) = \exp\left(\frac{-\| \bar{x}_i - \bar{x}_j \|^2}{2\sigma^2} \right) \qquad (18.6)$$

Maximizing the SVR objective function in Equation 18.2 by SVR training provides us with an optimum set of Lagrange multipliers α_i, $\forall i \in [1, M]$. The coefficient b of the estimated SVR function in Equation 18.1 can be computed by adjusting the bias to pass through one of the given training examples with nonzero α_i.

With the nonlinear kernel mapping, the regression function in Equation 18.1 can be interpreted as a linear combination of the input data in the feature space. Only those input elements with nonzero Lagrange multipliers contribute to the determination of the function. In fact, most of the α_i's are zero. The training data with nonzero α_i are called *support vectors*, which are the data points not inside the ε-insensitivity region as shown in Figure 18.2. Support vectors form a sparse subset of the training data. This type of representation is especially useful for high-dimensional input spaces.

18.4 FMRI DATA ANALYSIS AND MODELING THROUGH SVR

SVR has recently been applied to system identification, nonlinear system prediction, and face detection with good results (18,26,31). Comparisons of SVR with several existing regression techniques, including polynomial approximation, RBFs, and neural networks have been carried out (31). Initial attempts that directly use SVM have also been achieved for modeling hemodynamic response (5) and for comparing and classifying the patterns of fMRI activations (10,17,24). However, the application of SVR in the context of fMRI analysis has not yet been exploited, which is now introduced and developed in this work (39,40).

18.4.1 DATA REPRESENTATION

We formulate fMRI data as spatially windowed continuous 4-dimentional (4-D) functions. That is, the fMRI data is divided into many small windows, such as a $3 \times 3 \times 3$ region within which the entire time series is included. Each input (the training data) within a window is a 4-D vector equal to the row, column, slice, and time indices of a voxel. The output is the corresponding intensity. We approximate and recover all training data within the respective windows using SVR. The detailed formulation follows.

Let $y(u, v, w, t)$ be the fMRI signal of voxel $[u, v, w]^T$ at a given time point t, where u, v, and w are the respective row, column, and slice coordinates of the data. If the 4-D fMRI data size is $S_u \times S_v \times S_w \times S_t$, where S_t is the total number of time points, the corresponding input vector $\bar{x}$ is represented as

$$\bar{x} = [u, v, w, t]^T, \ u \in [1, S_u], \ v \in [1, S_v], \ w \in [1, S_w], \ t \in [1, S_t] \ . \qquad (18.7)$$

Within each spatiotemporal window of size $M_u \times M_v \times M_w \times S_t = M$, we have M input samples $\bar{x}_1, \bar{x}_2, \ldots, \bar{x}_i, \ldots, \bar{x}_M$, where $\bar{x}_i \in \Re^4$, and the respective scalar output $y_1, y_2, \ldots, y_i, \ldots, y_M$. SVR is used to restore the training examples within the window. Local intrinsic spatiotemporal correlations are accounted for during the regression by controlling function smoothness and training error through the parameter C (Equation 18.4). In order to compensate for the spatial correlation between neighboring windows, spatially overlapped windows are used (in all three dimensions) so that the recovered intensities over the overlapped voxels are averaged from the corresponding windows.

18.4.2 TEMPORAL MODELING

Without loss of generality, we assume an on–off boxcar function as our model variable corresponding to a simple block-design paradigm, which contains p zeros or ones during each off or on period and c repetitions or cycles of these two periods. The total number of time points S_t, should be equal to $c \times 2p$. The resulting boxcar function $m(t)$ is

$$m(t) = \begin{cases} 0, & \text{if } t = 1,\dots, p; \ 2p+1,\dots, 3p;\dots; \ S_t - 2p+1,\dots, S_t - p \\ 1, & \text{if } t = p+1,\dots, 2p; \ 3p+1,\dots, 4p;\dots; \ S_t - p+1,\dots, S_t \end{cases} \qquad (18.8)$$

where $t \in [1, S_t]$. That is,

$$m(t) = \underline{}\square\square\square\square\underline{} \cdots$$

An additional model entry, based on $m(t)$ in Equation 18.8, is added to each input data $\bar{x}$ and makes our spatiotemporal SVR (ST-SVR) a 5-D regression problem:

$$\bar{x} = [u, v, w, t, m(t)]^T \in \Re^5 \qquad (18.9)$$

whereas the output is still the corresponding fMRI signal $y(u, v, w, t)$.

The intuitive justification of our model-based formulation can be achieved by analogy with the GLM as it is typically used in fMRI (14). GLM is given by

$$Y = X\beta + e \qquad (18.10)$$

where Y is a fMRI data matrix, X is a "design matrix" specifying the time courses of all factors hypothesized to be present in the observed data (e.g., the task reference function, or a linear trend), β is a map of voxel values for each hypothesized factor, and e is a matrix of noise or residual modeling errors. Given this linear model and a design matrix X, the β maps can be found by least-squares estimation. The simplest example of the design matrix consists of a boxcar reference function (as in Equation 18.8) and a column vector with all entries being the constant 1 representing the mean value, without any other hypothesized factors (Figure 18.3). In this case, for each voxel, the time-series vector is regressed through fitting the boxcar function and the mean value μ (Figure 18.3). For this voxel, at a given time t, the fitting vector is the corresponding row of the design matrix and can be represented as

$$[m(t), \ 1]^T \qquad (18.11)$$

i.e., either $[0, 1]^T$ or $[1, 1]^T$. We extend this idea to SVR. SVMs have very good learning and generalization abilities. As long as we construct the input vectors with the essential features we would like the machine to learn, SVR can capture the complicated relationships (nonlinear or linear) hidden in the training examples. Therefore, for fMRI data representation, in addition to using the indices of the coordinates and time point as input vectors, we add extra model-fitting entries to the input vectors. For the model-fitting vector in Equation 18.11, the second

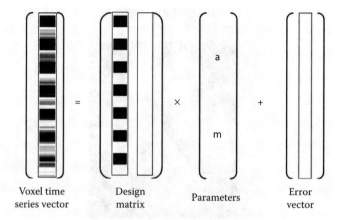

FIGURE 18.3 Diagram of GLM regression and boxcar function fitting.

entry is a constant 1, which is the same for all the input vectors and can be neglected in SVR learning.

With the input vector in Equation 18.9, temporal modeling is incorporated into the regression. Although $m(t)$ here is a simple boxcar function, a whole family of $m(t)$ could be used in the same way that the design matrix in the GLM is used to encode multiple experimental factors or confounds.

In order to test with access to the ground truth, we generate a 2-D time series (spatial size 52×63) of synthetic data that imitates a single fMRI brain slice in which four regions are activated. Three different amplitudes of activations are added to the gray matter to simulate weak, medium, and strong activations as in real fMRI data (see Figure 18.4a). For simplicity and easier intuitive visualization, the activations are temporally in the form of a boxcar function, with six images during each off or on period. Note that a more realistic and complicated reference function formed by convolving this boxcar with a gamma function (6) can also be used. The total number of time points is 72 (6 cycles). The generated data in Figure 18.4b is then used as ground truth for comparisons. Simulated noisy data (see Figure 18.4c) are obtained by adding Gaussian noise $N(0, 32^2)$, to the ground truth data. The recovered image by our SVR method ($W\text{-}model = 1$; Figure 18.4d) accurately restores the ground truth (Figure 18.4b). The image obtained using Gaussian smoothing of the original noisy data is shown in Figure 18.4e for comparison. Obviously, the ST-SVR method significantly improves the quality of the noisy data.

18.4.3 MULTIRESOLUTION SIGNAL ANALYSIS

With the aforementioned formulation, in order to capture the underlying relationship using ST-SVR for the windowed data and accommodate the differences in scale and training set size, the corresponding entries in the input vector are normalized over training examples within each window. After normalization, we

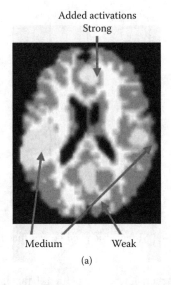

(a)

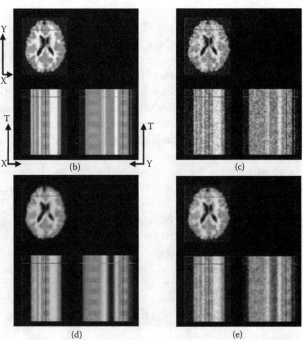

FIGURE 18.4 (See color insert following page 306). Simulated 2-dimensional time series (2-D + T data), visualized with 3 orthogonal slices (spatial axes: X, Y; time axis: T). (a) Added activations on a 2-D brain slice; (b) Ground truth data; (c) Simulated noisy data, with noise level $N(0, 30^2)$; (d) Restored data by the ST-SVR ($W\text{-}model = 1$); (e) Gaussian-smoothed data with Gaussian standard deviation 0.5 (s.t.d. = 0.5).

multiply all t_i by a coefficient *W-scale* and all $m(t_i)$ by a coefficient *W-model*. The notion of scaling different components of the input vector is critical to an understanding of the potential utility of SVR. The scaling can balance the relative explanatory power of the different components. Scaling has this effect because of the projection of the input to a higher-dimensional space using nonlinear kernel functions. It is this nonlinearity that renders the regression sensitive to scaling. In linear models, the scaling is irrelevant because the different components of the input vector do not interact. However, in SVR the scale of each explanatory variable can have a profound effect on the interactions.

The effect of temporal scale can be adjusted by varying *W-scale*, the coefficient for the time indices. Varying *W-scale* is equivalent to examining the temporal data at different scales and, therefore, achieves multiresolution signal analysis. A larger *W-scale* corresponds to a finer temporal resolution. We can restore the time courses at multiple resolutions and extract different frequency components by changing *W-scale*. Many voxel time series in fMRI exhibit low-frequency trend components that may be due to aliased high-frequency physiological components or drifts in the scanner. These trends can be removed in a variety of ways. In addition to using a simple high-pass filter in the temporal domain, a running-lines smoother has been proposed (28). However, most existing methods only aim to handle linear trends. In the spatiotemporal nonlinear SVR, with appropriate *W-scale* (usually relatively small), low-frequency noise can be extracted and removed and thus achieve nonlinear detrending.

The optimal *W-scale* for a specific frequency component is expected to be related to the total number of time points, the period of the stimulation, and the data noise level, whose value is currently determined empirically. A more rigorous formulation of *W-scale* determination is one of our future directions, which might be achieved in the frequency domain through spectrum analysis, etc.

For the data generated in Section 18.4.2 (shown in Figure 18.4), the ST-SVR window size used is $3 \times 3 \times 3 \times 72$. Figure 18.5 demonstrates the effects of *W-scale* by showing the recovered time courses for an activated pixel (Figure 18.5 left) and for a nonactivated pixel (Figure 18.5 right) of the simulated noisy data (Figure 18.4c) without model fitting (*W-model* = 0, data driven). As *W-scale* increases, higher-frequency temporal components are extracted. When *W-scale* = 5 (Figure 18.5a and Figure 18.5d), the restored signal captures the low-frequency component, which can be interpreted as a nonlinear trend.

18.4.4 MERGING MODEL-DRIVEN WITH DATA-DRIVEN METHODS

The coefficient associated with the model index, *W-model*, determines the degree of influence of the temporal-model term and the degree to which the approach is model-driven. A higher *W-model* (*W-model* = 1) is used when reliable temporal models are available. Otherwise, a lower or zero *W-model* is used, and the approach becomes more data driven. *W-model* can be interpreted as a model confidence or fitness

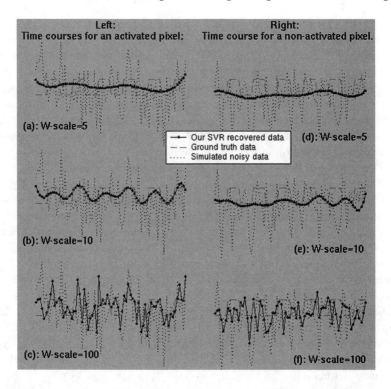

FIGURE 18.5 (See color insert following page 306). Effects on time course with varying *W-scale* for an activated pixel and a nonactivated pixel in the ST-SVR approach (*W-model* = 0).

measure, whose value could be empirically predetermined as a constant or estimated from regression residual analysis (40), although extra computation is needed.

For simulations on the data generated in Section 18.4.2, Figure 18.6 demonstrates the effects of varying *W-model* by showing the recovered time series when *W-scale* = 0, which corresponds to zero frequency (DC component). For the activated pixel, as *W-model* increases, the temporal model has stronger and stronger effects during the regression and data fitting (Figure 18.6 left). Note that because these are simulated data and no real physiological or neuronal activities are involved, the recovered time courses do not show any lag or undershoots in the activated pixel. In fact, the recovered time course accurately restores the ground truth time course (Figure 18.6c), i.e., the boxcar function. For the nonactivated pixel, as shown in Figure 18.6 right, the model term barely affects the data regression.

18.4.5 Generalization to Multisession Studies

Although so far only single-session analyses are discussed, typically fMRI experiments are run several times, either on the same subject (multirun) or with several

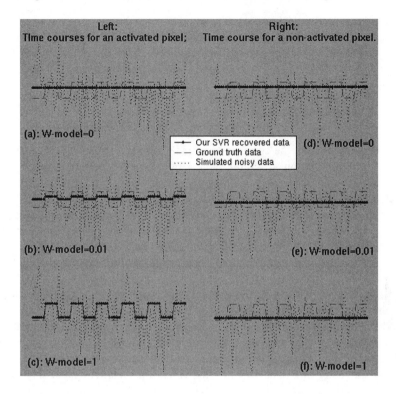

FIGURE 18.6 (See color insert following page 306). Effects on time course with varying *W-model* for an activated pixel and a nonactivated pixel in the ST-SVR approach (*W-scale* = 0).

different subjects (multisubject), or both. The additional data can increase the sensitivity of the experiment and allow the generalization of any conclusion to an entire population. A common technique for multisession analysis is to compute activation maps for each session and then combine them into a composite through *ad hoc* processing such as simple averaging. Limitations in this technique have led to difficulties in the measurement of group differences, especially for more subtle cognitive tasks. The ability of SVR to handle high-dimensional input data makes it ideally suited for extensions to multirun and multisubject studies. The ST-SVR formulation described in Subsection 18.4.1 and Subsection 18.4.2 allows easy incorporation of data from multiple sessions by expanding the input vectors and analyzing the data over multiple runs and multiple subjects together.

Similar to the spatial and temporal indices, now we have additional run and subject indices, r and s. Suppose we would like to process fMRI data on S subjects together and there are R runs for each subject, the input vector for data-driven ST-SVR is then

$$\bar{x} = [u, v, w, t, r_1, r_2, r_i, \ldots, r_R, s_1, \ldots, s_j, \ldots, s_S]^t \in \Re^{4+S+R} \qquad (18.13)$$

For model-driven ST-SVR, the corresponding input vector is

$$\bar{x} = [u, v, w, t, r_1, r_2, r_i, \ldots, r_R, s_1, \ldots, s_j, \ldots, s_S, m(t)]^t \in \mathfrak{R}^{5+S+R} \qquad (18.14)$$

where

$$r_i = \begin{cases} 1 & \text{if run index } = i \\ 0 & \text{otherwise} \end{cases} \qquad ; \qquad s_j = \begin{cases} 1 & \text{if subject index } = j \\ 0 & \text{otherwise} \end{cases}$$

The output is still the corresponding fMRI signal intensity $y(u, v, w, t)$.

After normalization, we multiply all r by a new coefficient W-*run* and all s by a new coefficient W-*subject*. These two associated coefficients will have the effect of emphasizing between-session (run and subject, respectively) differences when a higher value is used; lower values could pool common information over different sessions. This would have important applications on fMRI multisession inference, especially for analyzing and comparing control and subject group data. With the multiresolution effect of W-*scale* addressed in Subsection 18.4.3, the ST-SVR is expected to be able to handle and capture the possible sudden drifts between runs and subjects. Note that the multirun and multisubject group analysis scheme proposed here would have the advantage of accounting for between-run and between-subject variability, and could potentially increase statistical significance in regions in which temporal variations are shared by a group of runs and subjects even though the activations in these regions may be too low to be detected when conducting fMRI analysis over individual runs and individual subjects.

18.4.6 TESTING ON REAL fMRI DATA

The ST-SVR approach is validated by using the conventional t-test (25) on the SVR-restored fMRI data for activation detection, without additional presmoothing or postprocessing. Although the conditions of the t-test are not satisfied here (as is often true in fMRI analysis), we believe it provides a straightforward method of evaluation for the technique. The Gaussian RBF kernel function (Equation 18.6) is used in our experiments with σ set empirically to 0.1. Other SVR parameters are also set empirically $C = 1200$ and $\varepsilon = 20$ (Equation 18.2 and Equation 18.4).

The proposed approach is first applied to a block-design cognitive fMRI experiment performed at Yale (35), examining social attribution to geometric animations. T2*-weighted images were acquired using a single-shot echo-planar sequence. The pulse sequence parameters were TR = 1500 msec, TE = 60 msec, flip angle = 60°, and NEX = 1, providing an in-plane voxel size of 3.125×3.125 mm^2. Fourteen coronal slices were collected and were 10 mm thick (skip 1 mm). Corresponding T1-weighted structural images of the same thickness were collected in the same session (TR = 500, TE = 14, FOV = 200 mm, 256×192 mm matrix, 2NEX). The first four volumes of fMRI time series were discarded to discount T1 saturation effects. We have examined this dataset for a visuospatial task from one subject and one run. The window size used is $3 \times 3 \times 1 \times 160$, where 160 is the total number of time points. We did not use an isotropic window because the voxel shape is not cubic.

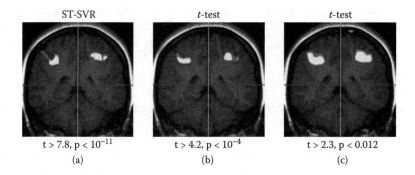

ST-SVR t-test t-test

$t > 7.8, p < 10^{-11}$ $t > 4.2, p < 10^{-4}$ $t > 2.3, p < 0.012$
(a)   (b)   (c)

FIGURE 18.7 (See color insert following page 306). Comparison of ST-SVR and t-test for real fMRI data from a visuospatial task (color activation maps).

The $3 \times 3 \times 1$ window covers a brain region whose physical size is almost isotropic ($9.4 \times 9.4 \times 10$ mm³). Visual comparisons in Figure 18.7 with results directly using t-test on presmoothed data (with empirically chosen FWHM = 6.25 mm × 6.25 mm) reveal that the SVR approach (W-$model$ = 1) leads to greater spatial extent in the intraparietal sulcus (IPS) with potentially better delineation and localization of the underlying spatial activation, in agreement with the underlying anatomy. Note that at the bottom of each slice in Figure 18.7 are the respective t values and p values for threshold. When the same t threshold used for SVR ($t > 7.8$) is used for the t-test, no activations are detected. For the t-test in Figure 18.7c, we intentionally further decreased the t threshold to $t > 2.3$ and tried to detect more IPS activation regions, which, however, lead to a more blurred spatial extent rather than localized spatial activation as in Figure 18.7a as well as some false activations outside of the cerebellum. The associated time course for an activated voxel from the SVR method for this data is shown in Figure 18.8.

The multirun SVR algorithm is tested using a block-design fMRI motor experiment on one subject (male). T2*-weighted images were acquired using

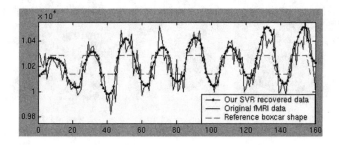

FIGURE 18.8 (See color insert following page 306). Time courses of an activated voxel for the real fMRI data in Figure 18.7. (Horizontal axis: temporal frame index; vertical axis: fMRI data intensity).

a single-shot, gradient-echo, echo-planar sequence. The sequence parameters were TR = 1500 msec, TE = 50 msec, flip angle = 80°, NEX = 1, providing an in-plane voxel size of 3.125×3.125 mm^2. Nine axial slices were collected and were 4 mm thick (skip 0 mm). Corresponding T1-weighted structural images of the same thickness were collected in the same session in the same slice locations (TR = 500, TE = 11, FOV = 200 mm, 256×192 matrix, 2NEX). Two experimental conditions were a finger-tapping task with left and right hands, respectively. During each run, the control (motionless and relaxed) and experimental conditions were alternated (11 scans active, 6 rest) and repeated so that each experimental condition would occur 4 times for a total of 136 volumes ($3.125 \times 3.125 \times 4$ mm^3). Four runs of data were collected in this experiment. For this data set, in order to examine the activation detection in the supplementary motor area (SMA), we focused on the task of right hand vs. the rest condition, with results shown in Figure 18.9. The window size used is $3 \times 3 \times 3 \times 136$, covering a brain region whose physical size is almost isotropic ($9.4 \times 9.4 \times 12$ mm^3). Although activations in the primary motor area are consistently detected by both the ST-SVR method and the traditional t-test (on presmoothed data with empirically chosen FWHM = 3.68 mm $\times$ 3.68 mm $\times$ 4.71 mm), ST-SVR on two-run data leads to activations in the SMA (Figure 18.9a), which are not shown by t-test on the same two-run data (Figure 18.9b). For the t-test on the two-run data in Figure 18.9b, if we intentionally further decrease the t threshold and try to detect the activations in SMA, spurious activations over regions other than the primary motor area and SMA, as well as false activations outside of the cerebellum, would appear. However, if the t-test is performed on four runs of data, the activations in SMA are able to show up as in

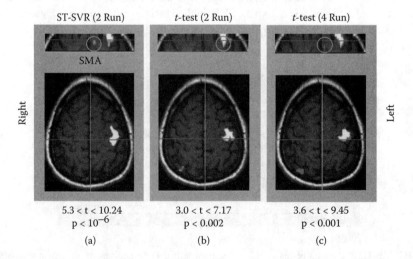

FIGURE 18.9 (See color insert following page 306). Multirun result comparison for a motor experiment of real fMRI data. Top: coronal slices; bottom: axial slices (color activation maps).

Figure 18.9c. This experiment indicates that the ST-SVR method could meet the need for sensitive and robust fMRI signal analysis.

18.5 CONCLUSIONS AND DISCUSSIONS

From a signal processing viewpoint, fMRI activation detection is a problem of nonlinear spatiotemporal system identification. We have presented a novel regression model involving spatiotemporal correlations using SVR, where nonlinear signal analysis is achieved. The framework allows the merging of model-driven with data-driven methods, as well as multiresolution signal analysis, by adjusting the associated model and scale parameters. Other advantages of the approach are embedded removal of low-frequency noise components and easy incorporation of multirun and multisubject studies into the framework. Experimental results on both simulated and real fMRI data revealed its effectiveness. Some comments on the particulars of the method are discussed in the following text.

A simple way to understand SVR is to consider it as a device that converts a highly nonlinear regression problem into a linear one by expanding the input variables using a series of nonlinear functions. This is closely related to the use of multiple basis functions of peristimulus time used in conventional analyses of fMRI time series. Another ubiquitous example is the use of discreet cosine functions of time that model low-frequency drift terms of smooth but arbitrary form. SVR goes further than simply expanding each component of the input. It also explicitly includes high-order interactions among the inputs. An analogy here might be the use of Volterra series expansions that use the same device to estimate high-order kernels (15) that model interactions among stimuli used in producing fMRI responses. These examples are used to provide an intuition into how SVR can model arbitrary and complicated nonlinear relationships between explanatory and response variables.

The size of the spatial window within which the SVR is performed is set to $3 \times 3 \times 3$ or $3 \times 3 \times 1$. We choose this size to allow for some spatial continuity while limiting the likelihood of heterogeneous activation within the same window. In addition, the window is intended to cover a brain region whose physical size is approximately isotropic. Alternatively, a Gaussian window can be applied to achieve more continuous weighting.

Correction for head motion involves rigid-body transformation estimation and resampling. In the multisubject case, the raw data need to be aligned through nonrigid transformation and resampling. Correcting differences in slice timing of 2-D acquisitions also needs data resampling. Because of the thick image slices typical of fMRI, intensity interpolation, required during the resampling process, can introduce significant artifacts (19). With the proposed ST-SVR approach, continuous variables are used for both the input vectors and output scalar. So, given the estimated motion parameters, nonrigid transformation parameters, and slice-timing differences, we could directly keep and use the floating-point continuous values for the spatial coordinates and time indices in the ST-SVR learning.

The output of the ST-SVR is a continuous high-dimensional regressed function, from which the discrete regressed fMRI data can be formed. Therefore, explicit interpolations required in motion correction, subject normalization, and slice-timing correction are embedded in the ST-SVR learning framework. The errors introduced by these interpolations are avoided. This desirable feature is not available for other methods.

The performance of the presented method is dependent on the validity of the explicit models. Without other prior temporal information, the on–off boxcar function in Equation 18.8 is used. Another way is to use a generic function, such as a gamma or Gaussian function, to model the time course (6, 33). In order to have reliable prior temporal models, we plan to learn the model functions (or hemodynamic response functions for event-related fMRI data) from the ST-SVR-restored data. Instead of relying on a generic parametric model (e.g., gamma, Gaussian), we would estimate the temporal models through statistical shape learning without assuming a specific shape of the hemodynamic response. The learned hemodynamic model can be incorporated to improve the specificity and sensitivity of fMRI signal detection.

Currently, the proposed ST-SVR method is validated by applying the conventional t-test on the SVR-restored fMRI data for activation detection. We are also interested in incorporating decision making (activation detection) into ST-SVR regression, by reformulating the SVR optimization objective function so that statistical clustering criteria can be optimized as well during data regression. In fact, validating the hemodynamic response estimation from ST-SVR-restored data is another way of evaluating the method. In addition, it would be interesting to further pursue the advantages of the nonlinear system analysis using this ST-SVR approach in exploring neuronal and hemodynamic responses as well as their interactions.

ACKNOWLEDGMENTS

The author would like to thank Drs. Lawrence Staib, Todd Constable, and Robert Schultz for their in-depth discussions and suggestions, and for providing the fMRI data concerning this work. The author is also grateful to Drs. Karl Friston, Will Penny, and John Ashburner for their valuable comments that improved the quality of this work.

REFERENCES

1. Backfriender, W., Baumgartner, R., Stamal, M., Moser, E., and Bergmann, H. (1996). Quantification of intensity variations in functional MR images using rotated principal components. *Phys. Med. Biol.* 41: 1425–1438.
2. Benali, H., Pelegrini-Issac, M., and Kruggel, F. (2001). Spatiotemporal covariance model for medical images sequences: application to functional MRI data. in *Information Processing in Medical Imaging,* 197–203.
3. Birn, R.M., Saad, Z.S., and Bandettini, P.A. (2001). Spatial heterogeneity of the nonlinear dynamics in the fMRI BOLD response. *Neuroimage* 14: 817–826.

4. Bonds, A.B. (1991). Temporal dynamics of contrast gain in single cells of the cat striate cortex. *Vis. Neurosci.* 6: 239–255.

5. Boulanouar, K., Roux, F., and Celsis, P. (2001). Modeling brain hemodynamic response in functional MRI using vector support method. in *Cognitive Neuroscience Society Annual Meeting* (Abstract).

6. Boynton, G.M., Engel, S.A., Glover, G.H., and Heeger, D.J. (1996). Linear systems analysis of functional magnetic resonance imaging in human V1. *J. Neurosci.* 16: 4207–4221.

7. Buxton, R.B., Wong, E.C., and Frank, L.R. (1998). Dynamics of blood flow and oxygenation changes during brain activation: the balloon model. *Magn. Reson. Med.* 39: 855–864.

8. Calhoun, V.D., Adali, T., Pearlson, G.D., and Pekar, J.J. (2001). Spatial and temporal independent component analysis of functional MRI data containing a pair of task-related waveforms. *Hum. Brian Mapp.* 13: 43–53.

9. Collobert, R. and Bengio, S. (2001). SVMTorch: Support vector machines for large-scale regression problems. *J. Mach. Learning Res.* 1: 143–160.

10. Cox, D.D. and Savoy, R.L. (2003). Functional magnetic resonance imaging (fMRI) "brain reading": detection and classifying distributed patterns of fMRI activity in human visual cortex. *Neuroimage* 19: 261–270.

11. Descombes, X., Kruggel, F., and von Cramon, D.Y. (1998). fMRI signal restoration using a spatiotemporal Markov random field preserving transitions. *Neuroimage* 8: 340–349.

12. Devor, A., Dunn, A.K., Andermann, M.L., Ulbert, I., Boas, D.A., and Dale, A.M. (2003). Coupling of total hemoglobin concentration, oxygenation, and neural activity in rat somatosensory cortex. *Neuron* 39: 353–359.

13. Fadili, M.J., Ruan, S., Bloyet, D., and Mazoyer, B. (2001). On the number of clusters and the fuzziness index for unsupervised FCA application to BOLD fMRI time series. *Med. Image Anal.* 5: 55–67.

14. Friston, K.J., Holmes, A.P., Worsley, K.J., Poline, J.-P, Frith, C.D., and Frackowiak, R.S.J. (1995). Statistical parametric maps in functional imaging: a general linear approach. *Hum. Brain Mapp.* 2: 189–210.

15. Friston, K.J., Josephs, O., Rees, G., and Turner, R. (1998). Nonlinear event-related responses in fMRI. *Magn. Reson. Med.* 39: 41–52.

16. Friston, K.J., Mechelli, A., Turner, R., and Price, C.J. (2000). Nonlinear responses in fMRI: the ballon model, volterra kernels, and other hemodynamics. *Neuroimage* 12: 466–477.

17. Golland, P., Fische, B., Spiridon, M., Kanwisher, N., Buckner R.L., Shenton, M.E., Kikinis, R., Dale, A., and Grimson, W.E.L. (2002). Discriminative analysis for image-based studies. in *Intl. Conf. on Medical Image Computing and Computer-Assisted Intervention*, 508–515.

18. Gretton, A., Doucer, A., Herbrich, R., Rayner, P.J.W., and Scholkopf, B. (2001). Support vector regression for black-box system identification. in *IEEE Workshop on Statistical Signal Processing,* 341–344.

19. Grootoonk, S., Hutton, C., Ashburner, J., Howseman, A.M., Josephs, O., Rees, G., Friston, K.J., and Turner, R. (2000). Characterization and correction of interpolation effects in the realignment of fMRI time series. *Neuroimage* 11: 49–57.

20. Hartvig, N.V. and Jensen, J.L. (2000). Spatial mixture modeling of fMRI data. *Hum. Brain Mapp.* 11: 233–248.

21. Huettel, S.A. and McCarthy, G. (2001). Regional differences in the refractory period of the hemodynamic response: an event-related fMRI study. *Neuroimage* 14: 967–976.

22. Iadecola, C. (2002). Intrinsic signals and functional brain mapping: caution, blood vessel at work. *Cereb. Cortex* 12: 223–224 (CC Commentary).

23. Katanoda, K., Matsuda, Y., and Sugishita, M. (2002). A spatiotemporal regression model for the analysis of functional MRI data. *Neuroimage* 17: 1415–1428.

24. LaConte, S., Strother, S., Cherkassky, V., and Hu, X. (2003). Predicting motor tasks in fMRI data with support vector machines. in *International Society for Magnetic Resonance in Medicine Annual Meeting* (Abstract).

25. Lange, N. (1999). Statistical procedures for functional MRI. in Moonen, C, Bandettini, P. Eds. *Functional MRI*. Springer-Verlag, pp. 301–335.

26. Li, Y., Gong, S., and Liddell, H. (2000). Support vector regression and classification based multiview face detection and recognition. in *Proc. Fourth IEEE Intl. Conf. on Automatic Face and Gesture Recognition,* 300–305.

27. Magistretti, P.J. and Pellerin, L. (1999). Cellular mechanisms of brain energy metabolism and their relevance to functional brain imaging. *Phil. Trans. R. Soc. Lond. B* 354: 1155–1163.

28. Marchini, J.L. and Ripley, B.D. (2000). A new statistical approach to detecting significant activation in functional MRI. *Neuroimage* 12: 366–380.

29. McKeown, M., Makeig, S., Brown, G., Jung, T., Kindermann, S., Bell, A., and Sejnowski, T. (1998). Analysis of fMRI data by blind separation into independent spatial components. *Hum. Brain Mapp.* 6: 160–188.

30. Miller, K.L., Buxton, R.B., Wong, E.C., and Frank, L.R. (1999). The linearity of cerebral blood flow response to brief motor tasks. in *Proc. ISMRM 7th Scientific Meeting.*

31. Mukherjee, S., Osuna, E., and Girosi, F. (1997). Nonlinear prediction of chaotic time series using support vector machines. in *Proc. IEEE Workshop on Neural Networks and Signal Processing* Vol. VII, 511–520.

32. Ogawa, S., Lee, T.M., Nayak, A.S., and Glynn, P. (1990). Oxygenation-sensitive contrast in magnetic resonance image of rodent brain of high magnetic fields. *Magn. Reson. Med.* 14: 68–78.

33. Rajapakse, J.C., Kruggel, F., Maisog, J.M., and von Cramon, D.Y. (1998). Modeling hemodynamic response for analysis of functional MRI time series. *Hum. Brain Mapp.* 6: 283–300.

34. Riera, J.J., Watanabe, J., Kazuki, I., Naoki, M., Aubert, E., Ozaki, T., and Kawashima, R. (2004). A state-space model of the hemodynamic approach: nonlinear filtering of BOLD signals. *Neuroimage* 21: 547–567.

35. Schultz, R.T., Grelotti, D.J., Klin, A., Kleinman, J., Van der Gaag, C., Marois, R., and Skudlarski, P. (2003). The role of the fusiform face area in social cognition: Implications for the pathobiology of autism. *Phil. Trans. of the Royal Society,* Series B, 358: 415–427.

36. Smola, A.J. and Scholkopf, B. (1998). *A Tutorial on Support Vector Regression.* NeuroCOLT Technical Report NC-TR-98-030, Royal Holloway College, University of London, U.K.

37. Vapnik, V.N. (1995). *The Nature of Statistical Learning Theory.* Springer-Verlag, New York.

38. Vapnik, V.N. (1998). *Statistical Learning Theory.* John Wiley & Sons, New York.

39. Wang, Y.M., Schultz, R.T., Constable, R.T., and Staib, L.H. (2003a). A unified framework for nonlinear analysis of functional MRI data using support vector regression. in *Human Brain Mapping Conference* (Abstract).

40. Wang, Y.M., Schultz, R.T., Constable, R.T., and Staib, L.H. (2003b). Nonlinear estimation and modeling of fMRI data using spatiotemporal support vector regression. in *Information Processing in Medical Imaging,* 647–659.

41. Vazquez, A.L. and Noll, D.C. (1998). Nonlinear aspects of the BOLD response in functional MRI. *Neuroimage* 7(2): 108–118.

42. Zarahn, E., Aguirre, G.K., and D'Esposito, M. (1997). Empirical analyses of BOLD fMRI statistics. I. Spatially unsmoothed data collected under null-hypothesis conditions. *Neuroimage* 5: 179–197.

19 Assessment of Cerebral Blood Flow, Volume, and Mean Transit Time from Bolus-Tracking MRI Images: Theory and Practice

Alessandra Bertoldo, Francesca Zanderigo, and Claudio Cobelli

CONTENTS

19.1 INTRODUCTION

In vivo noninvasive quantitative assessment of cerebral hemodynamics is of crucial importance for understanding brain functions in both normal and pathological states. Positron emission tomography (PET) offers a powerful tool, e.g., one can measure cerebral blood flow (CBF) and cerebral blood volume (CBV) by interpreting $[^{15}O]H_2O$ tracer activity images with suitable mathematical models [1,2]. PET methods are the gold standard for CBF and CBV quantification, but they have their own limitations, e.g., PET facilities are located only in specialized clinical centers, radioactivity tracers are employed, and arterial sampling is required. Recently, dynamic susceptibility contrast-enhanced magnetic resonance imaging (DSC-MRI) has emerged as an alternative and clinically appealing technique in *in vivo* assessment of cerebral hemodynamics. Briefly, in DSC-MRI, an intravascularly distributed paramagnetic contrast agent is rapidly injected into a peripheral vein. Once the bolus of the contrast agent reaches the region of interest, a short blood relaxation time because of the paramagnetic label leads to a decline in the MRI signal intensity acquired either by a spin-echo or gradient-echo method. Despite the inherent complexity of susceptibility contrast mechanisms, a theory to model the DSC-MRI information and several techniques to implement this theory correctly have been developed during the last 20 yrs in order to allow quantitative measures of CBF, CBV, and mean transit time (MTT).

The aim of this chapter is to review the theoretical fundamentals of the quantification of DSC-MRI signals and to discuss relevant issues in obtaining reliable estimates of CBF, CBV, and MTT.

19.2 THEORY

The model used for quantification of DSC-MRI images is based on the principles of tracer kinetics for nondiffusible tracers [3–5] and relies on the following assumptions:

1. The contrast agent is completely nondiffusible.
2. There is no recirculation of the contrast agent.
3. The contrast agent is confined to the intravascular space. In other words, the blood–brain barrier (BBB) is assumed to be intact; otherwise, tracer leakage can occur.
4. The system is in steady state during the experiment, i.e., the blood flow is assumed to be constant. As a consequence, only a stationary flow can be measured in a single experiment; however, flows that vary slowly compared with the duration of the experiment are still quantifiable by a series of consecutive experiments.
5. The contrast agent dose must not appreciably perturb the system.

19.2.1 TRANSPORT FUNCTION

Consider a bolus of amplitude q_0 of a nondiffusible tracer at time $t = 0$ in the feeding vessel to the volume of interest (VOI) of tissue. The amount of nondiffusible tracer leaving the VOI at a time t is given by

$$q_{out}(t) = q_0 \int_0^t h(\tau)\,d\tau \tag{19.1}$$

where h(t) is the *transport function*, i.e., the probability density function of the tracer transit time through the VOI. The transport function, h(t), is a characteristic of the system and has the dimensions of 1/time. In particular, h(t) is dependent on the flow and vascular structure of the VOI. When integrated over the interval 0 to infinity, the area is unitary and dimensionless. Therefore,

$$\int_0^\infty h(t)\,dt = 1 \tag{19.2}$$

19.2.2 RESIDUE FUNCTION

Following the definition of h(t), the amount of tracer remaining in the VOI is given by q_0 minus the amount that left the VOI:

$$q_{in}(t) = q_0 - q_0 \int_0^t h(\tau)\,d\tau = q_0 \left[1 - \int_0^t h(\tau)\,d\tau \right] \tag{19.3}$$

The function

$$R(t) = 1 - \int_0^t h(\tau)\,d\tau \tag{19.4}$$

is called *residue function* and describes the fraction of tracer still present in the VOI after a time t following an ideal bolus injection. R(t) is a dimensionless, positive, decreasing function of time for which

$$R(0) = 1 \tag{19.5}$$

Once h(t) and, consequently, R(t) are known for a given VOI, the concentration curve of the tracer at the exit and of that retained in the VOI can be predicted for any known input function to the VOI.

19.2.3 CEREBRAL BLOOD VOLUME

In case of an intact BBB, the amount of blood in a given VOI measures the central blood volume (CBV). From dynamic images acquired during bolus injection of a contrast agent, CBV can be determined from the ratio of the areas under the concentration time curve of the tracer within a given VOI, $C_{VOI}(t)$, and the concentration time curve of the tracer in the feeding vessel to the VOI, $C_{AIF}(t)$

(where AIF stands for arterial input function), respectively [3,4,6,7]. Normalizing CBV to the density ρ of brain tissue:

$$CBV = \frac{k_H}{\rho} \frac{\int_0^\infty C_{VOI}(\tau)\,d\tau}{\int_0^\infty C_{AIF}(\tau)\,d\tau} \qquad (19.6)$$

where k_H accounts for the difference in hematocrit (H) between large vessels (LV) and small vessels (SV) because only the plasma volume is accessible to the tracer, i.e., $k_H = (1 - H_{LV})/(1 - H_{SV})$. In fact, the CBV may be split into a cerebral plasma volume (CPV) and cerebral red cell volume (CRCV) [8], i.e.,

$$CBV = CPV + CRCV \qquad (19.7)$$

However, because hematocrit is defined by the ratio

$$H = 100\frac{CRCV}{CBV} \qquad (19.8)$$

we have:

$$CBV = CPV + H \cdot CBV \qquad (19.9)$$

$$(1-H)CBV = CPV \qquad (19.10)$$

and, consequently,

$$CBV = \frac{1}{\rho}\frac{amount \text{ of blood in a VOI}}{area \text{ under the blood input curve}}$$

$$= \frac{1}{\rho}\frac{amount \text{ of plasma in a VOI}}{1-H_{SV}}\frac{1-H_{LV}}{\text{area under the plasma input curve}}$$

$$\times \text{ (because the tracer is in the plasma only)} \qquad (19.11)$$

$$= \frac{1}{\rho}\frac{\int_0^\infty C_{VOI}(\tau)\,d\tau}{1-H_{SV}}\frac{1-H_{LV}}{\int_0^\infty C_{AIF}(\tau)\,d\tau} = \frac{1}{\rho}\frac{1-H_{LV}}{1-H_{SV}}\frac{\int_0^\infty C_{VOI}(\tau)\,d\tau}{\int_0^\infty C_{AIF}(\tau)\,d\tau}$$

The commonly used units for CBV are milliliters per 100 grams of tissue (ml/100 g) and microliters per gram (μml/g).

19.2.4 MEAN TRANSIT TIME

An additional parameter that characterizes the VOI is the mean transit time (MTT). MTT is defined as the center of mass of the distribution h(t) or, in different words, the average time required for any given particle of tracer to pass through the VOI. The MTT is given by

$$MTT = \frac{\int_0^\infty t \cdot h(\tau) d\tau}{\int_0^\infty h(\tau) d\tau} \qquad (19.12)$$

and from Equation 19.2 and Equation 19.4, one has:

$$MTT = \int_0^\infty t \cdot h(\tau) d\tau = \int_0^\infty R(\tau) d\tau \qquad (19.13)$$

MTT can also be calculated by using the central volume theorem of the indicator dilution theory [5–7]. According to this theory, MTT is the ratio of CBV to CBF in the VOI:

$$MTT = \frac{CBV}{CBF} \qquad (19.14)$$

MTT has dimension of time.

19.2.5 CEREBRAL BLOOD FLOW

In case of an intact BBB, following the preceding definitions, CBF can be obtained by the convolution of $C_{VOI}(t)$, $R(t)$, and the AIF $C_{AIF}(t)$:

$$C_{VOI}(t) = \frac{\rho}{k_H} CBF \int_0^t C_{AIF}(\tau) R(t - \tau) d\tau \qquad (19.15)$$

Equation 19.15 can be derived by starting from Equation 19.14:

$$CBF = \frac{CBV}{MTT} = \frac{k_H}{\rho} \frac{\int_0^\infty C_{VOI}(\tau) d\tau}{\int_0^\infty C_{AIF}(\tau) d\tau} \bigg/ \int_0^\infty R(\tau) d\tau \qquad (19.16)$$

from which

$$\int_0^\infty C_{VOI}(\tau)d\tau = \frac{\rho}{k_H} CBF \int_0^\infty C_{AIF}(\tau)d\tau \int_0^\infty R(\tau)d\tau$$

$$= \frac{\rho}{k_H} CBF \int_0^\infty C_{AIF}(\tau) \otimes R(\tau)d\tau$$

$$(19.17)$$

where $\otimes$ denotes the convolution operator and, thus,

$$C_{VOI}(t) = \frac{\rho}{k_H} CBF \cdot C_{AIF}(t) \otimes R(t) = \frac{\rho}{k_H} CBF \int_0^t C_{AIF}(\tau)R(t-\tau)d\tau \quad (19.18)$$

To obtain CBF, one needs to deconvolve (see following text) Equation 19.18 in order to calculate $R'(t) = CBF \cdot R(t)$ and, subsequently, the CBF from $R'(t)$ value at time $t = 0$:

$$CBF = R'(0) \qquad (19.19)$$

The commonly used units for CBF are milliliters per 100 grams of tissue per minute (ml/100 g/min) and microliters per gram per second (μml/g/s).

In conclusion, the algorithmic steps to assess cerebral hemodynamics are summarized in Table 19.1: first, CBF is obtained by deconvolution (Equation 19.18), and CBV from Equation 19.11, then MTT from Equation 19.14. Sometimes C_{AIF} is not measured. In this case, one can only measure a relative CBV, rCBV, from

$$rCBF = \int_0^\infty C_{VOI}(\tau)d\tau \qquad (19.20)$$

but CBF and MTT cannot be estimated.

TABLE 19.1
CBF, CBV, and MTT from DSC-MRI Images

Parameter	Formula
CBF Cerebral Blood Flow	$C_{VOI}(t) = \frac{\rho}{k_H} \int_0^t C_{AIF}(\tau)R'(t-\tau)d\tau \Rightarrow CBF = R'(0)$
CBV Cerebral Blood Volume	$CBV = \frac{1}{\rho} \frac{1-H_{LV}}{1-H_{SV}} \dfrac{\int_0^\infty C_{VOI}(\tau)d\tau}{\int_0^\infty C_{AIF}(\tau)d\tau}$
MTT Mean Transit Time	$MTT = \dfrac{CBV}{CBF}$

19.3 PRACTICE

19.3.1 FROM DSC-MRI SIGNAL TO TRACER CONCENTRATION

In DSC-MRI, the amount of contrast agent present within a voxel locally perturbs the total magnetic field, thus decreasing relaxation time constants and influencing the detected T2*-weighted signal, S(t), from the voxel, as follows:

$$S(t) = S_0 e^{-\Delta R_2^*(t) \cdot T_E} \tag{19.21}$$

where:

- $S_0 = S(0)$ is the signal value from water protons at time $t = 0$, when no contrast agent is yet present.
- $\Delta R_2^*(t) = R_2^*(t) - R_2^*(0)$ is the change in transverse relaxation rate, i.e., the difference between water proton T2*-relaxation rate $R_2^*(t)$ = 1/T2*(t) and its value at t = 0 $R_2^*(0)$.
- T_E is the echo time, i.e., a time parameter specific to the particular gradient-echo sequence adopted.

Within frequently used low-dosage ranges of contrast agents at common field strengths B_0, a linear relationship between the change in transverse relaxation rate and tracer concentration $C_{VOI}(t)$ within the voxel can be reasonably assumed [9–13] to be

$$C_{VOI}(t) = \kappa_{VOI} \Delta R_2^*(t) \tag{19.22}$$

in which κ_{VOI} is an unknown proportionality constant depending on the tissue, the contrast agent, the field strength, and the pulse sequence. From Equation 19.21 and Equation 19.22, one can derive

$$C_{VOI}(t) = -\frac{\kappa_{VOI}}{T_E} \ln\left(\frac{S(t)}{S_0}\right) \tag{19.23}$$

which is the fundamental equation of DSC-MRI, relating the tracer concentration profile within a voxel to the measured signal produced by the perturbed water protons spin-∫ system. Equation 19.21 is used to convert both arterial as well as tissue DSC-MRI-measured signals. Because of the complexity of the relaxation mechanism underlying the DSC-MRI signal generation and the consequent difficulty in retrieving the correct κ_{VOI} value for each voxel, the same proportionality constant ($\kappa = \kappa_{VOI}$) is usually assumed for both tissue and arterial concentration. However, this assumption can affect a correct quantification of CBF, CBV, and MTT [14,15].

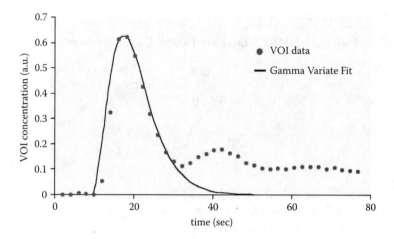

FIGURE 19.1 A typical concentration time course measured in a VOI (dot curve) in the presence of a tracer recirculation and the resulting gamma variate fit (solid line).

The dilution theory assumes that no tracer recirculation occurs. However, the measured $C_{VOI}(t)$ can include contributions from recirculation, which can be recognized as a second, smaller concentration peak or an incomplete return to baseline after the first pass (Figure 19.1). Therefore, the problem is to separate the first-pass tracer concentration profile from the recirculation peak. To separate this contribution of tracer recirculation, a gamma variate function is generally used to fit the $C_{VOI}(t)$ data [13,16–18] (Figure 19.1).

19.3.2 ARTERIAL INPUT FUNCTION

A correct measurement of the arterial DSC-MRI signal and, consequently, of the arterial concentration time curve, $C_{AIF}(t)$, is one of the most delicate steps in the quantification of DSC-MRI images. $C_{AIF}(t)$ is usually estimated from a large artery (e.g., the middle cerebral artery or the internal carotid artery), with the assumption that this represents the exact input to the VOI under examination [16,18,19]. However, several errors can effect this measurement and, consequently, introduce a bias in CBF estimates. Thus, presence of partial-volume effect, delay and dispersion, sequence type used, and wrong site of $C_{AIF}(t)$ measurement can concur, together or alone, to generate inaccurate CBF maps.

Because of the relatively low spatial resolution of DSC-MRI images, it is possible that the tissue surrounding the selected arterial vessel also contributes to arterial measured signal. In particular, the partial-volume effect can be due to vessel size, location, and orientation, and its presence introduces an overestimation of CBF. Recently proposed correction methods are based on the use of an appropriate scaling factor [20] or the use of *ad hoc* correction methods such as that developed in Reference 21. However, more work is needed to better understand the effect of partial-volume presence in $C_{AIF}(t)$ in the quantification of DSC-MRI images.

In addition to the presence of partial-volume effect, $C_{AIF}(t)$ may undergo dispersion during its passage from the point of measurement to more peripheral VOI, especially in pathology. If d(t) denotes the dispersion function, the deconvolved residue function does not represent CBF multiplied by the true residue function R(t), i.e., $R'(t) = CBF·R(t)$, but represents CBF multiplied by the convolution between the true residue function and the dispersion one, i.e.,

$$R'(t) = CBF \cdot R(t) \otimes d(t) \qquad (19.24)$$

with $\int_0^\infty d(\tau)d\tau = 1$, $R'(0) = 0$ and $\int_0^\infty R'(\tau)d\tau = CBF \cdot MTT$. Thus, in presence of dispersion, the actual residue function has a different shape and properties (Figure 19.2) in comparison with the true one. The measured $C_{AIF}(t)$ may also be affected by a delay. Delay and dispersion presence modifies the shape of the deconvolved curve and, in this case, CBF is usually estimated from the maximum of the deconvolved curve [13,22] instead of from the deconvolved curve at time t = 0, which is zero in presence of delay and dispersion. In this way, an error in quantification of CBF is introduced. In fact, recent studies have shown that delays of 1 to 2 sec can introduce a 40% underestimation of CBF and a 60% overestimation of MTT [23]. Of note is that these delays between the "measured" and "true" AIF are common in cerebral regions affected by cerebrovascular disease. Consequently, the quality of information provided by quantification of MRI images is reduced in important pathologies such as cerebral ischemia and carotid stenosis. As suggested in Reference 24 and Reference 25, $C_{AIF}(t)$ should be measured as close as possible to the true feeding artery to the VOI in order to minimize delay and dispersion effects on the CBF estimate. Although the use of

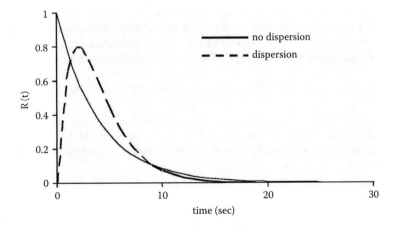

FIGURE 19.2 Residue function R(t) in the absence (solid) or presence (dashed line) of arterial dispersion.

a local $C_{AIF}(t)$ instead of that coming from an LV such as the carotid artery can reduce delay and dispersion, it can increase the presence of partial-volume effect. Thus, particular care is required in selecting the best place for $C_{AIF}(t)$ measurement, by evaluating all technical limitations and physiopathological conditions. In fact, of note is that the use of a local $C_{AIF}(t)$ could be important not only to minimize delay and dispersion but also in studying patients with cerebral ischemia or stenosis.

$C_{AIF}(t)$ is also dependent on the sequence type used. Usually both spin-echo and gradient-echo imaging are used to measure DSC-MRI signals because, at the moment, there is no clear evidence on which is the best method to reach accurate absolute CBF quantification. In Reference 12, the authors demonstrated that spin-echo functional images have great microvascular sensitivity resulting in images of good quality. But, as noted in Reference 13 and Reference 16, in this case $C_{AIF}(t)$ obtained with these sequences reflects more the situation of the SVs and, consequently, could be an underestimation of the "true" $C_{AIF}(t)$. On the other hand, gradient-echo sequence arises from both LVs and SVs, but $C_{AIF}(t)$ results are more affected by errors due to partial-volume effects [26].

19.3.3 DECONVOLUTION

In order to derive CBF from Equation 19.18, the function $CBF \cdot R(t)$ has to be determined by deconvolution. In general, no analytical solution is available, but several techniques allow one to compute an approximate numerical solution. These deconvolution techniques can be classified into two main categories, model-dependent and model-independent approaches.

In the model-dependent techniques, the function to be deconvolved is described by a parametric function, so that deconvolution loses its ill-posedness and ill-conditioning problems by having to solve a parameter estimation problem. This approach implies formulating *a priori* assumptions on the shape of the solution. Larson et al. [27] suggested an exponential residue model, which implies that the microvasculature is like that of a single, well-mixed compartment. More precisely, the following analytical expression was assumed for R(t) for $t \geq 0$.

$$R(t) = e^{\frac{-t}{MTT}}$$
(19.25)

Note that R(t) given by Equation 19.25 satisfies the residue function properties, i.e., $R(t) \geq 0$, $R(0) = 1$, and $R(t) \in [0; 1]$. Substituting Equation 19.25 in Equation 19.18, one has

$$C_{VOI}(t) = \frac{\rho}{k_H} CBF \int_0^t C_{AIF}(\tau) e^{-\frac{t-\tau}{MTT}} d\tau$$
(19.26)

and deconvolution reduces to a CBF and MTT parameter estimation problem. However, this approach, by introducing strong assumptions on R(t) behavior, is likely to introduce a bias in CBF estimates.

A more general model for R(t) was introduced by Ostergaard et al. [28], based on a model of macrovascular transport and microvascular retention in the brain. The model, originally introduced to describe tracer transport and retention in the heart [29], describes the vasculature as a major feeding artery in series with 20 SVs in parallel and allows to take into account delay and dispersion of the arterial input. However, the vascular model was found sensitive to data noise level.

The model-independent approaches owe their name to the fact that they make virtually no assumptions on the description of the unknown function to be deconvolved. These methods are more powerful and less biased than the model-dependent ones, but they have to deal with the ill posedness and ill conditioning of the deconvolution problem. One of the simplest methods to solve the inverse problem of Equation 19.18 is to use the convolution theorem of Fourier transform, which states that the transform of two convolved functions equals the product of their individual transforms:

$$F\{CBF \cdot R(t) \otimes C_{AIF}(t)\} = F\{C_{VOI}(t)\} \qquad (19.27)$$

From Equation 19.28, one obtains

$$CBF \cdot R(t) = F^{-1}\left\{\frac{F\{C_{VOI}(t)\}}{F\{C_{AIF}(t)\}}\right\} \qquad (19.28)$$

where F^{-1} denotes the inverse of the Fourier transform F. The Fourier transform approach has the attraction of being theoretically very easy to implement and insensitive to delays between the AIF and the tissue. However, its use is not without problems, and discordant results have been reported in the literature. For instance, Ostergaard et al. [13] showed that the Fourier transform approach biases CBF, in particular, underestimating it in case of high flow. They also showed that Fourier transform approach has an inherent problem in arriving at the actual CBF when the residue function has discontinuities. On the other hand, other researchers found satisfactory estimates of CBF in comparison with other, more sophisticated deconvolution techniques [30].

Another method to solve Equation 19.18 is to resort to a linear algebraic approach. More precisely, assuming that tissue and arterial concentrations are measured at equidistant time points, $t_i = t_{i-1} + \Delta t$, and choosing Δt so that CBF·R(t) is reasonably approximated by a staircase function, a discrete version of Equation 19.18 can be written in matrix form

$$C_{VOI}(t_j) \approx CBF \cdot \Delta t \cdot \sum_{i=0}^{j} C_{AIF}(t_i) \cdot R(t_j - t_i) \qquad (19.29)$$

which is equivalent to

$$
\begin{pmatrix} C_{VOI}(t_1) \\ C_{VOI}(t_2) \\ \dots \\ C_{VOI}(t_N) \end{pmatrix} = CBF \cdot \Delta t \cdot \begin{pmatrix} C_{AIF}(t_1) & 0 & \dots & 0 \\ C_{AIF}(t_2) & C_{AIF}(t_1) & \dots & 0 \\ \dots & \dots & \dots & \dots \\ C_{AIF}(t_N) & C_{AIF}(t_{N-1}) & \dots & C_{AIF}(t_1) \end{pmatrix} \cdot \begin{pmatrix} R(t_1) \\ R(t_2) \\ \dots \\ R(t_N) \end{pmatrix} \quad (19.30)
$$

In compact form:

$$
\mathbf{C}_{VOI} = CBF \cdot \Delta t \cdot \mathbf{C}_{AIF} \cdot \mathbf{R} \quad\quad (19.31)
$$

where $\mathbf{C}_{VOI} \in \Re^{Nx1}$, $\mathbf{C}_{AIF} \in \Re^{NxN}$, $\mathbf{R} \in \Re^{Nx1}$, and Δt is the length of the equally spaced sampling times ($\Delta t = t_i - t_{i-1}$). Equation 19.32 is a standard matrix equation that can be inverted to yield $CBF \cdot \mathbf{R}$ if $\det(\mathbf{C}_{AIF})$ 0:

$$
CBF \cdot \Delta t \cdot \mathbf{R} = \mathbf{C}_{AIF}^{-1} \cdot \mathbf{C}_{VOI} \quad\quad (19.32)
$$

This approach has been termed *raw deconvolution* in the literature [31]. Albeit appealing in its simplicity, it is known to perform poorly being extremely sensitive to noise. A widely used approach to solve Equation 19.18 that overcomes the limitations of the raw deconvolution is singular value decomposition (SVD). This was introduced as method to estimate R(t) by Ostergaard et al. [13,16]. The SVD constructs matrices $\mathbf{V}$, $\mathbf{W}$, and $\mathbf{U}^T$ so that the inverse of $\mathbf{C}_{AIF}$ can be written as

$$
\mathbf{C}_{AIF}^{-1} = \mathbf{V} \cdot \mathbf{W} \cdot \mathbf{U}^T \quad\quad (19.33)
$$

where $\mathbf{W}$ is a diagonal matrix, and $\mathbf{V}$ and $\mathbf{U}^T$ are orthogonal and transpose orthogonal matrices, respectively. Given this inverse matrix, $CBF \cdot \mathbf{R}$ is found simply as

$$
CBF \cdot \mathbf{R} = \frac{\mathbf{V} \cdot \mathbf{W} \cdot \mathbf{U}^T \cdot \mathbf{C}_{VOI}}{\Delta t} \quad\quad (19.34)
$$

SVD has been shown to be a reliable technique for deconvolution because it reduces the effect of noise on R(t) estimation. This is achieved by setting to zero the elements in the diagonal matrix $\mathbf{W}$ obtained by SVD when they are smaller than a threshold value given beforehand.

SVD represents the most used approach to quantify bolus-tracking MRI data. However, in the last years its limitations have been pointed out [32–37]. In particular, it has been shown that CBF values obtained by SVD largely depend on the threshold value selected to eliminate diagonal elements in $\mathbf{W}$ [32,33,35]. In addition, SVD introduces undesiderated oscillations and negative values in the reconstructed $CBF \cdot R(t)$, producing a nonphysiological R(t). This is far from ideal because there are situations in which the actual shape of the residue function, not just its maximum value, is of interest, i.e., when there is presence of bolus delay and dispersion and only an accurate determination of the shape of CBF · R(t) can allow an assessment and correction of this error. In such cases, the conventional SVD method is not suitable. In order to overcome SVD limitations, several deconvolution methods have been proposed during the last 10 yr. Thus, for instance, one of the disadvantages of SVD applied to DSC-MRI data is a tendency to underestimate the flow when the tissue tracer arrival is delayed relative to the AIF. This problem has been circumvented by the so-called block-circulant SVD proposed by Wu et al. [36]. This technique is made time-shift insensitive by the use of a block-circulant matrix W_c for deconvolution. Block-circulant SVD looks promising in providing tracer-arrival time-insensitive flow estimates and a more specific indicator of ischemic injury, but more work is necessary to better define its domain of validity. Andersen et al. [38] proposed the use of a Gaussian process to approximate the convolution kernel, i.e., the residue function. The method is termed *Gaussian process deconvolution* (GPD) and allows accounting for the smoothness (but not for nonnegativity) of the residue function by incorporating this constrain as *a priori* information. More recently, Calamante et al. [22] improved upon SVD by implementing the Tikhonov regularization method in order to overcome the presence of unwanted oscillations in the residue function. However, even if this new method was shown to provide an improved characterization (as compared to SVD) of the shape of R(t), it does not account for nonnegativity of R(t). Along this line, Zanderigo et al. [39] proposed the application of a nonlinear stochastic regularization (NSR) method, which is able to account for the smoothness of the residue function and handle possible violations of the nonnegativity constraint of CBF · R(t). NSR is a deconvolution method that exploits a model of the unknown residue function, only allowing nonnegative values. NSR considers CBF · R(t) composed by the exponential of a Brownian motion. This approach shows advantages over SVD in detecting only positive and smoothed (i.e., physiological) CBF · R(t) without fixing any threshold value and requiring only the knowledge of AIF and tissue data.

19.3.4 ABSOLUTE QUANTIFICATION ISSUES

The fundamental steps for CBF, CBV, and MTT quantification are summarized in Figure 19.3. The accuracy of CBF measures is strongly dependent on the values of the density ρ of brain tissue, and of the hematocrit in capillaries and large vessels, H_{SV} and H_{LV}, respectively. In particular, the frequently used values $\rho = 1.04$ g/ml, $H_{LV} = 0.45$, and $H_{SV} = 0.25$ [11] have been shown to generate in normal subjects CBF values that are in agreement with the flow values obtained with other techniques such as PET [11,40], but, on the contrary, for instance, in healthy smoker subjects, the same values of ρ, H_{LV}, and H_{SV} have been unable to provide reliable quantitative perfusion measurements [41]. In addition, nobody has tested the validity of the use of these values in pathologic conditions. To overcome these limitations, several other approaches have been proposed. In Reference 16, the authors obtained absolute CBF values assuming the microvascular hematocrit

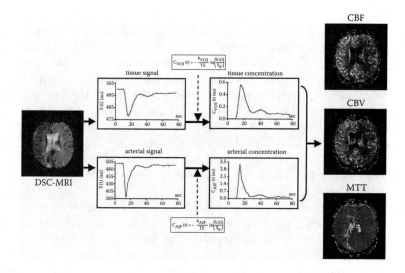

FIGURE 19.3 The quantification process of DSC-MRI image: from signal acquisition (left) to parametric mapping generation (right) of CBF, CBV, and MTT.

to be constant across the brain and by assigning the mean relative CBF value in white matter to a standard value of 22 ml/100 ml/min. The relative CBF values from each of the tissue gray regions were then multiplied by this individualized scaling factor to yield absolute CBF. The rationale for using normal cerebral white matter as an internal reference standard for generating absolute MRI CBF values is based on PET measurements that showed that in normal adult volunteers, white matter has a relatively uniform age-independent blood flow of 22 ml/100 ml/min. A different approach was introduced in Reference 42 and Reference 43 where in order to convert relative MRI CBF values to absolute ones, a conversion factor derived by comparison studies between relative MRI and absolute PET CBF measurements was proposed. A similar approach, with some dissimilarities, has been proposed in Reference 44 where CBF values were converted to absolute values by calculating the ratio of the CBF value measured by PET to the mean relative CBF measured by DSC-MRI in three subcortical white matter regions and multiplying by this scaling factor the relative CBF from each gray tissue to yield absolute MRI CBF values. However, even if this approach can provide a good qualitative index of CBF for patients with chronic carotid occlusion, these absolute CBF values are not accurate [44]. In summary, there is no gold-standard method to use in order to obtain accurate absolute CBF values in normal as well as in pathological subjects, and additional studies are needed in order to address the problem concerning the correct absolute quantification of CFB by DSC-MRI [23,45], especially in presence of pathologies such as chronic carotid occlusive diseases, tumors, and strokes.

19.3.5 CONCLUSION

DSC-MRI is becoming a fundamental tool for noninvasive quantitative assessment of cerebral functions and is regarded as a potential alternative to PET, which still is the gold standard for quantitative imaging. In fact, DSC-MRI, thanks to its minimal invasiveness and good spatial and temporal resolution characteristics, is increasingly being used to generate quantitative parametric maps of physiological processes related to brain function. Different from PET, conversion of MRI signals into measurement of physiological parameters is a less-developed field and only in recent years increasing attention has been paid to develop a methodology for correct quantification of DSC-MRI images. Numerous studies have already shown that CBF, CBV, and MTT quantifications from DSC-MRI images are possible, and the results obtained show DSC-MIR to be an extremely promising quantitative technique. However, further work is necessary to adequately address the problems highlighted in this chapter and to bring the robustness required in clinical applications.

ACKNOWLEDGMENTS

We thank Dr. Mirco Cosottini, Department of Neuroscience, University of Pisa, Pisa, Italy, and Dr. Gianluigi Pillonetto, Department of Information Engineering, University of Padova, Padova, Italy, for helpful discussions and contributions. This work was supported by NIH Grant EB-01975.

REFERENCES

1. Frackowiak, R.S., Lenzi, G.L., Jones, T., and Heather, J.D. (1980). Quantitative measurement of regional cerebral blood flow and oxygen metabolism in man using 15O and positron emission tomography: theory, procedure, and normal values. *J. Comput. Assist. Tomogr.* 4: 727–736.
2. Huang, S.C., Carson, R.E., Hoffman, E.J., Carson, J., MacDonald, N., Barrio, J.R., and Phelps, M.E. (1983). Quantitative measurement of local cerebral blood flow in humans by positron computed tomography and 15O-water. *J. Cereb. Blood Flow Metab.* 3: 141–153.
3. Zierler, K.L. (1962). Theoretical basis of indicator-dilution methods for measuring flow and volume. *Circ. Res.* 10: 393–407.
4. Zierler, K.L. (1965). Equations for measuring blood flow by external monitoring of radioisotopes. *Circ. Res.* 16: 309–321.
5. Axel, L. (1980). Cerebral blood flow determination by rapid-sequence computed tomography: theoretical analysis. *Radiology* 137: 679–686.
6. Stewart, G.N. (1894). Researches on the circulation time in organs and on the influences which affect it. Parts I–III. *J. Physiol.* 15: 1–89.
7. Meier, P. and Zierler, K.L. (1954). On the theory of the indicator-dilution method for measurement of blood flow and volume. *J. Appl. Physiol.* 6: 731–744.
8. Barbier, E.L., Lamalle, L., and Decorps, M. (2001). Methodology of brain perfusion imaging. *J. Magn. Reson. Imaging* 13: 496–520.

9. Rosen, B.R., Belliveau, J.W., Buchbinder, B.R., McKinstry, R.C., Porkka, L.M., Kennedy, D.N., Neuder, M.S., Fisel, C.R., Aronen, H.J., Kwong, K.K., Weisskoff, R.M., Cohen, M.S., and Brady, T.J. (1991). Contrast agents and cerebral hemodynamics. *Magn. Reson. Med.* 19: 285–292.

10. Weisskoff, R.M. and Kiihne, S. (1992). MRI susceptometry: image-based measurement of absolute susceptibility of MR contrast agents and human blood. *Magn. Reson. Med.* 24: 375–383.

11. Rempp, K.A., Brix, G., Wenz, F., Becker, C.R., Guckel, F., and Lorenz, W.J. (1994). Quantification of regional cerebral blood flow and volume with dynamic susceptibility contrast-enhanced MR imaging. *Radiology* 193: 637–641.

12. Boxerman, J.L., Hamberg, L.M., Rosen, B.R., and Weisskoff, R.M. (1995). MR contrast due to intravascular magnetic susceptibility perturbations. *Magn. Reson. Med.* 34: 555–566.

13. Ostergaard, L., Weisskoff, R.M., Chesler, D.A., Gyldensted, C., and Rosen, B.R. (1996). High-resolution measurement of cerebral blood flow using intravascular tracer bolus passages. Part I: Mathematical approach and statistical analysis. *Magn. Reson. Med.* 36: 715–725.

14. Kiselev, V.G. (2001). On the theoretical basis of perfusion measurements by dynamic susceptibility contrast MRI. *Magn. Reson. Med.* 46: 1113–1122.

15. Kiselev, V.G. (2004). Transverse relaxation effect of contrast agent: a crucial issue for quantitative measurements. in *Proceedings ISMRM Workshop on Quantitative Cerebral Perfusion Imaging Using MRI: A Technical Perspective*, Venice, Italy, March 21–23, pp. 6–9.

16. Ostergaard, L., Sorensen, A.G., Kwong, K.K., Weisskoff, R.M., Gyldensted, C., and Rosen, B.R. (1996). High-resolution measurement of cerebral blood flow using intravascular tracer bolus passages. Part II: Experimental comparison and preliminary results. *Magn. Reson. Med.* 36: 726–736.

17. Benner, T., Heiland, S., Erb, G., Forsting, M., and Sartor, K. (1997). Accuracy of gamma-variate fits to concentration-time curves from dynamic susceptibility-contrast enhanced MRI: influence of time resolution, maximal signal drop and signal-to-noise. *Magn. Reson. Imaging* 15: 307–317.

18. Porkka, L., Neuder, M., Hunter, G., Weisskoff, R.M., Belliveau, J., and Rosen, B.R. (1991). Arterial input function measurement with MRI. in *Proceedings of the Society of Magnetic Resonance Medicine,* 10th Annual Meeting, San Francisco, USA, p. 120.

19. Calamante, F., Thomas, D.L., Pell, G.S., Wiersma, J., and Turner, R. (1999). Measuring cerebral blood flow using magnetic resonance imaging techniques. *J. Cereb. Blood Flow Metab.* 19: 701–735.

20. Lin, W., Celik, A., Derdeyn, C., An, H., Lee, Y., Videen, T., Ostergaard, L., and Powers, W.J. (2001). Quantitative measurements of cerebral blood flow in patients with unilateral carotid artery occlusion: a PET and MR study. *J. Magn. Reson. Imaging* 14: 659–667.

21. van Osch, M.J., Vonken, E.J., Bakker, C.J., and Viergever, M.A. (2001). Correcting partial volume artifacts of the arterial input function in quantitative cerebral perfusion MRI. *Magn. Reson. Med.* 45: 477–485.

22. Calamante, F., Gadian, D.G., and Connelly, A. (2003). Quantification of bolus-tracking MRI: improved characterization of the tissue residue function using Tikhonov regularization. *Magn. Reson. Med.* 50: 1237–1247.

23. Calamante, F., Gadian, D.G., and Connelly, A. (2000). Delay and dispersion effects in dynamic susceptibility contrast MRI: simulations using singular value decomposition. *Magn. Reson. Med.* 44: 466–473.

24. Calamante, F., Gadian, D.G., and Connelly, A. (2002). Quantification of perfusion using bolus tracking magnetic resonance imaging in stroke: assumptions, limitations, and potential implications for clinical use. *Stroke* 33: 1146–1151.

25. Alsop, D.C. and Detre, J.A. (1996). Reduced transit-time sensitivity in noninvasive magnetic resonance imaging of human cerebral blood flow. *J. Cereb. Blood Flow Metab.* 16: 1236–1249.

26. van Osch, M.J., Vonken, E.J., Wu, O., Viergever, M.A., van der Grond, J., and Bakker, C.J. (2003). Model of the human vasculature for studying the influence of contrast injection speed on cerebral perfusion MRI. *Magn. Reson. Med.* 50: 614–622.

27. Larson, K.B., Perman, W.H., Perlmutter, J.S., Gado, M.H., and Zierler, K.L. (1994). Tracer-kinetic analysis for measuring regional cerebral blood flow by dynamic nuclear magnetic resonance imaging. *J. Theor. Biol.* 170: 1–14.

28. Ostergaard, L., Chesler, D.A., Weisskoff, R.M., Sorensen, A.G., and Rosen, B.R. (1999). Modeling cerebral blood flow and flow heterogeneity from magnetic resonance residue data. *J. Cereb. Blood Flow Metab.* 19: 690–699.

29. Kroll, K., Wilke, N., Jerosch-Herold, M., Wang, Y., Zhang, Y., Bache, R.J., and Bassingthwaighte, J.B. (1996). Modeling regional myocardial flows from residue functions of an intravascular indicator. *Am. J. Physiol.* 271: H1643–H1655.

30. Smith, A.M., Grandin, C.B., Duprez, T., Mataigne, F., and Cosnard, G. (2000). Whole brain quantitative CBF and CBV measurements using MRI bolus tracking: comparison of methodologies. *Magn. Reson. Med.* 43: 559–564.

31. De Nicolao, G., Sparacino, G., and Cobelli, C. (1997). Nonparametric input estimation in physiological systems: Problems, methods, and case studies. *Automatica* 33: 851–870.

32. Murase, K., Shinohara, M., and Yamazaki, Y. (2001). Accuracy of deconvolution analysis based on singular value decomposition for quantification of cerebral blood flow using dynamic susceptibility contrast-enhanced magnetic resonance imaging. *Phys. Med. Biol.* 46: 3147–3159.

33. Liu, H.L., Pu, Y., Liu, Y., Nickerson, L., Andrews, T., Fox, P.T., and Gao, J.H. (1999). Cerebral blood flow measurement by dynamic contrast MRI using singular value decomposition with an adaptive threshold. *Magn. Reson. Med.* 42: 167–172.

34. Wirestam, R., Andersson, L., Ostergaard, L., Bolling, M., Aunola, J.P., Lindgren, A., Geijer, B., Holtas, S., and Stahlberg, F. (2000). Assessment of regional cerebral blood flow by dynamic susceptibility contrast MRI using different deconvolution techniques. *Magn. Reson. Med.* 43: 691–700.

35. Sourbron, S., Luypaert, R., Van Schuerbeek, P., Dujardin, M., Stadnik, T., and Osteaux, M. (2004). Deconvolution of dynamic contrast-enhanced MRI data by linear inversion: choice of the regularization parameter. *Magn. Reson. Med.* 52: 209–213.

36. Wu, O., Ostergaard, L., Weisskoff, R.M., Benner, T., Rosen, B.R., and Sorensen, A.G. (2003). Tracer arrival timing-insensitive technique for estimating flow in MR perfusion-weighted imaging using singular value decomposition with a block-circulant deconvolution matrix. *Magn. Reson. Med.* 50: 164–174.

37. Smith, M.R., Lu, H., Trochet, S., and Frayne, R. (2004). Removing the effect of SVD algorithmic artifacts present in quantitative MR perfusion studies. *Magn. Reson. Med.* 51: 631–634.

38. Andersen, I.K., Szymkowiak, A., Rasmussen, C.E., Hanson, L.G., Marstrand, J.R., Larsson, H.B.W., and Hansen, L.K. (2002). Perfusion quantification using Gaussian process deconvolution. *Magn. Reson. Imaging* 48: 351–361.

39. Zanderigo, F., Bertoldo, A., Pillonetto, G., Cosottini, M., and Corbelli, C. (2004). Nonlinear stochastic regularization to characterize tissue residue function from bolus-tracking MRI images. in *Proceedings ISMRM Workshop on Quantitative Cerebral Perfusion Imaging Using MRI: A Technical Perspective*, Venice, Italy, March 21–23, pp. 77–78.

40. Schreiber, W.G., Guckel, F., Stritzke, P., Schmiedek, P., Schwartz, A., and Brix, G. (1998). Cerebral blood flow and cerebrovascular reserve capacity: estimation by dynamic magnetic resonance imaging. *J. Cereb. Blood Flow Metab.* 18: 1143–1156.

41. Carroll, T.J., Teneggi, V., Jobin, M., Squassante, L., Treyer, V., Hany, T.F., Burger, C., Wang, L., Bye, A., Von Schulthess, G.K., and Buck, A. (2002). Absolute quantification of cerebral blood flow with magnetic resonance, reproducibility of the method, and comparison with $H_2(^{15})O$ positron emission tomography. *J. Cereb. Blood Flow Metab.* 22: 1149–1156.

42. Ostergaard, L., Johannsen, P., Host-Poulsen, P., Vestergaard-Poulsen, P., Asboe, H., Gee, A.D., Hansen, S.B., Cold, G.E., Gjedde, A., and Gyldensted, C. (1998). Cerebral blood flow measurements by magnetic resonance imaging bolus tracking: comparison with $[^{15}O]H_2O$ positron emission tomography in humans. *J. Cereb. Blood Flow Metab.* 18: 935–940.

43. Ostergaard, L., Smith, D.F., Vestergaard-Poulsen, P., Hansen, S.B., Gee, A.D., Gjedde, A., and Gyldensted, C. (1998). Absolute cerebral blood flow and blood volume measured by magnetic resonance imaging bolus tracking: comparison with positron emission tomography values. *J. Cereb. Blood Flow Metab.* 18: 425–432.

44. Mukherjee, P., Kang, H.C., Videen, T.O., McKinstry, R.C., Powers, W.J., and Derdeyn, C.P. (2003). Measurement of cerebral blood flow in chronic carotid occlusive disease: comparison of dynamic susceptibility contrast perfusion MR imaging with positron emission tomography. *AJNR Am. J. Neuroradiol.* 24: 862–871.

45. Sorensen, A.G. (2001). What is the meaning of quantitative CBF? *AJNR Am. J. Neuroradiol.* 22: 235–236.

Index